HISTORY

OF THE

ROYAL IRISH RIFLES

GENERAL SIR CORNELIUS CUYLER, BART.

Colonel of Cuyler's Shropshire Volunteers (86th Regt.), 1793

HISTORY

OF THE

ROYAL IRISH RIFLES

By

Lieut.-Colonel GEORGE BRENTON LAURIE
Royal Irish Rifles

Dedicated

TO

MY REGIMENT

FOREWORD

I HAVE been asked to say a few words on the occasion of the publishing, in an amended or more popular form, of the Records of two distinguished Regiments, which taken from the line now go to constitute a separate corps, the Royal Irish Rifles.

As their Colonel, I have the pleasure of being able to claim, in common with every officer and soldier in it, a special interest in these papers, as after all it is our own history that is given in any annals of those famous old fighting Battalions, the 83rd and 86th of the Line.

The well-known General Order of the King, dated 1st January, 1836, which evinced an anxious desire that the meritorious act of any soldier in the Army should be carefully recorded does not seem to point to this having been always done, nor do I know that any system of registration was ordered. But this should only make the records we have more valuable and better known. The long periods in our own country of comparative peace have effected many changes, but amongst them all, the most satisfactory seems to me to be that the jealousy and coldness that I remember as once existing on the part of the public towards the Army is gone, and has given way to an entirely opposite feeling. Nothing now will suit the lighter literature of the day better than military news or story—even to anecdotes of the popular leaders of the past.

The little book that I have before me here this morning springs from the natural wish of both Battalions of the Royal Irish Rifles that their distinguished Family History so to call it, should be better known, in an inexpensive form, to every member. This very desirable end, I hope, has been accomplished, and the nature of a bequest of great value brought before the eyes of the younger generation.

I beg to thank Lieut.-Colonel F. J. H. Bell for his help, but especially Lieut.-Colonel G. B. Laurie, of the 1st Battalion, for his invaluable services.

WILMOT H. BRADFORD,
Lieut.-General.

Bournemouth,
3rd April, 1913.

PREFACE

THE One Hundred and Twentieth Anniversary of the raising of the 83rd and 86th Regiments sees this History of their doings published.

For about one quarter of that period I have had the honour of belonging to the Royal Irish Rifles.

I much appreciated the distinction conferred upon me when I was asked by my Regiment to write their History, and I deem it a fitting ending to my career in that Regiment.

I never realized before I went into the subject how very little I knew of the past history of my own Corps; but my task in writing it has been greatly lightened by the kindness of many who have helped me.

To enumerate their names only would make this preface assume undue length; but my most grateful thanks are due to them one and all.

Many a brave deed has been recorded in this story of the past; many equally courageous have been left untold, either through lack of space or because the deed itself has been forgotten.

When the sequel to this volume is written, I trust that the author of that sequel will be able truthfully to say that he found equal courage, honour and worth amongst those who have succeeded us, as I have found amongst our predecessors.

May the Officers and Men of the Royal Irish Rifles win yet more laurels for their Regiment by their staunchness, whenever their Sovereign calls for their services in war.

THE AUTHOR.

ERRATUM

Page 323, line 12. For "Lieutenant-Colonel Meurant," substitute "Lieutenant-Colonel Karslake."

CONTENTS

CHAPTER I.

FITCH'S AND CUYLER'S REGIMENTS OF FOOT (1793).

CHAPTER II.

EIGHTY-THIRD REGIMENT (1793-1802).

CHAPTER III.

EIGHTY-SIXTH REGIMENT (1793-1799).

CONTENTS.

CONTENTS.

CONTENTS.

CHAPTER VIII.

EIGHTY-THIRD REGIMENT (1805-1817).

CHAPTER IX.

2ND BATTALION 83RD REGIMENT.

Written by Professor C. Oman, Chichele Professor of Modern History, Oxford.

CONTENTS.

CHAPTER XII.

EIGHTY-THIRD REGIMENT (1857-1859).

CHAPTER XIII.

EIGHTY-SIXTH REGIMENT (1857-1859).

CHAPTER XIV.

EIGHTY-THIRD REGIMENT (1860-1912).

CONTENTS.

ILLUSTRATIONS

COLOURED PLATES.

MAPS.

COLONEL LLOYD
1801

Procured by Lieut.-Colonel Laurie from the late Mr. Lloyd Phillips, of Pent-y-Parc, Pembrokeshire, who was the son of the above Colonel Lloyd.

**DRESS OF 83RD REGIMENT OF FOOT—
PENINSULAR PERIOD**

From a sketch by E A. Campbell, Esq. COMPANY OFFICER SERJEANT PRIVATE

OFFICER OF 83rd REGIMENT WITH GORGET

Portrait of Major Thomas Somerfield, joined 83rd Regiment as Quartermaster, 25th December, 1796; received commission as Ensign, 4th December, 1798; became Brevet-Major 21st December, 1820; was severely wounded at Talavera; died, 1833.

Original picture in Officers' Mess, 1st Battalion Royal Irish Rifles.

83RD REGIMENT

LIEUTENANT, 1827

From an old print.

83RD REGIMENT

CAPTAIN, 1829

From an old print.

DRESS OF THE 86TH REGIMENT

CAPTAIN, 1796	SERJEANT, 1842	PRIVATE, 1875
(1)	*(2)*	*(3)*

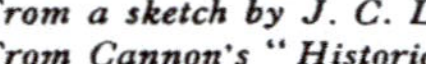

(1) From a sketch by J. C. Leask, Esq.
(2) From Cannon's "Historical Record of the 86th."
(3) From a sketch by Capt. Reynolds, Royal Irish Rifles.

DRESS OF THE 86TH ROYAL COUNTY DOWN REGIMENT
DURING THE INDIAN MUTINY

PRIVATE (During the Day)	PRIVATE (Early Morning or Night)	CAPTAIN

From a sketch by E. A. Campbell, Esq.

THE COLOURS OF THE 86TH REGIMENT, 1833-1867

THESE COLOURS WERE CARRIED IN THE INDIAN MUTINY, AND NOW HANG IN ST. PATRICK'S CATHEDRAL, DUBLIN.

DRESS OF THE ROYAL IRISH RIFLES

RIFLEMAN	SERJEANT AND RIFLEMAN	RIFLEMAN
(Marching Order, India, 1896)	(Review Order, 1910)	(Marching Order, Service Dress, 1913)

MOUNTED OFFICER, ROYAL IRISH RIFLES, 1913

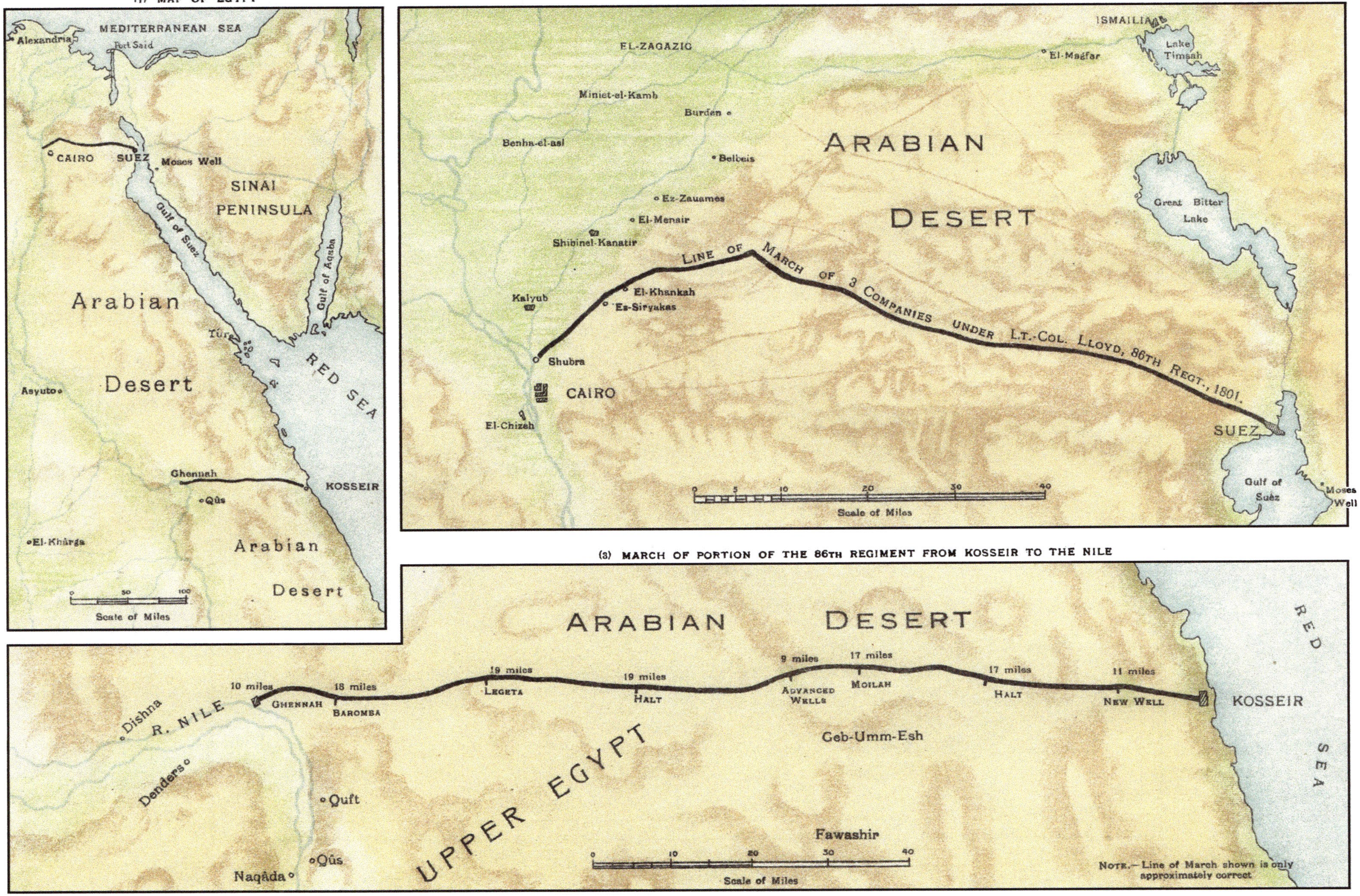
MEDITERRANEAN SEA
Alexandria
Port Said
CAIRO
SUEZ
Moses Well
SINAI
PENINSULA
Gulf of Suez
Gulf of Aqaba
Arabian
Desert
Tûr
RED SEA
Asyuto
Ghennah
Qûs
KOSSEIR
El-Khârga
Arabian
Desert
Scale of Miles
0 50 100
EL-ZAGAZIG
El-Magfar
ISMAILIA
Lake Timsah
Miniet-el-Kamb
Burden
Benha-el-asl
Belbeis
ARABIAN
DESERT
Great Bitter Lake
Ez-Zauames
El-Menair
Shibinel-Kanatir
LINE OF MARCH OF 3 COMPANIES UNDER LT.-COL. LLOYD, 86TH REGT., 1801.
Kalyub
El-Khankah
Es-Siryakas
Shubra
CAIRO
El-Chizeh
SUEZ
Gulf of Suez
Moses Well
0 5 10 20 30 40
Scale of Miles
(3) MARCH OF PORTION OF THE 86TH REGIMENT FROM KOSSEIR TO THE NILE
ARABIAN
DESERT
RED SEA
10 miles
18 miles
19 miles
19 miles
9 miles
17 miles
17 miles
11 miles
GHENNAH
BAROMBA
LEGETA
HALT
ADVANCED WELLS
MOILAH
HALT
NEW WELL
KOSSEIR
Dishna
R. NILE
Denderso
UPPER EGYPT
Geb-Umm-Esh
Quft
Fawashir
Qûs
Naqâda
0 10 20 30 40
Scale of Miles
NOTE.—Line of March shown is only approximately correct

EGYPTIAN CAMPAIGN OF THE 86TH REGIMENT. 1801.

John Bartholomew & Co., Edin.

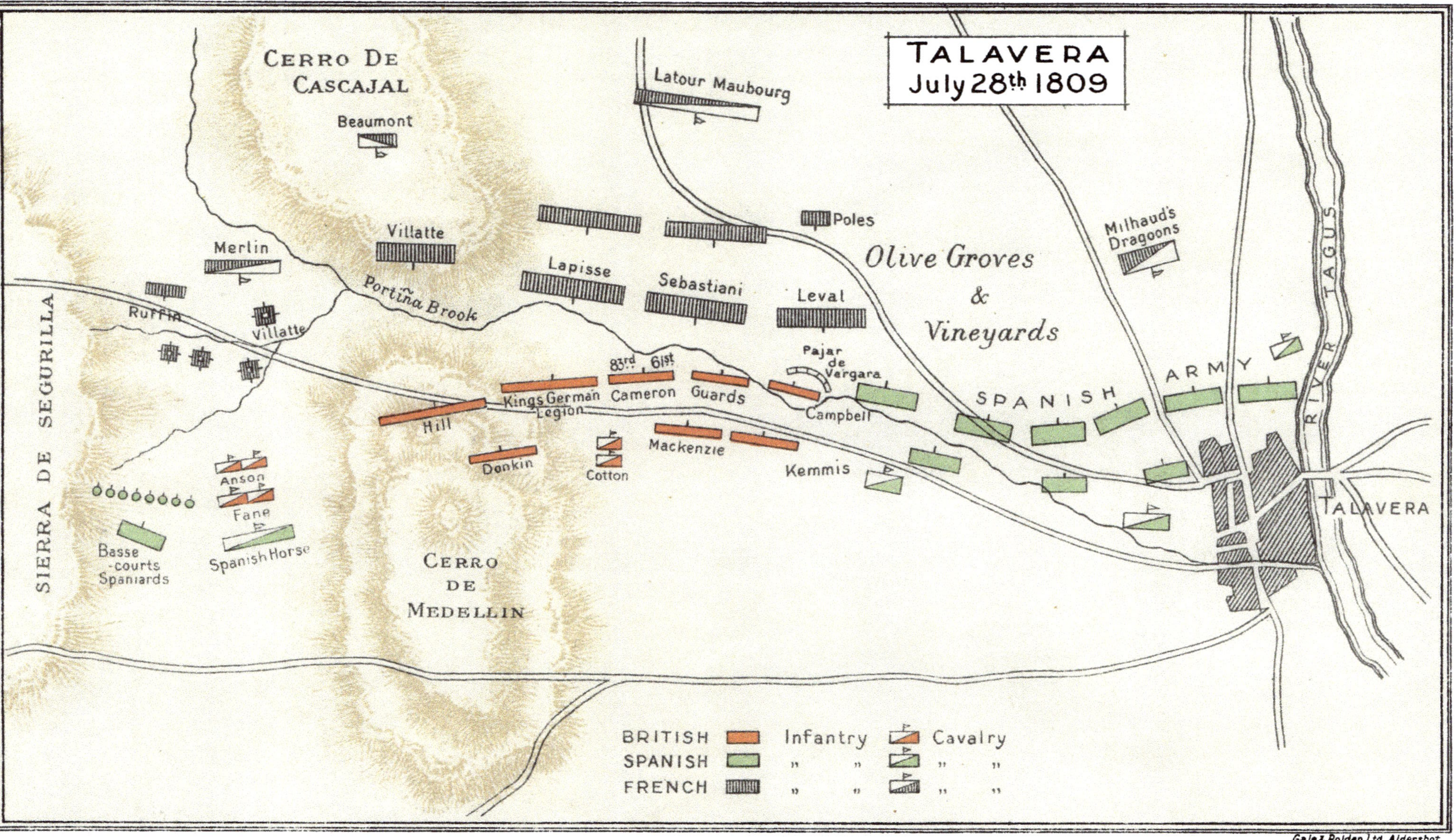
TALAVERA
July 28th 1809
Cerro De Cascajal
Beaumont
Latour Maubourg
Poles
Villatte
Merlin
Lapisse
Sebastiani
Leval
Olive Groves & Vineyards
Milhaud's Dragoons
Ruffin
Villatte
Portiña Brook
Pajar de Vergara
83rd 61st
Kings German Legion
Cameron
Guards
Campbell
Hill
Donkin
Cotton
Mackenzie
Kemmis
SPANISH ARMY
RIVER TAGUS
TALAVERA
Anson
Fane
Spanish Horse
Basse-courts Spaniards
SIERRA DE SEGURILLA
Cerro de Medellin
BRITISH Infantry Cavalry
SPANISH " " " "
FRENCH " " " "
Gale & Polden Ltd. Aldershot.

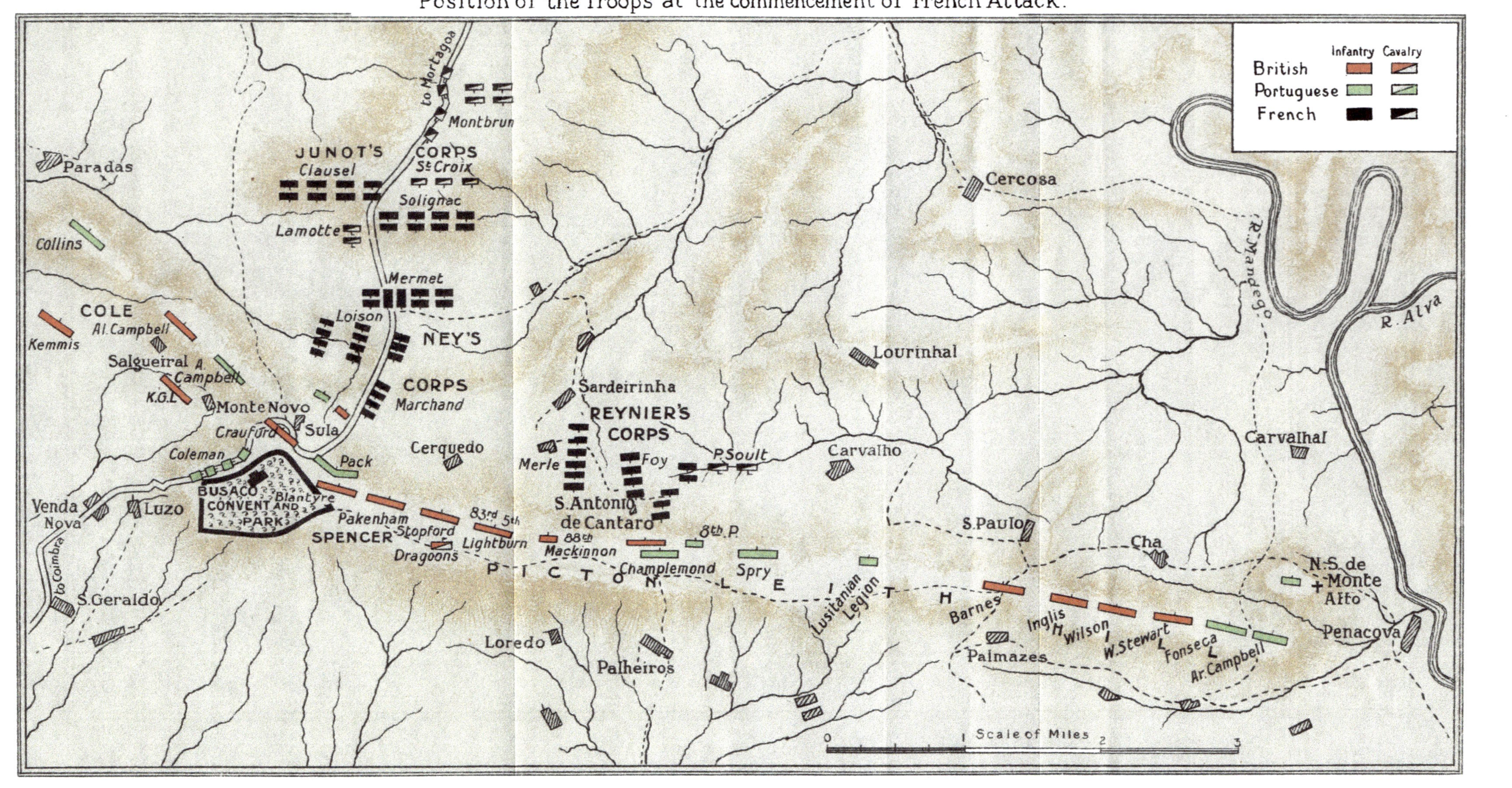

BATTLE OF BUSACO, Sept 27th 1810.
Position of the Troops at the commencement of French Attack.
Infantry
Cavalry
British
Portuguese
French
to Mortagoa
Montbrun
JUNOT'S CORPS
Clausel
St Croix
Solignac
Lamotte
Paradas
Collins
Cercosa
Mermet
Loison
NEY'S CORPS
Marchand
COLE
Al. Campbell
Kemmis
Salgueiral
A. Campbell
K.G.L.
Monte Novo
Sula
Crauford
Coleman
Pack
Cerquedo
Sardeirinha
REYNIER'S CORPS
Merle
Foy
P. Soult
S. Antonio de Cantaro
Lourinhal
Carvalho
Carvalhal
R. Mandego
R. Alva
Venda Nova
Luzo
BUSACO CONVENT AND PARK
Blantyre
Pakenham
Stopford
SPENCER
Dragoons
Lightburn
83rd
5th
88th
Mackinnon
8th P.
Champlemond
Spry
PICTON
LEITH
Lusitanian Legion
S. Paulo
Cha
N.S. de Monte Alto
Barnes
Inglis
Wilson
W. Stewart
Fonseca
Ar. Campbell
Penacova
Palmazes
to Coimbra
S. Geraldo
Loredo
Palheiros
Scale of Miles
0
1
2
3

Troops for false attack of Tete de Pont
St. Cristobal Fort
Tete de Pont.
RIVER GUADIANA
RIVER RIVILLAS
GEN LEITHS Assault 5th Division
San Vincente
CASTLE
3RD DIVISION
St. Roque
Trinidad
Sta. Maria
Pardaleras
La Picurina
LIGHT DIVISION
River Calamon
River Rivillas
5TH DIVISION
4TH DIVISION
ASSAULT OF BADAJOZ
APRIL 6TH 1812.
0 100 200 400 800
Yards.
Gale & Polden Ltd. Aldershot.

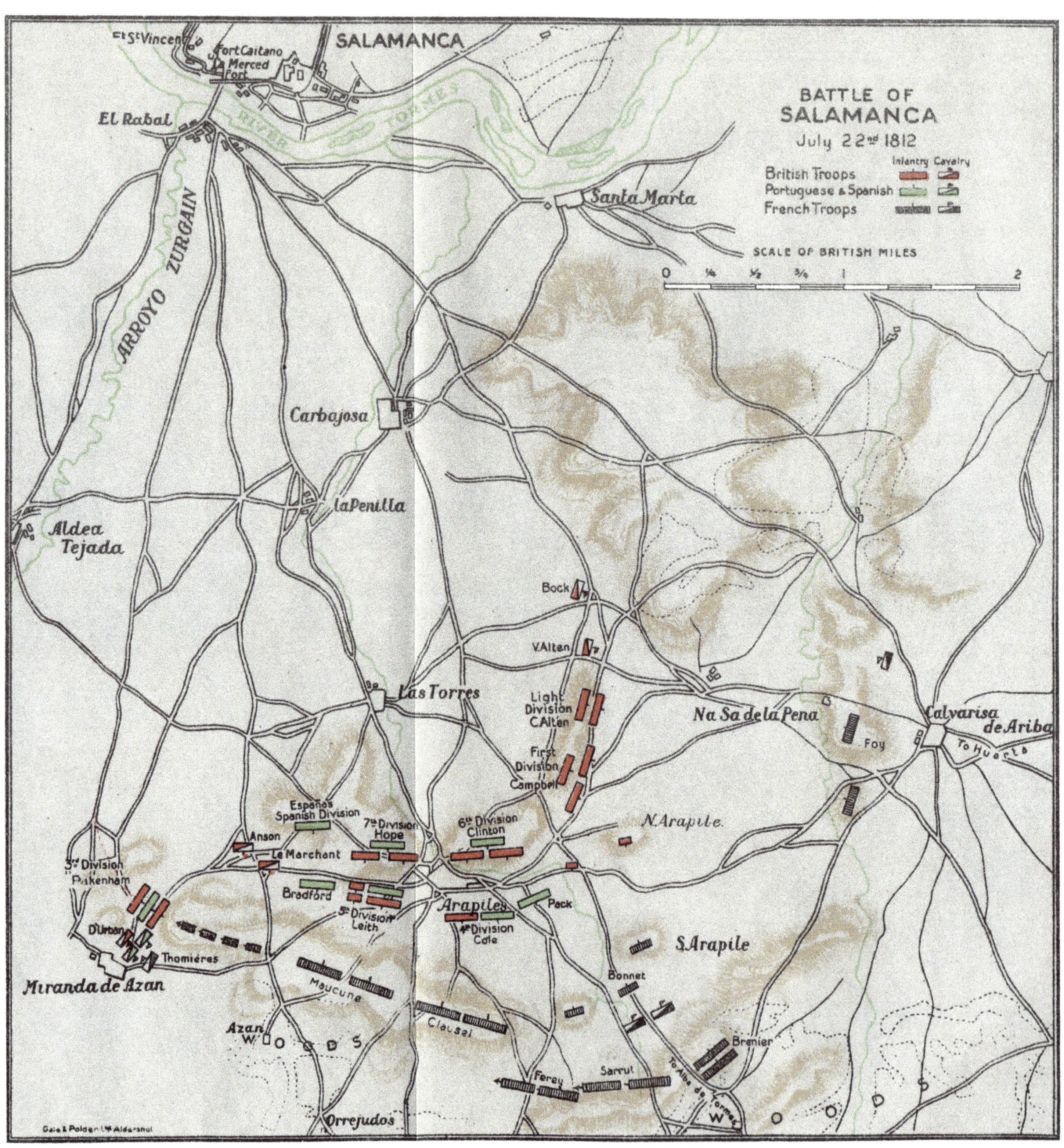
BATTLE OF SALAMANCA
July 22nd 1812
Infantry Cavalry
British Troops
Portuguese & Spanish
French Troops
SCALE OF BRITISH MILES
0 1/4 1/2 3/4 1 2
SALAMANCA
Ft St Vincent
Fort Caitano
Merced Fort
El Rabal
RIVER TORMES
Santa Marta
ARROYO ZURGAIN
Carbajosa
la Penilla
Aldea Tejada
Bock
V. Alten
Las Torres
Light Division C. Alten
First Division Campbell
Na Sa de la Pena
Calvarisa de Ariba
To Huerta
Foy
N. Arapile
Españos Spanish Division
7th Division Hope
6th Division Clinton
Anson
Le Marchant
3rd Division Pakenham
Bradford
5th Division Leith
Arapiles
Pack
4th Division Cole
S. Arapile
D'Urban
Thomières
Miranda de Azan
Maucune
Bonnet
Clausel
Azan
WOODS
Brenier
Ferey
Sarrut
To Alba de Tormes
Orrejudos
Gale & Polden Ltd Aldershot

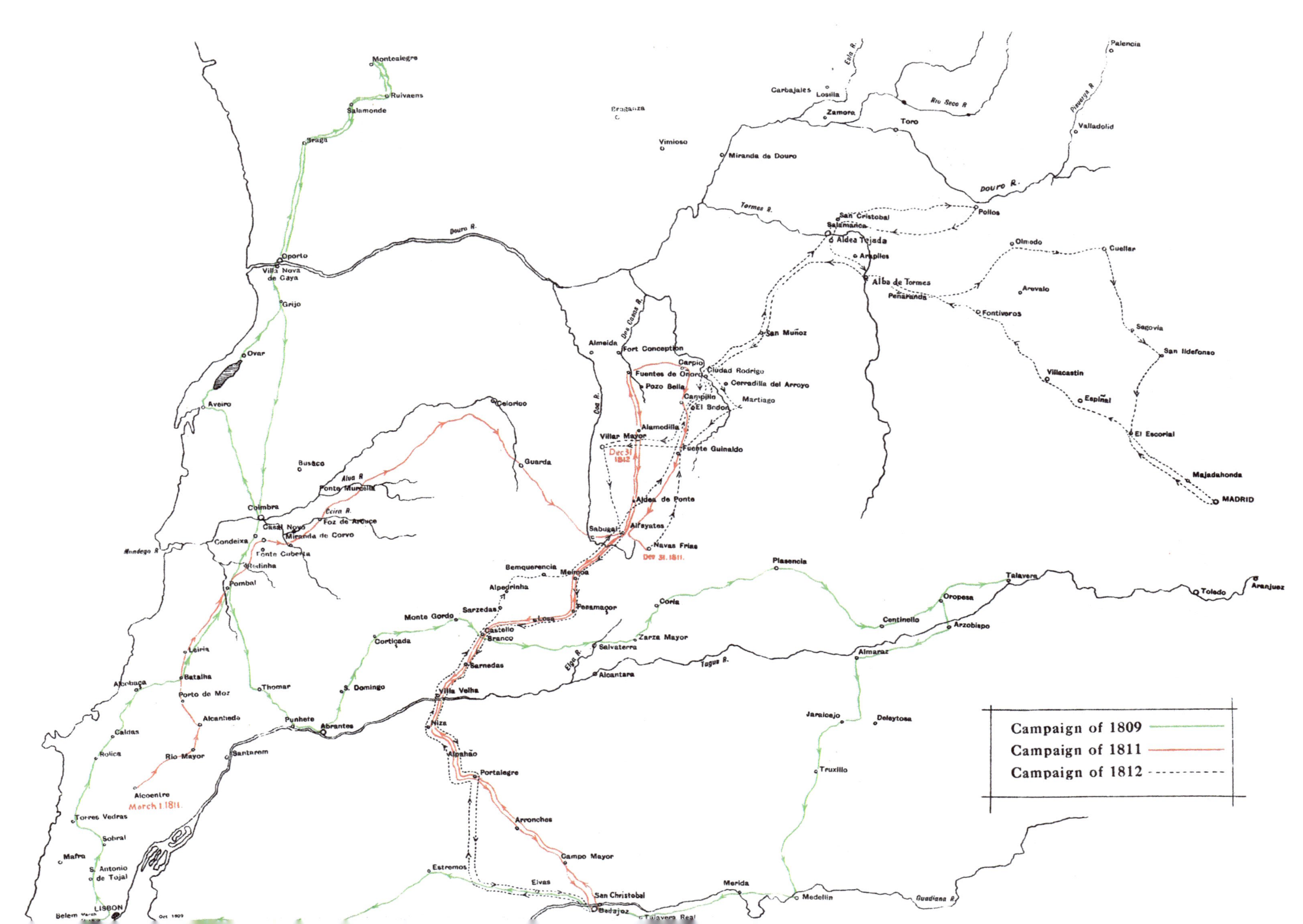

Campaign of 1809
Campaign of 1811
Campaign of 1812
Montealegre
Ruivaens
Salamonde
Braga
Oporto
Villa Nova de Gaya
Grijo
Ovar
Aveiro
Douro R.
Bragança
Vimioso
Miranda de Douro
Carbajales
Losilla
Esla R.
Zamora
Toro
Rio Seco R.
Palencia
Pisuerga R.
Valladolid
Douro R.
Tormes R.
San Cristobal
Salamanca
Pollos
Aldea Tejada
Arapiles
Alba de Tormes
Peñaranda
Olmedo
Cuellar
Arevalo
Fontiveros
Segovia
San Ildefonso
Villacastin
Espiñal
El Escorial
Majadahonda
MADRID
San Muñoz
Almeida
Dos Casas R.
Fort Conception
Carpio
Ciudad Rodrigo
Fuentes de Onoro
Pozo Bello
Cerradilla del Arroyo
Campillo
El Bodon
Martiago
Alamedilla
Villar Mayor
Dec 31 1812
Fuente Guinaldo
Coa R.
Aldea de Ponte
Sabugal
Alfayates
Navas Frias
Dec 31. 1811.
Celorico
Guarda
Busaco
Alva R.
Ponte Murcella
Coimbra
Ceira R.
Foz de Arouce
Casal Novo
Miranda de Corvo
Condeixa
Fonte Cuberta
Redinha
Mondego R.
Pombal
Bemquerencia
Meimoa
Alpedrinha
Sarzedas
Monte Gordo
Losa
Penamacor
Castello Branco
Corticada
Sarnedas
Coria
Plasencia
Zarza Mayor
Salvaterra
Elga R.
Alcantara
Tagus R.
Centinello
Oropesa
Talavera
Toledo
Aranjuez
Arzobispo
Almaraz
Leiria
Batalha
Alcobaça
Porto de Moz
Thomar
S. Domingo
Villa Velha
Alcanhede
Punhete
Abrantes
Caldas
Rolica
Rio Mayor
Santarem
Alcoentre
March 1 1811
Torres Vedras
Sobral
Mafra
S. Antonio de Tojal
LISBON
Belem
Oct 1809
Niza
Alpalhão
Portalegre
Arronches
Campo Mayor
Estremos
Elvas
San Christobal
Badajoz
Talavera Real
Merida
Medellin
Guadiana R.
Jaraicejo
Deleytosa
Truxillo

BATTLE
OF
VITTORIA
JUNE 21ST 1813.
0
2
VITTORIA
To Bayonne
To Salvatierra
To Estella
To Tudela
Durana
Longa
Gamarra
Reille
Abechuco
Grahams Column
Armentia
Zuazo de Alava
Gomecha
D'Erlon
River Zadorra
La Hermandad
Arinez
Gazan
Margarita
Mendoza
7th Division
Martiola
Tres Puentes
3rd Division
Villodas
Light Division
Subyana de Alava
Puebla Mountains
Hills Columns
Hills Columns
4th Division
Nanclares
Morillas Ridge
Gale & Polden Ltd

SOUTH AFRICA
Scale of English Miles.
Heights in English feet. Railways
Union of South Africa Red.
Other British Territory Purple.
INDIAN OCEAN
SOUTH ATLANTIC OCEAN
TRANSVAAL
ORANGE FREE STATE
NATAL
CAPE OF GOOD HOPE
UNION OF SOUTH AFRICA
BECHUANALAND PROTECTORATE
BECHUANALAND
GERMAN S.W. AFRICA
GREAT NAMA LAND
DAMARA LAND
KHAMA'S COUNTRY
MATABELE
BASUTOLAND
ZULULAND
SWAZILAND
GRIQUALAND WEST
GRIQUALAND EAST
KAFFRARIA
Kalahari Desert
Tropic of Capricorn
Cape Town
Port Elizabeth
East London
Durban
Pietermaritzburg
Bloemfontein
Kimberley
Pretoria
Johannesburg
Mafeking
Lourenço Marques
Delagoa Bay
Port Nolloth
Bacon's Geographical Establishment

68°
70°
72°
74°
34°
32°
28°
26°
24°
22°
AFGHANISTAN
KABUL
Jelalabad
Peshawar
Kohat
Attock
Murree
Rawalpindi
Ghazni
Kurum
Kilat i Ghilgee
Kandahar
Chaman
Quetta
Zhob Valley
Luni
Tal
Dadur
Bhag
Gundava
Shikarpoor
Sukkur
Bukkur
Rohree
Larkhana
Khairpoor
Mehar
Nowshera
Sewan
Hala
Hyderabad
Kotree
Karachi
Tatta
Mouths of the Indus
R. Indus
SIND
BOMBAY
PUNJAB
Shahpoor
Jhelum
Gujrat
Wazirabad
Sialkot
Goojeranwala
LAHORE
Amritsar
Ferozepore
Montgomery
Mooltan
Muzaffergurh
Bhawulpoor
Dera Ghazi Khan
Dera Ismail Khan
Bannoo
Khairpoor
Ooch
Bikaner
Jesulmer
Jodhpur
RAJPUTANA
Ajmere
Nusseerabad
Beawar
Pali
Jallor
Deesa
Mt. Abu
Oodeypoor
Neemuch
Mandisaur
Pertabgurh
Banswara
Rutlam
Ujjain
INDORE
Dhar
Mhow
Dohad
Baroda
Ahmadabad
Broach
R. Nerbudda
Kaira
Kaniby
GUJARAT
Deesa
Puttun
Western Runn
Runn of Kutch
GULF OF KUTCH
TROPIC OF CANCER
Dwarka
Porbunder
Junagurh
Beyt
Umerkote
Sirsa
Hissar
Jhoonjnoo
Sikar
Kishangurh
Nagore
Pulowdee
Barmer
Indian Desert
Srinuggur
Kashmir
Long. E. of Greenwich
SCALE OF ENGLISH MILES
0
50
100
150

MAP OF
NORTH-WESTERN
AND CENTRAL INDIA
Of the Railways shown in the Map the line between Allahabad and Cawnpore was available for the use of Government in 1858.
Names mentioned in the History are printed in Red.
78°
80°
82°
84°
34°
32°
30°
28°
26°
24°
22°
Rudok
Simla
Umballa
Dehra
Saharanpur
Muzaffarnagar
Meerut
Delhi
Moradabad
Bulandshahr
Bareilly
Aligarh
Budaun
Shahjahanpur
Fatehgarh
Agra
Etawah
Mainpuri
Cawnpore
LUCKNOW
Sitapur
Bahraich
Gonda
Fyzabad
Sultanpur
Azamgarh
Jaunpur
Ghazipur
Benares
ALLAHABAD
Fatehpur
Banda
Mirzapur
Gorakhpur
Chupra
Gaya
Jhansi
Gwalior
Jalaun
Kalpi
Lalitpur
Saugor
Damoh
Jubbulpore
Bhopal
Sehore
Hoshangabad
Rewah
Mandla
Seoni
Chhindwara
Bilaspur
Deogurh
Palkote
Kailas Mt
Gorkha
Khachi
Todum
Dhawalagiri Mt
Nabi Tal
Rampoor
Fategarh
Bhurtpore
Muttra
Dholpore
Kaimur
MIRZAPUR Hills
BUNDELKHAND AGENCY
N. W. PROVINCES
OUDH
NEPAL
R. Nerbudda
R. Soane
R. Gogra
R. Jumna
R. Ganges

THE HISTORY
OF THE
ROYAL IRISH RIFLES.

CHAPTER I.

FITCH'S AND CUYLER'S REGIMENTS OF FOOT (1793).

Foy on British Infantry—Size of British Army—Growth—Harold's house carls—Parliamentary Army—First regiment taken over from Dutch—Second and third regiments raised—17th to 78th Regiments raised—Review of state of Europe in 1793—Forces employed by Britain in India, Ceylon, Java, Cape of Good Hope, Continent of Europe, Ireland, West Indies, South America, Canada—Order to Major Fitch to raise a regiment—Order to Major-General Cuyler to raise a regiment—Lieut.-Colonel Fitch in "London Gazette"—Second Battalion of 83rd raised, 1794—Fifty regiments of foot raised—Other regiments numbered 83rd and 86th raised previous to 1793—Major-General Cuyler in "London Gazette"—First recruiting order to 83rd Regiment—Second and First recruiting order to 86th Regiment.

It is said that a gallant enemy of the United Kingdom once remarked: "The British infantry is the best in the world, fortunately there are very few of them." Whether this was really said or not, two brave French Generals who fought against Wellington's army gave the infantry of that army a very handsome testimonial. General Foy's opinion of them, freely translated, is as follows:—"In battle . . . . . there is no enemy more to be dreaded. Their infantry is steady under fire, well-trained, and fires perfectly. Their officers are the most brave and the most patriotic in Europe." *

At Waterloo General Reille said to Napoleon: "Sire, the English infantry in a fight is the devil."†

It is the history of two regiments of those sturdy fighters that is now written. The two corps commenced life originally as the 83rd Regiment of Foot, and the 86th Regiment of Foot, and it was not until each of these two infantry regiments had made their own modest reputations that

* *Note.*—"En bataille . . . il n'y a pas d'eunemis plus redoutables. Leur infanterie est solide au feu, exercée au métier, et tire parfaitement; leurs officiers sont les plus braves et les plus patriotes de l'Europe." (Page 277, "Vir Militaire du Général Foy.")

† "Sire, l'infanterie Anglaise en duel c'est le diable." (Ségur and Thiers.)

they were joined together in 1881, and became respectively the 1st Battalion and the 2nd Battalion of the Royal Irish Rifles.

The British are apparently not a military nation, for, whilst the British Empire stretches the world over, its army is scarcely equal to that of a fourth-rate Power like Roumania, and, by comparison with the size of the Empire, it really appears to be only a very weak police force, just sufficient to keep order on the frontiers, yet this nation of shopkeepers, as Napoleon called them, expect very great things of their army, and, what is passing strange, they do get marvellously good work out of their small armed forces. Of course, sometimes disaster overtakes a regiment, and this history of the Royal Irish Rifles forms no exception to the rule. The most surprising feature, however, is considering that the countries in which most of our wars take place are usually new to our troops, and that the tactics of the enemies vary in a bewildering manner, these operations are almost always successful.

Long before a single regular regiment of infantry existed in Great Britain, strong bodies of men were maintained by its rulers, it being their business to bear the brunt of any engagement, ably seconded by the civilian population, who were armed with anything that came handy, from the long bow to a bill-hook. The brave house carls of Harold, who perished to a man at the battle of Hastings, are a well-known example of household troops long before standing armies came into vogue in Western Europe.

Bodies of retainers were maintained with varying fortunes in England until the time of Henry VII., who put them down with a firm hand. England then passed through troublous times, including the Civil Wars between Charles I. and his Parliament. That Parliament, when it became victorious, was crushed by its own army, under Cromwell, and when Charles II. came to the throne, this fine army, of well over 30,000 men, was disbanded except a very few, who formed part of what is now the Foot Guards. But even at this time (1660) no British line regiment was in existence. In 1662, however, a small insurrection by some religious fanatics called the attention of the Government to its own great weakness in the face of armed opposition of any sort. Fortunately, a veteran regiment of infantry was near at hand, though not in the country. During the 30 years' war in Germany, when the Swedish King, Gustavus Adolphus, burst from the forgotten North on an astonished Europe, for his victorious though brief career, he was obliged to obtain troops from any and every Protestant country to make up for the paucity of numbers in his own kingdom of Sweden. Amongst other foreign troops, he had a Scots brigade. One regiment of this brigade, the Regiment of Douglas, after the death of Gustavus, passed into the service of the French King, and was borrowed from Louis XIV., by Charles II. in 1662. It thus became the 1st Regiment of Foot, and was called the Royal Scots, a name which it still retains. The 2nd Regiment of Foot was raised after King Charles the Second's marriage with Catherine of Braganza. This Royal lady brought, as part of her dower, Tangier, and the 2nd, or Queen's, Regiment was raised in 1661 to garrison this town, and to hold it against the Moors. The 3rd Regiment, or "The Buffs," again, was raised when the war with Holland broke out in 1665.

There were English regiments in the Dutch service. When war was declared, the Dutch, not unnaturally, asked these English soldiers to take the oath of allegiance to the Dutch Government. After consideration, these regiments refused, preferring to sacrifice their pay and prospects rather than be obliged to fight against their own countrymen. The English Government, therefore, formed "The Buffs" of officers and men who had retired from the service of Holland.

So things went on in England, regiments being formed when fighting was imminent and being disbanded when the danger was passed, Parliament always viewing any increase of the Army with disfavour for fear of the troops being used against themselves by the Crown. Still, some regiments had to be maintained, and in the last year of the reign of King James the Second (1688) the number had gone up to seventeen regiments of foot of the English Army; whilst several regiments had been brought over from Ireland, one of which—Lord Forbes's Regiment—was retained and formed the 18th, or Royal Irish, Regiment.

The Irish campaigns of William of Orange brought the number of infantry regiments of the standing army of Great Britain up to twenty-seven, the last formed regiment then being that recruited from the hardy farmers of Enniskillen, now the Royal Inniskilling Fusiliers. King William's wars continued, and in his last years the 39th Regiment of Foot was the latest permanent addition to his standing army.

As the years rolled by, and as the British Empire began to increase—though still very slowly—so did the sense of its responsibilities increase, and in 1756, of regiments which still existed, the junior infantry regiment was The "Royal Americans," recruited in Pennsylvania, and later known as the 60th Regiment.

About 1760 England took Canada from the French, and sixteen years later entered on the struggle with her American colonies, and regiments slowly increased, till at the beginning of 1793, when the long war with France, which really ended at Waterloo, was about to commence, the 78th Regiment (Seaforth Highlanders) was raised.

England, during her struggle with the revolted American colonists, also had as antagonists France, Spain, and Holland, and though she lost her American colonies, she made headway against her numerous foes. In 1793, however, things looked more promising for Great Britain. The state of Europe was much as follows: The French Revolution had broken out; King Louis XVI. had been guillotined on the 21st of January of that year, and the French Government was in the hands of somewhat incapable persons, whose tenure of office and life were synonymous terms to a great extent, as their loss of power was almost invariably followed by their decapitation. The French armies stood more or less on all the French frontiers, feeding themselves off the neighbouring countries, as France was suffering from famine. Austria had declared war on France. French troops had invaded what we now call Belgium, which country belonged to Austria. The lower classes in Holland were friendly to the new French regime, the upper classes opposed it. Spain was also attacked by France. The kingdoms of Sardinia and Naples also were marked down for plunder in succession by the French. Savoy and Nice were overrun at once, and Prussia, though also busily engaged in trying to steal a Polish province from Russia, who had recently taken charge of Poland, had a certain number of troops making war upon France, as a member of the German Empire, of which Austria was the head. Southern France was ripe for revolt against their own revolutionary Government. Finally in February, 1793, England, on hearing of the execution of the French King, asked the French Ambassador to leave the kingdom. So soon as he arrived in Paris, on 7th February, 1793, the French Government there declared war on both England and Holland.

What war with France meant is best seen by looking at what it led to. British troops were hurried off to the Low Countries. Orders were sent to occupy the French Windward Islands, in the West Indies; the British fleet moved to Toulon; Saint Domingo, in the Leeward Islands in the West Indies, was attacked; on the west coast of France, La Vendée rose in revolt, British troops had also to be sent there; Corsica, in the Mediterranean, also called for help, and more English troops were wanted there. As the war went on its sphere became still more vast. Holland made peace with France, and England then attempted to seize the Dutch Colonies and cripple the allies of France, the war spreading to South Africa. India also claimed attention, and French and Dutch possessions there had to be attacked, so Pondicherry and Ceylon were promptly seized, and Java, still farther away, was captured. Add to all this, various insurrections in Grenada, where the negroes captured the Governor and killed the white planters in all directions. A native tribe, called the Caribs, revolted in St. Vincent, and the Maroons in Jamaica, whilst yellow fever killed the British troops in the West Indies in thousands. At the same time Rebellion reared its head in Ireland, and French expeditions were thrown into that country, more troops being required to hold the Emerald Isle. The French seized Egypt, and British troops had to follow them there to turn them out. Expeditions were sent to Spanish South America, when Spain made peace with France. As the tide turned against the French, British troops were required in Spain and Portugal and in Italy; some even went to Germany. Add to all this the fact that the greater part of the population of Canada was French, and had only recently been acquired by conquest, and that no Government could really tell how long that French population would remain quiet, and remembering that the bitter struggle between England and her revolted American Colonies had only closed ten short years before, it will readily be seen that the demand for troops to undertake all these expeditions in all parts of the world, and to maintain regiments at war strength, was almost limitless. It will easily be seen, therefore, that seventy-eight infantry regiments were not enough for these numerous enterprises, so in 1793 regiments were raised apace. Orders for raising a regiment of foot, amongst others, were given to W. Fitch, Esq., who was already a Major in the Army.

The letter as kept at the Public Record Office is as follows:—

"WARRANT FOR RAISING A REGIMENT OF FOOT.

"George, R.

"Whereas we have though fit to Order a Regt. of Foot to be forthwith raised under your command, which is to consist of 12 Companys (?) with 4 Serjeants, 4 Corporals, 2 Drummers, and 66 Private men in each; with 2 Fifers to the Grenadier Company, beside a Serjeant-Major and Quarter-Master-Serjeant, together with the usual Commed. Officers.

"These are to Authorize you by beat of Drum or otherwise to raise so many men in any County or part of Our Kingdom of Great Britain, as shall be wanted to complete the said Regt. to the above mentioned numbers and all, etc.

"Given at Our Court at St. James, this 18th day of Dec., 1793, in the 34th year of Our Reign,

"By H. M. C.,

"GEO. YONGE.

"To Our F. & Wd. E. P. French, Esq., Lt.-Colo. Comdt. of a Regt. of Foot to be forthwith raised or etc., etc.

"Like Warrant of Same date to W. Fitch, Esq."

This regiment then became the 83rd Regiment of Foot. About the same time, Major-General Cornelius Cuyler was also ordered, to raise another regiment of foot by the following order, and when raised this regiment became the 86th Regiment of Foot:—

"George, R.

"Whereas we have thought fit to order a regiment of foot to be forthwith raised, under your command, which is to consist of ten companies, with three serjeants, three corporals, two drummers, and fifty-seven private men in each company, with two fifers to the grenadier company, besides a serjeant-major and quarter-master-serjeant, together with the usual number of commissioned officers, these are to authorise you, by beat of drum or otherwise, to raise so many men in any county or part of our kingdom of Great Britain as shall be wanted to complete the said regiment to the above-mentioned numbers.

"And all magistrates, justices of the peace, constables, and other of our civil officers whom it may concern are hereby required to be assisting unto you, in providing quarters, impressing carriages and otherwise as there shall be occasion.

"Given at Our Court of St. James's this 1st day of November, 1793, in the thirty-fourth year of Our reign,

"By His Majesty's Commands,

"(Signed) GEORGE YONGE.

"To our trusty and well-beloved C. Cuyler, Esq., Major-General in our Army and Colonel of a Regiment of Foot, to be forthwith raised, etc., etc., etc."

To shew the pressing demands made for troops, it may be added that a 2nd Battalion was raised in 1794 for the 83rd Regiment, but became immediately afterwards the 134th Regiment of Foot, or Loyal Limerick Regiment.

Some fifty regiments of infantry for the Regular Army were raised during the same twelve months, but most of them had, however, but a very short existence.

It should, perhaps, be added that other regiments numbered 83rd Foot and 86th Foot had existed before 1793, but as they had no connection with either of the two regiments raised in 1793, it would not be fair in any way to claim them as part of those regiments. These short-lived corps appear to have been as follows:—

"Fiemen's Regiments of British Army" states, under heading of "83rd Foot," that the first regiment numbered 83rd was disbanded in 1763. The second was the 83rd Royal Glasgow Volunteers, raised in 1778, and disbanded in 1783. Under the heading of "86th Foot," it says: "The first regiment numbered 86th was raised in 1759, and disbanded in 1763. The second was the 86th, or Rutland's Regiment, raised in 1780, and disbanded in 1783."

With reference to these regiments, another authority* states:—The 83rd was raised and retained on the Irish Establishment. Sir John Sebright, its first Colonel, when removed in 1760 to the 52nd, was succeeded by Bigoe Armstrong. The 86th was formed by drafts from other regiments. It was stationed in Senegal in 1760 and 1761, its Lieut.-Colonel, Commandant Worge, being governor of the colony.

From a consideration of these two warrants it will be seen that the 86th Regiment was raised before the 83rd Regiment, though in the 83rd

* By A. E. Sewell, "Journal R.U.S. Inst.," 1887, No. 138.

Regimental Records it is shewn as having been raised in September, 1793. The explanation of the discrepancy is as follows:—

In the "London Gazette" of Saturday, December 21st, 1793, the following extract appears:—

"Commissions signed by His Majesty for the Army in Ireland; dated, September 28th, 1793:—

"Lieut.-Col. Commandant Fitch's Corps: Major William Fitch, from the 55th Foot to be Lieut.-Colonel Commandant.

"Ditto: Captain Charles Handfield, from the 22nd Foot, to be First Major.

"Ditto: Captain William Sleigh, from the 23rd Foot, to be Second Major."

A previous "London Gazette," that of War Office, November 2nd, 1793, reads as follows:—

"A REGIMENT OF FOOT.

"Major-General Cornelius Cuyler, from the 55th Foot, to be Colonel.

"Major George Sladden, from the 67th Foot, to be Lieutenant-Colonel.

"Captain Richard Mark Dickens, from the 44th Foot, to be Major."

Thus it will be seen that though the order for raising the 86th Regiment was issued before that for raising the 83rd Regiment, and also that the officers were first gazetted to the 86th Foot yet by ante-dating the "Gazette" Colonel Fitch and his regiment were made the senior by some weeks. Whilst on this subject it might also be mentioned that the Letter Book, vol. 521, Secretary at War, gives 12th November, 1793, as the date of issue of order for raising Cuyler's Regiment of Foot, and two other infantry regiments, but this is probably simply the date it was passed through the Letter Book, as everything was very upset at the War Office then, owing to the extraordinary pressure put upon it by the raising of so many new infantry regiments, to say nothing of other troops.

It may not be out of place to give here the first recruiting orders issued to the 83rd Foot and the 86th Foot, though the actual date was naturally a little later than that of the warrant for raising them. It will be noticed that there is a slight difference between the two recruiting orders, owing to the fact that the 83rd Regiment was raised and stationed at that time in Ireland, whilst the 86th Regiment was raised and stationed in England.

ORDER FOR RECRUITING THE 83RD REGIMENT OF FOOT.

"George, R.

"These are to authorize you by beat of drum or otherwise to raise so many volunteers in any county or part of our Kingdom of Great Britain as are, or shall be, wanting to recruit and fill up the respective companies of our Regt. of Foot under your command in Ireland to the numbers allowed upon that Establishment, and you are to cause the said volunteers (to be raised or levied as aforesaid) to march under the command of such commissioned and non-commissioned officers in such numbers, at such times, and to such port as you shall think most convenient for their transportation to Our Kingdom of Ireland, and all etc., and for so doing this Our Order shall be and continue in force for 12 months from the 25th day of March next.

"Given at Our Court at St. James's, this 28th day of Febry., 1794, in the 34th year of Our Reign,

"By H. M. C.,

"GEO. YONGE.

"To Our F. & W[d]. W. Fitch, Esq.,
Lt.-Colo. Com[dt]. of Our 83rd Regt. of Foot,
or, etc., etc.,

"Like Order, dated 16th April, 1794,
C. Cuyler, Esq., Colo., 86th Foot."

But previous to this the following order had been issued to the 86th Regiment:—

WARRANT FOR AUGMENTING A REGIMENT OF FOOT UNDER THE COMMAND OF MAJOR-GENERAL CUYLER.

"George, R.

"Whereas We have thought proper to direct that each of the Companies of Our Regt. of Foot under your command shall be forthwith augmented with one serjeant, one corporal and forty-three private men.

"These are to authorize you by beat of drum, or otherwise, to raise so many men in any county or part of Our Kingdom of Great Britain as shall be wanted to complete the said augmentation, and all etc.,

"Given at Our Court at St. James's, this 28th day of March, 1794, in the 34th year of Our Reign,

"By H. M. S.,

"GEO. YONGE.

"To Our F. & W[d]., C. Cuyler, Esq.,
Major-Gen[l]. in Our Army, Colo. of a
Regt. of Foot, etc., etc."

The first Muster Rolls shew that no non-commissioned officers were taken by either Colonel Fitch or by General Cuyler from the 55th Regiment, as might have been expected, to form a disciplined nucleus for either the 83rd or 86th Regiments. All non-commissioned officers were simply recruits like the men.

EIGHTY-THIRD REGIMENT (1793-1802).

CHAPTER II.

EIGHTY-THIRD REGIMENT (1793-1802).

Recruited in Ireland—Essex Bridge—Royal Barracks—Second Battalion formed—Became 134th Regiment—Eighty-Third left Dublin—Bath, Poole, Southampton—Embarked for West Indies—One company landed in Isle of Wight—Situation in West Indies, etc.—Landed at Jamaica—Sailed for Saint Domingo—Recalled to Jamaica—Rebellion of Maroons—Description of Maroons, etc., from conquest of Jamaica—Lord Balcarres's plan of campaign—Its failure—Colonel Fitch attempted to carry it out—His death and character—Major-General James Balfour appointed Colonel—Colonel Walpole's successful conduct of the operations—Losses of the 83rd in Maroon War—Remainder of 83rd in Saint Domingo—Rejoined by company from Isle of Wight—83rd returned to Jamaica—State of affairs in Europe—83rd embarked for England—Losses of 83rd Regiment.

THE 83rd Regiment obtained most of its recruits in Dublin, and was allowed two months to teach them the rudiments of drill before being called upon to take over part of the garrison duties of Dublin. The first barracks it occupied was the Old Custom House at Essex Bridge.

Towards the end of the year a great number of the troops then in Dublin were embarked for service in the West Indies. This allowed of better quarters being given to the young regiment, and it was moved to the Royal Barracks, on the north side of the River Liffey. Ireland at this time was swarming with men, and was an almost inexhaustible reservoir for filling up British regiments. The Flight of the Wild Geese, which had begun at the close of James the Second's reign, and had continued to feed the Irish regiments in the French service, had ceased for some time. Most of the officers of the Irish Brigade had had to seek their personal safety by fleeing from France, and the Irish, an essentially aristocratic race, saw thus no reason for crossing in large numbers to bolster up the demagogues who controlled the destinies of France. Thus, though some few enthusiasts for Republican principles had gone to Paris, the great bulk of the Irish were quite pleased to join any British regiment that needed men. It was because of this plethora of recruits that the 83rd which had an establishment of 72 serjeants, 26 drummers, and 1,200 rank and file, had no difficulty in completing their numbers, whilst in October, 1794, a second battalion was raised, though the establishment of the 1st Battalion was reduced to 52 serjeants, 22 drummers, and 1,000 rank and file. This was the battalion already referred to which shortly became the 134th Regiment. After remaining for a year in Dublin, the 83rd Regiment embarked on its course of travel, which has never since ceased.

"Faulkner's Dublin Journal" of 8th November, 1794, has the following paragraph:—

"83rd Regiment—Lieutenant-Colonel Fitch—left Dublin between 8 and 9 a.m. on Friday, 7th of November, 1794, for England. The regiment consisted of 1,150 men, of whom not one was absent."

It landed at Pill and proceeded to the famous watering town of Bath. Having enjoyed itself at that centre of fashion for some months, the regiment was then sent to the seaside and Poole, in Dorsetshire, was the selected town. After a residence of five months there, the 83rd moved to Southampton, but whether by marching along the roads or by voyaging by sea in sailing vessels is not anywhere recorded.

Napier, the famous historian of the Peninsular War, writing of these times, mentions that it took three years to make an efficient British infantry soldier. Such a period of time, however, was not allowed to this young regiment before they saw their first active service.

On the 5th May, 1795, the 83rd Regiment embarked at Stokes Bay, and, after remaining on board ship in harbour for some ten days, finally set sail for the West Indies. The voyage was prosperous, except for one untoward incident. One transport which was carrying a company of the Regiment collided with another vessel just off the south coast of England, and the transport was so badly damaged that the company had to be landed in the Isle of Wight. This was in May, 1795. And the company did not rejoin its regiment until 1798, when it also went to the West Indies and picked up its comrades at Saint Domingo.

The situation in the West Indies at this time was rather curious. Fortescue, in his history of the British Army, points out that, out of eighty-one battalions of infantry in the British Army in the beginning of 1793, nineteen battalions were in the West Indies and nine in India, whilst twenty-five more battalions garrisoned what we now call Canada, and Gibraltar, and Ireland, leaving twenty-eight battalions of infantry to defend Great Britain and to form expeditionary forces when necessary, and also to act as marines on board the fleet. Moreover, he adds what anyone who has even only lightly skimmed through the history of the West Indies will admit to be true, that the climate there was so deadly to European troops that the Government had to count on actually replacing each of these nineteen battalions in the West Indies every second year, as the whole of the men, on the average, died in that time.

It can, therefore, be imagined that the British Government was very loth to keep such a large force in the West Indies.

However, at this time, they were obliged to keep an unusually large number of troops there for the following reason.

France owned a number of important islands in the West Indies. The white inhabitants were prosperous, important people, and when the French King was beheaded, and internal strife weakened the central government in France, these white inhabitants expressed strong disapproval of the course pursued by the National Assembly at Paris. This body, amongst other things, gave equal rights to all free mulattoes and negroes in the islands on the 4th April, 1792, and sent out commissioners to see that its decrees were duly carried out. In the meantime in some of the French islands the mulattoes and negroes had risen against the whites, and much plundering and massacring went on on either side. The French white residents, therefore, applied to the British Government, asking them to take over the islands and to either annex them or hold them in trust for the French King when he should come to his own again. In the meantime, the planters on the English West India Islands became alarmed at these negro risings on islands not a day's sail from their own, and begged for additional troops. Jamaica, for instance, agreed with the home authorities to pay for a regiment of dragoons (the 20th Light Dragoons), and also for two extra infantry regiments. The moment war broke out between England and France in February, 1793, orders were sent from England to occupy the French islands of Martinique, Guadeloupe, St. Lucia, and Mariegalante.

In September, 1793, the British Government accepted the capitulation of the French part of Saint Domingo from the planters there, who were suffering greatly from negro risings and the general folly of the commissioners from the French Assembly in Paris.

The part of Saint Domingo owned by the French was the western end of the island, and was roughly 125 miles from north to south, and 75 miles from east to west. The population was about 30,000 whites and over half a million of coloured people; 25,000 of both constituted the French armed forces in the colony, and pandemonium of the worst sort reigned supreme. One and a half regiments of infantry were sent from Jamaica to acquire and garrison this new possession. Some few of the French regular troops came over to the English, including the Irish Regiment of Dillon, and more troops were sent from Jamaica. By June, 1794, there were seven battalions of British infantry in French Saint Domingo, but disease was so rife that out of 3,700 men 1,700 were on the sick list. Desperate fighting with the French and their negro allies now ensued. For instance, at Tiburon, out of a British garrison of 480 men, 300 were killed and wounded, most of the wounded being afterwards massacred by the dusky enemy. In the meantime, Martinique was taken from the French, at the end of March, 1794. Saint Lucia capitulated after some stiff fighting on the 2nd of April, 1794, to the same expedition as had conquered Martinique, whilst General Grey, who commanded the expedition, pushed on for Guadaloupe, and by the 21st of May of the same year this island also surrendered. The French sent out an expedition to recover these possessions. Guadaloupe was recaptured by them. It is believed that hard fighting and disease accounted for 12,000 deaths amongst the British soldiers and sailors in the West Indies in the year of 1794 alone.

Whilst these things passed in the West Indies, the English were engaged elsewhere. In Holland the Duke of York and his gallant army had been driven out of the country after desperate fighting, not being too well supported by the Austrians and Prussians, who were really intent on obtaining from Russia a share of Poland, whilst the Dutch were affected by the French Republican ideas, and gladly allowed the enemy into their country, to their great regret afterwards. At Toulon, again, the English had been driven out, whilst only in Corsica had any measure of success attended their arms. Thus, as can be easily understood, reinforcements of troops were required everywhere, and the 83rd Regiment, though hardly as yet perfectly trained, was hurried off to Jamaica, with a view of being sent on to Saint Domingo.

Things had grown steadily worse from the British point of view in the West Indies. Guadaloupe had been retaken by the French; St. Lucia had been evacuated in June, 1795. Grenada was held by insurgent negroes, assisted by the French, the English being huddled into one corner of the island, having lost their Governor and forty of the principal white inhabitants, who were captured and massacred. St. Vincent was in a like bad state. There the Caribs, a tribe of aboriginal natives, had turned upon the British, egged on by French emissaries. Attempts by parties from Guadaloupe were made on the island of Dominica, but these were repulsed. Finally, to add to their troubles, the Maroons in Jamaica revolted in July, practically on the very day when the 83rd landed in Jamaica. In the ordinary course the Regiment would not have landed there at all, but would have gone straight to Saint Domingo, but the Government, no doubt for the sake of economy, had sent the troops out in loaded merchantmen, whose owners only contracted to carry the regiment as far as Jamaica. For the moment, it was not realised how very serious this Maroon rising was, and the troops were re-embarked on such vessels as were available and proceeded towards Saint Domingo, but Lord Balcarres, the Governor of Jamaica, sent a schooner after these ships, which overtook two of them, with rather over half the 83rd Regiment on board, and these troops returned at once to Jamaica, landed at Mondego Bay, and marched into the interior of the islands to assist in putting down this rebellion, which was serious not on account of the numbers of the enemy, for that was quite trifling, but on account of the inaccessibility of the country, and from the danger that the negro slaves in Jamaica would make common cause with the Maroons, and that a fierce servile war would devastate the whole island. A short account of the Maroons, of their haunts, and of the course of events will explain with what enemy and in what sort of country this regiment of recruits first saw service.

Jamaica originally belonged to Spain. The English tried continually to trade there with small success. During the reign of James I. the Spaniards caught some English sailors and sent them to the galleys. Complaint was duly made to the Court of Madrid by King James's Ambassador. There he received the haughty reply that inquiries should certainly be made into the matter, and the Governor of Jamaica should certainly be called to account for sending them to the galleys instead of hanging them, as he ought to have done.

Sir Walter Raleigh also complained to King James that the Spaniards had murdered twenty-six English who had been ashore trading with them for a month peaceably without a sword amongst them. The unfortunate men had been tied back to back and their throats had then been cut.

Such things might pass in the reign of James the First, but with Oliver Cromwell in power things wore a different aspect. An expedition was sent out in A.D. 1655, under General Venables in charge of the troops, and Admiral Penn in command of the fleet, and after a creditable defence Jamaica passed finally into British hands.

Some of the slaves deserted from their Spanish masters and fled to the mountains, where they maintained themselves by plunder, and received the name of "Maroons."

The planters were compelled to fortify their houses and buildings for protection. Government took action, and for a time succeded in keeping these savages in check, till a Maroon leader, by name Cudjoe, came to the front. He was a strong man and a strong leader of men. This remarkable savage—for he was nothing more—organised his fellows with a discipline of his own invention. His method of warfare was this: Never, if possible, to let one of his men be seen, and never, under any circumstances, to meet his enemy in the open. The country was such as to favour his tactics. It was intersected from east to west by a series of deep glens, rocky at the sides, invariably precipitous, and only to be approached by one narrow path, along which Indian file was absolutely necessary.

These glens were designated cockpits. The chain from east to west was connected by narrow passes, and there were certain others running in parallel lines from north to south. The fertile brain of Cudjoe invented a most elaborate system of signals, horns being used. How elaborate this was may be gauged from the fact that it is stated that each man had a separate call. Along the sides of the outlets to the cockpits Cudjoe would post his men in absolute concealment, and some of these outlets were as much as 800 yards in length. Woe betide any force which was caught therein. Each man was marked and shot down by an unseen foe. Still, British methods of warfare at length prevailed, and Cudjoe was driven out from the south. He took refuge in the north-west in what was called the Trelawney district. Here his fastness was wellnigh impregnable. Moreover, there was land there which could be cultivated and also water to be obtained. For ten years a desultory warfare was carried on with all the horrors of plunder, brigandage and butchery of wounded and prisoners on the part of the Maroons. At length Cudjoe practically compelled terms to be offered to him, and a treaty was entered into between that worthy and others of his crew, whose names have been handed down as Johnny Accompong, Cuffee and Quaco, and the Government of George II.

To the Maroons the terms were distinctly favourable—absolute independence and self-government, save that they were not allowed to inflict the death penalty. A complete amnesty was assured, hunting rights granted, and 1,500 acres of land awarded to them for ever.

All that the Government claimed was a right to send two white residents to live amongst the Maroons, and that all runaway slaves captured by the Maroons should be returned. The Maroons also covenanted to aid King George against all enemies.

The arrangement worked well till 1795, as the Maroons continued until then to carry out their obligations. For amusement they hunted pigs and runaway slaves. From the former the name Maroon, *i.e.*, Hog Hunter, is supposed to be derived. They were ruled by their chiefs with an iron hand. For religion—if religion it can be called—they practised what remained of their native African God or devil worship. They bred fowls, cattle and horses, and, though negroes, preferred the white man to those of colour, and, like white men, they both drank and gambled. In 1760 they joined forces with the Government when the blacks rose in revolt, and in 1779-80 made common cause against white invaders of the island. All seemed to point to security.

The restrictions on their intercourse with Jamaica in general gradually relaxed. The Maroon began to wander farther afield and inter-married with female slaves on the plantations. The chiefs lost their hold on the people, and the rod of iron rule was no more.

Some trifling trouble about a white resident whom the Maroons first wished to be changed, was the ostensible ground for the rising, and when the Government obliged them by appointing another resident in his place, they fiercely espoused the former man's cause.

Trouble was brewing, and the new Resident was driven from the Maroon town. A message came from the Maroons that they were ready for war, and that if the white men did not come to fight them they would come down and fight the white men.

The then Governor, Lord Balcarres, perceived the gravity of the situation. Already the Maroons had sent their women and children into their fastnesses. They threatened to kill their cattle and their women and children too if these should prove encumbrances.

No doubt French emissaries from the neighbouring islands were stirring up strife, and now a new menace appeared. On several plantations where the masters and slaves had hitherto subsisted in such harmony as was possible under the circumstances, unrest on the part of the negroes was apparent and discontent openly expressed.

Everything now looked most serious; Balcarres endeavoured by diplomatic means to remove the trouble, and, indeed, persuaded six chiefs to surrender. He then proclaimed martial law and prepared for military operations. The scheme of the Governor was to blockade the recalcitrant Maroons and starve them out. How he expected to blockade 40 square miles of terribly difficult country with the slender force at his disposal, effectually, is not easily to be understood.

However, on 9th August the main passes to the Maroon districts were seized. Between thirty and forty of the chiefs surrendered and were placed in confinement. This, however, did not bring in the Maroons. As an act of defiance they burnt their villages. Now, it is estimated that at this time the Maroons only totalled some 1,200 men, women and children. The force at the disposal of Lord Balcarres consisted of detachments of the 13th, 14th, 17th and 18th Light Dragoons, the whole of the 20th Dragoons (The Jamaica Regiment), the 16th and 62nd Foot (about 300 men), and part of the 83rd Foot. There was also a local Militia, officered by settlers, and the number of Major-Generals amongst these Militia officers was somewhat remarkable. The orders given by Lord Balcarres were as follow: Colonel Sandford, with the 16th Foot and 18th and 20th Dragoons, was told off to guard an outlet to the north. Colonel Walpole, with the 13th and 14th Light Dragoons (some 150 men), barred the outlet to the south.

Lord Balcarres, with the 83rd Regiment, closed the south-west, whilst Colonel Hull, with 170 men of the 62nd Regiment and one troop of the 17th Light Dragoons, guarded another northern place of egress.

First blood went to the Maroons, who attacked a militia post, killing and wounding several men. By the direct orders of Lord Balcarres on the same day (13th August), Colonel Sandford started to take "Old Town," a Maroon village. When well within the pass the column was met by a volley, though not a Maroon was to be seen. A second volley killed Sandford. The officer now in command seeing that a retreat in such a situation would be fatal, dashed right through at the head of his men and managed to rejoin Balcarres. Two officers and thirty-five men were killed and wounded. The local Militia, some of whom accompanied the column, fled. Six days later "New Town," or rather the site of a new town the Maroons were building, was occupied unopposed.

On the 23rd August a combined movement was made of three columns, under Colonels Fitch, Incledon, and Hull. Operations started at daybreak, and it was proposed to capture Old Town. The place was taken without opposition, having been evacuated by the Maroons for some days. Lord Balcarres now built a blockhouse on the "New Town" site, and determined to lay waste the Maroon provision grounds.

To his surprise the enemy, whom he thought he had penned in a cockpit, to which they had withdrawn, moved round some six miles in his rear, plundered a planter's house and burnt it. Colonel Fitch was now left in command, as Lord Balcarres returned to Port Royal. Fitch was expected to form a complete cordon round the Maroon position, which was, of course, a task beyond his powers.

The Militia now retired to their homes, leaving the Regular troops to finish the business. This Colonel Fitch essayed to do.

He had been kept very busy by Lord Balcarres, and the plan of campaign was none of his choosing. He had been ordered to cut a path through the Maroon country from end to end, and, not being able to see the direction owing to the close nature of the country, he placed buglers with the working parties at either end. The buglers continually sounded some call, and the men worked towards the sound. The brushwood was so thick that a Maroon named Dunbar, who was sent with a small party under a flag of truce to Colonel Fitch, said that he was standing with his party only six yards from Colonel Fitch, covering him with their muskets, and that they would have fired upon him and his party forgetful of their peaceful errand, if they had not observed two companies of infantry under arms close by, from whom they did not think they would be able to effect a safe retreat. In some way Colonel Fitch inspired confidence in the Maroons, and he had several interviews with them, first by the Maroons speaking to him from their concealment in the bush, and later on when they came in under a flag of truce, and offered to make peace with him but not with Lord Balcarres. He had to point out that he had no authority to make peace himself, and added that he must fight the King's enemies everywhere, and laughingly explained that even if the Maroons came to terms, he would have to go to Saint Domingo to fight the French.

Colonel Fitch had placed Captain Legh, one serjeant, two corporals, and thirty privates of the 83rd Regiment in a small pallisaded enclosure guarding a ravine. Captain Legh reported that he was under a heavy fire from the Maroons from some heights and wished permission to go forward to seize the high ground and hold it. On the 12th September, about 9 a.m., Colonel Fitch, attended by Lieutenant Brunt, Adjutant of the 83rd, and several other officers and men, went forward to Captain Legh's post. On arrival there he at once decided that Captain Legh was quite right, and he himself proceeded to select another place in advance for the post. He was remonstrated with for needlessly exposing himself, but answered in his usual placid way that he was only doing his duty, and ought to be as forward as any in doing that. He then allowed a small party to proceed 100 yards in advance, and this was fired upon at twelve paces' distance and most of the party were killed or wounded. Colonel Fitch was found by an officer, seated on a fallen tree, his arm supported by a projecting stump, and his head resting on his hand. Blood was trickling down from the middle of his waistcoat, and the short red and brown striped linen jacket which he wore stuck out behind appearing as if a rib had been broken.

Colonel Jackson, of the Jamaica Militia, seized his hand, and Colonel

Fitch turned his face towards him with a smile. The Maroons had reloaded and the click of their locks could be heard. Colonel Jackson gave the order to lie down, and tried to get Colonel Fitch to lie down too. Before the wounded man could move a bullet pierced his forehead just above the right eye, and he fell forward dead. Thus passed away, with courage and dignity, the first officer appointed to either the 83rd or 86th Regiments, the first of a numerous band who have been killed in the Regiment whilst serving their Sovereign and their country. Eight men killed and seven wounded was the result of this skirmish. Captain Legh was only apparently slightly wounded, but within four days he had died from the effects of the wound, whilst Lieutenant Brunt was also slightly wounded.

It is stated in history that no man was ever more lamented than Colonel Fitch. In person he was tall and graceful, easy and affable in his manner, and was never happier than when relieving the wants of his soldiers or providing some comforts for his younger officers from his own stores. He was brave, benevolent, and of a bewitching address.

However, Colonel Walpole, of the 13th Light Dragoons, who now assumed command, pressed the fighting, and the men began to beat the enemy at his own game. The Maroons commenced to lose heavily from the fire of marksmen concealed in the scrub like themselves.

By the 18th of December, 1795, these gallant savages were beaten into surrender. Colonel Walpole accepted the Maroons' submission, and gave them his word that they should not be sent out of Jamaica. So soon, however, as they had surrendered their arms the Jamaica Government ordered that they should be deported to Nova Scotia. The same Government also voted a sword of honour and £500 to Colonel Walpole, which he most properly refused, as his terms with the unfortunate Maroons had not been kept.

The 83rd Foot lost in this series of skirmishes 70 killed and wounded.

Major-General James Balfour was gazetted from the 13th September, 1795, to the Colonelcy of the 83rd Regiment, vice Lieutenant-Colonel Fitch, killed in action.

In the meantime, the remainder of the 83rd Regiment had proceeded to Saint Domingo, where they shared in the vicissitudes of the British forces in that island. There the English were only able to maintain themselves by the fact that the various parties for a long time spent most of their time fighting each other, only devoting their somewhat overworked energies occasionally to fighting the British. Mulattoes, negroes and French colonists all fought together, and the English would not have had much trouble in beating this divided enemy if it had not been for the scourge of yellow fever, to which we have already referred.

Fortescue places the actual burials of the English for 1794, 1795 and 1796 in the West Indies at over 35,000, and, further, from his works it may be assumed that wounds and infirmities accounted for an equal loss to the Services, or a grand total of 70,000 men alone in the West Indies.

The Generals in command in the West Indies continually called for more troops, and were sparingly supplied with what they called " Boys," and not men. It was this fact which accounted for the amazing mortality amongst the troops, as, though the recruits were full of pluck and dash, they easily fell victims to the local diseases.

In 1798 it was decided to withdraw from Saint Domingo, and in October of that year the evacuation of the island was complete. The part of the 83rd Foot which had been there, and which, it will be remembered, had been reinforced by the company which had been left in the Isle of Wight since 1795, as the result of the collision at sea when starting for Jamaica, returned to Jamaica reduced to a skeleton, and it was once more reunited as a regiment, being stationed on the northern side of Jamaica.

At this time the British had recovered the whole of Grenada, and maintained themselves in Martinique and Saint Lucia. Saint Vincent, though much devastated, was also in their hands completely again.

During 1798 the rebellion in Ireland occurred, and attempts at the invasion of that country were made by the French. In May, Napoleon sailed, with 30,000 men, to Egypt, taking Malta en route. Then the British fleet re-entered the Mediterranean, from which it had been absent for a year, fought and won the Battle of the Nile, and cut Napoleon's communications with France, and in August, 1799, he had to slip through that fleet in his remaining frigate, leaving his army cut off in Egypt behind him.

In September, 1799, the English, assisted by the Russians, again invaded Holland, but were driven out. Spain had made an alliance with France some time before this, and expeditions were sent to Ferrol, etc., with but poor results. An expedition was sent to Egypt in December, 1800, which met with the greatest success. Malta was captured by the British. War raged in India, fortune always favouring the British there. Denmark and Sweden formed an armed neutral alliance, aimed against British naval power, and the successful Battle of Copenhagen was the English reply, whilst the only comparatively peaceful spot in all this turmoil of war was the West Indies, where several attempts of the revolted Saint Domingo negroes to land in Jamaica were repulsed without difficulty, and where a small expedition captured all the Swedish and Danish Islands. On the 25th of March, 1802, the Peace of Amiens was signed, and in June, 1802, the 83rd Foot was embarked in men-of-war from the northern side of Jamaica, where they had remained since 1798. The Regiment embarked at the ports of Savannah le Mar, Falmouth and Mondego Bay, but only sailed as far as Port Royal, in the south-east of Jamaica, whence they marched inland a few miles to Spanish Town.

All preparations were now made for bringing the Regiment back to England, and, as a preliminary, any men who could be induced to do so were permitted to volunteer to extend their services by joining regiments which were remaining in the West Indies. A large number, therefore, joined the 60th and 85th Regiments, and a few went to the 2nd West India Regiment. On the 4th July, 1803, the 83rd Foot finally embarked for England. They had woefully shrunk in numbers, for the actual strength on board consisted of 1 lieutenant-colonel, 2 majors, 9 captains, 16 subalterns, 29 serjeants, 11 drummers, and 294 rank and file.

His Majesty's ship, "Delft," on which the regiment had embarked, had a most prosperous voyage, leaving Jamaica, as stated, on the 4th July, and landing the Regiment at Portsmouth on the 22nd of August.

During the seven years that the 83rd was in the West Indies they appear to have lost 26 officers and 870 non-commissioned officers and men from death. Besides small drafts from Great Britain, it had received some volunteers from other regiments, making a total of 410 rank and file, as reinforcements to the Regiment. The balance had volunteered to join other regiments or had been invalided home during that time.

Note.—The Regimental Records give the following names of the 26 officers who were killed or died in the West Indies, with the dates of their death from disease or wounds:—

(1) Dr. Weir, 6th Aug., 1795.
(2) Lieut. Reeves, 2nd Sept., 1795.
(3) Col. Fitch, 12th Sept., 1795.
(4) Capt. Legh, 18th Sept., 1795.
(5) Ens. Horridge, 24th Oct., 1795.
(6) Lieut. Armstrong, 27th Oct., 1795.
(7) Lieut. Moreton, 28th Oct., 1795.
(8) Lieut. Corr, 30th Oct., 1795.
(9) Capt. Hansard, 9th Nov., 1795.
(10) Capt. Hay, 14th Nov., 1795.
(11) Lieut. Wilton, 14th Nov., 1795.
(12) Surgeon's Mate Clancy, 14th Nov., 1795.
*(13) Ensign Byrne, 17th Aug., 1796.
(14) Ensign Morris, 20th Aug., 1796.
(15) Capt. Keane, 20th Aug., 1796.
(16) Lieut. Tramasse, 20th April, 1797.
(17) Ensign Lawton, no date.
†(18) (No rank stated) F. Smith, 8th Aug., 1800.
(19) Lieut. Ball, 20th Aug., 1800.
(20) Ensign Hill, 30th Sept., 1800.

* Believed to be Lieutenant Benson, Ensign Byrne being a clerical error in the Records.

† Probably an error for Lieutenant Thomas Smith.

(21) Major White, 27th Nov., 1800.
(22) Lieut. Gibson, 4th Oct., 1800.
(23) Lieut. Williams, 1st Dec., 1800.
(24) Capt. Wilson, 7th June, 1801.
(25) Lieut. Wright, 12th Dec., 1801.
(26) Lieut. Farrell, 26th Jan., 1802.

Note.—During the 83rd Regiment's tour of service in the West Indies, 640 non-commissioned officers and men were tried by Courts-Martial.

The Records of the Regiment show that 149,853 lashes were ordered to be inflicted by these Courts-Martial, of which number 49,283 were actually inflicted and 100,570 remitted.

OFFICER'S BREAST PLATE, 1793-1822, dead gilt plate, burnished gilt mounts.

CHAPTER III.

EIGHTY-SIXTH REGIMENT (1793-1799).

Major-General Cuyler's career—Colonel Maitland's report on General Cuyler—Roll of officers of the 86th Regiment whilst at Shrewsbury—Lord Hill's career—86th sent to Kilkenny—Lieut.-General Russell Manners succeeded General Cuyler—Regiment ordered on active service—Sent to Frome—Condition of Recruits sent to Army fighting in Holland—Bat and Blanket horses for 86th Regiment—Camp equipment for war in Holland—Rations of forage allowed to different ranks of Regimental Officers—Scale of Barracks accommodation and equipment—Parties embarked as Marines—H.M.S. "Boyne" burnt—Fleet action of Lord Bridport on 22nd June, 1795—Action of 13th July, 1795—Major-General Grinfield appointed Colonel of the Regiment—Inspection by the Duke of York—Order for augmentation by drafts from 118th Regiment—Establishment when so augmented—Transfers obtained from 121st Regiment—86th Regiment moved to Portsmouth—Augmented by two companies—Marched to Guildford and returned—Embarked for the Cape of Good Hope—Regiment remained at the Cape of Good Hope until 1799—Embarked for Madras—Appearance of Regiment on arrival in India—Sailed for Bombay.

THE 86th Regiment was raised by Major-General Cornelius Cuyler. At first they were not numbered, and were designated "General Cuyler's Shropshire Volunteers." General Cuyler had just returned from the West Indies, where he had commanded an expedition against the island of Tobago. Having stormed the principal fort on the 15th April, 1793, the rest of the island promptly submitted, and he came back to England a successful man, and was promoted Major-General. According to the records, General Cuyler was appointed to command a regiment of foot on the 30th of October, 1793, whilst the order for raising the regiment was dated 1st November, 1793. General Cuyler's services will be found at the end of the book in Appendix II. It may be of interest to add that whilst he was Commander-in-Chief in the West Indies, shortly after this—from 1796 to 1798—the Government sent out a Colonel Maitland to make confidential reports on various officers, etc. His report on General Cuyler, which has now come to light, was, that "it would not be to the disadvantage of the Service if ill-health or any other cause removed General Cuyler from the West Indies." This is evidently the first unsatisfactory confidential report on any officer of the Regiment that is recorded, and it may be interesting to remember that Lieutenant-General Cuyler was promoted General about the same time as this report was made, and was shortly afterwards nominated as Commander-in-Chief in Portugal (in 1799), and was created a baronet in 1814, so the report did not affect him adversely.

Though the regiment was formed at Shrewsbury, yet large numbers of men were obtained from Yorkshire, Lancashire and Cheshire; in fact, men were very scarce, and were sought wherever they could be found. It remained at Shrewsbury until April of 1794, when it embarked for Ireland, via Parkgate, and after landing at Cork, on the 4th of May, proceeded to march to Kilkenny. Its roll of officers in February, 1794, whilst at Shrewsbury was as follows:—

Colonel, Major-General C. Cuyler.

Lieutenant-Colonel, George Sladden.

Major, R. M. Dickens.

Captains, T. C. Hardy, W. H. Digby, Charles Byne, Edward Robinson, Alexander Campbell, Rowland Hill, and Robert Bell.*

Lieutenants, Thomas Nelson, Hugh Houstoun, W. S. Currey, Edward Barnes, Thomas Pickering, Charles Dod, George Middlemore, Charles E. Jolley, Daniel Gavey, William Semple, and J. C. Tuffnell.

Ensigns, William Murray, Thomas Thornhill,† Thomas Sims, W. C. Williams, James Burke, Daniel Macneil, Edward Fox,† and William St. Clair.

Adjutant, Daniel Coleman.

Quartermaster, Richard Jackson.

Surgeon, Hugh Dean.

Chaplain (?), Charles Austen.

The most successful man amongst these officers was Rowland Hill, who afterwards became Baron Hill, of Almanza, and Commander-in-Chief of the British Army, besides being the Duke of Wellington's right-hand man through the Peninsular Campaign. It is on record that the private soldiers under his command adored him, and he was always known amongst them as "Daddy" Hill, from the fatherly interest he took in his men. Captain Rowland Hill had family connections with Shropshire, and he brought a draft of local men to the Regiment when it was raised.

It was whilst the Regiment was at Kilkenny that they received their permanent number as 86th Regiment of Foot, and were also simply called "Shropshire Volunteers," as General Cuyler was transferred to the 69th Regiment on the 20th June, 1794.

Lieutenant-General Russell Manners succeeded General Cuyler as Colonel of the 86th Regiment.

Within ten weeks of its arrival in Ireland the Regiment received orders to proceed on active service. Fortunately for itself, the orders were cancelled, and the Regiment, instead of proceeding to the wars, was landed at Frome,

* George Cuyler is not shewn in this list, but was in the regiment at the time.

†Ensigns Thornhill and Fox are not shewn in list of officers of the Regiment, as it is not clear if they ever actually belonged to the 86th Regiment. Ensign F. Thornhill was Ensign in Army 17-4-1793, Ensign in 64th Regiment, 30-9-1793, but may have been antedated. Ensign E. Fox was Ensign in Army 5-4-1793, Ensign in 59th Regiment 5-2-1794.

in Somersetshire, and was then ordered to the Isle of Wight. This most fortunate circumstance gave the Corps time to learn its work thoroughly, and for the younger soldiers to grow into men. The result was that it became a well-matured regiment, and its sick list was always small. The frightful mortality caused in the West Indies, which principally arose from the youth of the soldiery, did not affect it, and from its first land campaign in Egypt to the capture of Bourbon, including the hard fighting in India, from 1802 on, the Regiment conducted itself with the greatest credit, and invariably carried out successfully the orders it received.

To our soldiers of the present day it appears strange that such things should be noticed. They are accustomed to everything being thought out in advance; to Army Reserve men of mature age being recalled to the Colours to fill up the ranks for war; and when on a campaign to receiving rations, etc., regularly. During the Duke of York's campaign in Holland in 1794 and 1795 things were not so well arranged. Parties of men were constantly sent over to the Netherlands who not only had never fired a musket, but for whom no muskets or accoutrements were issued. Sometimes they even had no uniforms given to them. Often they were served out with a canvas jacket and trousers, no shirts, drawers, stockings, or shoes, and sent direct to the Netherlands as soldiers! H.R.H. The Duke of York, who was in command of such an army in Holland as the Government allowed him to have, in vain protested against such scandals.

Curiously enough, no mention is made in the regimental records of the Regiment being ordered on service in July, 1794, though Cannon mentions its being embarked at Cork in readiness for active service, whilst in one of the warrant books in the Public Record Office an amount is paid out to provide bat and blanket horses for the 86th Regiment. This was in 1795, but probably was only passed through the accounts at that time. It appears that it was the Army of the Duke of York in Holland that the Regiment was to proceed to join, and its camp equipment for active service, as given in the orders of that time, is of interest to soldiers, and is as follows:—

Tents and camp necessaries for the following regiments going to join the Duke of York's Army on the Continent:—

Regiment——.	Bell Tents, Duke of York's Pattern, 12 men to a Tent.	Camp Colours.	Drum Cases.	Powder Bags.	Hatchets.	Kettles and Bags.	Wooden Canteens.	Haversacks.	Blankets.	Water Decks.	Pack Saddles.	ROBERT BISSET, Com'ers General.
	84	12	20	10	168	168	864	864	864	40	40	

" N.B.—No bell tents are wanted, as the arms are to be fixed in the tents with the men.

" The kettles are to be of iron, and fitted in wooden frames, five in a frame."

This order was dated 10th April, 1794, and held good when the 86th Regiment embarked at Cork, but was amended in August, 1794, by the additional order that round tents should also be taken and that dragoons were to have Hanoverian camp kettles, whilst the infantry were to use the old pattern kettles.

A further order issued with regard to forage is worth entering:—

" Number of rations of forage allowed . . . to officers of the line serving on the Continent:—

Lieutenant-Colonel		9
Major		7
Captain		5
Captain-Lieut., Commanding the Colonel's Company		5
Subalterns	each	1
Adjutant		1
Quartermaster		1
Surgeon and Mate	each	2
Companies, Captains absent	each	2
Captains without Companies	each	3"

Whilst the 86th Foot was at Kilkenny the following order was issued to them in common with other regiments stationed at home with reference to domestic arrangements. The order is dated the 30th day of May, 1794, and proceeds to relate that:—

" Each Field Officer shall have 2 rooms.
Each Captain shall have 1 room.
Each 2 Subalterns shall have 1 room.

But if there are plenty of rooms, each subaltern may have one, with due allowance of candles and coal.

Twelve N.C.Os. and Private men, 1 room.
Officers' Mess, 2 rooms.

Where kitchens are provided, soldiers are not to cook their victuals elsewhere.

The weekly deliveries of coals and candles for every room occupied as above are not to exceed the following quantities, viz., three bushels, one quarter of coals, and one pound and one quarter of candles in winter. . . One bushel, three quarters of coals, and three-quarters of a pound of candles to the infantry for the summer.

N.B.—When sea coal is not used, one hundredweight of coal is considered equal to a bushel."

The order then goes on to consider the furniture, etc., in officers and men's quarters.

" SCHEDULE OF FURNITURE AND UTENSILS.

Commissioned and Warrant Officers' Rooms.

Closet, 1; table, 1; chairs, 2; coal box, 1; coal tray, 1; bellows, 1; fire irons, 1; fender, 1.

Non-Commissioned Officers' and Private Men's Rooms.

Bedsteads, mattresses, palliasses, blankets (pairs), sheets, rugs—12 single or 6 double.

Round towel closet or shelves, table, rack for arms, set of fire irons, fender—1.

Forms, 3.

Iron pots, wooden lids, pair of iron pot-hooks, iron trivets, wooden ladles—2.

Iron flesh-fork, frying pan, gridiron—1.

Large bowls or platters—2.

Small bowls or porringers, trenchers, spoons—12.

Water bucket, coal tray, pair bellows, candelstick, tin can of 3 gallons, large earthern pan for meat, basket for carrying coal—1.

Drinking horns, 2; earthern chamber pots, 6.

Broom, mop—1."

Evidently usually all cooking was expected to be done in each barrack room.

About the 21st January, 1795, three captains, 5 subalterns, 14 serjeants, 5 drummers, and 395 rank and file were embarked as marines. They were placed on four line of battleships, viz., "Prince of Wales," "Triumph," "Brunswick," and "Hector." Though it is now unusual to use line regiments as marines in the fleet, it had then long been the custom to do so. The fleet at the time was largely manned by men who had been seized by the press gang and forced to serve as sailors in the men-of-war. About the years 1795 and 1796 large numbers of disaffected Irishmen, called "Defenders," had been forcibly drafted into various ships, and in April, 1797, most formidable mutinies broke out in the fleet in different places.

The marines, or troops acting as marines, were berthed in every man-of-war, between the bluejackets' and their officers' quarters, and armed guards of marines were always mounted about the ship. Looking at the matter from our present standpoint the wonder is not that marines had to be employed for such purposes and that mutinies occurred, but that these ships, manned by these poor pressed men, almost invariably went so gallantly into action, and sustained such dreadful losses as they did without flinching.

On the 17th and 24th February more men were called for to serve as marines, and 1 captain and 6 subalterns, with 10 serjeants, 5 drummers, and 261 rank and file went on board H.M. ships "Prince," "Saturn" and "Boyne."

All the above ships mounted 74 guns, except the "Prince of Wales" and "Prince," which were 98-gunned ships.

The grenadier company of the 86th was on the "Boyne," and they are supposed to have been the cause of her loss. The "Boyne" was anchored off Spithead on the 1st May, 1795, and was commanded by Captain George Gray, and carried the flag of Vice-Admiral Taylor. Though it has never been correctly ascertained how the fire originated, it is known that the marines were engaged at musketry practice on the windward side of the ship. It is believed that some lighted paper from their muskets must have blown back into the portholes of the Admiral's cabin, setting fire to his papers and any other inflammable articles which may have been there. The flames burst through the poop before the fire was discovered, and in spite of every exertion to extinguish the conflagration, the ship was ablaze fore and aft within half an hour. All the fleet sent boats to the "Boyne's" assistance, and all the crew were saved except eleven of the sailors. To add to the danger of the salvage work, the "Boyne's" guns, which were loaded, were discharged as the iron became heated, and two men in the fleet were killed by a chance shot. About 1.30 p.m. the ship broke from her anchors and drifted away on the tide until she grounded, still burning furiously, on the spit opposite Southsea Castle. Here she blew up at 6 p.m., when the flames reached her magazine, but, fortunately, without doing any further damage.

The grenadier company lost their arms, accoutrements and baggage in this disaster.

At the Public Record Office there is a note to the effect that compensation was granted as follows:—

"Pay Book No. 3, 1797, 59: Warrants to Lieutenants Eyre, Mapey, and Mr. Campbell, for loss of luggage when the 'Boyne' was burnt at Spithead." This is included in the warrants for payment of money sent to the 86th Regiment, and evidently the names are really those of Lieutenants Eyre Massey and John Campbell, and it is to be hoped that these two gentlemen obtained payment despite these clerical errors on the part of the War Office clerk.

On the 16th June, 1795, part of the Regiment went into action for the first time. Five British line of battleships were ordered to reconnoitre three French men-of-war, which were lying at anchor off Belle Isle. The British ships were respectively, H.M.S. "Triumph" and H.M.S. "Brunswick," on which the men of the 86th were serving as marines, and H.M.S. "Royal Sovereign," H.M.S. "Mars," and H.M.S. "Bellerophon." On getting close to the enemy's vessels they suddenly discovered nine more line of battleships and 15 frigates, all French.

The English ships promptly drew off, with the French fleet in hot pursuit. Next morning the "Bellerophon" and "Brunswick" cut away their anchors, and threw some of their provisions overboard, but still were overhauled by the enemy. About 9 a.m. the action commenced. The "Mars" was very badly knocked about, and would have been taken by four French ships, but the "Triumph" and "Royal Sovereign" bore boldly down to her assistance, and, after a certain amount of fighting, by 7 p.m. the very greatly superior enemy had withdrawn. The "Triumph" and "Mars" were the only two ships which suffered any damage, the "Mars" losing twelve men killed and wounded, whilst the "Triumph" did not lose a man, all her damage being confined to the masts and rigging, and the whole of her stern being smashed in. It is recorded that this ship fired over 5,000 lbs. of powder from her stern chasers alone in this fight. This action is known in Naval annals as "Cornwallis's Retreat."

The withdrawal of the French fleet can only be accounted for by its dread of being led into a trap, and its retreat being cut off by the British fleet, which might be using these five ships only as a decoy. Though not intentionally arranged, this very nearly happened. The French had brought their action with H.M.S. "Mars," "Triumph," etc., to a close about 7 p.m. on the 17th June, 1795, and about 3.30 p.m. on the 22nd June the regimental records state that a fleet of 14 sail, under Rear-Admiral Henry Harvey, fell in with the same French fleet, commanded by Admiral Villaret. As a matter of fact, Admiral Harvey's squadron was part of Lord Bridport's fleet, and the latter's total strength was twenty-one line of battle ships.

The British squadron at once made all sail for the French fleet, and carried on after them until about 6 o'clock on the morning of the 23rd June, when, after some fighting, three French ships were captured—"Alexandre," "Formidable," and "Tigre," whilst the remainder of the French fleet ran into shelter at Port L'Orient.

The Admiral's ship, H.M.S. "Prince of Wales," and H.M.S. "Prince" had detachments of the 86th Regiment on board, acting as marines, as already stated.

The total loss of the fleet in this second action was 31 killed and 113 wounded. It is only fair to the gallant enemy to add that before they struck their flag, these three French ships lost 670 of their complement killed and wounded.

The "Hector" and "Saturn," on the 13th July, 1795, had an indecisive engagement with the French near Cape Roux, but again the enemy's ships escaped by running into Frejus Bay.

On the 25th March, General Russell Manners was transferred to the 26th Light Dragoons, and Major-General Grinfield was appointed Colonel of the 86th Regiment, having before been Lieutenant-Colonel in the Third Foot Guards. His Royal Highness the Duke of York, who had just returned from his unsuccessful campaign in Holland, inspected the headquarters of the

Regiment, which remained in the Isle of Wight on the 22nd August, and expressed his approbation of their good order and appearance.

In October the 86th Foot was ordered to be augmented to the strength of 100 rank and file per company, and the letter accompanying the order is given in full, as being of a certain amount of interest, shewing what bounties were paid at the time for transferring and how the arms and accoutrements of the men were dealt with :—

"War Office,
"14th October, 1795.

"SIR,

"In the absence of the Secretary at War, I have the honour to acquaint you that the King has been pleased to order that the 86th Regiment of Foot, under your command, shall be completed to the augmented establishment of 100 rank and file per company by drafts from the 118th Regiment, now on marine duty.

"His Majesty has further directed that in consideration of the high establishment now ordered, the Regiment shall have an additional lieutenant per company, to be taken from amongst the ensigns of your Regiment, as His Majesty shall be pleased to direct. The serjeants and drummers are to be transferred as such.

"The drummers, beyond those allowed in the augmented establishment of the Corps, are to be paid as supernumeraries and mustered accordingly until vacancies shall occur in the 86th, to which they can be appointed.

"The drafts will be allowed one guinea and a half each, to be accounted for to them by your Regiment. Their regimental debts (if not exorbitant and beyond what the usual custom of the Army admits of) are to be transferred with them.

"It has been the practice to transfer the rank and file of new corps as private men and with the same bounty, but on the present occasion the corporals transferred as privates are to receive one guinea additional bounty.

"The drafts are to take with them their clothing, etc., and those on board the men-of-war are to retain their arms and accoutrements, the latter to be paid for by you, unless you should prefer exchanging them by accoutrements of your own, in which case the accoutrements of the 118th are to be carefully returned to the Colonel.

"The officers on marine duty (whose names are specified in the letter above referred to) are to continue on that service until releived [relieved] by officers of your Regiment which should be done when the ships come into port.

"Enclosed I send you a state of the numbers of the Regiment, according to its augmented establishment.

"I have, etc., etc., etc.,

"(Signed) M. LEWIS.

"Major-General Grinfield,
"Colonel of the 86th Foot."

The establishment, as augmented, appears to have been as follows:—

1 Colonel and Captain.
1 Lieutenant-Colonel and Captain.
1 Major and Captain.
7 Captains.
1 Captain-Lieutenant.
21 Lieutenants.
8 Ensigns.
1 Chaplain.
1 Adjutant.
1 Quartermaster.
1 Surgeon.
2 Mates.
52 Serjeants.
50 Corporals.
20 Drummers.
2 Fifers.
950 Private men.

1120 Total.

To complete the Regiment, besides men from the 118th Regiment, others were brought in from the 121st Regiment, which had been recently released from prison in France.

In December the Regiment moved to Portsmouth and Hilsea, and its detachments acting as marines were gradually withdrawn from the men-of-war as they came into port, until in January, 1796, the whole Regiment was again together, and it was still further augmented by the addition of two extra companies (the 11th and 12th), which were added to the establishment of the Regiment as recruiting companies.

In April, the 86th Regiment marched to Guildford, but marched back to the Isle of Wight in June, and embarked from there for the Cape of Good Hope, where they landed on the 22nd of September, six days after the surrender by the Dutch Governor of the Colony to the expeditionary force, under the command of General Sir Alured Clarke.

The 86th Regiment was stationed at the Cape of Good Hope during 1797 and 1798, being augmented by drafts from the 95th and from other corps.

On 20th February, 1799, it embarked for India and landed at Madras on the 10th of May.

The appearance of the men was especially commented on. Well grown, well set up, and 1,300 strong, the Regiment excited universal admiration. The Regimental Records add: "Perhaps the finest body of men that ever came to that country."

The 86th Regiment remained at St. Thomas's Mount, near Madras, until the 8th June, when it sailed for Bombay, as the troops were no longer required on the east side of India, owing to the fall of Seringapatam on 4th May, 1799, and the death of Tippoo Sahib.

The Regiment landed at Bombay on the 22nd July, 1799, where it remained for the time being, sending small detachments by sea to Tannah and Surat, under Major Bell and Captain James Richardson respectively. Those detachments rejoined in December, and now a short summary of the situation in India will not be out of place.

CHAPTER IV.

EIGHTY-SIXTH REGIMENT (1799-1802).

Tippoo Sahib—Situation of affairs with regard to France in the Indian Ocean and Egypt—Abercromby's Army sent to Egypt—British Expedition from India ordered to Java—Three companies of 86th Regiment sent—Destination changed to Isle de France—Again changed to Egypt—Expedition went to Ceylon—Returned to Bombay—Sailed for Egypt—Arrived at Cosseir—Marched to Ghennah on the Nile—March of 11th Soudanese Regiment compared—Force moved down the Nile to near Cairo—Three companies of 86th Regiment also sent to Suez—Ordered to march across the desert to Cairo—Details of march—Description of British campaign in Egypt—Great moral effect of desert march of the 86th Regiment—The French capitulated at Cairo and Alexandria—Strength of French—Regimental states of the 86th—Rewards given to the British Troops—The 86th returned to India—Route followed.

1800

THE 86th Regiment had been hurriedly brought across from the Cape of Good Hope, with a view to taking part in the campaign against Tippoo Sahib, the warlike ruler of Seringapatam. Though by no means such a capable warrior as his father, Hyder Ali, yet Tippoo had courage and a certain amount of generalship. He attempted to make use of his central position to concentrate against one of the two British forces converging on his capital. His attack failed, yet he bravely resisted the combined forces of his foes in Seringapatam, and when the breach was finally stormed, he fell, as became him, at the head of his gallant native troops. This occurred on the 4th May, 1800, and the 86th landed in India on the 10th May. One great danger to British existence in India had passed away with the fall of Tippoo Sahib, for there is no doubt that he was in league with the French. As mentioned earlier in this history, Napoleon had occupied Egypt with an army of roughly 30,000 veteran troops, whilst England's only way to send reinforcements or assistance of any sort to her Indian possessions was by a tedious voyage round the Cape of Good Hope in sailing ships. This took at the least some months, and if the ship was not a strongly armed one it was liable to be attacked by French privateers from the Islands of Mauritius and Bourbon, which lie in the Indian Ocean, well to the eastward of Madagascar.

The situation has been well summed up during 1800 and the early part of 1801 as follows:—It is said: " Very different was the position in Hindustan at the commencement of the present century (19th) from that which the British now enjoy. The fierce Mahrattas stood in unbroken strength; a mere rumour of invasion either by Sikhs or Afghans struck deep alarm into the East India Company's Council at Calcutta. To the southward an arduous and costly campaign had been but recently concluded by the conquest of Mysore (Tippoo Sahib), nor were tranquility and undisputed government established as yet in the provinces which now form part of the Presidency of Madras.

" In Europe another consideration failed not to excite some apprehension in the minds of British statesmen. If France should retain a long and undisputed possession of Egypt it was but too probable that they might acquire a dangerous influence.

" At Constantinople still stronger were the grounds for uneasiness when Paul, the hot-brained Czar, broke wrathfully from his alliance with England and appeared likely to accept the proffered friendship of Napoleon Bonaparte. Such a union might have been based on mutual guarantees of the Danubian provinces to Russia and of Egypt to France.

" But there was yet another and more urgent motive for desiring to expel the French from Egypt, and to press forward the attempt without loss of time.

" England had lost all her Allies. Napoleon was suddenly making the greatest efforts to equip the fleets of France, Spain, and Holland. Calculating on the United Kingdom being obliged to keep a strong fleet in the Baltic to make head against the Northern Powers, he was sanguine and strong in the hope that he would acquire a naval superiority in the Mediterranean."

The British people had come to desire peace, and to feel that a peace was necessary. The ministers of the Crown, whilst acquiescing reluctantly in this conviction, were anxious that they should be in a position to negotiate on strong and advantageous ground. Abercromby, with 15,000 men, was sent to invade Egypt, held by the French with more than 28,000 veteran soldiers; whilst the Government of India determined to surprise the Dutch Settlements in Batavia or Java, Holland being in the hands of the French.

1801

In the beginning of 1801 troops were secretly collected in Ceylon for this purpose, and on 1st January, 1801, three companies of the 86th Regiment sailed from Bombay and landed at Point de Galle, in Ceylon.

By this time, however, the Indian Government had changed its mind and had determined to send the expedition to capture the French settlements of Isle de France, or Mauritius, and Bourbon, which as already stated, were used as a rendezvous for French fleets and privateers, and, being on the direct sea route round the Cape of Good Hope, were a standing menace to all British ships and most destructive to British trade in general. The expedition consisted of 10th, 80th and 88th Regiments, seven companies of the 19th Regiment, three companies of the 86th Regiment, a battalion of Bengal Volunteer Sepoys, with detachments of Bengal, Madras, and Bombay Artillery. The whole under command of Colonel The Hon. Arthur Wellesley, of the 33rd Regiment.

Again the Indian Government changed its mind, this time on account of urgent orders from the Home Government that they had decided to turn the

French out of Egypt, and they called upon the Indian Government to co-operate by sending a force up the Red Sea.

On the 14th of February the whole force, except the seven companies of the 19th Regiment, sailed from Ceylon to Bombay to refit and take in provisions. The transports were escorted by two gunboats.

On 30th March Major-General Baird arrived and took over command of the expedition, Colonel Wellesley remaining at Bombay.

The expedition sailed out on the 1st of April. The last transport, in which was Major Bell, of the 86th Regiment, did not leave Bombay until the 13th April. Three companies of the 86th sailed with this expedition. The force—strength, 5,226 rank and file, of which 2,838 were King's white troops—landed at Cosseir on the 18th of June. Cosseir is stated in the regimental records to be in Upper Egypt, but is more accurately described as a port on the west side of the Red Sea, north of Suakin.

On the 28th the three companies commenced their march by detachments across the desert to Ghennah, on the Nile, distance 120 miles.

It was divided up into marches as follows:—

Cosseir, to the New Well	11 miles; water probable.
Half-way to Moilah	17 miles; no water.
Moilah	17 miles; water.
Advanced Wells	9 miles; water, but very bad.
Half-way to Legeta	19 miles; no water.
Legeta	19 miles; water and provisions.
Baromba	18 miles; no water.
Ghennah	10 miles; the Nile.

Though the troops only marched on eight days, seventeen days were taken to perform the whole distance, as numerous halts were made, and water was found every other day, so that no very great inconvenience was suffered by the men.

On this march some of the men of the 86th Regiment, whilst halted at Moilah, decided that water could be found at the foot of a small hill near the camp. Captain Middlemore, in charge of the detachment, urged them on, and they proceeded to dig a hole four yards square and three yards deep, and were rewarded by discovering a fine spring of water.

It may be of interest to notice here that when, in July, 1889, the 2nd Battalion Royal Irish Rifles proceeded up the Nile on the Toski Expedition, the 11th Soudanese Regiment marched across the desert to the Nile to join in the campaign by exactly this route, taking only 5½ days to perform this march, at least one of the Soudanese soldiers died from sunstroke on the way. Captain Macdonald, of the Gordon Highlanders, afterwards Major-General Sir Hector Macdonald, K.C.B., who commanded this Soudanese Battalion, stated to the Author that they had found all the wells exactly as laid down in the above records, with English names still given to them by the Arabs, and that the one dug by the troops was especially pointed out to them by their guide.

On the 6th of August the force was assembled at Ghennah, on the Nile, and then proceeded in boats down the river for 300 miles, landing finally on the Island of Rhonda, near Cairo, on the 29th of August.

On the 26th of December, 1800, three companies of the 86th Regiment, being the two flank companies and the General's company, embarked at Bombay, under Lieutenant-Colonel Lloyd, and sailed in the "Leopard" (50 guns), "Fox" (36 guns), "Bombay" (32 guns), and two store ships, the "Adam Smith" and the "Babel Mandel." On the same ships was a detachment of the Bombay Artillery, a battalion of Bombay Sepoys, etc. Admiral Blanket (sometimes spelt Blanquett), in the "Leopard," was the naval officer in charge of the expedition.

On arrival at Mochoa, on the 12th of January, the "La Forte," frigate, joined the expedition, which then made sail, Jeddah being reached on the 26th of February. Here the "La Forte" struck on a bank and became a total loss, but the crew and the guns were removed from the wreck by the boats of the remaining ships.

The expedition sailed again on the 5th of March, and arrived at Suez on the 16th of May. Its slow progress was accounted for by the shallowness of the water, which compelled the ships to anchor and to take advantage of the tides to cross the shoals. The Admiral's ship took the ground on one occasion, but no harm was done to it. The troops were not finally landed until the 4th of June.

On the 6th of June, at 6 p.m., the three companies of the 86th Regiment commenced their march on Cairo. The exact distance is 78 miles from Suez to Cairo. It was considered necessary to make a detour of some 12 miles to avoid a party which the French in Cairo were reported to have sent out to intercept these three companies. The total distance was thus 90 miles to be marched, and on the way at only one place could water be found, viz., at El Hanka (or El-Khankah), 12 miles from Cairo.

Three pints of water per man was carried on camels, but, unfortunately, the water-skins were old and some leaked, and the water became a putrid liquid, with maggots in it in a very short time.

The thermometer had stood during the daytime at 109°. Gradually it fell during the night to 86°. Before nine o'clock that evening three officers—Captain Cuyler, Lieutenants Morse* and Goodfellow—became so ill that the two lieutenants were sent back to Suez.

*Lieutenant Morse is believed to have been attached from the 8th Regiment. No Lieutenant Goodfellow is shewn in the Army List of this period, so it may be a clerical error for Lieutenant Goodlad, of the 33rd Regiment, who may have been attached to the 86th Regiment.

At 11 o'clock, after a march of twelve miles, Lieutenant-Colonel Lloyd halted the party for two hours, and moved off again at one o'clock in the morning and marched until 7 a.m. on the morning of the 7th of June.

The day then became so intolerably hot that the Commanding Officer ordered tents to be pitched, with a view of sheltering the men from the sun until evening. This entailed unloading the baggage camels, and gave a good deal of trouble generally, but as 26 miles of the march had been accomplished everyone was in good spirits.

Unfortunately, however, other circumstances had to be considered. The Arab guides now came to Colonel Lloyd and told him that as the day promised to be excessively hot, if the camels remained resting on the ground they would soon become so stiff in their limbs that they would be unable to advance without water. On the contrary, if the camels were once on the move, despite the heat, they would be able to keep advancing. The guides also added that they suspected that the camel drivers would steal the water from the skins whilst the soldiers slept, and the skins again could be better watched whilst actually slung on the camels on the march.

After due consideration, Colonel Lloyd determined to push on at once, especially as he considered that he was already short of water, and also on account of his lack of food for the men. They had been duly supplied with salt pork rations at Suez, but, as the men well knew that eating these would only increase their thirst, they, with cheerful improvidence, had thrown the meat rations away.

At 11 a.m. the tents were struck and the march was resumed. The thermometer again stood at 109°. Captain Cuyler, who had rejoined in a weak state, now fainted, and was left with a camel and two men to look after him. He was finally able to return to Suez.

About one o'clock the heat began to tell on the men, and they straggled so terribly that Colonel Lloyd, seeing that many would fall exhausted by the way and die of thirst, cut his own baggage off the camels and put as many men as possible up on these ships of the desert. All the officers followed their Colonel's example.

At two o'clock the thermometer rose to 116°, and the Khamsen, or south wind, commenced to blow. The detachment struggled forward manfully, despite the hot wind, hunger and thirst, until four o'clock in the afternoon, when a halt was made.

The officers and men had been seized with dreadful sensations during this last five hours' march. Some were affected with giddiness and loss of sight; others fell down gasping for breath and calling for water; others, including Lieutenant-Colonel Lloyd, had seen the mirage—clear, cool lakes which always vanished away just before they were reached.

The water-skins were now found to be cracked by the sun, and the water had become an evil-smelling, thick compound. The men who drank it were seized with vomiting and violent pains. The officers had brought some Madeira with them, and they divided this amongst the men, and then mixed a little rum with what remained of the water and threw the remainder of the rum away. They then issued this concoction to the three companies, accompanied by the warning that every drop was now in their possession, that half the journey had not been performed, and that it was entirely on their own prudence and self-restraint they had to rely as to whether they would, by judiciously using this water, cover the remainder of the journey; as if they drank it at once they must inevitably die of thirst before the wells of El Hanka (or El-Khankah) were reached.

Between six and seven o'clock the wind ceased, and at seven the party moved off, leaving 17 men unable to follow, but some camels remained to bring them on so soon as they could move.

At 11 o'clock on the night of the 7th of June the detachment halted, and all the Europeans fell asleep exhausted by their efforts. The Arabs seized the occasion to break into a small trunk carried by one of the camels, which contained the money of the 86th. At four o'clock, on the 8th, the party again fell in, this time suffering from cold, as a raw and heavy dew had fallen upon them, rendering their limbs stiff and benumbed. The south wind again commenced to blow at two o'clock in the afternoon, but the effect on the men was not so great. Between four and five o'clock in the afternoon of the 8th of June the springs of El Hanka (or El-Khankah) were reached.

The troops had thus marched 78 miles in less than 48 hours, with only three pints of water per man and no food. On arriving at the wells great self-restraint was shewn by the men in using the water, but two of the officers' horses broke loose and drank until they died on the spot.

Eight of the seventeen men left behind joined at El Hanka (or El-Khankah); the others had died in the desert.

During the whole of the march no vegetation, bird, or beast had been seen. After halting at the springs until the evening of the 9th of June, the detachment commenced its march in the dark, to prevent being discovered by the enemy, and at 11 o'clock on the 10th of June it joined the Turkish Army, encamped at Chobra (or Shubra), under the Grand Vizier, the British, under Lieutenant-General Hutchinson, being encamped on the other side of the Nile.

The Turkish and British armies had arrived thus far on their march to Cairo, after a campaign which may briefly be described as follows:—The English Government had thought that the French Army in Egypt only numbered some 16,000 sickly troops. General Sir Ralph Abercromby was, therefore, sent with an army of 16,000 men, which disembarked after a fight at Aboukir Bay, near Alexandria, on the 8th of March, 1801. On the 13th of March it fought a drawn battle, losing 1,300 of all ranks, and on the 21st of March yet another combat took place in front of Alexandria, in which the French lost about 1,000 killed and 900 prisoners, most of the latter being wounded, whilst the English lost 1,700 killed and wounded, including their General, Abercromby, mortally wounded, but the French were driven into Alexandria, and, leaving a party to contain them, Lieutenant-General Hutchinson, marched up the Nile towards Cairo, calling up the Turkish army to his assistance. This army had marched up from Palestine. The Mamelukes who were the late rulers of Egypt, and who formed magnificent light cavalry, also helped in the advance, though they suspected treacherous dealing so much from the Turks, their suzerain Power, that they always kept the British forces between them and the Grand Vizier's army.

Such, then, was the army which the three companies of the 86th joined at the end of their arduous desert march. Cannon's History states that the men had been obliged to abandon their knapsacks on the march, and that their uniforms had been burnt in consequence of the plague, but that, despite all these small drawbacks, the men, being of exceedingly fine physique, attracted "great interest." They joined General Stuart's Division, and the whole force marched on Cairo.

Now, the full effect of the daring march of the three companies of the 86th Regiment was seen. The French at Cairo, over 14,000 strong, knew that an expedition from India was landing at Suez and Cosseir, and they had no idea of their numbers beyond the fact that they were very numerous, and their numbers were not reduced by Dame Rumour. Now, however, these forces from India were actually joining the Army from Europe, and the French saw that they were in an evil plight. It was not for them to know that only 160 of the 86th Regiment had essayed to cross the desert, that nine of them had perished in the attempt, and that the others had just managed to struggle through and no more. All they knew was that a regiment from India had joined Hutchinson, and that many more were following in their steps. This fact weighed largely with the French General, Belliard, who commanded at Cairo, as he afterwards acknowledged, and he made a convention with the British that on condition of his surrendering Cairo, he would be permitted to march his army and guns to a seaport and from there be taken to France by British ships.

The 86th regimental records proudly note that the three companies marched to the Citadel at Cairo on the 10th of July, when several companies of the French *Imperial Guards* grounded their arms and surrendered as prisoners of war. The mistake is obvious, as no French Imperial Guard existed at that time, and also the French did not ground their arms. As a matter of fact, the British arranged to take over the Citadel on the 11th of July. The French, by some mistake or by some design, marched out on the night of the 10th of July, and for some time no English troops could be obtained to take over the Citadel. When the 89th Regiment, which had been hastily fallen in for the duty, did arrive they found the gates of the Citadel closed, and no one to open them, and it was three o'clock in the morning of the 11th of July before an entrance could be effected. The regimental records and Cannon's History both say that the 86th Regiment took over the Citadel. Other records, however, do not corroborate this, but all agree that the 86th Foot did take over the charge of Fort Ibraham (a most important post) from the French Army on 14th of July.

The total strength of the French troops from Cairo, as shewn by their embarkation returns at Rosetta, was 12,912 European soldiers, 760 Greeks, Copts and Mamelukes, with 82 civilians (librarians, mathematical teachers, etc.). Five hundred more French soldiers deserted on the march from Cairo to Rosetta, not wishing to leave the country.

The French at Alexandria surrendered on the 30th of August, and were also shipped off to France, numbering 10,528 of all ranks.

Captain Cuyler and Lieutenants Morse and Goodfellow, who, it will be remembered, had to return to Suez, rejoined from there on the 16th of June, having crossed the desert with a caravan, which was proceeding into the interior of the country.

The parade states shew that the following officers and men of the 86th Regiment landed at Suez and Cosseir:—Lieutenant-Colonels, 2; Majors, 0; Captains, 4; Lieutenants, 8; Ensigns, 1; Assistant Surgeons, 1; Serjeants, 20; Drummers, 11; rank and file, 308; whilst the parade state of the 5th of October, 1801, is as follows:—

"At Camp, El Hamed, 5th October, 1801, 86th Regiment.

Lieut.-Colonels.	Majors.	Captains.	Lieutenants.	Ensigns.	Assist. Surgeons.	Serjeants.	Drummers.	Rank and File Fit for Duty.	Rank and File, Sick, Present.	Rank and File Sick in Hospital.	Total Rank and File.
1	0	4	8	1	1	20	11	247	41	4	292

It is known from the India Office Records that 63 men of the Regiment died in Egypt during this campaign, whilst 8 men are reported to have deserted and 24 men were sent back sick to Bombay, so the Regiment must have received a draft.

The total French forces in Egypt when Abercromby landed with his 16,000 men, was 32,180 soldiers, and 768 civilian establishment.

The Eighty-Sixth received, in common with the other corps which served in this enterprise, the thanks of Parliament and the Royal authority to bear on their colours the Sphinx, with the word "Egypt," to commemorate the share taken by the Regiment in this splendid achievement.

To perpetuate the remembrance of the services rendered to the Ottoman Empire, the Sultan established an order of knighthood, which he named the Order of the Crescent. The superior officers of the Army and Navy were constituted members of this order, whilst the regimental officers received gold medals, varying in size from large ones for the Lieutenant-Colonels, to small ones for the Subalterns, also presented by the Sultan, which they were permitted by King George III. to accept and wear.

The 86th were now ordered to return to India, and they performed this journey by sailing up the Nile in boats to El Hamed, already mentioned, where they remained during October, and on the 30th of that month proceeded to Gheeza, where they remained several months, after which, in May, they crossed the desert to Suez, went thence to Moses' Well, on the Arabian side of the Gulf of Suez, from which place they finally embarked, landing in Bombay on 4th of July, 1802. Whilst at Moses' Well several men died of the plague, and all the men's clothing, bedding and tents were burnt, which appears to have completely stopped the disease. Whilst waiting at Gheeza 207 volunteers joined the Regiment from the 20th, 35th, 48th and 63rd Regiments, presumably being anxious to return to India with the Shropshire Volunteers.

1802

CHAPTER V.

EIGHTY-SIXTH REGIMENT (1802-1806).

Political division of India—Mysore—Travancore—Mahrattas—Armies of these States—Quarrels with the Mahratta Confederation—Two companies of 86th Regiment sent to Dieu, one company sent to Surat—Feuds in Gaikawar of Baroda's family—Capture of Khurri—Losses in the storming of Khurri—Returned to Bombay—Headquarters of Regiment returned from Egypt—Regiment ordered to Baroda—Captain John Grant stormed first position—Enemy surrendered unconditionally—Despatches from Baroda—Losses at Baroda—Field Orders by Lieut.-Colonel Woodington—Operations against Khanoji in Gujarat—Combat on the Myhe River—Major Holmes's despatches—Losses at Mhye River—Holmes reinforced by five companies of the 86th—Combat at Gural and losses—Holmes's despatches—Skirmish at Champria and losses—Combat at Karella and losses—Events leading up to second Mahratta War—Berar—Scindia—Holkar—Marquis of Wellesley—Lord Lake—Colonel Harcourt—Major-General Arthur Wellesley—Operations in Gujarat—Colonel Murray appointed to command—Governor-General's orders to civil authorities—Woodington's further operations against Khanoji—Captain Grant, with volunteers from 86th Regiment, captured a native village—Combat under Captain James Richardson on the banks of Nerbudda—Major-General Arthur Wellesley's forces and dispositions—Murray's return of troops on 18th July, 1803—Woodington ordered to seize Baroach—His forces—Description of Baroach—Woodington's orders for the storming of Baroach—Losses at siege and storming of Baroach—Complimentary order by the Governor-General—Incidents during the assault—Woodington's official despatches describing assault—Situation in Mahratta Confederacy—Further operations of Major-General Wellesley and Colonel Murray—Capture of Powanghur—Return of Murray's force, 17th December, 1803—Monson's and Murray's advance and retreat—Flank companies of 86th marched to Mhow—Battle of Deig—Lake's campaigns—Major-General Jones superseded Murray—State of his troops—Long march from Ralawan to Bhurtpore—Appearance of troops—Captain Grant stormed an outworks and captured eleven guns—Enemy attempted to recapture guns—Assault on Bhurtpore repulsed—Losses—Siege raised—Maharajah of Bhurtpore sued for peace—Holkar retreated to the Punjab—Lake pursued him—Holkar sued for peace—Lord Lake's orders—86th returned to Bombay—List of the Regiment's services from 1801 to 1806—Items from India Office Records.

Note.—All reported occurrences in this Chapter were verified by Captain Norman (late 90th Regiment) with the original despatches at the India Office.

It is difficult, nay, impossible, in this year of grace, to realise the conditions of service in India at the commencement of the nineteenth century. The British settlements then were few and far between, the garrisons scattered and deplorably weak, whilst the position then held *vis-à-vis* the native princes must have been gall and wormwood to the British officer. Now England practically governs the whole of the country, and, though the so-called independent princes do rule over a fourth of Hindustan, they know full well that if they overstep the rules laid down as to what is, and what is not, good government, they will soon be called to order by the British Resident at their Court. In the opening years of the last century matters were on a far different footing. The East India Company had just emerged from a long and costly war with the Moslem usurper of Mysore, and, after a campaign of doubtful results with Hyder Ali, had overthrown his son Tippoo Sultan and restored the Hindoo Prince to the throne. In spite of this, British relations with the ruler of Mysore were strained. The Nizam of Hyderabad, a Moslem chief, ruling over some millions of Hindoos, nominally as the soubah or viceroy of the Emperor of Delhi, looked upon himself as independent, at any rate of the East India Company. The Maharajah of Travancore had on more than one occasion tried conclusions with the Company, and was destined in the years to come to feel the weight of its arms.

To the north lay the great Confederacy of Mahratta Princes, whose territories stretched across the Peninsula from the Bay of Bengal in the east, to the Gulf of Cambay in the west, a belt some 1,400 miles from east to west, and 900 miles from north to south. The chief princes in the Confederacy were the Ruler of Berar in the east, whose territories marched with the frontiers of Bengal, and whose armies had even threatened Calcutta. Scindia, whose headquarters were at Gwalior, geographically speaking, came next. Then Holkar, the ruler of Indore. The Peishwa of Poona and the Gaikawar of Baroda made up the chief quintet. There were many minor chiefs, masters of armies by no means to be despised but owning allegiance to one or other of the five. Then, beyond the frontiers of Bengal, lay Oude, then absolutely independent, and the "*Roi faineant*," the Emperor of Delhi, the descendant of the Great Mogul, at that time a pensioner of the Mahratta Prince, Scindia of Gwalior. Further to the north-west lay the Punjab, ruled then by the Lion of the Punjab, Runjeet Singh.

All of these States possessed armies, formidable in numbers, formidable in organization. There were many European adventurers in India at that time, French officers for the most part, Royalists who, having been proscribed by the Revolutionists, had sought an outlet for their talents in the East. These men were readily and warmly welcomed by the independent princes of India, who entrusted them with the training and organization of their armies on the European model. They taught the chiefs of India to cast cannon, to make gunpowder, to improve the trace of their fortresses, and, generally, to fit their armies for war, and not for mere predatory raids. From the geographical, as well as from the military point of view, the most formidable of all the powers was the great Mahratta Confederacy. Its nominal head was the Peishwa of Poona, and yet he was the one from whom the least was to be feared, whilst Scindia, Holkar and the Bhonslas of Berar, though nominally looking up to the Peishwa as their chief, were far more formidable as foes.

The Mahrattas were originally a pastoral people, who united the profession of war and plunder to the more peaceful avocation of agriculture. Until the year 1773 British relations had been purely mercantile, and English merchants in Surat had been treated with contumely by the Mahratta Chiefs, and had to purchase the privilege of trading at the cost of many insults. In that year one of the ever-recurring internecine quarrels broke out in the dominions of the Peishwa, and the British espoused the cause of one Ragonaut, a claimant to the throne, receiving in return the cession of the Islands of Salsette, and Caranjah, near Bombay, and the Port of Bassein, with extended liberties of commerce. The campaign was fraught with disaster. The Company's troops met with a rude reverse at Baroach (which was afterwards to be captured by the 86th). The Supreme Government at Calcutta disapproved of the action of the Bombay authorities, but, to retrieve the situation, marched a force across India from Bengal into Gujarat, finally conquering that province and restoring our protégé to the throne.

It was said that the Mahratta forces numbered some 260,000 cavalry and 96,000 foot, and it was their boast that they had watered their horses in every river in India—from the Cauvery to the Indus. Be that as it may, they had met with one crushing defeat at the hands of the Afghans on the Field of Paniput, in 1761, and Goddard, with his Bengal Sepoys, unaided by any British troops, had shown that the Sepoy, led by British officers, was more than a match for the best of the Mahratta hordes. Still, the Mahrattas of Berar and of Central India were loth to accept tamely the superiority of the Feringhee, and it was foreseen after Goddard's campaign and the subsequent reverses on the Coast of Malabar that the day was not far distant when Mahratta and Englishman would have to measure swords. British policy then was to win to its side one or more of the chiefs and to fight it out with the rest. Two hundred and sixty thousand cavalry, who carried their all on their saddle-bow, men who would plunder a village at dawn and by sundown be fifty miles away, were not a foe to be despised. British cavalry, so far as shock tactics were concerned, could ride through and over them, but for mobility Europeans were not in the same category. Oddly enough, the Government of Bombay, who had felt the need of cavalry in their previous campaigns with the Mahrattas, had made no attempt to grapple with the evil, and when the Second Mahratta campaign was undertaken, in 1802, there was not a troop of cavalry, British or native, in the Bombay Presidency!

In the year 1802 the headquarters of the 86th Regiment was in Egypt with Sir David Baird, and four companies, under Captain Richardson, had been left behind in Bombay, which had been greatly denuded of troops in order to counteract Napoleon's designs upon India. Not only had the East India Company to take care of their own possessions, but also of those of our Allies, the Portuguese, which also were threatened by the French. The Portuguese then, as now, owned Goa to the south, and Damaun to the north of Bombay, with Dieu still further to the north in Cambay. Two of the companies of the 86th Regiment were sent to garrison Dieu, and one was quartered in the fort at Surat, an important port at the mouth of the Nerbudda River, whilst a fourth was left in Bombay.

Matters at this time were assuming a serious aspect in Gujarat and the Government of Bombay determined to strengthen their forces at Surat in order to be prepared for any eventuality. One of the many and ever-recurring feuds had broken out in the family of the Gaikawar of Baroda. We espoused the cause of the legitimate heir to the throne, one Anunt Rao. He was a man of weak intellect, whilst his illegitimate brother Khanoji was a man of iron will and indomitable energy. Khanoji had the support of the Arab mercenaries, who formed the better part of the Gaikawar's army, and with a number of these he, early in 1802, took up a strong position outside the fortress of Khurri, some thirty miles north-west of Ahmedabad. It is said that Khanoji had at his disposal 28,000 men (of whom 12,000 were infantry), and 68 guns. Colonel Anderson, the Commandant at Surat, was asked as to whether the force at his disposal was sufficient to cope with Khanoji, and he expressed himself in favour of an immediate attack. Major Walker, an officer of the Bombay Army, who had just returned from Egypt, was in garrison at Cambay, with his regiment, the First, now the One Hundred and First Grenadiers, and he at once moved forward towards Khurri, with his own corps and the companies of the 86th. On nearing the fortress, he was met by emissaries from Mulhar Rao, the Mahratta Commander, expressing a desire to arrange terms. Walker, who was entrusted with full political powers, sent his Chief Staff Officer, Captain Williams, to discuss the matter, but Williams was treacherously seized, and whilst Walker was waiting the outcome of the negotiations the Arabs sallied out and attacked his force. Considerable loss was incurred before they were driven into the fort, and in the course of the fighting *Lieutenant Creagh, of the 86th, and two officers of the Bombay Grenadiers, were killed. Walker now saw that the place was far too formidable to be carried by his little force. In addition to the fort itself, which was surrounded by walls thirty feet high, the Arabs had thrown up a formidable entrenchment, flanked by lofty cavaliers; in these were no less than sixty-eight guns, whilst the total garrison was estimated at 32,000, of whom 12,000 were cavalry. Walker himself had not a single sabre! Retreat was not thought of, so Walker sat down before the place and reported fully to Bombay. The Governor rose to the occasion. An express boat was sent to Goa, then held by the 84th Foot and the 3rd Bombay Infantry, its Commandant being Sir William Clarke, the Colonel of the 84th. He immediately collected sufficient boats to move the force up to

* Lieutenant Creagh is always shewn in the Regimental and Indian Records, and also in his family records as belonging to the 86th Regiment. As a matter of fact, he belonged to the 74th Regiment, and is shewn in all Army Lists as belonging to it. He was either attached to the 86th Regiment or had exchanged into it just before his death. He was an elder brother of Sir Michael Creagh, and was promoted Lieutenant 14-6-1800.

Cambay, to which place the 75th Foot, the 1st Battalion 6th and 2nd Battalion 7th Regiments of Bombay Infantry had been ordered. Disembarking at Cambay, Sir William Clarke reached Walker's camp at Budason, three miles from Khurri, on the 23rd of April, and at once assumed command. He found the Mahratta position strongly entrenched, and the enemy, encouraged by their success on the 17th of March, full of fight. On the 24th his artillery having arrived, he threw up batteries, and subjected the Mahrattas to a close bombardment until the morning of the 30th April, when he determined to assault.

Clarke's force was told off into three columns, the actual assault being made by the 75th, supported by the Bombay Grenadiers, with the flank companies of the 84th, Colonel Coleman, of the 75th, being in command. Lieutenant-Colonel Woodington, of the Bombay Grenadiers, was told off to move to the right with the 84th and the whole of the battalion guns, whilst Major Grummont, of the Bombay Army, commanded the reserve—consisting of the four companies of the 86th, the 1st Battalion of the 3rd Regiment, and 2nd Battalion of the 7th Regiment Bombay Infantry. Woodington's fire had the desired effect, and though the Arabs fought gallantly, they were soon driven out of their entrenchments and compelled to take refuge in the fort, which was evacuated on the 5th of May. Sir William Clarke determined to retain possession of the place, and leaving two companies of the 75th, some Bombay Artillery, and a native regiment as a garrison, he retraced his steps to Surat, thence resuming his own position as Envoy to the Portuguese Governor of Goa, and Commandant of a British Garrison in a foreign state! Khurri had not been captured without loss, and, though the 86th had not been actively engaged in the final assault, it had to deplore the loss of a promising young subaltern in the person of Lieutenant Price. The casualties in the force were as under:—

	Officers.		Men.			Officers.		Men.	
	Killed.	Wounded.	Killed.	Wounded.		Killed.	Wounded.	Killed.	Wounded.
Bengal Artillery ...	—	—	5	23	Bombay Grenadiers ...	—	1	2	32
Bombay Artillery	—	—	—	4	1/3rd Bombay Infantry	—	—	—	8
75th Foot	—	—	8	36	1/6th	—	1	—	—
84th	1	—	10	41	2/7th	—	—	1	2
86th	1	—	—	—					

The trophies of the day consisted of six guns, captured in the outworks on the 30th of April, whilst no less than fifty-two were surrendered with the fort on the 5th of May.

The companies of the 86th Regiment which had taken part in the operations at Khurri had a long and arduous march back to Surat. Roads there were none, the many streams were all unbridged, and, owing to its being the rainy season, the whole country was under water. Even at Surat there was but little comfort. The troops were quartered in the fort, where they were exposed to the pestilential exhalations from the river, with the result that the corps became fever-stricken. In October, to the joy of all ranks, orders were received for the companies at Surat to proceed to Bombay, there to rejoin the headquarters, which had just arrived from Egypt.

The stay in Bombay was short. The position of the Gaikawar at Baroda was once more most critical, and Major Walker, who, after the capture of Khurri, had proceeded to Baroda as Chief Political Officer, urgently demanded reinforcements. In aiding Anunt Rao to curb the rebellious propensities of his illegitimate brother Khanoji, the Company had stirred up a hornet's nest, but it had acted with method. It was not working merely for the *beaux veux* of Anunt Rao. The Government wished to regularise their position on the northern side of the Nerbudda, and in return for their assistance, the Company received grants of considerable tracts of land, bringing in a net revenue of close on a quarter of a million per annum. For this annual income it agreed to maintain a garrison of two native regiments with a proportion of artillery in Baroda.

The presence of these troops had anything but a soothing effect on the Arab mercenaries who hitherto had played fast and loose in the Gaikawar's territory. They had shown a strong disposition to favour the claims of Khanoji, and Major Walker, the Resident, thought it advisable that the native regiments at Baroda should receive a "stiffening" of British soldiers. The 86th was chosen to give this " stiffening," and in the middle of November, 1802, it embarked for Cambay, whence it marched to Baroda, the heavy baggage and guns which accompanied the Battalion being taken up the river in country boats. On the 4th of December, 1802, the much-needed reinforcements arrived at Baroda, and then the Resident deemed himself strong enough to demand the cession of the Fort. Both the Resident and the Gaikawar felt that it would be more conducive to the peace of the State if the fort were garrisoned by Sepoys, rather than by Arabs. The Arabs, however, were of a contrary opinion, and refused to listen to the terms of cession. On the 16th of December Major Walker called upon Colonel Woodington, the senior officer in the garrison, to take the necessary measures to secure possession of the fort, and on the same day Woodington shifted his camp to the near vicinity of the fort, outside of which was a walled enclosure, which had been strongly entrenched.

The force at his disposal consisted of the 75th, the 86th, a detachment of Bombay Artillery, the Bombay Grenadiers, and the 2nd Battalion 7th Bombay Infantry. With each battalion were two 6-pounder guns, and with the detachment of Bombay gunners were three 18-pounder and four 12-pounder guns. Three days later the batteries opened, but the Arabs had established themselves in a masonry house covered by a deep tank, from which they kept up a galling fire, or, in the language of Colonel Woodington, "sniped" at the British. The 86th were called upon to carry this house, which they did in the most gallant style, led by Captain John Grant; they were supported, with much dash, by a company of the Bombay Grenadiers. The Arabs now, as at Khurri, signified their willingness to negotiate, and the Resident desired Colonel Woodington to suspend hostilities. However, the Political Officer soon learnt that the main object of the Arabs was to secure a suspension of fire in order that they might repair their works. On the 23rd Woodington re-opened his cannonade and made all preparations for the assault, when, on Christmas morning, the enemy capitulated unconditionally, and Woodington at once occupied the fort with the Bombay Grenadiers. The official despatches on the capture of Baroda afford a connected narrative of the operations, and these are here appended:—

"To Major-General Nicholls,

"Commanding Officer of the Forces,

"SIR,—I have the honour to acquaint you that on the morning of the 18th inst. I moved down from the encampment, leaving a force for the protection of the camp, with the forces under my command and took possession of that part of the Pora, or suburb, of Baroda, situated to the west of the fort. During the night a battery of three 18-pounders was erected within three hundred yards of the fort, at the north-west angle of the part intended to be breached.

"From the difficulties and obstacles usual on such occasions, it has required the labour of a second night to bring the battery to such a state of perfection that I judged necessary. I hope to-night to reinforce it with the additional 18-pounder and to-morrow everything will be ready to breach.

"The loss in officers and men on the day we entered the Pora has been considerable, but I have certain intelligence that it has not been so great as that of the enemy.

"As soon as circumstances admit, I shall have the honour to transmit a detail of the operations, and in the meantime transmit a return of the killed and wounded.

"I have, etc., etc.,

"H. WOODINGTON.

"Camp near Baroda,

"20th December, 1802."

Two days later Colonel Woodington addressed a second despatch recounting the progress of the operations:—

"To Major-General Nicholls,
"Commanding Officer of the Forces.

"SIR,—Since my letter to you of the 20th I have the honour to inform you that our attention has been principally engaged in establishing ourselves in the Pora and completing our batteries, which was done last night, with the loss of only four or five Europeans slightly wounded. On the morning of the 21st a battery of three 18-pounders and one for a howitzer and a 6-pounder was reported complete with their guns in them. About 8 o'clock they opened on the tower before-mentioned with considerable effect, and the Arabs apparently evacuated it for a few moments. However, they soon returned and kept up so smart a fire from the upper part of the tower that we were obliged to discontinue our fire. At dusk I ordered the battery to re-commence, and am happy to state that we did not meet with a single casualty. During the night Captain Hoffman expressed a desire to be relieved from duty, and I was very happy to avail myself of the opportunity, and have ordered him to proceed to Bombay, and Captain Warden is appointed in his room.

"This morning a considerable body of Arabs were dislodged from the houses which they occupied across a tank on our right, and from which they had been sniping at us. This service was performed by the grenadier and light companies of the 86th, together with some sepoys from the Flank Corps, and executed under my eyes in a manner which did credit to the officers engaged; they came up with and bayonetted eight or ten of the enemy; our loss was one European killed and four wounded.

"I have, etc., etc.,

"HENRY WOODINGTON.

"Camp near Baroda,
"22nd December, 1802."

On the early morning of the 27th Colonel Woodington received official intimation from the Resident, Major A. Walker, that the negotiations between the British nominee to the Throne of the Gaikawars and the rebels in the fort had been brought to a successful termination. The fort was accordingly handed over to the British troops, and the new Gaikawar installed with befitting pomp and ceremony by the two native regiments in Woodington's force, the 1st Bombay Grenadiers and the 2nd Battalion of the 7th Bombay Infantry. All the officers of the British regiments attended to give éclat to the entry of Anunt Rao into his capital.

The fighting, such as there was, in this short siege of a week's duration fell on the two British regiments, the 75th and 86th, their casualties being approximately the same, whilst the artillery suffered heavily in proportion.

Casualties at the siege of Baroda, 18th to 27th December, 1802:—

	Officers.		Men.	
	Killed.	Wounded.	Killed.	Wounded.
Artillery	—	2	9	18
75th Regiment ...	1	—	6	30
86th Regiment ...	—	—	7	28
Bombay Grenadiers ...	—	1	4	7
2/7th Bombay Infantry... ...	—	2	—	7

Colonel Woodington's final orders to the troops under his command emphasise in a most striking manner the esteem in which the 86th Regiment was held by their commander.

"FIELD FORCE ORDERS BY LIEUTENANT-COLONEL WOODINGTON.

"Camp, Baroda,
"27th December, 1802.

"Whilst Lieutenant-Colonel Woodington laments the loss of the gallant men who fell before Baroda, he congratulates the troops on a successful termination of hostilities by compelling the enemy to evacuate the Fort of Baroda, and accept the terms prescribed to them by Government, and he entreats that the officers and men employed during the siege will accept his unfeigned thanks for the willing and ready support he received from them, and although the enemy gave the army in general but few opportunities of distinguishing themselves, still they did not fail to avail themselves of such as offered, as was instanced in the attack and defeat of a considerable body of Arabs in the White House and Pagoda by a party of His Majesty's 86th Regiment, under Captain Semple, . . .

"By Order,
"GEORGE WILLIAMS, Brigade-Major."

The capture of the fort at Baroda and the ejectment of its Arab garrison was of material benefit to the Gaikawar, rendering his position in his capital absolutely secure, for under the terms of the recent treaty the British were compelled to maintain an adequate force for his protection. Beyond the range of those troops his writ no longer ran. His illegitimate brother Khanoji was still in arms, and, at the head of a very considerable body of troops (for the most part cavalry), ravaged the Province of Gujarat, levying contributions where he willed, and destroying those villages which dared to say him nay. Khanoji's forces received a welcome reinforcement in the shape of the Arabs from the Fort of Baroda, and, as his depredations prevented the collection of the Gaikawar's revenue, Colonel Woodington, at the request of the Resident, detached a force under Major Holmes, of the Bombay Army, to bring him to reason. The task was no easy one. The whole force at Woodington's disposal numbered 918 British soldiers, made up of a detachment of the 75th, the headquarters of the 86th and some Bombay Gunners. His native troops were just 2,400 strong. Not a single sabre in the force, yet it was well-known at Bombay that the strength of the Mahrattas lay in their horse, of which Khanoji possessed several thousands.

Woodington had to provide, not only for the security of the Gaikawar's capital, Baroda, but also for the many forts scattered over the country, so that he was unable to spare Holmes more than 442 British and 720 native soldiers —among the former three weak companies of the 86th, under Captain James Richardson. Khanoji, knowing the weakness of his opponents in cavalry, was by no means averse to trying conclusions. All the chances were in his favour; roads there were none—such tracks as then existed were unmetalled— and the streams unbridged; they ran either across sandy wastes, through which the European troops, in heavy boots, painfully toiled, or through dense jungles, in traversing which Khanoji's horse were able to hover on the flanks, wearying the British with ceaseless attacks on the baggage. The few guns with Holmes's detachment were dragged by bullocks, thus increasing the immobility of the force. A small body of the Gaikawar's cavalry had been attached to the column. Of their fighting qualities no very high opinion was entertained, but at any rate they furnished Holmes with information as to the whereabouts of the enemy, and on the 6th of February Holmes learnt that Khanoji, with 12,000 men, had taken up a strong position between Savali and Pralhampur, on the Mhye River, and, procuring some local guides, he determined to attack. Whether the guides played false is not known. The 75th Regiment were leading, and apparently marched into the thick of the enemy, and were received with a heavy fire, followed by a dashing charge of the Arabs, who had flocked to Khanoji's standard. The road only permitted the men to march in fours. On either side were lofty banks from twelve to fifteen feet in height, and in the dense jungle beyond lay, ensconced, Khanoji's men, who now swarmed down on the 75th sword in hand. The regiment fell back in confusion, leaving one of its guns in the hands of the enemy. In falling back they threw the 86th into some disorder, for the road was too narrow to admit of deployment, and Arab and 75th were in one disorganized mass. All that was possible was for little groups of men to try the effect of British bayonet versus Arab tulwar, Holmes, ably seconded by Captain

Richardson, setting a fine example to all. The 86th, with the Bombay Grenadiers charging up in support, dashed forward with a cheer, swarmed over the banks and drove the Arabs helter-skelter through their camp, re-capturing the lost gun and bayoneting large numbers of the enemy.

DESPATCH FROM MAJOR HOLMES, COMMANDING "THE DETACHMENT TO THE NORTHWARD," TO LIEUTENANT-COLONEL H. WOODINGTON, COMMANDING IN GUJARAT:

"Camp near Sowley,

"7th February, 1803.

"SIR,—I have to inform you that I marched to attack Canojee [Khanoji] at eight o'clock yesterday morning. Owing to some difficulty in procuring the guides, we were not able to march off sooner.

"The whole of the baggage was left behind with a guard and one of the guns to protect the camp during our absence; perhaps it would have been better to have left the whole of the guns behind.

"Canojee [Khanoji] was encamped in a very strong situation, having the River Mahe [Myhe] in his rear, and the side we were obliged to attack was protected by a very thick jungle and a number of defiles through which there was no other passage than a narrow road with high banks on each side; by it both men and guns had to advance.

"This enabled the enemy to attack our advanced guard with great advantage. For a considerable time we were unsuccessful, owing to the narrowness of the road and the thickness of the jungle into which it was impossible for our men to enter. We lost a great number of men and one gun which was supported by the 75th Regiment. This corps was forced to retire in some disorder. The enemy immediately advanced in great force and pressed hard upon us, when I sent orders for the 86th to move forward, and . . . for the first time in my life made use of my sword. We re-took the gun immediately.

"The 75th soon returned to the attack and acted with their usual spirit.

"A great number of the enemy was killed, the remainder fled and crossed the river in the utmost confusion, leaving their camp standing. Whilst they were attempting to cross the river a great number were killed or drowned. I can give no correct account of the number of the enemy, but from their having made no preparation for moving off it would appear that they expected to have been able to stand their ground, or that they were surprised or nearly so.

"I have, etc., etc.,

"G. HOLMES, Major,

"Commanding the Detachment to the Northward."

EIGHTY-SIXTH REGIMENT (1802-1806).

Return of killed and wounded in the detachment under the command of Major G. Holmes, on the 6th of February, 1803:—

	Officers.		Men.	
	Killed	Wounded.	Killed.	Wounded.
Bombay Artillery	—	—	1	—
75th Regiment (1st Gordon Highlanders)	—	3	31	51
86th Regiment (2nd Royal Irish Rifles)	—	—	2	6
2nd Battalion 1st Bombay Infantry (102nd Grenadiers)...	—	—	2	3
1st Battalion 3rd Bombay Infantry (105th Mahratta Light Infantry)	—	—	2	—
2nd Battalion 7th Bombay Infantry (114th Infantry) ...	—	3	—	—

Colonel Woodington, almost as soon as he had despatched Holmes on the arduous task of hunting down Khanoji, realised that the force was far too weak to carry out the desired effect, and on the very day that Holmes was attacking Khanoji, on the banks of the Mhye River, Captain Cuyler, with five companies of the 86th, was ordered from Baroda in his support, Captain Grant, with two companies, remaining behind to afford the necessary stiffening to the native troops in the Gaikawar's capital. Holmes was not the man to allow the grass to grow under his feet. He thoroughly understood the secret of Oriental warfare, " When once you have got your enemy on the move keep him running." The whole province was seething with revolt. To use the expression of the late Lord Salisbury when alluding to the Crimean War, " England had put her money on the wrong horse." Anunt Rao was slothful, dissolute and bordering on the imbecile. He was despised by the Mahrattas. Khanoji, on the other hand, though illegitimate—and that counts for little in the East—was bold, energetic, and of great personal gallantry. The people flocked to his standard and held aloof from our own protégé. Danger lurked near the capital. Holmes continued pressing hard on Khanoji's footsteps. On the 26th of February he attacked some of his adherents at the village of Gural, our losses being three men wounded. His despatch on this occasion is given below.

"Gural Camp, Coppergunge,"
"27th February, 1803."

"To Lieutenant-Colonel Woodington,
"Commanding the troops in Gujarat,

"SIR,—I have to inform you that we have taken and destroyed the village of Gural, which was occupied by some of Canojee's [Khanoji] adherents (about 200 Arabs and 700 collys). The village was to have been taken entirely by Setteram's men (the Gaikawar Contingent). I accompanied them with the 86th and 2nd Battalion 1st Bombay Infantry, but we were intended only to look on at the attack.

"Setteram's men had very nearly succeeded in taking the place, when the Arabs, perceiving, that none of our men were employed, rushed forward and drove Setteram's men back. This circumstance obliged me to order our men to attack, and the advanced guard, under Captain Richardson, advanced so rapidly that the Arabs who were following Setteram's men found themselves nearly at the point of our men's bayonets before they discovered them. This was just what our men wished for, and they took care to take every advantage of it by dashing up to the Arabs with their bayonets and killed and wounded a large number.

"The Arabs, immediately on discovering our men, ran back as fast as they could, without attempting to fire at them. Our men followed them so closely that we were in possession of their village very shortly. We had on this occasion only three men of the 86th wounded.

"I have, etc., etc.,
"G. HOLMES, Major,
"Commanding the Detachment to the Northward."

Holmes continued to follow up the enemy with all the vigour and rapidity of which his force was capable; in fact, having got the forces of Khanoji on the move, he was determined to give them no rest. On the 3rd March he addressed a semi-official note to Woodington, in which he writes:— "I am just now returned from giving Khanoji a second dressing.

"We marched at five o'clock this morning, leaving a guard and one gun to protect the camp and baggage during our absence. At about seven we arrived in sight of Canojee's [Khanoji] position at Champria, which appeared to be exceedingly strong, much more so than the one we attacked him in on the 6th ult. After several rounds from our gun we commenced the attack in two columns; each consisted of 150 Europeans of the 86th Regiment, and 200 sepoys from the 2nd Battalion of the 1st and of the 2nd Battalion of the 7th Bombay Infantry, the one commanded by Major [?] Cuyler, the other by Captain Richardson. I remained with the reserve, consisting of about 100

Europeans and 300 natives. The two columns, although they had to fight their way through the thick jungle for a considerable distance, arrived at Canojee's [Khanoji] camp at the same time. The success was complete, the enemy fled in the greatest confusion, our loss being confined to a couple of men of the 86th wounded."

On the following day Colonel Woodington, who had remained behind at Baroda, was requested by the Resident to disperse a body of Arabs, who had taken up a threatening attitude in the village of Karella, a short distance from Baroda. Woodington immediately moved out with the 75th, a company of the 86th made up to 100 rank and file, and a wing of the 1st Battalion of the 3rd Bombay Infantry. At 2 a.m. on the 5th he arrived opposite the Arab position and at once attacked it. The surprise was complete, and the enemy were driven out of their position and the village was destroyed with the loss of five men of the 75th, Lieutenant Grant and one man of the 86th wounded. The Paymaster of the Force, who had applied for leave to accompany the troops on this little expedition, met the fate of so many volunteers, a stray shot from one of the Arab fugitives killing him on the spot.

In the meantime signs were not wanting that the Company's policy of supporting the Rulers of Baroda and Poona was exciting the wrath of the other Mahratta Princes, Scindia of Gwalior, Holkar of Indore, and the Maharajah of Berar. The cessions of territories in return for this support were viewed with particular disfavour, and Holkar and Scindia, who for many years had been at daggers drawn, now entered into a compact to drive the British out of Gujarat, if not into the sea. The Governor-General, foreseeing that war was inevitable, prepared to meet the coming storm. A campaign against the Mahrattas was fraught with big issues. Khanoji and his followers sank into insignificance when compared to the greater issue, and Woodington and Holmes received orders to refrain from further active operations pending instructions for the new campaign. Holmes now threw his men into the forts of Dholkar and Neriade, the while that Woodington concentrated the bulk of his troops at Baroda.

The Second Mahratta War.

1803

Three of the five ruling Mahratta Chiefs were opposed to the British Government, being the Maharajahs of Berar, Scindia, and Holkar. The two westernmost Chiefs were British Allies. The territories of the latter were open to attack, and in all probability would be attacked, so that it became necessary to provide a sufficient force, not merely for their defence, but also to strike a counter-blow when the suitable moment should arrive.

In the north the Mahratta fortresses of Allighur, Agra, and Delhi threatened the East India Company's own proper frontiers, and the armies of the Maharajah of Berar were a menace to the peaceful Bengali ryots on that side of India. There were strong rulers in India in those days. The Governor-

General, the Marquis of Wellesley, unlike some of his successors, believed that British rule in India rested on British bayonets, and that without their powerful advocacy British justice would have but short shrift. His Commander-in-Chief, Lord Lake, had shown his fitness for command in the dashing fight at Lincelles in the Low Countries and in the suppression of the Rebellion in Ireland, where he had acted with vigour and determination. He was now entrusted with the whole conduct of negotiations, civil as well as military, for the Governor-General was of the opinion that when once the sword has been drawn no interference should be permitted until the enemy had confessed himself beaten. Lake's objectives were the fortresses of Allighur, Agra, Deig, and Delhi; when once these rallying points were in his possession he was to deal with the forces of Scindia and the Maharajah of Berar as he thought fit.

A second column, under Colonel Harcourt, the Military Secretary of the Governor-General, was to march to the south and to seize the territories of the Maharajah of Berar, after making itself master of the fortresses of Balasore and Barabuttee.

General the Honourable Arthur Wellesley, the brother of the Governor-General, who had distinguished himself at the storming of Seringapatam, and later at the capture of Copenhagen, was back in India in command of the Mysore Division, and to him was entrusted the whole charge of the operations in the south, with supreme military and political powers. This was somewhat resented by the Government of Bombay, for, even at the age of thirty-three, the future Duke of Wellington was what may be termed a "masterful man," and his language on occasions could be forcible if polite; the Governor-General was fully prepared to support his brother even if it necessitated rapping provincial governors over the knuckles.

The forces in Gujarat came under the direction of General the Honourable Arthur Wellesley, who at first selected Sir William Clarke, the Colonel of the 84th, for the command of the troops north of the Nerbudda River. Sir William, however, was acting in the double capacity of Envoy to the Portuguese Governor at Goa and commandant of the British garrison in that fortress. This was a post demanding special qualifications, and as the Portuguese Governor officially expressed a wish to retain Sir William by his side, Wellesley conferred the command upon the junior lieutenant-colonel of the 84th, Colonel Murray, an officer who had performed the duties of Deputy-Adjutant-General to Sir David Baird in the Egyptian campaign, and whose reputation at that time stood high.

In announcing this appointment to the Bombay authorities, the Governor-General let it be clearly understood that he would brook no interference from the local authorities, and that his brother was to be supreme. The terms of his despatch (given below) were sufficiently forcible to ensure compliance with all the requirements of the young Major-General.

"Colonel Murray, of His Majesty's 84th Regiment, will be invested with the chief local military authority in the Province of Gujarat, subject only to the control of the Honourable Major-General Wellesley, or of the General Officer Commanding in the Dekkan, and all orders of a contrary nature are to be immediately revoked. . . . The skill, judgment, and heroic valour of our officers and troops must not be frustrated by vexatious counter-action in subordinate officials or by the minute and unserviceable pretensions of inferior civil authorities."

In the meantime, while Wellesley was perfecting his plans and awaiting the arrival of the bullocks, without which his troops were unable to take the field, Woodington, with his acquiescence, made a series of efforts to drive Khanoji out of Gujarat. His whole force now amounted to 918 British Infantry, including the headquarters of the 86th and 3,071 native troops, of whom perhaps one-half were with Woodington at Baroda, the rest being distributed between the various strategic centres in the neighbourhood, Holmes having the control of these. Early in May, despite the terrible heat of the weather, Holmes was ordered to occupy the fort of Keira, which had been ceded to us by the Gaikawar. This was successfully accomplished, and Captain Cuyler was placed in command of the fortress, whilst the Regiment moved on to Nerriade, where it remained under canvas, suffering terribly from the rains. On the 12th of June Captain Grant volunteered to attack a body of the enemy, who were reported to be occupying a village some thirty miles distant. This raid was most successful, the enemy were surprised just before dawn, their position carried at the point of the bayonet, and a number of prisoners, and, what was more valuable, some forty horses were captured, and this with the loss of Lieutenant Procter and six men wounded.

On the 14th of July the Regiment was again distinguished. A large body of Khanoji's followers threatened to cross the Nerbudda above Baroach, thus cutting off communications with the coast. Captain Richardson, with four companies of the 86th and the 1st Bombay Grenadiers, was detached in search of the enemy. Two days later Richardson's guides located Khanoji's men, but the difficulty was to reach them. The unmetalled roads were hidden tracks across dismal swamps, but the men cheerfully harnessed themselves to the guns, which the bullocks were unable to draw, and by almost superhuman efforts the detachment came upon the enemy waiting for the river to subside before they could venture to cross to the opposite bank, where a small body had already established themselves.

Richardson's plans were soon formed. A few rounds from his guns gave his men time to deploy, and then the companies of the 86th stormed the enemy's entrenchments with the bayonet. The success was complete, the Mahrattas were driven into the river, many being drowned, and the few who reached the shelter of the farther shore were disagreeably surprised on the morrow to find that Richardson had secured a few boats, crossed the stream under cover of the darkness, and was on their heels before the sun was up. In his despatch

Richardson spoke in the highest terms of Lieutenant Lanphier, of the Regiment, who was slightly wounded in the action of the 16th July. Richardson, with his four companies, now returned to Baroda, whilst 15 officers and 587 men remained with Major Holmes at Dholka, Nerriade and Keria.

The future Duke of Wellington had been appointed to the supreme command of the troops in Southern and Eastern India on the 26th of June, 1803, and he at once ordered Colonel Murray up from Goa to take over the command in Gujarat. The choice was not a good one, as Wellesley soon discovered, but there was no King's officer of sufficient seniority to put in his place, and Wellesley had not yet gauged the value of the officers of the Bombay Army, though ultimately Major-General Jones, of the Bombay Artillery, did supersede Murray. The senior Lieutenant-Colonel of the 86th, William Dowdeswell, was at the time Military Secretary to Lord William Bentinck, the Governor of Madras. He had commanded a company of the First Guards during the last campaign in Holland, and had there attracted the attention of General Lake, who commanded the Brigade of Guards, and as soon as the Mahratta War broke out Lake offered Dowdeswell the command of a division of the Grand Army, which he naturally accepted, and so Murray was left to mismanage affairs in Gujarat.

In addition to his active army, which comprised the 19th Light Dragoons, the 74th and 78th Highlanders, and the Scots Brigade, with a large force of native troops of the Madras Presidency—cavalry as well as infantry—Wellesley had to provide for the safety of the Provinces of Cambay and Gujarat, since incorporated with the Bombay Presidency, and to be prepared to act against the Mahrattas from the South. In a despatch dated the 2nd August, 1803, he recapitulated the measures he proposed to adopt, and as the regimental history is bound up with the military history of those days, extracts from the future Duke of Wellington's instructions and despatches will prove of interest.

On the 6th August, 1803, when the final rupture with Scindia and the Berar Maharajah took place the forces in Gujarat were distributed as under:—

1. Echeloned between Surat and Songhur, to act on the defensive south of the Nerbudda:

Artillery	217
61st Foot	45
65th Foot	763
75th Foot	573
84th Foot	272
88th Foot	148
1/6th Bombay Infantry	1218

2. At Baroda, and to act north of the Nerbudda:—

61st Foot	109
86th Foot	815

First Bombay Grenadiers	625
2/1st Bombay Infantry	697
1/6th Bombay Infantry	106
2/6th Bombay Infantry	763
2/7th Bombay Infantry	696
1/9th Bombay Infantry	780
Artillery	268

Of these, the 1/9th was a newly-raised regiment, and the ranks of the other regiments were largely filled with recruits. Murray inveighed strongly against being compelled to take the field with such a force, more especially did he dwell on the absence of cavalry, and announced his determination to raise at least a troop which he proposed should be brought on the establishment of the Bombay Army, and should be officered with a couple or more selected officers. To this the Bombay Government gave a ready assent, and an application was made to General the Honourable Arthur Wellesley for a dozen sets of arms and accoutrements of the Madras pattern, in order that others might be made up in Bombay to equip the troop.

The General then proceeded to lay down the composition of the striking force which he wished to be maintained, as follows:—

"Having thus provided for the principal garrisons and stations in Gujarat and the stations dependent on Surat, there will remain for service in the field:

Artillery	99
65th Foot	753
86th Foot	815
1/1st Bombay Infantry	625
2/1st Bombay Native Infantry	697
1/6th Bombay Native Infantry	519
2/6th Bombay Native Infantry	763

"These corps ought to be divided into two detachments, one consisting of 50 artillery, the 86th, and two native battalions, to be stationed in a convenient situation north of the Nerbudda, and in front of Baroda, and the other, consisting of 49 artillery, the 65th, and two native battalions, south of the Taptee, between Songhur and Surat.

"In the cantonments, with these two detachments, ought to be the necessary proportion of ordnance and stores, viz., two 6-pounders for each corps, and two 12-pounders and two 5½-inch howitzers for each detachment. Camp equippage for each detachment ought to be in readiness in Baroda and Surat."

General Arthur Wellesley then issued minute instructions for the guidance of the officers commanding these detachments, as well as for those left in charge of the various posts, and he finally suggested that Colonel Murray should move up to Baroda to assume command without delay.

On the 18th of July, 1803, Murray sent in a return of the troops north of the Nerbudda destined not only to secure the safety of the province from attack, but also to furnish a moveable column to act against Scindia's forces.

	Officers.	Men.
Cavalry	2	58
Artillery	8	368
65th Foot	19	500
84th Foot	3	10
86th Foot	20	600
1st Grenadiers	12	854
2/1st Bombay N.I.	13	978
2/2nd Bombay N.I.	12	871
1/3rd Bombay N.I.	12	614
1/9th Bombay N.I.	14	687
Detachment, 2/6th Bombay N.I.	2	185
Gaikawar's Cavalry	0	266

In the artillery there were just 90 British gunners, and the cavalry had only been enlisted since Murray assumed the command. Saddlery and accoutrements had been borrowed from Madras, and a couple of young officers from the Bombay Native Infantry had been attached to the newly-raised troop. The Gaikawar's cavalry was without discipline or drill, but Murray, in order to infuse a little of both into these wild horsemen, attached Captain Williams, who had been on Woodington's staff, to act as commandant.

The rest of the troops were brigaded as under:—First Brigade: The 65th Foot, Bombay Grenadiers, and 2/1st Bombay Infantry. Second Brigade: The 86th Foot, 2/2nd Bombay Infantry, 1/3rd Bombay Infantry, and 1/9th Bombay Infantry. Captain Cuyler, of the 86th, was placed in command of the First Brigade eventually, with Lieutenant Marston, of the 86th, as his brigade-major.

The Government of Bombay demurred to the instructions of the young General, but the orders of the Governor-General were explicit, and, needless to say, that Arthur Wellesley, young as he was, carried his point.

On the 6th of August General Wellesley issued orders to Woodington, then in command at Baroda, to seize Baroach, and at the same time he suggested to the Government of Madras to push forward a force towards Ganjam, on the Eastern Coast, whilst he himself advanced to Ahmadnagar. On the 21st of August, in conformity with Wellesley's instructions, Colonel Henry Woodington, of the 2/1st Bombay Infantry (now the 102nd King Edward's Own Grenadiers) advanced on Baroach with 500 of the 86th, under Major (?Captain) Cuyler, a detachment of the Bombay Artillery, and his own corps. At the same time the Bombay authorities despatched the "Fury," of the Bombay Marine, round by sea, with orders to ascend the Nerbudda and afford such aid as was possible to the land attack. Woodington reached Baroach on the 23rd, and after a short skirmish drove the enemy into the fort, and threw up batteries armed with the guns of the infantry of his force (only four 6-pounders and two 12-pounders), with which he commenced to batter the place. On the 25th a sortie from the walls was repulsed with the regrettable loss of Captain Semple, of the 86th, and on the 29th the breach having been declared practicable, Woodington issued his orders for the assault.

The fortress was indeed a formidable one, and although the Engineers had reported the breach as practicable, it was clear that if the enemy retired in anything like order to the inner keep, all the work of breaching would have to be done over again. Baroach itself is built on an elevation some 180 feet above the surrounding country and is surrounded by a massive masonry wall eighteen feet in height and twelve feet in thickness. On this stands a solid crenellated and loop-holed parapet, six feet high and three feet thick. At various parts circular bastions had been erected, which afforded the defenders ample flanking fire. Up this steep hill the assaulting columns had to mount under a sharp fire from the enemy's musketry, and the still more deadly rays of the August sun, whilst the ditch was knee deep in mud.

Woodington's orders, which are here reproduced, were a model of clearness, and the men of the 86th had for their leaders officers who had passed their whole service in the Regiment and had earned the respect and confidence of the men—Cuyler in Egypt, Richardson in the many affairs in which the companies left behind in India had been engaged during the past two years.

" Baroach,

" Monday, August 29th, 1803.

" FIELD FORCE ORDERS BY LIEUTENANT-COLONEL WOODINGTON.

" Parole, ' Britannia.' Countersign, ' Success.'

" The breach being reported practicable, the fort will be stormed at three o'clock.

" Storming party, under the command of Major Cuyler.

" Forlorn hope, a serjeant and twelve men of His Majesty's 86th Regiment.

" First party to lead, under Captain Richardson: 100 rank and file of His Majesty's 86th Regiment, including one flank company, and 100 rank and file of the Grenadier Battalion, including one flank company.

" Second party, under Major Cuyler: 150 rank and file of His Majesty's 86th Regiment, including one flank company and 150 rank and file of the Grenadier Battalion, including one flank company. Each party to have hoes, pick-axes and crows with them, carried by soldiers, the scaling ladders (to be used if necessary) to be also carried by soldiers.

" Reserve, under Captain Bethune (Bombay Artillery): 100 rank and file of His Majesty's 86th Regiment, 100 rank and file of the Grenadier Battalion.

"These parties will parade and be formed in the streets the troops occupy, at two o'clock, and then to be completed to thirty-six rounds, after which they will be marched up to the rear of the battery under cover from the view of the fort and wait in it until the signal is given from two 6-pounders fired in quick succession one after the other, which will direct the advance to storm the breach. The reserve will follow the storming party, and after having entered the breach Captain Bethune will immediately form his men to act as circumstances may direct. Twenty boxes of musquet ammunition to be taken with Captain Bethune's party.

"On entering the breach Captain Richardson will turn to the left and march by the works [walls ?] to take possession of the Cutterpore Gate. Major Cuyler's party will follow and push on also by the works [walls ?] to take possession of the Jarrasar Gate. When these gates are obtained, the works [walls ?] are to be cleared of any parties of the enemy in their vicinity, and the men to be kept under arms to act at a moment's warning."

It appears from the official return of the casualties at Baroach during the short siege, as well as in the assault, that the losses of the 86th amounted to Captain Semple, 2 serjeants and 3 rank and file killed; Captains James Richardson, Laughlan, McLaurin, 3 serjeants, and 12 rank and file wounded.

Captain Semple had been severely wounded at the capture of Baroda in December, 1802.

On the 14th of October, 1803, the Governor-General, The Marquess of Wellesley, published a General Order eulogising the conduct of Colonel Woodington and of the force employed in these operations. In this Order the 86th is thus referred to:—

"The Governor-General in Council signifies his particular approbation of the valour and judgment manifested by Major Cuyler, of His Majesty's 86th Regiment throughout the services at Baroach and in commanding the storm of the fort, and by Captain J. Richardson, in leading the assault.

"The Governor-General in Council laments the loss of Captain Semple, of His Majesty's 86th Regiment, killed on the 25th August, and of the brave men who fell at Baroach."

In a subsequent despatch, Colonel Woodington reported the capture of fifteen stand of colours, adding: "I have, at Major Cuyler's particular request, permitted the 86th Regiment to keep two."

Colonel Woodington's despatch bears ample testimony to the gallantry displayed by the officers and men of the 86th, and the fact that the future Duke of Wellington wrote in the warmest praise of the services of Major Cuyler and Captain Richardson, must ever be a source of pride to the successors of the old 86th.

The despatches which are reproduced further on give the bald official account of a feat of arms which has never been adequately rewarded, and which even in the Regiment has been allowed to fall into oblivion. To storm a fortress surrounded by a masonry wall eighteen feet in height, in the full glare of an August sun, was but the forerunner of what the 86th were to do half a century later in the memorable campaign in Central India, under Sir Hugh Rose.

Previous to advancing to storm this fortress, the 86th fixed their bayonets more securely on the muzzles of their muskets by wedging them on with pieces of cloth. This prevented a well-known feat of the Arab swordsmen of striking the bayonet off the muzzle by a blow of the sword on the locking ring.

Captain McLaurin, who was wounded, would have been made a prisoner or have been slain by the natives, but he was rescued with great gallantry by Private John Brierly. Serjeant Bills was conspicuous for his heroic courage, and was appointed serjeant-major in recognition of his soldierly conduct, whilst Serjeant John Moore, who led the "Forlorn Hope," received £50 from the Government, and Private Brierly was promoted to the rank of corporal.

"Baroach.

"To The Hon. Major-General Arthur Wellesley,

"On the Western Side of India.

"SIR,—I wrote to you yesterday evening, after we had stormed and taken possession of the Fort of Baroach. I have now the honour of acquainting you more fully on the subject.

"The breach was reported practicable by the engineer at eleven a.m., when I determined to storm, but delayed until three o'clock, not only that I might benefit by the assistance of the "Fury" and an armed boat, which I expected would arrive in time to take their stations opposite the fort, but as I thought it a very likely hour to find the enemy off their guard.

"The vessel and armed boat, however, did not arrive in time to offer any assistance.

"The enclosed order for the storm will inform you of the dispositions I made, as will the accompanying profile and elevation of the western front of the fort.

"The enemy made a desperate attack in opposing our entrance into the breach, but, by the valour and spirit of the troops, were speedily repulsed, and my orders were carried into full execution. After Captain Richardson had obtained possession of the first gate, Major Cuyler pushed on so rapidly that he over took the Arabs before the greater part, both horse and foot, could get out of the gate, and put to death about two hundred of them; many horses also were killed.

"I beg leave to submit to your notice the ready co-operation of Major Cuyler. Throughout every part of the service his gallantry and conduct in commanding the storm, as also that of Captain Richardson, who led, were most noticeable.

* * * * * * * * * *

" I have great pleasure in informing you that our loss is small. Captain McLaurin, of His Majesty's 86th Regiment, is the only officer wounded, and not badly. A return of the killed and wounded and a general return of the killed and wounded during the siege is enclosed.

" I have, etc., etc.,

" HENRY WOODINGTON, Lieutenant-Colonel.

" Baroach, 30th August, 1803.

" (The ' Fury ' was coming up the river with two 18-pounder guns, but owing to want of water failed to arrive in time.)"

The column commanded by Colonel Woodington, acting in the neighbourhood of Baroda, was under the general supervision of General the Honourable Arthur Wellesley, and in forwarding Woodington's despatch relating the storming of Baroach to the Governor-General, the future Duke of Wellington, pointedly alluded to the gallant services of the 86th Officers.

" Camp,

" September 12th, 1803.

" MY LORD.—I have the honour to enclose copies of papers which contain detailed accounts of the attack upon, and capture of, Baroach. I beg to draw your Excellency's notice to the conduct of the troops employed on this service, particularly to that of Lieutenant-Colonel Woodington, who commanded, to that of Captain Cuyler and Captain Richardson, of the 86th Regiment, and Captain Cliffe, of the Bombay Engineers.

" I have, etc., etc.,

" ARTHUR WELLESLEY, Major-General."

No doubt now existed that the Maharajah of Scindia and the Maharajah of Berar would oppose a serious resistance to our arms, but Holkar appeared disposed to temporise. Fortunate it was for the British that he did not throw his armies into the field at the commencement of the war. In order to appreciate the difficulties, it will be well to recapitulate the situation when Arthur Wellesley was appointed to the command of the forces in Western India.

Lake had concentrated about 10,500 men in the neighbourhood of Cawnpore, and was preparing to operate against the fortresses of Allighur, Agra, Delhi and Deig. His cavalry comprised the 8th, 27th and 29th Light Dragoons; his infantry, at first, had but one British battalion, the 76th, now the Second Battalion of the West Riding Regiment. He was afterwards reinforced by the 75th Foot (which was sent round from Bombay), the 22nd Regiment, and 1st Bengal European Regiment, which, after the operations in Cuttack, marched to the north-west to join him.

EIGHTY-SIXTH REGIMENT (1802-1806).

Colonel Harcourt, of the 12th Foot, Military Secretary to the Governor-General, with the 22nd, 1st Europeans, and 4,000 native troops, were massed at Ganjam prior to a descent on Cuttack and Orissa.

General Wellesley, with 11,000 men, was in the vicinity of Ahmadnuggur; his force included the 19th Hussars, the 74th and 78th Highlanders. A separate division, under General Stevenson, some 9,000 strong, was near Aurungabad. This included the Scots Brigade (now the 2nd Battalion of the Connaught Rangers).

Murray, with less than 5,000 men, was in Gujarat. His was the only column unprovided with cavalry, for the Governor-General's Body Guard was attached to Harcourt's force; whilst Stevenson, who was acting in conjunction with Wellesley, had three regiments of Madras Cavalry attached to him. Murray has been blamed, and justly blamed, for indecision, but it must always be borne in mind that without cavalry his army was deprived of its eyes, and he dependent on native information for every item of intelligence.

The different phases of the campaign may now be briefly recounted chronologically. War was declared on the 6th of August (the very hottest season of the year) against Scindia and Berar, and the same day all the columns put themselves in motion.

8th August.—Wellesley carried Ahmadnuggur by storm.
29th August.—Woodington captured Baroach.
2nd September.—Stevenson captured Jalnapore.
3rd September.—Lake carried Allighur by storm.
10th September.—Lake defeated Mahrattas outside Delhi, which he entered, releasing the King from confinement.
17th September.—Woodington captured the fortress of Powanghur.

The fortress of Powanghur, which the natives considered impregnable, situated, as it was, on an almost inaccessible rock, capitulated to Woodington's force after a short but very effective bombardment on the 17th September, 1803. In reporting the capture of Powanghur, Colonel Murray, of the 84th, then commanding all the troops to the north of the Nerbudda River, wrote as follows to the Commander-in-Chief at Bombay:—

"Headquarters, Baroda,
"21st September, 1803.

"SIR,—I have the honor to enclose a letter which I received this morning from Lieutenant-Colonel Woodington.

"Colonel Woodington highly praises the zeal and activity of the troops under his command, and, to judge by their success, this praise is well merited.

"Colonel Woodington has, in a most particular manner, requested that I should lay the meritorious services of Sergeant Moore, of His Majesty's 86th Regiment, before you. He led the "forlorn hope" at the assault of Baroach, and behaved with the utmost gallantry on that occasion. Major Cuyler speaks highly of his general character.

"I have, etc., etc.,

J. MURRAY, Colonel,

"Commanding the Forces in Gujarat."

18th September.—Stevenson, who had advanced from Ganjam, occupied the important position of Juggurnaut.

23rd September.—Wellesley defeated the Mahrattas at the memorable battle of Assaye.

9th October.—Colonel Wallace, of the 19th Hussars, who had been detached from Stevenson's column with the Scots Brigade and two native regiments, captured Lussaulgaum after a sharp fight.

10th October.—Lake defeated the Mahrattas outside Agra, and Harcourt occupied Cuttack.

12th October.—Wallace occupied the fort of Chandore.

14th October.—Harcourt carried the fortress of Barabuttee by storm, and on the same day, on the opposite side of India, Wallace seized Dhoob.

18th October.—Lake occupied the fortress of Agra, which from that date became a permanent British station.

21st October.—Stevenson carried Assirghur by storm.

26th October.—Wallace occupied Jalna, thus securing all the passes into the Bombay Presidency from the East.

1st November.—Lake defeated the enemy at Laswarri.

29th November.—Wellesley defeated the Berar Army at Argaum.

15th December.—Wellesley carried Gawilghur by storm, a fortress hitherto considered impregnable.

The campaign had been admirably conceived and executed with the most consummate skill. Not a flaw marred any of the proceedings save in Gujarat, where Murray had shown indecision and had failed to do more than wander aimlessly about the country. His subordinate, Woodington, on the contrary, had done all that was asked of him. It is true that Murray, so far back as the 24th of September, the day after Assaye, had reported his force as too sickly to move, and sent the major part of the 65th back to Bombay to recuperate, retaining the men of the other regiments. He wrote to Wellesley:—

"To defend my position is all I can expect, and I am not sanguine that I shall succeed in this. I fear that to ask for a reinforcement from Bombay will be useless unless circumstances will permit a battalion being drawn from Goa."

Wellesley, however, ordered Murray to advance to the north-east, and to form the various details of the 61st, 65th (of which corps 100 men had been retained), 75th and 88th into one battalion. Murray recommended Major Cuyler, of the 86th, for the command of this battalion: "I have much pleasure in recommending an officer of Major Cuyler's merit to your notice."

On the 9th of October, in obedience to the reiterated orders of General Wellesley, Murray moved forward to Godra, where he halted to establish magazines. A month later he occupied Loonawarra, having passed through Barcaon on the 6th November. He made another prolonged halt at Loonawarra, ostensibly to organise a troop of cavalry, as he professed the utmost distrust of the Gaikawar's cavalry, which he reported as worse than useless. On the 10th of December he was at Arcotah, and on the 17th he marched into Dahud, where he again halted to organise a fresh depôt and magazines.

The following return of Murray's force on the 17th December, 1803, the day he marched into Dahud, the very healthiest season of the year, shows the hardships to which the troops had been exposed and how unfit they were to undertake operations in the field:—

Artillery	363 of all ranks,	with 104 sick.
65th Foot	105 ,, ,,	,, 39 ,,
75th Foot	552 ,, ,,	,, 218 ,,
86th Foot	759 ,, ,,	,, 229 ,,
1st Grenadiers	610 ,, ,,	,, 166 ,,
2/1st Bombay N.I.	745 ,, ,,	,, 123 ,,
2/6th Bombay N.I.	820 ,, ,,	,, 82 ,,
2/7th Bombay N.I.	741 ,, ,,	,, 78 ,,

Here, therefore, was a force of 4,694 men about to take the field against an enemy whose horsemen had spread desolation from the frontiers of Mysore to the Punjab, and yet it was not thought worth while to furnish it with a single reliable trooper. Much as Murray must be blamed for indecision, it must be borne in mind that Lord Lake had at his command a cavalry division, which included three regiments of British Dragoons, and that Wellesley had the 19th Dragoons and three regiments of Madras regular cavalry.

A week later peace was declared with the Maharajahs of Scindia and Berar, and Murray's tribulations were at an end; the force fell back from Dahud first to Godra, and then to Jerode, where it went into winter quarters.

1804

Scarcely was the cold weather over than hostilities broke out with Holkar, and Murray was ordered once more to take the field, this time minus the 75th, which was sent round to Calcutta by sea, to reinforce Lake, his own regiment (the 76th) having been cut to pieces in the preceding campaign. A terrible famine was raging in Central and Southern India, and Wellesley reported his own force as unable to move; at the same time, he did not hesitate to urge Murray forward with all the energy of which he was capable. Lake, who had been in winter quarters near Cawnpore, had pushed forward a column under Colonel Don, which had carried the fortress of Rampoorah, one of Holkar's strongholds, on the 15th May, and Holkar himself had retired before Don towards Ujjain—that is to say, towards Gujarat. Don at once ordered Colonel Monson, an officer of great personal gallantry, to follow up Holkar's Army, whilst Murray was directed to advance on Ujjain in the hopes that Holkar, caught between the two columns of Monson and Murray, might be crushed, as it were, between the upper and the nether millstone. On the 12th of June Murray re-occupied Dahud, and slowly advanced to the north-east, the while that Monson was pressing forward to meet him.

Passing through Sheoghur on the 25th of June, and Pitlaud on the 28th, Murray crossed the Myhe River on the following day, and on the 1st of July was in Badnawar; a week later he reached Ujjain. Then his troubles began. His force was, it must be confessed, very small, and on the line of march it was exposed to constant attack on the part of the Mahratta Horse. His one troop was scarcely fitted to cope with the many thousands Holkar could bring into the field, and the Gaikawar's troopers had little relish for hand-to-hand fighting with their own co-religionists. Monson and Murray never met. Hearing that Holkar, with fifty thousand men, was about to attack him, Murray fell back. Monson, learning that Murray had retired, fell back also, when Holkar threw himself with all his force on Monson, pursuing him to the very gates of Rampoorah. With the details of that terrible disaster we have no concern. Suffice to say that Monson reached Rampoorah on the 29th of July, with the loss of more than half his force.

As soon as Murray felt the immediate danger of attack from Holkar was at an end, he at once retraced his steps, and occupied Ujjain on the 8th of July. Here he constructed a depôt and collected supplies; whilst doing this he detached the 86th, with the Bombay Grenadiers and the 2nd Battalion 7th Bombay Infantry, to occupy Indore, one of Holkar's strongholds, and at the present time (1913) his capital.

Indore was occupied on the 24th of August, and the following day Captain Richardson, with the flank companies of the Regiment and the Bombay Grenadiers, made a forced march to Mhow, which, it was reported, was strongly held. On reaching the place it was found to be deserted, and the 86th, leaving a garrison in Indore, retraced their steps to Ujjain.

It was physically impossible for the column to penetrate further into Holkar's territory. The men were shoeless, their clothes in rags, for weeks they had been on half rations, and arrack was only issued on the doctor's orders. The country to the eastward was mountainous, intersected by many streams, which in the rainy season became raging torrents, and of bridging equipment there was none. Had Murray continued his march, in all probability his force would have met the fate which overtook Monson. Lake himself realised that a second hot weather campaign was more than the regiments could stand, and he deferred his own advance until the close of the rains. On the 3rd of October he re-occupied Muttra, which had been abandoned by its native garrison a fortnight previously, and on the 15th of the same month he relieved Delhi, which had been besieged by Holkar for some days. Holkar now realised that his unbroken career of victory was at an end, and he fell back in some haste, calling on the Rajahs of Bhurtpore and Dholpore to come to his aid, and despatching emissaries in the hope of inducing Scindia again to take the field. Lake, with the cavalry, now followed up Holkar's main force, while the infantry, under General Fraser, proceeded to attack the infantry of the Mahrattas at Deig. Here, on the 13th November, General Fraser won a signal victory, falling, however, in the very hour of success. Four days later Lake overtook Holkar's cavalry at Furrackabad, and routed the famous Mahratta Horse with the most consummate ease. From Furruckabad Lake retraced his steps to Deig, which was carried by assault on the 24th of December, when the Army advanced to Bhurtpore and commenced that ill-fated siege.

In the meantime Lake had been urging on Murray the necessity of advancing to the eastward, and on the 19th of October, his men having received fresh provisions of clothes and stores, he broke camp at Ujjain. On the last day of the month he had only reached the entrance to the Muccandrah Pass, and a fortnight later, passing through Hallode, Jaorah, and Rutlam, he entered the town of Mundaseer. Murray's dilatory proceedings during the preceding campaign had led General Wellesley to suggest his supersession, and at last the Government of Bombay, worn out by his querulous letters, ordered Major-General Jones, of the Bombay Artillery, to take command of Murray's force. Jones at the time was in command at Baroda, but he immediately called on the Resident, Major Walker, of Khurri and Baroda fame, to furnish him with an escort of the Gaikawar's cavalry, and he at once set out to overtake the moveable column. This he did on Christmas Day, at Ralawan, some 25 miles west of Shahabad. Jones did not mince matters; he wrote in the strongest terms to the Bombay Government as to the condition of the little force under his command. It consisted of the headquarters of the 65th and 86th, the 1st Bombay Grenadiers, the 2/2nd Bombay Infantry, 1/3rd Bombay Infantry, and 1/9th Bombay Infantry, with fourteen 6-pounders, four 12-pounders, and four 5½-inch howitzers. The force was honeycombed with sickness, and the native regiments contained a large proportion of recruits. The present state of the force on the day he assumed command makes but sorry reading:—

	Total.	Sick.	Recruits.
Artillery	97	13	—
British Infantry	651	164	—
Native Infantry	2,616	278	676
Native Cavalry	52	9	—

The two British battalions, the backbone of the force, between them could muster something short of 500 bayonets!

Unfortunately, no records exist of the exact route followed by the regiment on its march from Ralawan to Bhurtpore.

The great distance marched by the Regiment on its way to assist at the siege of this place may be judged from the map.

After a long and harassing march, the troops approached the fortress of Bhurtpore on the 10th of February, when a large body of hostile horsemen surrounded the column and impeded its movements across level country. Half of the force was employed in protecting the baggage, and the guns were repeatedly unlimbered to keep the hostile cavalry at a distance. On the following day Major-General Jones's division joined the army before Bhurtpore, and was inspected by Lord Lake, who expressed his satisfaction at the bearing of the troops.

The soldiers of the Sixty-Fifth and Eighty-Sixth Regiments presented a motley appearance; their worn-out uniforms were patched with various colours, or replaced by red cotton jackets; many of the men wore sandals in the place of shoes and turbans instead of hats; but beneath this outward war-worn appearance the innate courage of Britons still glowed.

The siege of Bhurtpore had, unfortunately, been undertaken without a battering train of sufficient weight necessary to ensure the reduction of so strong a fortress. The siege, however, was persevered in. At three o'clock on the 20th of February, 250 men of the Eighty-Sixth Regiment and two companies of sepoys, commanded by Captain Grant, of the Regiment, stormed an outwork covering one of the principal gates, with the bayonet, driving the Arabs, who fought with their usual determination, into the city, and capturing eleven brass guns, in which service Lieutenants Lanphier and D'Aguilar distinguished themselves, the former receiving a spear wound in the neck. As the Arabs fled to the gate, Captain Grant followed, in the hope of being able to enter with them, but he found it closed, and, after destroying the fugitives who were shut out he retired to the Pettah, to await the results of other attacks, which did not succeed. While the soldiers of the Eighty-Sixth Regiment and the sepoys were dragging the captured guns to the camp, they were attacked by a numerous body of the enemy, who issued from the fortress to re-take the guns, but were repulsed by the steady valour of the soldiers. Captain Grant formed a square round the guns, and, under a heavy fire from the fort, succeeded in bringing them to camp. This was the only successful part of the attack.

The enemy's numerous cavalry also attacked the British camp, but were defeated by the troops not engaged in the trenches or in the assault. Lord Lake commended the determined bravery of the storming party of the Eighty-Sixth Regiment in orders and directed the captured guns to be placed in front of the camp of the Regiment, a mark of distinction highly prized by the corps and by the Bombay Division of the army to which they belonged.

1805

At three o'clock on the following day the flank companies of the Sixty-Fifth and Eighty-Sixth, supported by the Seventy-Fifth and Seventy-Sixth Regiments, commanded by Lieutenant-Colonel Monson, stormed a large and high bastion. After passing the ditch, the "forlorn hope" was destroyed in attempting to ascend the breach, which was extremely steep and knee-deep in mud and loose stones. Every effort was made, the men climbing over the dead bodies of their comrades and struggling to gain the ramparts, but in vain. Some strove to climb by the shot-holes made by the British guns, and others drove their bayonets into the mud walls to ascend by, while the enemy above hurled upon their heads large stones, logs of timber, packs of flaming oiled cotton, and jars filled with combustibles, with a terrible destruction. The killed and wounded lay by hundreds, crushed beneath the falling timbers, or burning under the flaming oiled cloth, when Lieutenant-Colonel Monson, seeing the impossibility of succeeding, ordered the survivors to return to camp.

The casualties in the Bombay column were most severe.

	20th February.				21st February.			
	Officers.		Men.		Officers.		Men.	
	Killed.	Wounded.	Killed.	Wounded.	Killed.	Wounded.	Killed.	Wounded.
65th Foot	—	3	11	44	—	8	10	99
86th Foot	—	4	19	42	—	2	6	33
Bombay Grenadiers	1	2	33	11	1	5	9	88
2/2nd Bombay Infantry	—	2	?	?	?	?	?	?
1/3rd Bombay Infantry	—	1	3	29	1	1	8	27
1/9th Bombay Infantry	—	2	?	?	—	3	?	?

The officers wounded were Captain Morton, Lieutenants Travers, Baird (five wounds), Lanphier (twice wounded), and D'Aguilar.

These two last-named officers particularly distinguished themselves, as did Serjeant George Ibertson, and Corporal Crawfurd, who was promoted serjeant for "gallantry at Bhurtpore."

Lake now found that with the forces at his disposal the capture of the fortress was out of the question. He had a totally inadequate siege train, and the British soldiers of the Bengal Brigade were utterly dispirited. He, therefore, reluctantly determined to abandon the siege, and to prosecute the war with

Holkar by other means. If Lake was dispirited, the Maharajah of Bhurtpore was far more so. He considered that he had not received adequate support from Holkar, that his men had been compelled to bear the brunt of the fighting, and he sent emissaries to Lord Lake, offering to conclude peace. Terms with him were speedily arranged, and the army broke up; the Bombay column, under General Jones, taking up their quarters between Tonk and Rampoorah, whilst the Bengal troops were echeloned between Agra, Muttra, Secundra, and Futtehpore. Here they spent the hot weather, whilst Holkar, not knowing quite what was in store for him, wended his way towards the Punjab, hoping to obtain the support of Runjeet Singh, the Ruler of the Punjab, and of the Amir of Afghanistan. Lake awaited the break-up of the rains before following him up.

At the end of October, Lake, feeling that his men had recovered from the fatigues and privations incidental to their long marches and the siege of Bhurtpore, determined to continue the pursuit of Holkar, who was known to be beyond the Sutlej, in the Dominions of Runjeet Singh. That monarch had evinced no disposition to assist Holkar. At the same time it was a rash proceeding to cross the frontiers of an independent State in search of a foe. Still, Lake never hesitated. His one fear was that Holkar might elude him by doubling back from the Punjab to Central India. He therefore determined to carry on the pursuit with two independent columns. He himself, with the 8th, 24th, and 25th Dragoons (the 27th and 29th had recently been re-numbered), the 22nd Cheshire, and First Bengal European Regiments, with two regiments of native cavalry and two of native infantry, were under his own command, whilst the Bombay column, under Major-General Jones, formed the second column. Lake's own route lay through Delhi, Paniput, Pattiala, Ludhiana, where he crossed the Sutlej, through Jullundur, to the River Beas. Here he learnt that Holkar had taken refuge in Umritsar, the headquarters of the Sikh religion, though not the capital of the country. Lake halted on the banks of the Beas, to allow Jones to come up. That officer, leaving Rampoorah on the 5th of November, had marched through the territories of the Maharajah of Jeypore, then through the Shekawati country, and, skirting the Bikanir State, reached the Beas on the 10th of December. Pressure was now brought to bear on Holkar. Lake informed the Punjab authorities that unless Holkar accepted the terms offered, he would cross the Beas and attack him, even in the Temple of Umritsar itself. Lake was a man of his word, and both Holkar and Runjeet Singh knew full well that what he threatened he would carry out. Holkar was at his last gasp. His treasury was empty, his troops dissatisfied, and Runjeet Singh refused him any assistance. He was advised to make friends with his adversary quickly, lest a worse thing befel him, and on the 24th December, 1805, he accepted the terms laid down by Lake. These included the cession of large tracts of territory and the disbandment of his army.

EIGHTY-SIXTH REGIMENT (1802-1806).

In orders dated Riapoora, Ghaut, on the left bank of the Hyphasis, 13th December, 1805, General Lord Lake returned thanks to Major-General Jones, the officers and men of the division of the army from Bombay, for the important services rendered by them during the war; and, alluding to the period they had been under his immediate command, added: "His Lordship has been proud to witness on every occasion on which they have been employed the steady conduct and gallantry in action of all the troops composing the division."

1806 From the banks of the Hyphasis, near the spot where Alexander the Great crossed that river when he invaded India, the Regiment commenced its march to Bombay early in January. It arrived at Bombay on the 29th of March, and embarked for Goa, landing on the rock of Aguada on the 3rd of April. Here the Eighty-Sixth was at last destined to have some repose, after practically more than five years of active service. During those five years it had sailed up the Red Sea, crossed the desert twice, served a campaign in Egypt, and traversed the Northern and Western Provinces of India —from Bombay to Bhurtpore. It had also sustained a loss of ten officers and over a thousand non-commissioned officers and men.

Whilst the Regiment was engaged in these Indian wars, their Honorary Colonel, Lieutenant-General Grinfield, had died, and he was succeeded by Lieutenant-General Sir James Henry Craig, K.B., who was transferred from the Forty-Sixth Foot on the 5th of January, 1804. Sir James was removed to the Twenty-Second Foot in October, 1806, and Lieutenant-General Sir Charles Ross, Baronet, was appointed to the vacant colonelcy from the Eighty-Fifth Regiment.

Many small details connected with the Eighty-Sixth Regiment at this time may be found in the records at the India Office. The following are mentioned as examples of the internal life of the Regiment during these five busy years:

In 1803, 400 rupees each were allotted to Lieutenants Creagh and Marston and to Ensign McQuarrie, for the purchase of a tent apiece. This would appear to be a liberal allowance for this purpose.

Lieutenant Marston also received 1,417 rupees in September of this year for acting as Brigade-Major to a "detachment of troops to the northward." The same officer was appointed Brigade-Major by Major-General Jones, when he marched his Bombay Division across India to join Lord Lake.

In 1805 there was a general subscription raised in Bombay to assist the Home Government to prosecute the war with France. All ranks of the Eighty-Sixth joined with alacrity in this subscription, and sent a handsome donation of 8,750 rupees from their scanty pay to this most patriotic object.

CHAPTER VI.

EIGHTY-SIXTH REGIMENT (1806-1812).

Establishment reduced—Draft from Cape of Good Hope campaign—Lieut.-Colonel Fraser assumed command—General Sir Charles Ross appointed Honorary Colonel —General Order re Captain John Grant—86th created Leinster Regiment—Establishment further reduced—Regiment marched to quell Madras Army Mutiny—Account of Mutiny—Movements of Regiment during this Mutiny—Regiment returned to Goa—Charitable Fund formed—Declaration of Portuguese Viceroy—86th embarked for Bourbon—Description of French island of Bourbon—Voyage to Bourbon—Composition of force—Account of the capture of Bourbon—Lieut.-Colonel Fraser's despatch—Losses of the Regiment—Major Edwards's complimentary order to the 86th—Corporal William Hall—Private John Moore—Major Edwards's career and death—Monument to Lieut Monro—Anecdote about Captain Lanphier's grave, etc.—Further despatch by Lieut.-Colonel Fraser—Lieut.-Colonel Keating's order—Rewards for Bourbon—Losses at Bourbon — Major-General Needham appointed Honorary Colonel — H.M.S. "Africaine's" action—H.M.S. "Ceylon's" sea-fight—Regiment proceeded to Mauritius —State of the Regiment—86th returned to Bombay—Title of Regiment changed to Royal County Down Regiment—The words " India " and " Bourbon " ordered to be placed on colours.

1806

THE establishment of the Regiment was reduced early in 1806 to 54 serjeants, 22 drummers, and 1,000 rank and file, but this was only a matter of form, as the Regiment itself was a perfect skeleton, for, after receiving a draft of 213 volunteers, in October, from the 77th Regiment, which was returning to England, and a further draft of 30 men, also in October, under Lieutenant Michael Creagh, the records state that the strength was under 500 rank and file.

This detachment, under Lieutenant Creagh, had been attached to the 93rd Highlanders, and had been serving at the Cape of Good Hope, with the Army under General Sir David Baird, which had recaptured this colony from the Dutch, to whom it had been returned at the Peace of Amiens in 1802. Three other officers arrived with this party, viz., Ensigns Blackhall and Hillhouse and Paymaster Cope.

Lieutenant Michael Creagh had received a grape-shot wound in the head at the battle of Blueberg, and one other officer (Ensign Heddrick) appears to have been so badly wounded that he could not continue the voyage to India with this party and never joined the Regiment.

On 29th March, 1806, the Regiment left Bombay and embarked for the Portuguese possession of Goa, this time, however, not on conquest bound. It arrived at Goa on the 3rd of April, and remained there until the 17th of August, 1809.

EIGHTY-SIXTH REGIMENT (1806-1812).

In 1806, whilst at Goa, Lieutenant-Colonel Hastings Fraser assumed command, having come out from England, as the records state, overland, which, of course, means only that he came by way of Egypt.

General Sir Charles Ross, Bart., was appointed Colonel of the Regiment, vice Lieutenant-General Sir J. H. Craig, K.B.

The latter was removed to the 22nd Foot, whilst Sir Charles Ross was transferred to the 86th Regiment from the 85th Foot.

*On 24th August, 1809, His Majesty was pleased to direct the Regiment to be called The 86th or Leinster Regiment of Foot. No reason was given for this change of designation.

Lieutenant-Colonel Peter Carey, of the 86th Regiment, commanded the force which captured the Danish settlement of Serampore, near Calcutta, on the 28th of January, 1808. His prize-money on that occasion amounted to £16,420 (sixteen thousand four hundred and twenty pounds). He had been brought in from another regiment as a Major, and went to the 84th Regiment as Lieutenant-Colonel in 1811. His rank was therefore either "Local" Lieutenant-Colonel or "Brevet" Lieutenant-Colonel.

On the 16th of February, 1808, Captain John Grant, who had been granted leave of absence, had the following most complimentary order, with reference to his services, published by the Government of the Bombay Presidency:—

"Bombay, 16th February, 1808.

"GENERAL ORDERS BY GOVERNMENT.

"It having been reported that Captain John Grant, of H.M. 86th Regiment, has obtained permission to proceed to Great Britain on leave of absence from the Regiment, The Honourable The Governor in Council cannot allow him to depart without being accompanied by this testimonial to the credit which that meritorious officer has done to the British arms during his service in India—at the siege of Baroda, the capture of the important fortress of Baroach, and Powanghur, and particularly in the arduous and successful attack of the columns which Captain Grant commanded on the 20th of February, 1805, at the siege of Bhurtpore, on which memorable occasion he most gallantly carried the enemy's post and captured the whole of their guns, being eleven in number, etc., etc."

1808 — In March, 1808, the establishment was further reduced to 44 serjeants, 22 drummers, and 800 rank and file.

1809 — In August, 1809, orders were received by the 86th Regiment to join a force which was being assembled in the Presidency of Madras.

* The Regimental Records and Cannon's History both give October, 1806, as the date of the change of title to Leinster Regiment, but the actual warrant is in existence at the War Office bearing the date 24th August, 1809, with "Approved G.R." written in by King George the Third himself.

This was the largest force of European troops up to that time which had ever been brought together in India, and the reason of the assembling of this body of troops was most serious, being no less than the Mutiny of the Madras Native Army.

There were various reasons for this mutiny, but, in very general terms, it may be said that any trouble that arose was caused by the unsympathetic conduct of the Lieutenant-Governor of Madras, Sir George Barlow. He was noted for most autocratic conduct, which, generally, was pressed to an unreasonable extent. The Commander-in-Chief in Madras, Lieutenant-General Hay Macdowall, took exception to various acts of the Government, but his complaints were not heeded, and even when the use of the troops became necessary in the Province of Travancore, he was not consulted, and a plan of military operations was decided upon by the Government, without calling him in as their military adviser. Lieutenant-General Macdowall, naturally, became very much incensed, resigned his appointment, and set sail for England, making some unnecessary remarks about the Government in his farewell orders.

The Government, another name for Sir George Barlow, immediately directed that the order should be expunged from the public records, and publicly dismissed the Commander-in-Chief. Not stopping at this, the Government then suspended from the Service Major Boles, D.-A.-General, who had signed the order of the Commander-in-Chief. The officers of the Madras Army undertook to subscribe for the defence of Major Boles, as it was quite apparent that he had acted under orders. Lieutenant-General Macdowall's case was easily settled, for the poor old man was lost at sea on his way to England.

Sir George Barlow, having heard of the assistance rendered to Major Boles, determined to punish various senior officers without trial, and this order was agreed to by the Council on the 1st of May, 1809. As if to make sure that everything should be done to exasperate the troops, certain allowances had been abolished in 1808, which abolition pressed very heavily on the senior officers, and this may shortly be described as follows. In 1802 it was found that heavy expense was incurred by the Government for tents and transport for the troops, and by captures of both tents and transport made by the enemy, which the Government had to make good. They transferred these charges to the Colonels of regiments, by paying them certain allowances, they to make profit or loss out of the transaction. During the native wars up to 1806, the losses predominated. In the peace that followed these officers hoped to recoup themselves, as, naturally, the wear and tear of tents and animals was not so great as in war. Here, however, the Government stepped in, and, being anxious to economise, decided to stop these allowances in 1807; this was carried into effect in May, 1808. Thus these unfortunate officers lost heavily. On the 18th of June, 1809, the officers of the native troops at Hyderabad issued an address to the Army, in which they condemned the action of the Government and announced their resolution to contribute towards the support of the suspended officers, as well as to join in any legal measures calculated to remove the cause of the existing discontent. After unsuccessfully trying to induce the officers to withdraw from their position, the Resident at Hyderabad then on the 3rd of August, spoke to the 16th Madras Regiment on parade, when the men having begun to load their muskets, the Resident, Colonel Close, prudently withdrew. Practically all the other Madras troops had thrown in their lot with the Hyderabad Garrison, including the Madras European Regiment, at Masulipatam, and a committee of officers was formed to direct affairs and to obtain a redress of grievances.

At Seringapatam, the revolted troops asked for more assistance to hold the fortifications, and a battalion of the 8th Madras Native Infantry and another of the 15th Madras Native Infantry proceeded to march to that place. They were attacked, by orders of the Acting Resident, by 3,000 Mysore Horse and 1,500 armed footmen, but repulsed them and moved on, but, being again attacked by the same troops, reinforced by His Majesty's 25th Dragoons, they were completely broken and dispersed. This was accounted for by the fact that the Native Infantry were surprised by the charge of the European Dragoons, whom they looked upon as supporters. The Native Infantry lost 442 men in killed and wounded, whilst the Mysore Horse lost 125 men, and one European officer of the revolted troops died of fatigue and another was wounded and taken prisoner; the remaining Native Infantry effected their escape into the fort at Seringapatam to the number of 20 European officers and about 850 natives.

On the 11th of August the officers at Hyderabad placed themselves and their interests in the hands of the Governor-General of India, Lord Minto, and the fort at Seringapatam, after an investment of some days—from the 10th of August to the 23rd of August—also surrendered.

Throughout Madras the native troops were brought to their allegiance again by the wise use of His Majesty's British troops, supported by any native soldiers whose officers had remained loyal to the Company without any further bloodshed, the force, of course, being the one to which the 86th Regiment from Goa was attached.

Despite the fact that Lord Minto had supported the Madras Government in all its acts, yet there was such a general feeling of confidence throughout the Army in the justice and moderation of His Lordship, that it is probable that if he had been able to arrive in Madras earlier no action would have been taken by the officers beyond the representations of their grievances to him. Everything was, happily, settled after his arrival by the publication by the Governor-General of a general amnesty to all concerned with few exceptions.

Leaving Goa, the 86th Regiment proceeded up the river in boats to Candiaparr, from whence it ascended the Ghauts.

The monsoon having set in, the men suffered much from the incessant rains and inundations, and after a long march through the Mahratta territories the Regiment arrived at Bellary on the 15th of September.

The First Brigade was composed of the 86th Regiment, the 2nd Battalion of the 1st Foot, and two battalions of sepoys. Lieutenant-Colonel Fraser acted as Brigadier and Lieutenant Michael Creagh was appointed Brigade-Major.

The force moved forward against the opposing native troops, who retired for several marches. Negotiations were carried on and a satisfactory settlement arrived at between the Government of Madras and this dissatisfied party of native troops.

The Army was now re-distributed, and the 86th formed the garrison of Bellary and Gooty.

The effects of the march through the flooded cotton fields, etc., now became apparent, and Captain James Burke and many non-commissioned officers and men died as the result of the fatigues they had undergone.

The regiment shortly returned to Goa, which was a Portuguese possession. The French had driven the Portuguese Regent from Portugal and had seized that country. The Regent retired to Brazil, also a Portuguese colony. As Britain was in command of the sea, and in alliance with Portugal, the French could not follow him to Brazil, and this point was emphasized to them by the fact that Britain could, and did, land a force in Portugal to turn the French out of that country. More will be heard of that expedition when this History describes the doings of the 2nd Battalion of the 83rd Regiment in the Peninsula. But it suffices to say that the English had to furnish a garrison to guard their Allies' Indian territory, and it was thus that the 86th Regiment found themselves doing garrison duty in a foreign country's possession.

It was at this time, at the friendly little station of Goa, that the small but most useful charitable fund of the Regiment was started under the following circumstances:—

The origin of the Benevolent Fund of the 86th Regiment was shortly as follows:—Whilst the Regiment was stationed at Goa, in 1806, Lieutenant-Colonel Fraser, then in command, established a canteen for the men, which was for the purpose of supplying a better description of liquor from Bombay than would ordinarily be supplied by a local native contractor. Up to the year 1808 the profits from this canteen amounted to some £1,412, and by the existing regulations and customs of the Service in India this sum was the personal property of the Lieutenant-Colonel. Lieutenant-Colonel Fraser, however, refused to keep the money, but placed it in Government funds, in the joint names of the Lieutenant-Colonel of the Regiment and the two next senior officers, the interest accruing thereon to be expended as the Commanding Officer might direct in relieving such cases of distress as might be brought to his notice.

A fuller account of the formation, etc., of the fund will be found in Appendix III.

The fund still continues to exist, and at the present date (1913) amounts to the sum of £3,160. Its usefulness to serving soldiers and to those in old age and poor circumstances cannot be over-rated.

The Regiment made many friends during its stay in Goa, and when they left it to join the Madras Field Force, in August, 1809, the following "declaration" was issued by the Viceroy and Captain-General of the Portuguese possessions in Asia, dated Palace of Panjan, 16th August, 1809:—

"On the departure of His Britannic Majesty's Eighty-Sixth Regiment from Goa, His Excellency the Viceroy and Captain-General of the Portuguese possessions in Asia avails himself of the opportunity to express his sentiments of praise and admiration of the regular order and conduct which Lieutenant-Colonel Fraser and the officers and soldiers of that Corps have so honourably observed during a period of three years, while they have been employed in the territories subject to his authority, so highly creditable to the discipline of that Corps.

"His Excellency the Viceroy will never forget the invariable harmony which has always subsisted between the subjects of His Royal Highness the Prince Regent of Portugal and all ranks of His Britannic Majesty's Eighty-Sixth Regiment, whose remembrance will always be grateful to him, and he doubts not they will continue to acquire, in whatever part of the world their services may be called for, glorious claims on the rewards of their Sovereign and the admiration of their country."

The final part of the good Governor's declaration soon came true. Whether the 86th Regiment "acquired *glorious* claims on the rewards of their Sovereign," is, perhaps, open to question; but from Goa, they sailed away to do their duty manfully and to be rewarded by His Majesty George III. in gracious approval of that well-done duty, by the title of "Royal," which they have ever since carried.

On the 5th of March, 1810, the 86th embarked on the transports "Minerva" and "Lowgie Family," escorted by H.M. ship "Diomede," and sailed on the following day to form part of an expedition against the French island of Bourbon. The total strength of the 86th Regiment who embarked was 420 of all ranks. Bourbon, as has already been pointed out, lay well to the eastward of the East African Coast, about the latitude of Madagascar. It commanded the trade route from England by the Cape of Good Hope, and in those days a trade route could not be changed with the same facility that it can be, to a certain extent, in these days, because sailing ships were practically bound to follow the trade winds. The French were not slow to avail themselves of their happy position on the flank of this sea road, and from the ports of Bourbon there poured forth privateers of all sorts besides any French men-of-war or frigates that had been able to escape the attentions of the English fleet. Despite the fact that all trading vessels went armed in those days, very heavy losses were inflicted on British shippers. It was time that the Mistress of the Sea should see to this island, and it was effectually seen to.

The following description of the island, of its inhabitants, its physical features, etc., may be of interest, and soldiers will at once see the point of immediately landing the troops and capturing the forts and towns with their garrisons, before the latter, seeing that the defence of the towns was hopeless, should take to the mountains and woods, and commence a guerilla warfare in a most difficult country:—

Réunion, known also by its former name of Bourbon, is 400 miles south-east of Tamatave, in Madagascar, and 130 miles south-west of Port Louis, Mauritius. It is elliptic in form. Its greatest length is 45 miles, and its greatest breadth is 32 miles. The coastline, about 130 miles long, is little indented, has no natural harbours, and there are no small islets round the shores.

The narrow coast lands are succeeded by hilly ground, which, in turn, gives place to mountain masses and table-land, which occupy the greater part of the island. It is of volcanic origin, and one mountain, Piton des Nieges, is 10,069 feet high, whilst another, Grand Bernard, is 9,490 feet high. Part of these hills is cut off from the rest of the island by two ravines 500 or 600 feet deep. The thirty miles of mountain-wall round the volcano is, perhaps, unique in its astonishing regularity.

Eruptions constantly take place, but nothing of any great consequence, though a village was overwhelmed in 1875. Hot mineral springs are found.

Sugar plantations and forests abound. Its population towards the end of the 18th century was 35,000, descended from the first French settlers, chiefly Normans and Bretons, who married Malagsy women. About A.D. 1872, the population was 212,000. St. Denis, in 1902, had a population of 27,392. Bourbon was discovered in A.D. 1513 by the Portuguese navigator, Pedro Mascarenhas. In A.D. 1643 it was annexed by France.

On the fleet entering Cannanore Harbour, to embark the 12th Native Infantry, one transport struck on a rock and was lost, but all aboard were saved. Calling at Quillon and Point de Galle, the transports duly arrived at Madras on 13th of April, where the troops landed and encamped at St. Thomas's Mount.

Leaving Madras on the 4th of May, the fleet sailed for the island of Rodriguez. The white troops consisted of part of the 56th Foot and the whole of the 69th and 86th Regiments, with several native corps.

This force was divided into three brigades. The First Brigade was again commanded by Lieutenant-Colonel Hastings Fraser, with Lieutenant Michael Creagh as his Brigade-Major, and the 86th formed part of this brigade.

CAPTURE OF BOURBON BY THE 86th REGIMENT
1810

By kind permission of the Artist
Miss Marjory Watherston

The Second Brigade was also commanded by an officer of the 86th Regiment, Lieutenant-Colonel Drummond, and Lieutenant Richardson, of the same Regiment, was his Brigade-Major. Lieutenant-Colonel Keating, of the Fifty-Sixth Regiment, was in command of the whole force.

1810

The van of the expedition, consisting of the Eighty-Sixth Regiment, 180 rank and file of the Sixth Madras Native Infantry, a small detachment of artillery, and fifty pioneers, under Lieutenant-Colonel Fraser, of the 86th, with difficulty, effected a landing at Grand Chaloupe about one o'clock on the 7th of July. On gaining the shore, the light company of the 86th, under Lieutenant Archibald McLean, supported by the grenadiers, under Captain Lanphier, dashed forward to drive back parties of the enemy's riflemen, who kept up a harassing fire, and to secure possession of the heights, which service was performed with great gallantry. The other part of the Regiment, having landed, pressed forward to the heights above St. Denis, and, as the sun was setting, approached to within range of the enemy's batteries. The violence of the surf had become so great that the other brigades could not land, and the 86th, in consequence, fell back to the heights, where they were joined during the night by the sepoys, pioneers and artillery, with one 4½-inch howitzer.

At four o'clock, on the morning of the 8th of July, the Eighty-Sixth commenced descending the mountain, leaving the sepoys on the summit to defend the rear; their advanced guard was soon discovered by the enemy's post, and at daylight the Regiment was assailed by a heavy fire of cannon, mortars, and musketry; at the same time some of the enemy's riflemen attempted to gain the road on its right.

The light company, supported by the grenadiers, and followed by the Regiment, descended the mountain at a running pace. Two columns of the enemy having each a field piece, and being supported by the heavy guns of the redoubt, opened a sharp fire of grape and musketry, but as the Eighty-Sixth arrived on the plain they closed on their adversaries with the bayonet. This spirited conduct decided the contest; the opposing ranks, unable to withstand the shock of steel, faced about and fled. Their Commandant, M. de St. Luzanne, escaped with difficulty, and their Second-in-Command was wounded and taken prisoner by Captain Lanphier.

The enemy attempted to re-form behind the parapet of the redoubt, but they were pressed so closely by the grenadiers that they abandoned it, leaving a brass 6-pounder behind, which was immediately turned against themselves.

The halyards of the flagstaff in the redoubt were shot away, but Corporal William Hall, of the Eighty-Sixth, climbed the staff under an incessant fire of round shot and musketry, and fixed the King's Colour of the Regiment to it. The French soldiers viewed this daring feat with admiration, and as he descended the staff, unhurt, they raised a loud shout. The seamen of the fleet off the shore, who had witnessed the gallant charge, hailed the well-known flag of the Regiment, which waved on the redoubt, by a loud huzza, which rang from ship to ship as they passed; at the same time, the grenadiers of the Regiment stormed two batteries, capturing nine 24-pounders, a 12-inch mortar, and a furnace of red hot shot.

The position seized was held by the Regiment, the guns of the captured redoubt answering the incessant fire of the enemy until the arrival of additional troops.

At four o'clock the enemy attempted to retake the redoubt, but were repulsed with the loss of their Commanding Officer, who was taken prisoner. About this time Lieutenant-Colonel Drummond's Brigade arrived, and the enemy sent out a flag of truce: Lieutenant-Colonel Keating having joined, the surrender of the Island of Bourbon to the British Arms was concluded by him.

Colonel Fraser stated in his despatch: "I cannot conclude without requesting permission to offer my humble tribute of praise to the noble spirit which animated every individual of my detachment. From Major Edwards, who commanded the Regiment, I received the greatest assistance. Captain Lanphier, Lieutenant Archibald McLean, and every officer and soldier of the Corps displayed the most ardent valour, which must have been conspicuous to the whole force off the coast, who witnessed their conduct. To Lieutenant Creagh, my Brigade-Major, I was highly indebted, for his unremitting exertions and attention to the duties of his station, from the beginning of the service I was sent on, to the moment when he was struck by a cannon-ball while he was in the act of encouraging our artillery men in the redoubt, which, I fear, will deprive his Sovereign and his country of the services of a most promising officer."

The Regiment lost on this occasion Lieutenant John Monro, of the grenadier company, killed during the charge, Major Edwards (commanding the Regiment), Captain Lanphier, Lieutenants Michael Creagh (Brigade-Major), Archibald McLean, Blackhall, Webb, and White wounded, Serjeant Millan, two other serjeants, two drummers, and seventy-five rank and file killed and wounded.

On the 9th of July Major Edwards published the following regimental order:—

"It affords Major Edwards great pleasure in having received Lieutenant-Colonel Fraser's directions to confirm the appointment of Captain Lanphier to the grenadier company, a distinction to which that officer is well entitled for the gallant manner in which he led the brave grenadiers to the assault of the redoubt and batteries.

"The conduct of the light infantry, under Lieutenant Archibald McLean, has on all occasions been equally distinguished, and Major Edwards sincerely laments the severe wound that gallant officer has received, but trusts he will soon be restored to the Service.

" Major Edwards has great pleasure in confirming Lieutenant Blackhall's appointment to the grenadiers; he regrets the wound received by that officer, but hopes it will not prevent his joining that company, which he animated by his zeal and example.

" The conduct of the officers, non-commissioned officers and soldiers of the Eighty-Sixth Regiment is above all praise; they have fought the enemy with every species of disadvantage and deprivation; they have borne the latter without a murmur, and their determined valour has achieved victory. In spite of every obstacle, they have nobly sustained the character of their country, and it will be gratifying to their feelings to know that their gallant exertions have been witnessed and applauded by the whole of the British force off the shore.

" Major Edwards cannot conclude without expressing his regret for the loss of Lieutenant Monro, of the grenadiers, and the brave men who have fallen on this occasion; their memory, however, will long survive and be held dear in the recollection of the Regiment.

" Most sincerely does he regret the severe wound received by Lieutenant Michael Creagh, but which, he hopes, will not deprive the Service of that valuable and gallant officer; and he trusts the wounds received by Lieutenant White will not long prevent him joining the Corps.

" Corporal William Hall, who hoisted the King's Colour on the redoubt, is appointed serjeant, for his gallant conduct, in the room of Serjeant Millan, killed.

" Private John Moore, of the light infantry, is appointed corporal, for his gallant behaviour on the 8th instant."

So ends the regimental order. Major Edwards, who dictated this order, appears to have been as ready with his pen as he was to command his regiment efficiently in every trying circumstance. He was finally promoted into the 14th Foot, and was killed in the storming of Bhurtpore, under Lord Combermere (Stapleton-Cotton) in 1826, whilst commanding that regiment. On the same day that Major Edwards wrote the above sonorous regimental order the Regiment, preceded by their light and grenadier companies, marched into St. Denis, the capital of the island, and, entering the principal battery, struck the tri-coloured flag of France and hoisted the King's Colour of the Eighty-Sixth.

The French garrison surrendered, and was sent as prisoners to the Cape of Good Hope. The island was handed back to France in April, 1815, after Napoleon's first abdication. Captain Michael Creagh's diary contains the following entry:—" 1810, 8th July. Captured the island—at the storming of St. Dennis I received a wound from an 18lb. shot which carried away my collar bone and all the muscles of the left shoulder. The same shot took off the heads of six of our men."

The Regiment erected a monument with the following inscription upon it: " Lieutenant John Graham Monro fell near this spot on the 8th of July,

1810, while charging the enemy, at the head of His Britannic Majesty's 86th Grenadiers. The officers of the Regiment have erected this monument as a mark of respect for his memory."

Some years after this, the monument in question suffered by a hurricane. The island had then been handed back to France, and the French officers then stationed in the island had the monument thoroughly repaired out of their own scanty pay.

The following anecdote is narrated regarding Captain Lanphier, who commanded the grenadier company on this occasion:—A detachment of the 86th Regiment, on marching through Tipperary in 1823, halted at the village of Middleton. In the evening the Commanding Officer observed the soldiers assembled round a tomb in the burial ground, with their caps off. On inquiring the cause, a soldier of the grenadiers replied: "Your honour, we are come up to see our old captain." On joining the group, he observed the tomb of his old and respected comrade, Lieutenant-Colonel Lanphier, and the following words, which had been scratched by the soldiers, beneath the inscriptions of the tombstone: "A Brave Soldier." "Please, your honour" (the soldier continued) "the 'boys' of the company would like to fire three rounds over the grave, and would be glad to pay for the powder, if your honour will let them fire." On the following morning the grenadier company, which the deceased had gallantly commanded for a number of years, paid the last tribute of respect to their late Captain's remains, which was duly appreciated by his surviving relatives and also by the villagers. Lieutenant-Colonel Lanphier entered the Army as Ensign in the 10th Foot in 1798, and was promoted to be Lieutenant in the 86th Regiment in 1800, to be Captain in 1806, to the rank of Brevet-Major in 1810, and of Brevet-Lieutenant-Colonel in 1819. He retired from the Service by the sale of his commission on the 30th of January, 1823, being then the senior Captain of the 86th Regiment, and he died very shortly after he retired.

In his report on the capture of Bourbon,* Colonel Fraser, who commanded the 1st Brigade, states that, as to the landing, he transferred the men of the 86th from the transport "Minerva" to H.M. ship "Syrius," with a few native troops. Captain Pym, who commanded the "Syrius," was enabled to make sail about eleven o'clock p.m. (meaning a.m.), and about one o'clock p.m. on the 7th inst. lay to off Grand Chaloupe, and immediately commenced the debarkation. Colonel Fraser then adds that when he first pushed forward, after capturing the heights and driving back the enemy's skirmishers, he found

* The general idea of the attack on the island appears to have been to land the 1st Brigade at Grand Chaloupe, some six miles west of the town of St. Denis, whilst the other troops landed near Riviere de Plevies, three miles east of St. Denis, thus partly cutting off the main town from any assistance. Only 150 men landed at Riviere de Plevies before the surf rose and prevented any more landing, despite the expedient adopted of running a small transport ashore and landing under her lee in boats.

he had only 350 bayonets with him all told, and that he fell back after dark to the heights, so that his retirement was not known by the enemy, whilst his position prevented any reinforcements being brought forward during the night to the French redoubts.

Colonel Fraser commanded the whole operations almost to the end, including the negotiations, and was most firm in his refusal not to grant a twenty-four hours' truce, as requested by the enemy, simply replying that the surrender must be practically unconditional.

Besides all the other orders issued on this occasion, the regimental records give an extract from the orders of Colonel Keating, who commanded the whole expedition.

EXTRACT FROM DETACHMENT ORDERS BY LIEUTENANT-COLONEL KEATING.

"Headquarters, Saint Denis,

"9th July, 1810.

"It is impossible for the Commanding Officer to be sufficiently expressive in returning his thanks to the several brigades, for their steadiness and discipline since their landing under most trying circumstances.

"It is, however, particularly to the 1st Brigade, under the command of Lieutenant-Colonel Fraser, to notice their gallantry before the enemy yesterday morning, in taking possession of the most important posts on the west side of the river of Saint Denis, and in maintaining that position against the enemy with all the advantages which he possessed."

In connection with an order published by Major Edwards, promoting Private John Moore to be corporal, for his gallant behaviour on this occasion, there is a note in the regimental records, which reads as follows:—"Private John Moore, promoted in the latter part of this order for his gallant conduct, was the same person who as serjeant on 20th August, 1803, led the "forlorn hope" at the storming of Baroach, and for his gallantry on that occasion was presented with 500 rupees by the Bombay Government."

Comparing the respective losses, the 86th Regiment, which practically took the island, lost 11 killed and 58 wounded, out of 420 of all ranks; the remainder of the force lost 7 killed and 21 wounded, out of a total of over 3,200.

The French surrendered 145 pieces of ordnance and great military stores. The total French garrison appears to have been 600 regulars and 2,700 militia. The enemy were found to have recruited their forces largely from any Irishmen they could take out of captured British ships. At the capture of Mauritius over 500 troops of the enemy were found to be thus made up.

In the terms of capitulation one of the conditions laid down was "that at 12 o'clock to-morrow the French troops which occupy the arsenal and Imperial battery shall evacuate their post, and the grenadier company of His Majesty's

86th Regiment . . . will take possession of them, when the French flag will be struck, and that of His Britannic Majesty displayed."*

Lieutenant-Colonel Fraser was created a Companion of the Bath for his conduct on the 8th of July, and the officers of his Regiment (the Eighty-Sixth) presented him with a sword, and the officers of the remainder of his brigade with a valuable piece of plate.

On the 25th of September General Sir Charles Ross was transferred as Colonel to the Thirty-Seventh Foot, and was succeeded in the colonelcy of the Eighty-Sixth Regiment by Lieutenant-General the Honourable Francis Needham, who was gazetted from the Fifth Royal Veteran Battalion. He was noted as having taken a serious interest in his regiment, and was an energetic person himself. Having, in 1818, succeeded his brother as Viscount Kilmorey, he was created Earl of Kilmorey in 1822.

On the same date the flank companies of the 86th Regiment were ordered to proceed on the expedition which captured the neighbouring island of Mauritius. They marched to join the force at Saint Paul, but were finally sent back to their Regiment, as it was considered that the 86th was too weak numerically to spare these two companies.

In the meantime some of the Regiment seized the opportunity of again becoming marines for a short time. The occasion arose as follows:—The British fleet had for some time been blockading Mauritius and Bourbon. So soon as Bourbon had passed into English keeping, the Navy proceeded to harry Mauritius, and, venturing inside the Isle de Passe on the south-west side of Mauritius, came to grief, losing one ship, which was sunk, and several others were temporarily disabled. This unfortunate circumstance gave the French numerical superiority, and their gallant sailors at once put to sea, and proceeded to Bourbon, where their frigates threatened to engage the British batteries.

The " Africaine," frigate, had arrived from England short of hands, so Lieutenant W. Home, a serjeant, and twenty-five men of the Regiment were permitted to go aboard, and the " Africaine," accompanied by the " Boadicea," immediately put to sea, all under the command of Commodore Rowley. The " Boadicea " was becalmed nine miles astern, but the " Africaine " managed to get up to two French frigates and promptly engaged them both. After a most severe action, of two and a half hours' duration, in which Captain Corbett, Commander of the " Africaine," and 2 officers, and 47 men were killed, whilst 8 officers and 105 men were wounded, the " Africaine," also becalmed and unable to escape, surrendered to the two French frigates, but Commodore Rowley then managed to get up in the " Boadicea," when the two hostile vessels fled and the British frigate was taken in tow and brought in to Saint Paul's.

Out of the little party of Eighty-Sixth men on board, Lieutenant Home

* "The Naval Chronicle," vol. 24, 1810.

and seventeen men were wounded, six were killed, and only three escaped unhurt. Verily hard knocks were freely given and received in those days.

The relative forces in this smart little engagement were as follow:—

English frigates engaged were the "Africaine" (38 guns), and the "Boadicea" (38 guns).

French frigates, "Astrée" (44 guns), and "Iphigenia" (38 guns), the last having been captured from the English.

The "Africaine" had 295 people aboard, and lost 163 killed and wounded; whilst the French lost 9 killed and 33 wounded aboard the "Iphigenia," and 1 killed and 2 wounded in the "Astrée."

The English loss was out of all proportion to that suffered by their enemies, and is accounted for, first by the latter's great superiority of strength, and secondly by the rumour current throughout the Navy at the time that Captain Corbett was such a strict disciplinarian that his men refused to fight for him and purposely took poor aim with their cannon. If it is true it was a clear case of "biting off one's nose to spite one's face."

The "Boadicea" shewed great pluck in standing up to engage these two French frigates after the "Africaine" was captured, as every moment more French ships were expected to appear.

This fight could be heard on shore, and, naturally, the Regiment was wild with excitement until the return of the two British frigates.

Again, in September, the "Ceylon," frigate, having Major-General Abercromby and staff, also Lieutenant Clarke and twenty-five men of the Eighty-Sixth Regiment on board, was captured by "La Venus," French frigate, after a severe action, both ships being dismasted. The indefatigable Commodore Rowley again appeared, in the "Boadicea," recaptured the "Ceylon," and took the "La Venus." The description of the combat is as follows:—

"The 'Ceylon,' frigate, 32 guns, had been despatched from Madras to join Rowley. Looking in at Port Louis on September 17th, she saw what appeared to be a considerable French force in the harbour, and, bearing up, made all sail for Bourbon. Commodore Hamelin, with the 'Venus' and 'Victor,' promptly put to sea in chase of her.

"The 'Ceylon' discerned her enemies at 2 p.m., and at a few minutes past midnight, observing that the 'Venus' was far ahead of her consort, shortened sail to begin action. Nominally a 32-gun frigate, she actually carried twenty-four long 18-pounders, two long 9-pounders, and fourteen 24-pounder carronades, or forty guns in all; while the 'Venus' had twenty-eight long 18-pounders, four long 8-pounders, twelve 36-pounder carronades, or forty-four guns in all; so that the broadside weight of metal of the British ship was only 343 lbs. to the 'Venus' 484 lbs. Moreover, the 'Ceylon' had on board but about 295 people, including 100 men of the 69th and 86th Regiments, and the Frenchman probably had nearly her full complement of 380. In spite of the disparity of force, Captain Gordon, commanding the

'Ceylon,' maintained a hot fight for an hour, at the expiration of which time the 'Venus' dropped astern and gave him an opportunity of repairing damages and of endeavouring to escape ere the 'Victor' should get up. But at 2.15 a.m. the 'Venus' again overtook him, and the battle was renewed, until both frigates became unmanageable. At 4.30 a.m. the 'Victor' arrived and placed herself athwart the 'Ceylon's' bows, prepared to rake her, whereupon Gordon struck. At 5.10 a.m. his ship was taken possession of. She had lost ten killed and thirty-one wounded, including Gordon himself, and Master William Oliver. The losses of the 'Venus' cannot be specified. . . Had the 'Ceylon' realised in time that the 'Victor,' though a three-masted vessel of imposing appearance, was only a mere shell of a craft, less formidable than the ordinary eighteen-gun brig, she might have sunk her with a broadside, and perhaps have kept her flag flying for a few hours, when, as will be seen, she would have been relieved.

"At 7.30 a.m. on the 18th Rowley, who was then at anchor in St. Paul's Bay, saw the French ships and their prize at a distance of about nine miles from the shore. The 'Boadicea,' reinforced with 50 volunteers from the 'Africaine,' at once got under way, with the 'Otter' (sixteen guns), and the 'Staunch' (fourteen guns), and made sail in chase. The 'Victor' took the 'Ceylon' in tow, and the three endeavoured to make the best of their way to Mauritius, but they were delayed first by the tow-rope breaking and then by the disproportion in size between the 'Ceylon' and the 'Victor.' At 3.30 p.m., therefore, the prize was cast off, the 'Venus' lay by to protect her, and the 'Victor,' in accordance with orders, stood away to the eastward. Scarcely was the corvette out of range ere the 'Ceylon' hoisted her colours, Lieutenant Philip Gibbon having temporarily taken command of her in the absence of his seniors, who had been removed to the 'Venus.' At 4.40 p.m. the 'Boadicea' ran alongside the 'Venus,' and in ten minutes obliged her to strike, with a loss of nine killed and fifteen wounded.

"The 'Boadicea' had only two wounded. Rowley then put back to St. Paul's Bay. The 'Venus' was a fine frigate of 1,105 tons. She was added to the Navy as the 'Nereide.'"*

1811

The Regiment remained in garrison in Bourbon until the 14th of March, 1811, when it was ordered to embark for Mauritius, which had been captured by Colonel Keating's expedition in 1810.

The Eighty-Sixth duly landed at Port Louis, in Mauritius, on the 20th of March.

The right wing and headquarters proceeded to Mahebourg, and the left wing occupied the detached posts on the south and east sides of the island. Here a small draft joined the Regiment directly from England, under the command of Ensign James Creagh; it consisted of Ensign John Grant, Assistant-Surgeon Bell, and a few recruits, and formed part of a large draft, of which the remainder had proceeded to India.

* (Extract from "The Royal Navy," A History.)

EIGHTY-SIXTH REGIMENT (1806-1812).

The "state" of the Regiment is now given to show how an infantry battalion was organised at this time:—

NAMES OF OFFICERS AS POSTED TO COMPANIES, AND EFFECTIVE STRENGTH ON THE 24TH NOVEMBER, 1811.

Company.	Rank and Name.	Serjeants.	Corporals.	Drummers.	Privates.
Grenadier.	Capt. Lanphier ... ... Lieut. McLaurin. ,, Home. ,, Clarke.	4	4	4	50
No. 1.	Capt. Impey (Europe) ... ... Lieut. Campbell. ,, Hodson. Ens. Munro.	4	4	2	44
No 2.	Capt. Morrice ... ... Lieut. McLean (Europe). ,, McIntosh ,, Ens. Grogan.	4	4	2	41
No. 3.	Capt. Richardson ... ... Lieut. Jacob (Europe). ,, Mercer. Ens. McLean (Europe).	4	4	2	44
No. 4.	Capt. Shearman (Europe). Lieut. Vanspall. Ens. McQuarrie.	4	4	2	45
No. 5	Capt. Travers (Europe)... ... ... Lieut. Deane. ,, Stuart (Europe). Ens. Carroll, absent.	4	4	1	40
No. 6.	Capt. Marston... ... ... Lieut. McQuarrie. ,, Sandon. Ens. Grant.	4	4	2	44
No. 7.	Capt. M. Creagh ... ... Lieut. White. ,, Birkitt. Ens. I. Creagh.	4	4	2	42
No. 8.	Capt. Nicholson ... ... Lieut. Hilhouse. Ens. Wilkins (Europe).	4	4	2	41
L. Infantry.	Capt. Williams ... ... Lieut. Arch McLean. ,, John Webb.	4	4	2	46
	Total Effective	40	40	21	437

Orders were received from England directing that the Regiment should return to India. Therefore, on the 27th of December, 1811, the headquarters marched from Mahebourg, and the whole assembled at Grand Riverrie, near Port Louis, and on the 9th of January, 1812, the headquarters of the Eighty-Sixth and six companies embarked and sailed on the ship "Sir William Burroughs," under the command of Lieutenant-Colonel Fraser, the remaining four companies following in the ship "Helen," under Major Marston.

1812 On the 21st of February the Regiment landed at Madras, and encamped on the south beach until the departure of the 33rd Regiment for England, when it moved into barracks at Fort St. George.

Here it found the remainder of its draft awaiting it, consisting of sixty-eight recruits, under the command of Captain Impey.

Captain Michael Creagh now embarked for England, on leave, granted him for the recovery of his wound from a cannon-ball received in action at Bourbon.

This chapter of the most successful expedition to Bourbon can now be closed with the following extract from the "London Gazette," which reads as follows:—

"War Office,

"18th May, 1812.

"His Royal Highness the Prince Regent, in the name and on behalf of His Majesty, has been pleased to approve of the 86th Regiment being in future styled the 86th or Royal County Down Regiment, and bear the 'Irish harp and crown' upon their buttons."

The name of the Regiment has since been changed, and the number has been done away with, but for over 100 years the "Irish harp and crown" has been worn upon the buttons, and the title of "Royal" has distinguished the Regiment for their gallant conduct on the 8th of July, 1810.

This distinction entailed the changing of the facings from yellow to blue, and of the lace from silver to gold; whilst on their Colours the Regiment bore the "Irish Harp," in addition to the "Sphinx," and the word "Egypt." The names of "India" and "Bourbon" were ordered to be inscribed on the Colours some years later, as witnessed by the following letter sent to the Colonel of the Regiment, Lord Kilmorey:—

"Horse Guards,

"15th October, 1823.

"My Lord,—

"I have had the honour to lay before the Commander-in-Chief your Lordship's letter of the 21st ultimo, with its enclosures, and am now directed to acquaint your Lordship that His Majesty has been graciously pleased to approve of the 86th Regiment bearing on its Colours and Appointments, in addition to any other Badges and Devices which may have heretofore been granted to the Regiment the word 'India,' in consideration of the distinguished conduct of the Regiment during the period of its service in India, from the year 1799 to the year 1819.

"His Majesty has been also pleased to approve of the 86th Regiment bearing on its Colours and Appointments the word "Bourbon," in consideration of the distinguished conduct of the Regiment in the attack and capture of the Island of Bourbon, in the month of July, 1810.

"I have, etc., etc.,

"(Signed) C. TORRENS, Adjutant-General.

"To General Earl of Kilmorey,
"Colonel, 86th Regiment."

Monument erected at Bourbon by the 86th Regiment to the Memory of Lieutenant Monro and the N.C.Os. and Men who fell on that occasion.

CHAPTER VII.

EIGHTY-SIXTH REGIMENT (1813-1819).

Eighty-Sixth marched from Madras to Goa—Destination changed to Vellore—Marched 400 miles to Masulipatam—Second Battalion formed and disbanded—Left wing moved to Hyderabad—Description of the Pindarees—Their customs and destruction—Wing returned by trying marches to Masulipatam—Insurrection in Hyderabad previous to left wing leaving—Regiment ordered to Madras—Insurrection in Ceylon—H.M.S. "Orlando" takes the Light and Grenadier Companies to Ceylon—Description of that island and history—Operations of the Light and Grenadier Companies in Ceylon—Rebellion crushed out—The two companies returned to Trincomalee—Congratulatory order of General Sir Robert Brownrigg—Voyage from Ceylon to India on the "Forbes" —One other company's experiences in Ceylon—Volunteering for other regiments—Complimentary order from Madras Headquarters—Three months batta issued—Voyage to England on the "Golconda."

1813

IN January, 1813, the Regiment was ordered to march from Madras to its old station of Goa. Whilst on the march and engaged in ascending the Pada-Naig-Droog Ghauts, it received amended orders to march to Vellore, so the Regiment retraced its steps.

On arrival at Vellore, a draft, consisting of Lieutenants Jacob and Kirkland, Ensigns Munro, McLean, McQuarrie, Kennedy, and Maclachlan, and 217 men joined from England. Captain Impey, who was a Brevet-Major, and Lieutenant White died in May.

At the end of August, the Eighty-Sixth left Vellore and marched four hundred miles to the fort of Masulipatam. The route lay through the Calistry country, and when near the Kistna River the monsoon broke. For several days all ranks had to wade through the inundated cotton grounds waist deep in mud and water.

They arrived in Masulipatam in October, and suffered much from illness as the result of the march, losing several men. Whilst stationed at this fort, a number of officers arrived from England, bringing the new Colours with them. The officers were Captain Michael Creagh, Lieutenants Home and Perry, and Ensigns Gould, Bradford, Caddell, Henry and Moreton.

The strength of the 86th was now much above the establishment, including the recruits in England, this being due principally to the large number of men from the Militia who had volunteered to join the line.

The authorities decided to raise a second battalion, which was to bear date from the 25th of December, 1813, though it was not ordered to be raised until February, 1814. The order for raising it is as follows:—

EIGHTY-SIXTH REGIMENT (1813-1819).

"War Office,

"18th February, 1814.

"SIR,—I have the honour to acquaint you that as the effectives of the 86th Regiment of Foot, under your command, considerably exceed the Establishment in consequence of receiving Volunteers from the Militia, His Royal Highness The Prince Regent has been pleased, in the Name and on behalf of His Majesty, to order that a 2nd Battalion shall be added to the Establishment of the said Regiment, to consist, in the first instance, of One Field Officer and Four Companies of 100 Rank and File each. Agreeably to the detail specified in the margin hereof.* When the effective strength of this Battalion shall exceed 400 Rank and File, the Establishment will be augmented to Six Companies of 100 Rank and File each, with another Field Officer, and on the completion of that number the Battalion will be further augmented to the Establishment of Ten Companies, consisting of 800 Rank and File, with the usual proportion of Officers and Non-Commissioned Officers.

"I am also to acquaint you that the Establishment of the said Regiment is to be reduced from its present Establishment of 11 Companies and 1,208 Rank and File, to 10 Companies of 100 Rank and File, and according to the numbers specified in the margin† and that the Officers, Non-Commissioned Officers and Drummers now in this Country who, by the reduction in the 1st

* 2ND BATTALION 86TH REGIMENT.

4 Companies.

1 Major.
4 Captains.
4 Lieutenants.
4 Ensigns.
1 Paymaster.
1 Adjutant.
1 Quartermaster.
1 Surgeon.
1 Assistant-Surgeon.
1 Serjeant-Major.
1 Quartermaster-Serjeant.
1 Paymaster-Serjeant.
1 Armourer-Serjeant.
1 Schoolmaster-Serjeant.
20 Serjeants, including
4 Colour-Serjeants.
20 Corporals.
1 Drum Major.
7 Drummers.
380 Privates.

455 Total.

† 1ST BATTALION 86TH REGIMENT.

10 Companies.

1 Colonel.
2 Lieutenant-Colonels.
2 Majors.
10 Captains.
22 Lieutenants.
8 Ensigns.
1 Paymaster.
1 Adjutant.
1 Quartermaster.
1 Surgeon.
2 Assistant-Surgeons.
1 Serjeant-Major.
1 Quartermaster-Serjeant.
1 Paymaster-Serjeant.
1 Armourer-Serjeant.
1 Schoolmaster-Serjeant.
50 Serjeants.
50 Corporals.
1 Drum-Major.
21 Drummers.
950 Privates.

1128 Total.

Battalion, would become Supernumeraries, are, of course, to be posted to Appointments in the Establishment of the 2nd Battalion, agreeably to the rule laid down in the General Order dated 31st March, 1812.

"The arrangements are to take effect from the 25th December, 1813.

"You will be pleased to make the earliest communication upon the subject of the above-mentioned transfer of the Reduced numbers of the 1st Battalion to the Officer Commanding that Battalion.

"You will also acknowledge the receipt of this letter, and at the same time report the number and Ranks of the Officers and men transferred to the 2nd Battalion.

"I have, etc., etc.,

"(Signed) PALMERSTON.

"To General The Honourable F. Needham."

Major Baird was placed in charge, and the Second Battalion was duly formed at Hythe. It consisted of only four companies. In March, 1814, it was ordered to proceed to Colchester, with a view to embarking for Holland, but the abdication of Napoleon ended the war, and the Second Battalion never left England. It returned in October of the same year from Colchester to Hythe, and then marched to Deal, where it was disbanded.

Major Baird, Captain Edwards, Lieutenants McLaurin, Webb, Leach, and Hodson, Ensigns Stuart, Law, Russell, Holland, and Home, with 16 serjeants and 230 rank and file, being the effectives of the 2nd Battalion, embarked for India and joined the 1st Battalion at Masulipatam on the 11th of September, 1815. Previous to this, however, the left wing of the 86th Regiment, under the command of Captain Williams, had marched, in January, 1815, for Hyderabad, to join a force which was subsidized by His Highness the Nizam. This force arrived on the 2nd of February, at the cantonment of Secunderabad.

1815

In January, 1816, the remainder of the Regiment moved to Hyderabad, but returned to Masulipatam in October of the same year, where they found that a draft of forty-six men, under Captain Chadwick, had arrived from England in August to join.

During the period that the Regiment was stationed at both Masulipatam and Hyderabad, it was constantly employed against barbarous hordes of plunderers, called Pindarees.

The Pindarees were not a tribe, but a military system of bandits. They were largely used by the Native States to make war on their neighbours, for the purpose of ruining those with whom they were probably at variance, and also for the purpose of saving themselves from being plundered by this intractable organization. General Wellesley wrote of them in February, 1804: "I think that we run a great risk from the freebooter system. It is not known to the Governor-General, and you can have no idea of the extent to which it has gone, and it increases daily . . . No inhabitant can, nor will, remain to cultivate unless he is protected by an armed force stationed in his village."

The evil attained to such dimensions that in 1817 a great army had to be assembled for the destruction of these freebooters.

These rascals were augmented by military adventurers from every State, and frequently amounted to as many as 30,000 men, though their numbers fluctuated. No vagabond who had a horse and sword at his command needed to be at a loss for employment.

The Pindarees, however, were generally armed with spears, in the use of which they were very expert; a proportion of them were provided with match-locks, and all were mounted.

Their one thought was plunder. Bringing but little with them, they strove to carry away as much as possible.

Their ordinary marches were thirty or forty miles a day, and their secrecy was as great as their activity. Moving in bodies up to 4,000 strong, they rode to their destination divided into smaller parties, plundered all, burnt the houses, and drove off the cattle. Such work is, in truth, bad enough, though sometimes necessary, in regular war, but to their iniquities they added the torture and murder of their helpless victims. Red-hot irons were applied to the soles of their feet, a bag filled with hot ashes was tied over the mouth and nostrils of the sufferers, who were then beaten on the back to make them inhale the ingredients; oil was also thrown on their clothes, which were then set on fire. The hands of children were frequently cut off, as the most easy way of obtaining the bracelets which adorned their wrists. Many other modes of torture equally frightful were resorted to.

Having extracted the last jot from their victims, the Pindarees then rode for some pre-arranged point on the frontier, far distant from where they had entered, and reunited and divided up the spoil. No redeeming virtue marked the character of the Pindaree. Those who knew him best said that his courage was beneath contempt.

In 1814, 1815, and 1816 they raided the Madras Presidency, committing widespread depredations. On the 25th of December, 1816, Major Lushington, with 350 men, marched nearly fifty miles and attacked some Pindarees, killing and wounding about 800 of them. From that time on no quarter was given to these people, who, in 1817, raided over a very wide extent of territory, including the Nizam's country and the Honourable East India Company's domains.

Seven thousand inhabitants were killed and wounded by one small band of these robbers in twelve days in the Company's territory alone.

In 1817 and 1818 they were effectually destroyed by the simple process of attacking them with cavalry, etc., wherever they assembled in large numbers, and they were always thus beaten and dispersed. They therefore resorted to the practice of remaining dispersed, which would have entailed a never-ending guerilla war, but, fortunately, the native villagers, cheered up by seeing the Government really taking steps to destroy their enemies, gladly sallied out to attack the dispersed robbers, and dealt firmly with them. This ended the career of these turbulent scoundrels. The Pindarees are now forgotten even in name.

1817

On the 5th of January, 1817, Major Baird, with three officers and 120 men, marched to reinforce the left wing at Hyderabad. At Masulipatam there were detachments of from 100 to 150 men each employed in the field, under Captains Williams, Morrice, and Creagh, in the Ganjan district, on the banks of the Kistna, and towards Vizigapatam, leaving little more than the band and the sick in charge of the colours.

The left wing at Hyderabad was continually on the alert and frequently called out, but invariably failed to come up with the Pindarees, though they occasionally picked up a few sick men and horses.

These predatory hordes, having passed the frontiers and generally escaped unhurt into the Company's territories in various parties, caused considerable alarm, and, naturally, all disposable troops were in consequence fully employed.

On the 2nd of June, 1817, the left wing was ordered to march to Masulipatam, to join headquarters. The troops suffered greatly from the heat; several men died from the effects of it, and it is noted that great difficulty was experienced in burying an officer, Lieutenant Moreton, and two men in the same grave, on account of the decomposed state of their bodies, though they had only been dead a very few hours.

On another occasion the guides led the column astray, and the troops did not arrive in camp until after sunset, though they had started at one a.m. It is not, therefore, surprising to learn that on this occasion only ten men marched into camp representing the left wing, out of 450 men who had started just after midnight.

Several men died after being brought in, and three were found dead by the natives sent in search of them. A three days' halt was then made to allow of the stragglers coming in. They complained that lack of water brought on delirium.

A short time before leaving Hyderabad two of the Nizam's sons rebelled and threatened to depose their father, seizing and putting to death several of his adherents.

The British Resident having been appealed to by the Nizam, a column, consisting of the wing of the 86th Regiment, a battalion of sepoys, and two guns, were ordered to the city for his support. The force was preceded by two battalions of the Nizam's regular infantry, under European officers, with two 6-pounders. The adherents of the Princes attacked this force in one of the narrow streets and overpowered them, with the loss of one of their guns, one officer killed and two wounded, and fifty men killed and wounded.

The advanced guard of the British forces, consisting of the Light Company of the 86th and sixty pioneers, with hand grenades and entrenching tools, under Lieutenant Creagh, marched forward towards the sound of the firing in the town, recovered the gun, covered the removal of the wounded, and drove off the enemy. The troops remained under arms in the city until the next morning, by which time the Nizam's two rebellious sons had surrendered and had been sent to the fort of Golconda.

The force then returned to the Residency and remained encamped there until order was thoroughly restored in the city.

1818

On the 27th of January, 1818, the flank companies were sent out on an expedition, under Captain M. Creagh, with some native cavalry, etc., to try and intercept some Pindarees on their return from the East India Company's territories through those of the Nizam's. The force remained out until the middle of April, when the flank companies left it and marched to join the Regiment, which was then moving to Madras, picking them up at Guntoor on the march.

On the 15th of April the Eighty-Sixth left Masulipatam for Madras. When within one march of the latter place it was ordered to Wallaghabad, being under orders to return to England. It had already been under orders in 1816 to sail for home, but the orders had been countermanded on account of the Pindaree War.

Since leaving Vellore the following officers had died: — Captain Nicholson, Lieutenants Mercer and Henry, and Ensign Maclachlan, at Masulipatam; Lieutenant Hodson, at Hyderabad; Lieutenant Moreton, on the line of march; Brevet-Majors Williams and Shearman, at Calcutta and the Cape of Good Hope, where they had gone for the recovery of their health.

Again the Regiment was prevented from sailing by an urgent demand for troops to be sent to Ceylon. The "Orlando," frigate, brought the despatches from the Governor of Ceylon, General Sir Robert Brownrigg. The flank companies of the 86th were at once completed to 100 rank and file each and were embarked in the "Orlando," frigate, on the 8th of September, 1818, and sailed for Ceylon, under the command of Major Marston.

The following officers composed the flankers: —

Grenadier Company.	*Light Company.*
Capt. Michael Creagh.	Capt. Archibald McLean.
Lieut. W. Home.	Lieut. J. Creagh.
Lieut. D. Bradford.	Lieut. P. Gould.
Lieut. A. Russell.	Lieut. E. Caddell.

And Assistant-Surgeon R. H. Bell.

On the 12th of September they arrived and landed at Trincomalee, and commenced preparations for moving into the interior of the Island.

Before following the movements of the troops, it may be worth while to consider for a moment the physical features of the Island of Ceylon, and the origin both of its inhabitants and of the rebellion which was now in full swing in this British possession.

Ceylon is an island to the south of India. It resembles somewhat a pear in shape, and is some 270 miles in length and averages 100 miles in breadth. The sea-shore is, generally, level, but the greater part of the interior is diversified by mountains. These approach on the southern, eastern, and western sides to within about forty miles of the sea; whilst to the north the level district extends for about seventy miles. Owing to its configurations, few countries enjoy the same variety of climates as Ceylon. On the mountains the thermometer frequently falls to 50° Fahrenheit and is seldom above 65°, whilst on the coast 80° may be taken as an average. Its highest mountains reach an elevation of from 6,000 to 8,000 feet, and form a circular barrier to the interior, which rendered it impregnable to European arms for nearly 300 years. Several rivers take their rise in the elevated district, and, although there is not a natural lake of any extent in the entire Island, few countries are better provided with water.

The inhabitants are called Cingalese, and are variously supposed to be descended from the Chinese or from the Rajputs of the neighbouring Continent, with a strong mixture of Malay blood. Ceylon is supposed to have been quite well-known to the ancient Greeks and Romans, under the name of Taprobane. They always describe it as abounding in elephants. The Portuguese commenced their settlements there in 1505 and the Dutch followed and displaced them about 1602, or at the end of Queen Elizabeth's reign. In 1795 the Dutch had joined with the French in the war prosecuted against England. The latter replied by seizing the Dutch colonial possessions, and General Stuart commanded a force which captured Ceylon from the Dutch. It is probable that internal dissensions amongst the Dutch troops facilitated Stuart's task. It was now necessary to deal with the native King, who held a court at Kandy, and ruled over a collection of more or less subordinate Chiefs. In 1798 the King died, and his half-nephew, by marriage, was elevated by the Prime Minister to the throne. Later on he killed the Prime Minister, and shortly after decided to inflict the same fate on the then Prime Minister. The latter hastily fled to British protection. The King thereupon seized his wife and family, and had all the children beheaded one by one, whilst the unfortunate mother was compelled by torture to crush their dissevered heads in a mortar, even to the baby in her arms, and afterwards was killed herself. This personage was finally caught by the British native allies, but, unfortunately, instead of hanging him to the nearest tree, he was handed over to the English, who sent him to India with a pension. Previous to his being caught, however, there had been some fighting. A garrison, some 1,000 strong, had been left at Kandy, under a Major Davies. Being starved out, they had to make a treaty with the Cingalese and retreated towards Trincomalee. Being stopped in their retreat by a river, which they ineffectually attempted to cross, they finally surrendered their arms, and were marched off as prisoners, when, of course, they were all massacred, excepting three officers, who, shortly afterwards, died in captivity. Major Davies must bear the blame for the whole of this sad and unnecessary business. Having, however, cleared out the King, the country appeared to settle down, but in October, 1817, a spirit of insubordination first exhibited itself in Uva, a hilly district in the south-eastern part of the Island. The Government Agent was killed and a Buddhist priest was installed as king. In a short time the whole country was aflame, excepting one or two inconsiderable districts. The only way of settling the affairs was to clear the country thoroughly, killing all those found with arms in their hands. As the following diary shews, this was well carried out, and the country settled down, and has been peaceable ever since on the whole.

To return to the two flank companies of the Eighty-Sixth Regiment, who were left preparing for their advance into the interior.

The following account of the operations in Ceylon is preserved in the regimental records, being principally an extract from a journal kept by the companies:—

1818 On the 18th of September, 1818, part of the two flank companies embarked at Trincomalee in four dhonies, or native rice boats. Lieutenant Home and fifty men had to be left behind, owing to the crowded state of the boats, whilst Major Marston also remained behind, sick, thus leaving Captain Michael Creagh in charge. Three days' cooked provisions were carried by the men, and the boats sailed in the evening for Batticolee.

During the night of the 19th-20th of September one of the boats struck on a sunken reef and remained there. The weather was, fortunately, calm, but a certain amount of surf broke on the reef. This boat carried the ammunition of the force, and the occupants attracted attention to their situation by firing musketry during the night. Only one of the other boats heard them and came to their assistance. As it could not approach, on account of the surf, Lieutenant Creagh made a raft and went to the assistance of the shipwrecked crew—a gallant act, in view of the darkness and surf, not to mention the possibility of sharks.

Fortunately the surf was the salvation of the party, for it battered the boat over the reef at last, and it was found that it was but little damaged. At daybreak a breeze sprang up, and on the next day the little flotilla, after having crossed the harbour bar with some difficulty, anchored in the harbour of Batticolee, and the troops landed and encamped near the fort.

Here the Regiment found the sick of the 19th and 73rd Regiments. These men had been sent from the interior of Ceylon and were dying fast, at the rate of from three to six a day, whilst over 100 had been buried beside the tents of the 86th alone.

Lieutenant Home and his fifty men joined on the 23rd, and on the 25th the two companies embarked in open boats and proceeded up the Mandoor Lake and landed at Mandoor on the following day.

Here the force found coolies in readiness to carry the ammunition, provisions, and light baggage, and the troops at once marched off to the Mangalar River, about thirteen miles away. On the way they had a foretaste of what they might expect in Ceylon—heavy rain, a little cultivation, much jungle, and a gradual ascent. The companies halted here until the 29th of September, when they marched twelve miles to a place named Chinna Kandy. Here there was a post of the 19th Regiment, commanded by a Lieutenant Robinson. On the march the troops passed a party of sick soldiers proceeding to the coast. Some of the party consisted of women, and all were carried in hammocks, fastened to poles, and each hammock was carried by two bearers. All the invalids appeared to be in a dying state, and additional pathos was lent to the sad scene by the little children crying piteously over their weak and exhausted mothers. The journal notes that all cultivation had now ceased, and that there were no inhabitants, whilst, naturally, there was much jungle, and several rivers had to be forded even in this short march. A halt had to be made at this post for one day for the following strange reason, as stated by the writer, viz., that the halt was made to enable Lieutenant Robinson to pack up a supply of provisions, all his assistants having died, and all his men suffering from fever and ague.

On the 1st of October the march was resumed to Kataboa, a post some fifteen miles further on. Here Captain Ritchie, of the 73rd Regiment, was in charge, and he and his surgeon were the only Europeans, out of a force of seventy, free from fever. The severity of the fever may be judged from the simple statement that Captain Ritchie had buried nearly a hundred of his light company at this place.

On the march several herds of wild elephants lent interest to the weary way through thick jungle, without even a road, and with the usual rivers to wade through. On the way a halt was made to allow of a party of seventy men of the Ceylon Corps catching up. This detachment was commanded by Lieutenant Noonan, who added to that duty the important functions of interpreter and guide to the whole force.

A guide was certainly most necessary in the jungle, but Lieutenant Noonan does not appear to have been very successful at his first attempt to conduct the 86th companies through the jungle, for the day after his arrival the force marched, lost their way in the jungle, and at nightfall found themselves, as the writer naïvely says, much to their surprise, within a mile and a half of Kataboa, which place they had left that morning. The rain still continued, but the troops marched again next morning some ten miles through more open country, passing herds of wild elephants, buffaloes, hog-deer, pea and jungle fowl, but saw no inhabitants.

On the 5th the party arrived at Etenawallee, and encamped beside the place that a Mr. Wilson, with his attendants, had been killed a short time before by the enemy. Here the detachment's numbers were increased by the arrival of Lieutenant Stuart, of the Ceylon Corps, with a friendly Chieftain and his native followers.

On the 6th a further march of ten miles was made, and several of the enemy were seen on the hills in the Palwatee country. These were the first inhabitants that had been seen since leaving Maddoor. They were armed and abused the British troops calling them the not very opprobrious name of beefeaters.

Captain Michael Creagh now proceeded to business, and in the evening sent out various parties to scour the country and the neighbouring hills.

One detachment, under a Malay serjeant, of the Ceylon Corps, discovered a mountain village which was held by its inhabitants, but which, after a little difficulty, was stormed and destroyed, seven Kandians being killed. Another party, under Lieutenants Stuart and Russell, destroyed some houses and captured several prisoners; whilst Lieutenants Home and Creagh, after unsuccessfully chasing some of the enemy, discovered a large store of grain and household utensils. After supplying the force with the grain they required from this hoard, the remainder was destroyed. One of the prisoners stated that a rebel Chief was in the neighbourhood, so Captain Creagh, with Lieutenant Bradford and a mixed force of Europeans and Malays, marched away at once and reached the place in the mountains, where he resided, at daybreak next morning. The Chief himself had fled, but a few prisoners and some arms were taken and a few houses were destroyed.

1818

That day the force halted to rest themselves after their fatigues and marched, on the 8th of October, some seven miles, to Goadiagamme.

The country was found to be hilly, with much jungle, and several rivers had to be crossed. The Kandians watched the march from the hills, but offered no opposition.

On the next day, the writer of the journal notices with much satisfaction that a buffalo was shot, as up to now the whole party had been living entirely on bad salt beef and rice, and he adds that everyone was beginning to suffer from fatigue and exposure to alternate heat and cold and heavy rain.

The force, this day, remained halted whilst Lieutenant Noonan went with a detachment to Badula, returning next day with a small supply of ammunition, provisions and arrack and with a cohorn, in charge of four of the Royal Artillery.

On the 10th of October the force still remained halted. Difficulties were now cropping up with the carriers, who were much reduced in number by desertions and sickness. The allowance to officers of carriers was now reduced to two carriers to each officer, so as to set free enough to carry the more important items of provisions and ammunition.

On the 11th of October the first casualty occurred, Lieutenant Noonan being sent in sick to Badula, suffering from fever.

Four parties were now sent out that evening into different parts of the country to harass the enemy. The most important detachment was the light company of the 86th Regiment, under Captain McLean, with fifty-seven Malays and the cohorn, with its four artillery men attached. Captain McLean was seized with fever on this march. Captain Creagh arrived next morning and took over command, sending Captain McLean to hospital at Badula, where he died on the 20th of October, very much regretted by all ranks of the Regiment.

On the 12th of October the troops marched to Nindia Gamme. There was no road through the jungle, but there were vestiges of a path, which had been improved by the enemy according to their view of the situation. This improvement took the form of cutting ditches across the road; also trees and bushes were cut down and placed across the path to impede the march. But the most trying device of the Kandians to harass the British was very simple and yet annoying. They frequently cut paths off from the main path, and the force would first be in doubt which of the two to take. In case they took the wrong one it ended in a blind alley, and the whole force, strung out in single file along the path, then had to retrace its steps with all the attendant confusion, and start to work along the other and right path. The advanced guard was surprised and thrown into confusion this day by a wild elephant, and the writer of the diary adds that the jungle that day was very thick, that there was no cultivation, and that officers and men alike were suffering from sore feet and leech bites.

Captain Creagh afterwards ascertained from prisoners that the whole of this day a rebel force of some four hundred or five hundred men marched on his right flank, waiting for an opportunity, which never came, of attacking him at a disadvantage, or of cutting off stragglers.

Despite all these fatigues and worries, the troops were full of fight, and Captain Michael Creagh appears to have been quite a commander after their own heart. On bivouacking that night, a party was immediately sent off, under the command of Lieutenant Creagh, with Lieutenants Gould and Stuart, the latter to act as interpreter. It was reported that a particularly obnoxious rebel chief was somewhere in the neighbourhood. The Government of Ceylon had offered five hundred dollars for his capture. The chief's name was Polyahagodera. After a harassing night march, his village was found to be situated on the side of a mountain. It was surrounded and stormed; the chief was killed, and the village burnt, and all the remaining inhabitants, including his family, were brought in as prisoners.

On the 14th and 15th the two companies halted to rest European and native troops, all of whom were in a very bad state of health, from indifferent food, overwork and exposure.

On the 16th Lieutenant Creagh was ordered out for some duty, with forty rank and file. Whilst actually parading for this duty, he and several of his men were seized with fever and had to be carried off parade.

On the 17th of October parties were sent out by night, to endeavour to secure rice and prisoners. The diary adds that the rice was always found in husk, and the prisoners were utilised to beat it out.

On the 18th various parties that had been out since the 11th of October rejoined. They had been in charge of Lieutenants Home, Russell, and Caddell, and last, but not least, of the Dessaue, which was the title of the loyal chief, who had joined the force early in its march.

All these detachments had been universally successful in any encounters with the enemy, and brought in some prisoners, but, better still, had persuaded many of the Kandians to come in and make their formal surrender. It is noted that in one place that a small detachment of the grenadier company was ordered to occupy, eighteen men were taken ill within twenty-four hours, after which the commander of the party took it upon himself to quit the place.

On the 19th Lieutenant Bradford and a party made a long march and captured a rebel chief, with a price of five hundred dollars on his head. Several Kandians were killed in the scuffle.

The rebels replied to these successes by attacking the Dessaue's party and wounded several of his followers.

Lieutenants Home and Caddell were sent the same day with the sick to Badula, whilst on the 20th Captain Creagh and Lieutenant Bradford marched towards the great Kewligedera jungle in search of the rebel chief of the same name, leaving Dr. Bell and a party in charge of some of the sick, who could not be moved on account of the scarcity of coolies, as most of those brought by the force had already died.

The force marched 30 miles that night, and next morning fell in with a body of the enemy, who were immediately attacked and several were killed and several taken prisoners. After this, alarm signals were passed along the hills by the Kandians, and, as nothing further could be done for the moment, Captain Creagh concealed his party and sent out some spies and also some Malays disguised as inhabitants of the country, to try and get into touch with a small party, under a Lieutenant Burns, of the 19th Regiment, who had been ordered to join Captain Creagh here. The spies brought back information that Dr. Bell's camp had been attacked during the night and that an officer and some soldiers had been killed.

The Malays came back with two prisoners, who confirmed the above statement, and caused much anxiety to everyone. The two prisoners were now induced to lead the party to the " wanted " chief's dwelling. This turned out to be a cave very difficult of approach, and situated some distance up the mountain side. Kewligedera had observed the British as they toiled up the slope, and he and his followers wisely decamped, leaving official letters addressed to the Pretender, some stands of arms, and their wardrobes behind.

The two prisoners who were acting as guides, feeling that they could not return to their friends, now made the sporting offer to Captain Creagh to conduct him to where the chief's treasure was hidden, which they stated was only a few miles distant. The general anxiety about the fate of Dr. Bell's party was so great that the tempting offer was declined. The party then proceeded to march back to Dr. Bell's camp, and trudged steadily along all night in torrents of rain with not a firelock or cartridge amongst all the men which could by any chance have been made to go off. Hyenas, jackals, tigers, hogs and elephants roared round the marching column unceasingly in a terrifying manner. Private Stanton, of the light company, met a wild elephant face to face in the narrow path. He tried in vain to fire, but his firelock refused to go off on account of the wet. The elephant seized him, tore off his belt, threw him some distance into the jungle and broke his useless firelock to pieces. The poor fellow died from the injuries he sustained. At daybreak Captain Creagh ordered the bugles to sound occasionally, and their calls were soon answered by Lieutenant Burns's party, which had been delayed by the heavy rain of the previous day, which had swollen some rivers that he had to cross. He was, however, able to furnish Captain Creagh's party with a much wanted dram of arrack (native spirit), and a handful of rice per man. On arrival in camp it was found that the enemy had attacked Dr. Bell's party at night, but that sensible man had moved his party quietly out of the tents and away from the camp fire, at which two points the Kandians directed most of their fire. No one in the whole party was injured. A shell was then fired from the cohorn mortar, which exploded behind the rebels, who decided that they were surrounded, and fled from the field.

1818

Captain Creagh, in the meantime, with wise foresight, had sent Lieutenant Gould and a small party to the heights to erect a shed and collect grain and cattle, and, as all the coolies (who were all natives of India) were now on the sick list, the whole force moved from the pestilential low grounds to the heights near Hatpeira. From this place Lieutenant Gould took as many sick into Badula on the 25th of October as their means of transport would permit of.

1818

This post at Hatpeira was by no means a haven of rest. Though let severely alone by the enemy, it was continually annoyed by wild elephants at night. These unwelcome visitors destroyed the tents and sheds and killed several native followers, though, fortunately, no soldiers were injured by them.

From this time to the 15th of November the detachment was employed every night in pursuit of the rebel Kandian chiefs and in collecting prisoners and provisions.

The captures made were various, 200 head of cattle, a large store of paddy, or unhusked rice, 80 stand of arms, 300 bows, with many arrows, four "grasshopper" field guns, well mounted, several loads of Dutch money, many elephants' tusks, and, finally, in conjunction with other detachments, they captured the King's crown of the Pretender, his jewels, his favourite horse and elephant. All these last Royal belongings were forwarded by Captain Creagh to the Governor of Ceylon. Most of the arms taken were those which had belonged to the unfortunate party commanded by Major Davies, which had been massacred.

But by this time the good work of the flank companies and many other similar detachments was beginning to tell. Many rebel chiefs had been killed and many taken prisoners, and the Pretender himself was believed to have fled from the island of Ceylon to the mainland of India. The remaining Kandians who had joined in the insurrection decided to return to their allegiance, and they and their followers began to come in in large numbers to make their submission.

Evidently so many European troops were not required, and as the 86th "Flankers" had been hastily borrowed, it was decided that they should return as soon as possible to the Regiment, which was being detained from embarking for England; so they were called in to Badula about the middle of November. A complimentary order, thanking the two companies for their successful work and good behaviour was published by the Commandant of Badula. On the 19th and 20th of November the two companies set out on their return march to Trincomalee.

The composition of the force was given at the beginning of the expedition on the 8th of September as nine officers and two hundred rank and file, with the usual proportion of serjeants and drummers. The companies had sadly shrunk in size; five officers and sixty-six rank and file, with two serjeants and two drummers only, marched away from Badula on the 20th of November.

The rainy season had now set in, and all the rivers were full and rapid. No bridges or boats were available, and the crossings were attended with much danger and discomfort. With much difficulty Kataboa was reached, and Captain Ritchie's detachment was found almost annihilated by disease. Here an elephant was obtained and the march was resumed with much more comfort, as the elephant was used to convey the stores and men across the rivers.

At Chinna Kandy all was in a state of ruin. Lieutenant Robinson had been sent away sick, and his three serjeants had died. The few remaining men were unable to move, and the defences of the place had been broken down by wild elephants. Worse than all, the buildings containing the stores had been washed away by the tropical rain, and with great difficulty enough was recovered to supply the detachment of the Eighty-Sixth.

After ten days' marching, under almost incessant rain, the troops reached

Mandoor, where they halted two days to recover, and then sailed down the lake to Batticolee.

The party remained at this place until further orders were received. When these orders arrived, to everyone's surprise they contained directions to proceed by land instead of by sea. At this time, as was well known, most of the country to be traversed was covered by water. However, there was no help for it. The orders were perfectly distinct, and the troops proceeded to carry them out.

Captain Creagh, with his usual foresight, prepared two canoes, and these, with the invaluable borrowed elephant, were sent along the road for some forty miles. This distance was usually performed in three harassing marches by the infantry, but a friendly Collector had supplied boats for the men, which conveyed them by sea these forty miles. Then the elephant and canoes were picked up and the whole party set their faces steadily towards Cottiar Bay, a place near Trincomalee. The time taken for the march was ten days. The country was nearly covered by water, and often there was scarcely any dry ground to rest upon. The column moved in the following order:—The elephant marched in front, feeling for the numerous rivers; then came the canoes, dragged by the remaining coolies, assisted by those soldiers who were still able to march. In the canoes all the sick were placed. The towing party were generally up to their waists in water and frequently swimming. The tents were by now destroyed and were abandoned, as there was, generally, no ground above water to pitch them on, so they were only useless encumbrances.

On the 18th of December the dishevelled party reached Cottiar Bay and found boats awaiting them, in which they at once embarked for Trincomalee. The boats were open ones and leaked badly, and the bar was with difficulty crossed. Once outside, rough weather came on, and it was found too dangerous to remain in the open sea and impossible to re-cross the bar with safety. In this unfavourable position Captain Creagh again shewed much resource, and ordered the boats to pull for some small low rocks. Here the party landed and remained the whole night, huddled together, and in constant danger of being swept away as the tide rose, and the sea still remained rough. To add to their privations the boats had neither provisions nor water, and, drenched, shivering and starving, the little party remained waiting for the dawn.

The following morning the rain ceased and the sea went down; the party re-embarked and pulled for the harbour. Here they were met by the men-of-war's boats, which had been cruising, looking out for them for the last two or three days, and had known of their perilous plight the evening before, but, owing to the extreme roughness of the sea, had been unable to proceed to their assistance. The "Orlando," frigate, which had brought the two companies to Ceylon, at once invited all to breakfast, the Naval officers inviting the infantry officers, and the crew all the men, but the kind invitation was refused. The frigate's boats towed the soldiers' boats to the shore, whilst the crew of the frigate crowded to the side of the ship and cheered the little party. A feeble cheer was raised in reply, and the soldiers landed and immediately carried off twelve of their small number seriously ill to the hospital, whilst one captain, four lieutenants, one serjeant, two drummers, and forty-seven rank and file marched into the fort, the remains of two flank companies of just two hundred rank and file.

During this short campaign all—officers and men—marched barefooted, with their trousers cut off at the knees, to enable them to discover and remove the leeches with which every pool, river, bush, and blade of grass abounded. The appearance of the whole detachment on landing was heart-rending. They were dressed in rags, unshaved, emaciated, tottering from exhaustion and hunger, with their feet and legs ulcerated.

Lieutenant Home found his brother, Ensign Home, dead in the hospital when he landed. The latter had belonged to one of the Battalion companies; that is, one of the eight remaining companies which were neither grenadier nor light companies. Part of his company had been sent to a most unhealthy station, called Minnery, and Ensign Home had contracted a fever there, from which he died on the 19th of December. Most of the non-commissioned officers and men who had proceeded there with him also died.

The flank companies had to remain at Trincomalee until the 15th of January, 1819, when they embarked on a ship named " Forbes."

Previous to their sailing, the following order was published by the Commander of the Forces in Ceylon, which read as follows:—

" GENERAL ORDER.

" Headquarters, Colombo,

" 22nd December, 1818.

" The embarkation of the detachment of H.M. 86th Regiment at Trincomalee for Madras, having been directed by the General Order of the 15th instant, the Commander of the Forces is desirous of expressing his public acknowledgments to Major Marston, Captain Creagh, and to all the officers and soldiers of that distinguished Corps, for the gallant and important services rendered by them in suppressing the Kandian rebellion; at the same time, he cannot conceal the deep regret he feels at the severe loss which has fallen on the detachment, and the death of a gallant officer, Captain Archibald McLean, and many brave soldiers, and the sickness under which several of the men still labour, but whom, he trusts, will ere long be restored to health and to the service of their country."

1819

The General (Sir Robert Brownrigg) also stated that Captain Creagh's gallant and most useful services in the command of the flank companies of the Eighty-Sixth Regiment during the Kandian war, under very trying circumstances, were such as to entitle him to the favourable notice of His Royal Highness the Commander-in-Chief, and to the best commendation that he could bestow upon him.

The ship "Forbes," with the flank companies aboard, met with contrary winds, and ran out of provisions, and had to put back to Trincomalee, where she arrived on the 2nd of February, but put to sea again immediately and reached Madras on the 5th of February, where the detachment landed and joined the Regiment at Poonamalee the same evening.

The "Forbes" also brought back the Battalion's company to which poor Ensign Home had belonged. It had left for Ceylon seventy strong, but only brought back thirty men, whilst the flank companies only brought back fifty of all ranks, so deadly had the Ceylon climate proved. Assistant-Surgeon Bell and Lieutenant Caddell, and several of these died after their return to India.

The Battalion company, before referred to, had been ordered to Ceylon as follows: So soon as the grenadier and light companies had gone to Ceylon, the remainder of the Eighty-Sixth was ordered to follow, and they embarked on the 17th of September, 1818, on H.M.S. "Fowey" and on two merchant ships named "Neptune" and "Beckwith." They sailed to Colombo, and were ordered to leave one company at Trincomalee and to take the remaining seven companies back to Madras, which place they reached on the 17th of October, and disembarked and marched to Wallaghabad.

On the 22nd of December the Eighty-Sixth marched to Poonamalee, where the three companies found them on their return.

1819

The custom in India at that time was for any corps which was expecting to go to England to allow their men to volunteer for other corps serving in India. By this means the transporting of large bodies of troops round the Cape of Good Hope was avoided, as regiments went home reduced, excepting for the number of officers necessary to the strength of a company. The Indian Government gave a bounty of so much to each man volunteering to remain behind in India, which bounty was, of course, a great deal less than it would have cost them to bring out a man from Great Britain to replace him, and in place of a recruit they obtained a seasoned soldier. The man received the bounty, which was for him a large sum of money, and the only people who came really badly out of the transaction were the officers of the regiment going home, who had to lose all their old men and renew practically the whole regiment with recruits. So the first signs that the Eighty-Sixth was really going home at last came in the issue of General Orders of the 9th of February, 1819, issued at Calcutta, giving permission to the men to volunteer for other regiments serving in India.

EIGHTY-SIXTH REGIMENT (1813-1819).

The results of the volunteering are as shewn on the attached table.

	Serjeants.	Corporals.	Drummers.	Privates.		Serjeants.	Corporals.	Drummers.	Privates.
Royal Scots	—	—	—	1	47th Regiment ...	—	—	1	7
17th Regiment ...	—	3	—	79	69th Regiment ...	1	1	8	92
24th Regiment ...	—	—	—	42	87th Regiment ...	—	2	4	190
34th Regiment... ...	—	—	1	—	Honourable Company's Service.	13	9	2	101

	Serjeants.	Corporals.	Drummers.	Privates.
Total	14	15	16	512

The following most complimentary order was then issued by the Lieutenant-General Commanding in Madras:—

"Headquarters, Madras,

"6th March, 1819.

"The Lieutenant-General has traced with gratification the well-earned tributes of applause bestowed by successive Governments and Commanders in commendation of the numerous instances of gallantry and efficiency displayed during the active and varied service in which His Majesty's Eighty-Sixth or Royal County Down Regiment has been engaged since its arrival in the East Indies; and Colonel Fraser, the officers and men, are requested to accept His Excellency's thanks for the correct and orderly conduct of the corps. . . .

"By order of Lieutenant-General,

"SIR THOMAS HISLOP, Bart, K.C.B."

This was followed by a still more interesting order from the Government, which shows in what a practical manner the Honourable East India Company recognised merit. It reads as follows:—

"GENERAL ORDERS BY GOVERNMENT.

"Fort Saint George,

"20th March, 1819.

"The meritorious conduct of His Majesty's Twenty-Fifth Light Dragoons and Eighty-Sixth Regiment of Foot in every situation where they have been employed during the long period of their services in India has been brought under the notice of Government in a particular manner by His Excellency the Commander-in-Chief, as establishing for them strong claims to its consideration, and the honourable mention which has been made by His Excellency the Commander-in-Chief has been coupled with the expression of His Excellency's wish that their services may meet with the same acknowledgment on their approaching departure as in General Orders dated 27th of August, 1805, and 11th of October, 1806, marked the close of the distinguished career of His Majesty's Seventy-Fourth Regiment and Nineteenth Light Dragoons then returning to England; concurring entirely in the commendations which His Majesty's Twenty-Fifth Light Dragoons and Eighty-Sixth Foot have received from the Commander-in-Chief, and entertaining the same high sense of their merits and services, the Governor in Council is pleased to extend to the officers of those corps the indulgence recommended by His Excellency, and accordingly directs that a donation of three months full batta be passed to them on the occasion of their embarkation for Europe.

"By order of The Right Honourable The Governor-General in Council."

On the 1st of April the Regiment received its orders to move to Madras and to embark from that port. It duly marched to the fort at Madras for the last time, and on the 15th of April embarked on the "Golconda."

Twenty-seven officers and 110 non-commissioned officers and men constituted the actual strength of the Eighty-Sixth at this time. On the 17th of April, the Regiment were all on board, with the exception of the Colonel and five other officers, when it came on to blow so heavily that the "Golconda" had to cut her cables and get clear of the Madras roadstead for fear of being blown ashore; so the Regiment went home leaving these six officers behind. They followed immediately afterwards in the ship which brought the Eighty-Fourth Regiment home. It is interesting to note the time occupied on this voyage compared with the time now taken to bring a regiment home from India.

The "Golconda" left Madras in a storm on the 17th of April, and anchored in Simon's Bay, Cape of Good Hope, on the 6th of July. Sailed on the 19th of July and anchored at Saint Helena on the 9th of August; sailed from there on the 17th of August, and anchored off the Isle of Wight on the 18th of October, 1819. The ship reached the Nore on the 20th of this month and was placed in quarantine, as, besides losing Lieutenant John Campbell on the 31st of August, several of the few remaining non-commissioned officers and men had died on the way home. The quarantine did not last more than three days, and the Regiment disembarked at Gravesend, after a voyage of twenty-seven weeks.

The Eighty-Sixth had been abroad for twenty-three years and four months. Of all those who embarked in June, 1796, only two individuals returned to England with the Regiment, namely, Major Marston, who brought it home, in the enforced absence of Lieutenant-Colonel Fraser (Major Marston had gone abroad with the Regiment as one of the junior Lieutenants), and Quartermaster R. Gill, who had embarked with the Regiment as a private. The Regiment now settled down to the peaceful routine of home and colonial garrison life, which was not fated to be disturbed by the rude alarms of war for just under forty years.

OFFICER'S BREAST PLATE, 1793-1822,
silver polished plate, silver mounts.

CHAPTER VIII.

EIGHTY-THIRD REGIMENT (1805-1817).

Cape of Good Hope returned to the Dutch in 1803—1805, the British Government determined to re-take the Cape—Reasons for this course—Strength of force employed—Embarked from Falmouth and Cork—Voyage to the Cape of Good of Hope—Dutch forces at the Cape under General Janssens—General Janssens's information and preparations—British attempted to land—First landing failed—Force sent to Saldanha Bay under Beresford—Second attempted landing successfully carried out—Combat at Blueberg—Casualties of both sides and of 83rd Regiment—March on Cape Town resumed—Cape Town surrendered—List of stores, etc., surrendered—83rd Regiment sent to Mossel Bay—General Janssens signed a capitulation—Lieut.-Colonel of the 83rd Regiment selected to take the despatches to England—Pay lists of the 83rd Regiment—Dr. Cowan, 83rd Regiment, sent on an exploring expedition and lost—Major Collins, 83rd Regiment, also sent to explore and discovered the Caledon River—83rd Regiment ordered to India to suppress a mutiny—17 natives ordered to be enlisted into 83rd Regiment—Native risings checked by 83rd Regiment—2nd Battalion 83rd Regiment disbanded and officers and men drafted to 1st Battalion—83rd Regiment ordered to Ceylon.

1805

THE Dutch colony at the Cape of Good Hope had been in British hands from the 16th of September, 1795, until the 20th of February, 1803. In 1802 there had been a general peace arranged between various European Powers, called the Peace of Amiens, and in consequence of that peace orders had been sent from England, dated the 16th of November, 1802, directing that the colony should be handed over to the Dutch again. This was duly carried out on the 20th of February, 1803, though war had been again declared between Great Britain and other European Powers.

Early in 1805 the British Government determined to recapture the Cape. There were various reasons for this decision; amongst others, that it would be a convenient station for their vessels in the East India trade to refit and refresh their crews, whilst at the same time the enemy—French and Dutch—would be denied the use of the place. At this time the Cape was much used as a harbour for hostile privateers, as is proved by constant despatches in the "London Gazette" of that period, detailing captures by British men-of-war. The seas were so efficiently patrolled by the English fleet that the Dutch in Cape Colony found that the quickest way to get to Holland or France from the Cape was to embark in any vessel belonging to the United States and to sail to America, there changing to any American vessel bound for European ports. The United States, of course, took no part in these European wars.

Having determined on this expedition, the Government had to decide upon the strength of the force to be employed.

In July they ordered the 59th Regiment and the 20th Light Dragoons to embark at Falmouth in transports belonging to the East India Company, and to these two regiments were added 320 artillery men, and 546 recruits, some of whom belonged to the 86th Regiment, and were under the command of Ensigns Heddrick and Creagh. These transports were escorted by H.M.S. "Espoir" (eighteen guns), "Encounter" (fourteen guns), and "Protector," and their destination was announced to be the East Indies.

Shortly afterwards the 24th, 38th, 71st, 72nd, 83rd and 93rd Regiments were embarked at Cork, ostensibly for the Mediterranean, and, being joined by victuallers' tenders and merchantmen, sailed under the protection of the "Diadem," "Raisonnable," and "Belliqueur" (all of sixty-four guns), and the "Diomede" (fifty guns), and the frigates "Narcissus" and "Leda" (thirty-two guns).

Both expeditions had secret orders to join at Madeira, and then proceed in company to the Cape of Good Hope.

Commodore Home Popham commanded the fleet, and the troops were under the command of Major-General Sir David Baird.

Sir David Baird's return of the strength of his force, dated the 27th of August, is as follows:—

	Whole Force.	83rd Regt.
Colonels	1	—
Lieutenant-Colonels	6	1
Majors	12	2
Captains	57	7
Lieutenants	75	11
Ensigns	33	7
Adjutants	6	1
Paymasters	6	1
Quartermasters	6	1
Surgeons	8	1
Assistant-Surgeons	12	1
Serjeants	213	43
Bombardiers	22	—
Drummers	128	19
Rank and file	4416	701
Women	366	Not noted.
Children	278	Not noted.
Civilians	21	—
Total	5666	796

The above does not include 19 officers and 229 non-commissioned officers and men of the 20th Dragoons, nor the troops on board the East Indiaman sailing from Falmouth, which consisted, according to Sir David Baird's complaint of the 1st Battalion 59th Regiment, only 900 strong, instead of 1,000, and some 546 recruits, very young, unarmed, unequipped, and not acquainted with the use of firearms.

He very strongly urged that the 8th Regiment, stationed at Cork, should also be sent with him, but the Government did not comply with his request.

So engrossed was the world in the nearer and greater events, that this formidable expedition left the shores of Great Britain almost without attracting any attention. When the 83rd Regiment sailed on this enterprise the French Army, nominally for the invasion of England, was encamped at Boulogne, and could practically be seen from the English coast at Folkestone, and whilst the fleet was still sailing to its destination, Trafalgar was fought, and the Austerlitz campaign, the first of the great victories of the French Empire, was won. When it returned from the Cape, in 1817, the French Empire had vanished, and the great deeds of the Peninsular War and Waterloo were but memories.

Setting sail on the 31st of August, from Cork, the fleet reached Funchal, in Madeira, on the 28th of September without incident, save a collision on the first day between two transports, and sailed again on the 3rd of October, and reached Bahia, in Brazil, on the 10th of November. This will appear nowadays a roundabout way to go to South Africa, but in the days of sailing ships it was the shortest way in point of time, on account of the prevalence of the regular "trade" winds, which carried the vessels to the coast of South America, and could then be counted on to take the ships across to the Cape.

In the bay at Bahia (also called San Salvador) a transport, the "King George," and storeship, the "Britannia," were wrecked, and Brigadier Yorke, in charge of the artillery, was drowned, together with two of his men. Sailing again on the 26th of November, the expedition arrived near Table Bay on the 4th of January, having left forty men in hospital at the friendly port of Bahia.

1806

The white inhabitants of the Cape numbered at this time 25,757, exclusive of soldiers, and had 29,545 slaves and 2,006 indentured natives. Cape Town, the only real town, was fortified, and had a population of over 6,000 Europeans and some 11,000 blacks. A fair number of the white inhabitants were British subjects, mostly men who had settled there as traders during the British occupation. They had been ordered to leave the colony in neutral ships, but the order was not enforced.

General Sir David Baird, who knew the Cape well, having served there for nearly a year, in 1798, over-estimated the strength which the enemy could bring against him, and states in his despatches of the 12th of January that the force opposed to him was some 5,000 strong. As a matter of fact, all the excellent and energetic Dutch Governor, General Janssens, could muster for a field force, was as follows:—

The 5th Battalion of Waldeck	400
The 9th Battalion of Jagers (or Sharpshooters)	202
The 22nd Regiment of the Line	358
Dragoons	138
European Artillerymen	160
French sailors	240
Javanese Artillerymen	54
Hottentot Infantry	181
Slaves from Mozambique, Artillery Train	104
Mounted Burghers	224
Total	2061

To this strength must be added sixteen guns, according to General Janssens, and 23, according to Sir David Baird's despatches.

Also, there was a large force of burghers both in the country and in Cape Town, besides a few soldiers, not reckoned in the above returns, left in Cape Town to stiffen up the burgher force.

The arrival of the English fleet was not unexpected, for on the 25th of December, 1805, the French privateer, "Napoleon," which had recently brought some fifty English prisoners of war from the Mauritius to the Cape and had then gone to cruise in the route of homeward-bound ships was chased ashore by an English frigate, near Cape Town, and it was suspected that the frigate had companions. A day later another vessel ran into Cape Town with the news that she had passed a great fleet steering south; whilst on the 28th of December yet another vessel arrived with the news that a large number of English ships had sailed from Madeira on the 4th of October.

General Janssens took his measures accordingly. He arranged a system of signal guns, which worked so well that the British having arrived off the coast on the 4th of January, by daybreak on the 5th all the country within 150 miles of Cape Town knew that the enemy was close at hand. Unfortunately for him, the English had arrived at quite the worst season of the year from his point of view. The country burghers—farmers to a man—were threshing, and the grapes were beginning to ripen. Everything conspired to keep them at home, whilst the weather was exceptionally hot and travelling was very uncomfortable. On reflection, the countrymen decided to let military matters stand for the present, and attend to more profitable and safer business nearer home.

This took away a very large part of the Dutch forces, and left their General only those already detailed. As to these, the Waldeck Battalion was not well disposed; they had no national dislike to the British, and foresaw that the Dutch would be beaten, and did not see any point in being killed for honour and glory to no purpose. The Jagers were of all nationalities, but were full of fight. The crews of the French ships were collected from the "Atalante," a frigate, driven ashore and dismasted in Table Bay a short time before this, but floated and then being repaired, and of the privateer, "Napoleon," driven ashore by a British cruiser on the 26th of December. They were commanded by Colonel Beauchêne, the commander of the marines in the "Atalante."

Such was the state of the Dutch ashore. The English afloat in the transports were in none too enviable a position. It was one thing to sail to the shores of South Africa; it was quite another thing to land there on arrival. General Baird had intended to put his army ashore at a curve in the coast, north of Melkbosch Point, some sixteen miles' march north of Cape Town, and to march straight upon the town, trusting to his fleet to keep him supplied with provisions, and afterwards to land his siege train, if he should require to employ it, near Salt River.

Using his power of rapid movement by sea, General Baird made a feint on the morning of the 4th of January, with the 1st Battalion 24th Regiment against Campo Bay (Green Point), guarding the transports with the "Leda," frigate. The wind failed, and the transports were not able to get close enough to land their troops. The essence of the whole expedition was surprise; therefore, no time could be wasted, so on the morning of the 5th the First Brigade, under Brigadier Beresford (afterwards Marshal Beresford, of Portuguese and Albuera fame), proceeded in boats to land in the small bay north of Melkbosch Point. However, it was found that the surf was breaking in too heavily to permit of a landing, so the troops, including the 83rd Regiment, had to re-embark. Then that evening Brigadier Beresford was sent, with the 38th Regiment and the 20th Light Dragoons, to Saldanha Bay, about seventy miles north of Cape Town, to effect a landing, and try and seize cattle and horses for the force. The "Diomede" protected this landing party. The remainder of the force was to have followed on the morning of the 6th, but as it was observed that the surf was beating much less violently on the shore than on the previous day, it was determined to make another attempt to land. The "Diadem," "Leda," "Encounter," and "Protector" were moored so as to cover the beach, and then a small transport was run aground, to make a sort of breakwater to the part of the beach on which it had been decided to land the troops. The Highland Brigade (71st, 72nd and 93rd Regiments) was then landed, without any further casualties than one boat, containing thirty-five of the 93rd Regiment, which tried to land a little wide of the breakwater, being upset, and all its occupants drowned.

The Dutch opposed the landing with one company of Burgher Militia, but the guns of the ships kept them at a respectful distance, and only one man of the Highland Brigade was killed, and two officers (including the Brigade-Major) and three rank and file were wounded, on the 6th of January.

On the 7th the 83rd Regiment landed, accompanied by the 1st Battalion 24th Regiment and 1st Battalion 59th Regiment, all under the command of Lieutenant-Colonel Baird, of the 83rd Regiment, who was brother to the General. No time was lost now by the British. Formed in two brigades, they marched away very early on the morning of the 8th of January towards Cape Town. At 3 a.m. on that morning the Dutch force was advancing towards Blueberg, from Cape Town. They had passed the night near Rietvlei, and at that hour their scouts brought word that the British were approaching. At five o'clock General Baird's army could be seen descending the shoulder of the Blueberg, formed in two columns. The column on the right—the First Brigade—contained the 83rd Regiment, the 1st Battalion 24th Regiment and 1st Battalion 59th Regiment of Foot. Besides the two infantry brigades, Sir David Baird had two howitzers and six light guns. The Dutch deployed into line, and their General rode down the front, saying a few encouraging words to his men.

General Baird states that the enemy's force then looked to him 5,000 strong, mostly cavalry, with 23 guns "yoked to horses," and he considered that they intended to refuse their right wing, and with their left wing to turn the British right, so he frustrated this attempt by sending the Second, or Highland Brigade straight up the Cape Town road to attack their front, whilst he ordered the First Brigade to advance through a defile of the mountains to attack the Dutch left. Deploying as they advanced, the Highland Brigade received a very severe fire of round shot, grape and musketry. The cannonade was kept up by both sides as the two infantry brigades moved forward to the attack, but the Waldeckers had no heart for fighting, and turned and ran. The 22nd Regiment of the Line followed suit, was rallied by General Janssens, and again broke and fled as it caught sight of the Highlanders coming on with the bayonet. The remainder of the Dutch troops fought well, but they were defeated, and by General Janssens's orders retreated on Reitvlei, where they were rallied.

In the meantime, the First Brigade, not without loss, had cleared the irregulars out of the heights on the British right flank, hastening the Dutch retreat.

The total British loss on this day was 1 officer and 14 rank and file killed, 9 officers and 180 rank and file wounded, and 8 rank and file missing.

The loss of the 83rd Regiment was reported to be two serjeants and two privates wounded, and three privates missing. All were really wounded. Their names are as follow:—

Serjeants Crane and William Scott, Privates John Elvin, James O'Bern, Joseph Griffen, John Foster, Dennis Foley.

The last-named died on the 11th of January.

The Dutch loss has been usually stated as being 700 killed and wounded, and was so reported in Sir David Baird's despatch, but it appears to have been much over-estimated; 337 was the actual number returned by the Dutch as killed, wounded, and missing, of whom 110 came from the gallant French sailors, 240 strong.

The Eighty-Sixth Regiment's officers and men were in the charge of the Highland Brigade, attached to the 93rd Regiment, and both Ensigns Heddrick and Creagh (misspelt Craigh) are returned by General Baird as wounded.

General Janssens sent the Waldeck Battalion from Reitvlei, excepting one company, back to Cape Town, and also the French sailors. With the remainder of his force he moved to the east, to the mountains of Hottentot Holland, where he had laid in a store of provisions, whilst General Baird, suffering much from want of water, moved on to Reitvlei and halted there for the night. The sailors, by great exertions, landed sufficient provisions for all the troops, excepting the 1st Battalion 59th Regiment, who had nothing to eat. Next morning (the 9th of January) the march was resumed, and about two miles north of Cape Town Brigadier Beresford's force, of the 1st Battalion 38th Regiment and the 20th Dragoons, joined Sir David Baird, and a flag of truce came out from the town. The Commandant of Cape Town demanded a forty-eight hours' suspension of arms, in order to negotiate a capitulation. General Baird, very properly, gave him six hours to surrender the outworks of the town, and then allowed thirty-six hours for the capitulation. The 1st Battalion 59th Regiment at once took possession of Fort Knokke, and next day all the military posts of the town were surrendered and occupied by the British.

1806

The inhabitants suffered no ill-treatment, and the Regulars were sent to Europe as prisoners of war; 450 pieces of cannon fell into the hands of the victors. The total value of military stores captured was £120,877 15s. 11d., which was afterwards divided amongst the troops as prize money. The Dutch inhabitants of Cape Town immediately sent in a petition to General Baird, asking that some of the prize money might be paid at once, so that it could be spent in the colony.

General Baird sent Brigadier Beresford, with two regiments, towards General Janssens's party, pointing out to him, in a courteous letter, the misery which would be caused to the inhabitants if he pursued a guerilla war.

To hasten his decision, the 83rd Regiment was embarked on the 14th of January, with orders to sail round eastward to Muscle Bay (*Mossel* Bay now), for the purpose of occupying the defiles of the mountain of Zellendam, called the Attaquas Pass, from which position it could advance to attack General Janssens's force from the rear, or could, by holding the mountain, cut off his retreat.

On account of the extreme violence of the wind, only one of the ships of war, with the light and grenadier companies of the 83rd Regiment on board, was able to effect the passage. They reached False Bay. The ships conveying the remainder had to relinquish the attempt and return to Table Bay.

But the threat had due effect. The Dutch General knew full well that he could expect no help from Europe for a long time, and on the 18th of January he signed the final capitulation, which transferred the Cape to British rule.

The Waldeck Battalion at once enlisted into the British service, whilst the trained Hottentot infantry with the Dutch also transferred into the British service as the Cape Regiment.

During this short and successful campaign Major W. H. Trotter, of the 83rd Regiment, acted as the Deputy Adjutant-General, as the Chief Staff Officer to the force had been prevented by severe illness from landing. General Baird notes in his despatches that the " zeal and activity manifested by Major Trotter, of the 83rd Regiment, happily precluded all deficiency in that department."

In Lieutenant-General Sir David Baird's despatch from Funchal, dated 2nd of October, 1805, he states that he had appointed Captain Sorell, of the 86th Regiment as Assistant Adjutant-General to the Cape Expedition. He was sent from Madeira with despatches to the Governor at St. Helena, and did not arrive at Cape Town in time to take part in the operations.

The Lieutenant-Colonel of the 1st Battalion 83rd Regiment was specially selected to take home to the King the despatches announcing the conquest of the Cape of Good Hope.

The Regiment now settled down to garrison duty with few exciting incidents.

Some of the most interesting events that occurred are as follow:—

On 11th of September, Sir David Baird, writing to the Government at home, complains that Richard L—— McCarthy, of the 83rd Regiment, has been gazetted to an ensigncy, and it is his duty to state that the character of this man is exceptional. He has been repeatedly punished in the 83rd Regiment, and was lately brought before a general court-martial and sentenced to a severe corporal punishment for desertion, and it was accordingly inflicted upon him, and he (Sir David Baird) very much fears that there is no reasonable ground for expecting so radical a reform in his conduct as to render him deserving of a commission in the Service either now or at a future period. A pencil note on the letter states that McCarthy was recommended for the commission by Colonel McMahon.

Some troops were sent during this year under General Beresford to South America, for the reduction of Buenos Ayres. Reinforcements were sent to him of some 2,000 men from the Cape, and Major Trotter, of the 83rd Regiment, was again selected as Chief Staff Officer to accompany them.

As a point of interest to all officers and men, it may be noted that the 83rd Regiment received their January pay in June. That of privates amounted to 15s. 6d. for the thirty-one days. The company officers were paid on the same sheet as the men, and the Colonel and Adjutant had to certify on their honour as officers and gentlemen that the list was correct; whilst the Paymaster had to proceed to a notary and then make an oath that he had "mustered" the Regiment, i.e., personally called the roll, and found everything as stated on the pay-lists.

1808 In October, 1808, an expedition was fitted out by order of Lord Caledon, Governor of South Africa, with the object of exploring the country between Lithako and the Portuguese province of Mozambique. It was in charge, as head explorer, of Dr. Cowan, assistant surgeon of the 83rd Regiment. With him was sent Lieutenant Donovan, with twenty Hottentots, of the Cape Regiment. A white man, named Kruger, was sent as guide. Dr. Cowan's last letter was dated the 24th of December, from latitude 24° 30′ S., longitude 28° E. (The latter is supposed to be incorrect.) Nothing certain is known of the fate of the explorers. They were supposed to have been murdered by the natives, or to have died of fever.

1809 Lord Caledon used Major Collins, of the 83rd Regiment, in 1809, for the purpose of exploring the interior. He took with him some of the Regiment, including Dr. Cowdery, the assistant-surgeon. They travelled to the Orange River, and then went up the river, discovered the Caledon tributary, naming it after the Governor, and also the Kraai, which they first named "Grey," after Lord Grey, the South African Commander of the Forces, and returned via the Eastern Coast, rendering their report to the Governor on the 9th of August, 1809.

Lord Grey ordered a force to be ready to proceed to Madras, consisting of 21st Dragoons, 72nd Regiment, and 83rd Regiment; total, 2,278. The 83rd Regiment furnished 1,032 of this brigade. The order was dated the 18th of October, and was only cancelled on the 20th of November, in consequence of further information received from India stating that their services were not now required. The pressing need for their services had arisen from the breaking out of a mutiny of the native troops in India in support of their European officers' grievances.

This was the same mutiny that brought the 86th Regiment from Goa. The advices at the Cape stated that over twenty-five battalions of the Madras Army had mutinied, and the Bombay Army was expected to follow suit.

1810 General Orders, dated 8th March, 1810 (Cape of Good Hope), direct "that the 83rd Regiment is to take seventeen negroes on the strength of the Regiment, and the pay of soldiers is to be drawn for them from the 25th of February. Each negro is to be provided with and kept complete in the following necessaries: 1 forage cap, 1 jacket, 1 waistcoat, 1 pair of pantaloons, 3 check shirts. The Quartermaster of the Regiment is responsible that they are not ill-used, unnecessarily harassed, or employed for any private purpose, but they are to be used for carrying fuel, cooking, etc., etc."

Battle of Blueberg.

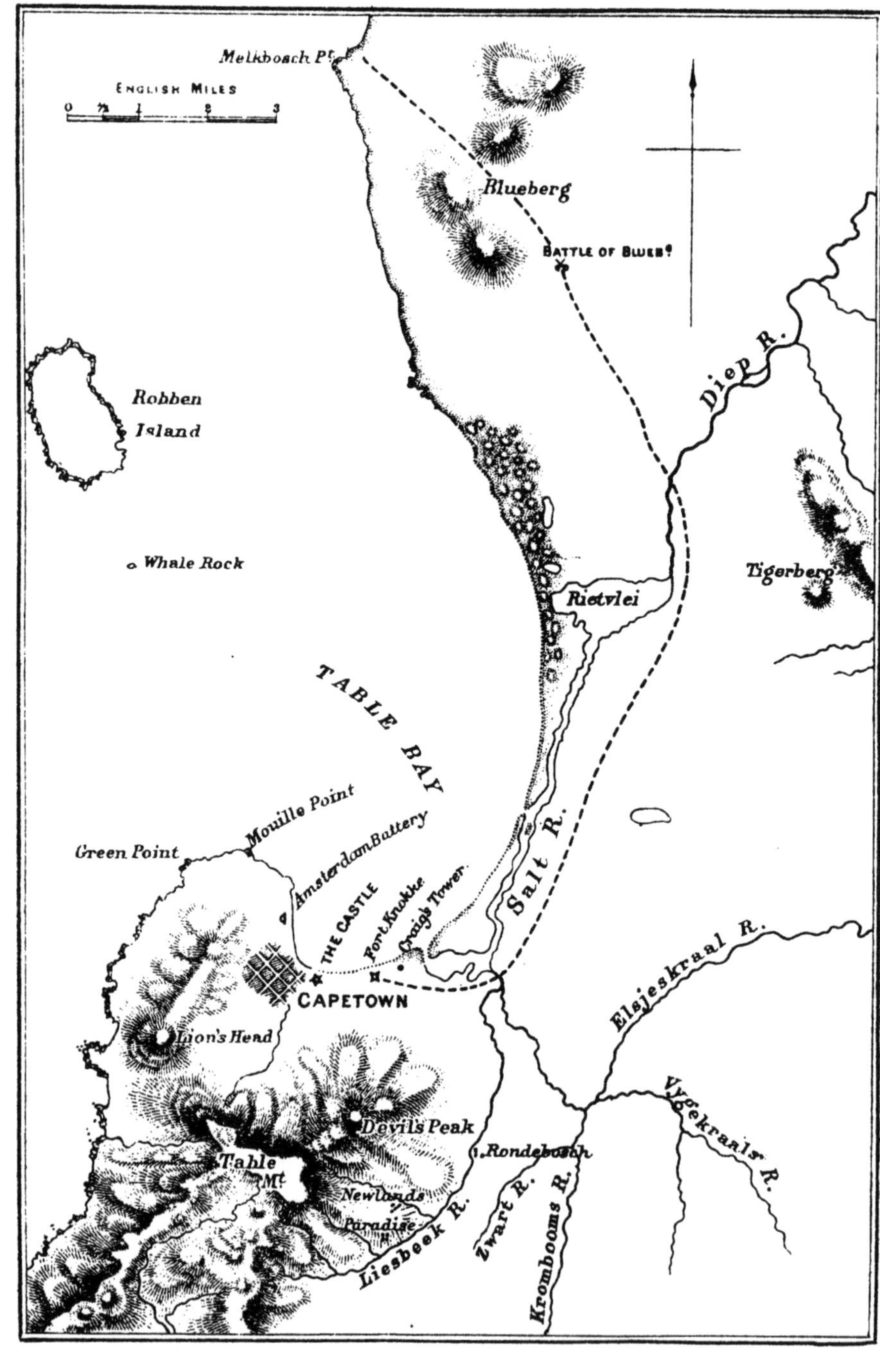

The War Office mildly protested against these new recruits, but did not insist on their objections, and they were duly enlisted.

1811 In 1811 troubles with the natives brought 1 captain, 2 subalterns, and 106 non-commissioned officers and men to Algoa Bay, and a further detachment was sent to reinforce them of 4 officers and 107 other ranks.

Lieutenant Vereker and 100 men of the 83rd Regiment were sent in pursuit of some "cattle lifters" from Grahamstown. They traced the spoor of the cattle to the kraal of the Imidange Captain, Habana, and took sufficient of his cattle to replace the deficiency, and remained there until the morning, having sent a message to the chief that he could have his cattle back so soon as he brought the others in. At dawn the heights round the kraal were covered with armed Kaffirs. Habana refused to restore the looted cattle, and the Lieutenant withdrew towards Grahamstown with the cattle he had taken possession of; he was followed by the Kaffirs, who rushed in on him in a narrow pass in the valley of the Kat River, hurling their assegais at him and his men. Three of the soldiers were wounded. The order to fire was given; five Kaffirs were killed and many more wounded. Those who were alive fled. Lieutenant Vereker was not again molested on his way to Grahamstown.

On the 22nd of September the draft, comprising all the serviceable men of the 2nd Battalion 83rd Regiment, arrived in Simons Bay, under command of Lieutenant-Colonel Cother, C.B., in the transports "Adamant" and "Eliza," and joined the headquarters of the 1st Battalion 83rd Regiment in Simons Town, where it was under the command of Lieutenant-Colonel Brunt.

1816 Five companies of the 83rd Regiment were sent into the interior of the Colony in the year 1816. Brevet-Major Somerfield was in command, and the destination of himself and his detachment is vaguely described in the regimental records as "The Frontiers of Africa." As a matter of fact, this detachment really went by sea to Algoa Bay, and marched inland for 350 miles to Graaf Reinet, on the banks of Sunday River. The country is described as being "barren." The reason for this march was a revolt of the Boers, Hottentots and Kaffirs. When matters were duly settled the five companies returned to Algoa Bay in October, and remained there until the September of the following year.

1817 The Regiment had been under orders to proceed to Ceylon for some time, and finally it embarked at Simons Bay on the 1st of October, presumably in the same two transports, the "Adamant" and "Eliza," and arrived at Colombo on the 16th of November and 3rd of December.

Note.—During the 83rd Regiment's tour at the Cape of Good Hope, the regimental records show that 1,020 Courts-Martial were held; 177,055 lashes were ordered to be inflicted by those Courts-Martial, also that 94,399 lashes were so inflicted and 83,656 remitted.

CHAPTER IX.

2ND BATTALION 83RD REGIMENT.

Written by Professor C. Oman, Chichele Professor of Modern History, Oxford.

Military forces in United Kingdom, 1808—Fifty 2nd Battalions raised in 1803-04—Services, etc. of 2nd Battalions—Actual strength of 83rd Regiment, 1809—2nd 83rd ordered to Cadiz—Embarked 11th January, 1809, Ireland—The battalion again embarked in March, 1809—Names of officers landed in Peninsula—Sir Arthur Wellesley inspected 2nd 83rd Regiment—2nd 83rd Regiment marched to the Douro—Passage of the Douro—Pursuit of the French—Marched to Talavera—Battle of Talavera—Charge of the 2nd 83rd Regiment—Its losses—The battalion ordered to Lisbon——General Cameron's complimentary order—Strength of the battalion at Lisbon—2nd 83rd Regiment ordered to rejoin Wellesley's army, September, 1810—General Picton in command of its division—Marched to Busaco—Battle of Busaco—Retirement on the lines of Torres Vedras—Description of lines of Torres Vedras—Massena's retreat from the Lines—Wellington's pursuit—Combats at Redinha, Casal Novo, Foz de Arouce, Sabugal—Battle of Fuentes de Oñoro—2nd 83rd Regiment marched to Badajoz—Ineffectual siege of Badajoz—British forced to raise siege by Marshals Soult and Marmont—The march to Ciudad Rodrigo—Combat of El Bodon—Halt of three months—Preparations for the siege of Ciudad Rodrigo—The siege—Strength of 2nd 83rd Regiment in January, 1812—The march to Badajoz—The third and final siege of Badajoz—The Regiment's losses in storming the Picurina—The storming of Badajoz—Losses of the 2nd 83rd Regiment—Marshal Marmont's raid into Portugal—Wellington's march to cut him off—Minor operations—Battle of Salamanca—Losses of the 2nd Battalion—The advance on Madrid—Garrison duty in Madrid—The retirement from Madrid to Portugal—Courts-martial in the 2nd 83rd Regiment—Reasons that the campaign of 1813 was delayed until May—The operations during the advance to Vittoria—Battle of Vittoria—General Colville directed names of certain soldiers of the 83rd to be noted, and sent them a guinea apiece next day—Losses of the Regiment—Combat at Roncesvalles—Losses in battles of the Pyrenees—The crossing of the Bidassoa, October 7th—Nivelle—The losses of the 2nd 83rd Regiment—The battles of the Nive—Combat of Sauveterre—Combat at Orthez—Losses of the battalion—The advance to Toulouse—Losses of the 2nd 83rd Regiment—The Battle of Toulouse—End of Peninsula campaign—March to Bordeaux—The 2nd 83rd Regiment embarked for England on the 7th of June, 1814.

THE PENINSULAR WAR, 1809-1814.

PART I., 1809-1811.

SINCE the Napoleonic War began, in 1803, the United Kingdom had never been so completely stripped of battalions fit for immediate service

[*Note.*—Professor Oman's spelling of Fuentes de Oñoro and Orthez has been adopted in this chapter.]

abroad as at the mid-winter of 1808-09.* The dispatch of the army which won Vimeiro and drove Junot to sign the Convention of Cintra had reduced the available force at home to a low figure. When the supplementary expedition, under Sir David Baird, had sailed for Corunna, in October, the number of units ready for foreign service shrank almost to vanishing point. There were, indeed, putting aside the Guards, only six first battalions left within the four seas. The others were all second battalions.

The difference in efficiency between a first battalion and a second battalion was in those days very marked. When the French War re-commenced, after the rupture of the Peace of Amiens, the whole of the British infantry of the Line was found organised in single-battalion units. The first device which was used for raising it to a strength fitted to face the expected French invasion was the embodiment of the new second battalions, called the "Army of Reserve," of which fifty were created in 1803-4, and eleven more a little later. These units were originally intended purely for home service, but their men were to be encouraged to volunteer into the senior battalions, which were intended to provide all over-seas garrisons and expeditions. It was only after the crowning mercy of Trafalgar had removed any danger of invasion, that the terms of engagement of the second battalions were changed, and the whole of them made available for general service. But it was only a limited number of them which were ever sent out of the kingdom; many spent the whole time of their existence, from 1804 to 1814 or 1816, within the British Isles. Some score of regiments had their first battalions in the West Indies, and their home battalions could do no more than send out perpetual drafts to fill up the gaps in the senior units made by the deadly climate: they were (and always remained) little more than depôts. The situation of another twenty units, or a trifle over, which had their first battalions in the East Indies, was only a little better, though the annual wastage which had to be filled up in that quarter was not quite so appalling. Others, again, finding special difficulties in recruiting, were so weak that they could not put more than 350 or 400 men on parade—a figure so low that it would have been useless to send them to the front. It must be remembered that whenever a first battalion went abroad, it discharged into the junior unit all its non-efficients, men too young, or too old, or too weakly for the field, draining off efficients in their place. Hence the second battalion, left at home, always had in its ranks not only its own ineffective men, but all those belonging to its sister unit. It needed to be very strong if it was to provide a number of men sufficient to constitute a second field-service battalion.

When wellnigh every available first battalion in the Army, and a certain number of the better second battalions, had already gone off to the Peninsula, a sudden demand was made for yet more troops from home. Napoleon had

*In November, 1808, the only first battalions, or single battalion regiments, left at home were the 74th, 75th, 77th, 85th, 1/88th, and 94th.

crossed the Pyrenees in person, and the first disasters to the Spanish patriot armies had begun. It then occurred to the British Government that it would be well to establish a base of operations in Southern Spain, in the impregnable isle of Cadiz, in case the army, now under Sir John Moore, should be driven back, and no longer be able to depend on Lisbon. On December 9th Canning wrote to the British Minister accredited to the Supreme Junta, bidding him ask for leave to land troops at Cadiz, and, without waiting for an answer (time forbade the delay), ordered 4,000 men to be shipped for that port. It was difficult to procure even this small force.

The 2nd Battalion 83rd Regiment was at this time one of the few second battalions which was so strong that, although it had sent out a draft of 214 men to the 1st Battalion at Cape Town in February, 1808, it could show, in the following December, a strength sufficient, after all non-efficients had been deducted, to put nearly 1,000 men into the field. Indeed, the Estimates of 1809 show that at this time the 83rd was actually the strongest two-battalion regiment in the British Line. Its numbers are stated at 2,461 of all ranks, the next most numerous corps being the 56th, with 2,301. There had been an immense volunteering into the Regiment from the Militia, during the spring and summer. Most of these recruits were Irish, largely from the North Mayo, Wexford, and Carlow battalions, but there were about 100 from Scotland, of whom sixty-seven came from the Aberdeenshire Militia, and some thirty or so from English corps, mainly in Lancashire.

Hence, after allowing for a full first battalion at the Cape, the Regiment had some 1,400 men in the Palatine Barracks at Dublin, with the colours of the 2nd Battalion: and even when it had set aside 400 non-efficients, it could show over 900 rank and file fit for foreign service.

It was natural, therefore, that the 2nd Battalion 83rd Regiment should be one of the few battalions warned to make ready to sail at once on the Cadiz expedition at mid-winter. The others were the 2nd Battalion 87th Regiment, the 1st Battalion 88th Regiment, and the first battalions of the Coldstream and Scots Guards. The corps marched from Dublin to Fermoy in December, and waited in that town for some days, while shipping was being prepared in Cork Harbour. On January 11th it marched down to Cove and there embarked, with a strength of 45 officers and staff and 986 men.

1809

Contrary winds prevailing, the transports were unable to beat out of the harbour for some ten days; but just as the weather abated and the start became possible, orders were received to land the battalion without delay, while the ships were to sail for Corunna. The cause of this change was that Sir John Moore's despatch of December 31st had just come to hand at London, in which that General stated his intention of bringing his whole army away from the Peninsula, and abandoning the attempt to defend it from the French. If the expedition was to return, it would need no base at Cadiz, and the 2nd Battalion 83rd Regiment need not sail; while the empty transports were required to join those already assembled at Vigo, on which the retreating army was to be taken off. The other battalions originally designated for the Cadiz service had already got away, and after a fruitless appearance before that place, where the Spaniards refused to admit them, retired to Lisbon.

Accordingly, the 2nd Battalion 83rd Regiment and its baggage were hastily re-landed, apparently about January 22nd, and marched back to Fermoy. The ships went on, but were far too late to be of any use at Corunna, where the troops had embarked, after the victorious battle of January 16th, and the glorious death of Moore, and were all on their homeward way before the transports had even started from Cove.

1809

The stay of the 2nd Battalion 83rd Regiment at Fermoy, however, was not to be a long one. After some weeks of hesitation, the British Government had determined that the attempt to defend the Peninsula should not cease with the return of Moore's army, but that the small force which had been left behind in Portugal, under Sir John Cradock, should receive prompt reinforcement, and be raised to a strength of some 25,000 men. This resolve was taken, as it appears, about March 7th. On April 2nd Sir Arthur Wellesley, the conqueror of Vimeiro, was formally appointed to the command of the army in Portugal. Wellesley sailed on April 14th, but long ere his departure the troops which were to be given over to him had already been started off. None of the regiments which had just returned from Corunna were yet available; they had all landed in such a debilitated condition, and with ranks so thinned by sickness, that it was not before June that any of them were again fit for active service.

Wellesley's reinforcements had to be scraped together out of the few second battalions at home, which were not altogether too weak for the field. Of these the 2nd Battalion 83rd Regiment was far the strongest and most efficient; the others, all decidedly weaker in numbers, were the 2nd Battalions 7th, 30th, 48th, 53rd, and 66th Regiments. These troops, together with the thirteen battalions which Cradock had commanded in Portugal during the winter, and the four battalions which had originally been sent to Cadiz, but had been refused admission there, composed the infantry of Wellington's original Peninsular army; in addition he had five regiments of cavalry.

The 2nd Battalion 83rd Regiment made its second start with ranks not quite so full as in January. At the inspection before it embarked Brigadier-General A. Campbell rejected some scores of men as too young or too weak for active service, and the battalion embarked with 40 officers and about 930 men. Lieutenant-Colonel Alex. Gordon was in command; the major was John Napper; there were 9 captains, 19 lieutenants, and 10 ensigns (an unusually large proportion of the juniors), and 6 staff. The adjutant was Lieutenant Henry Brahan. The battalion sailed from Cove on transports which formed part of a considerable fleet, there being five other corps from

Irish quarters—the units whose numbers are cited above—as well as some 300 artillery horses and a great accumulation of commissariat stores and provisions. The whole was under the charge of Lieutenant-General Rowland Hill, who, though he had only just returned from the Corunna retreat, had been put under orders to command a division in the new Army of Portugal. The convoy set sail with a fair northerly breeze on March 29th and was favoured with excellent weather during the whole of its transit. Making an exceptionally rapid passage, it had passed Cape Finisterre by April 4th, and ran into the mouth of the Tagus on the evening of April 5th.

The pay book of the battalion gives the following as the names of the officers who landed in Portugal in April:—

Lieutenant-Colonel Alexander Gordon.

Major John Napper.

Captains William Geddes (Brevet-Major), Thos. Somerfield, Jas. Reynolds, Samuel Hext, Chas. Laird, Francis Creagh, Geo. Mansergh, Geo. Noleken, and Hon. Henry Powys.

Lieutenants Francis Abell, Connel Jas. Baldwin, Wm. Cotter, Nicholas Colthurst, Jas. Cruttwell, Wm. Dahman, Thos. Elliott, Geo. Fry, Francis Flood, Jas. Hingston, James Jackson, John Montgomery, Gilbert Elliott, Wm. G. Cummins, Jones Ferris, John Nicholson, Robert Pyne, John Ramsay, Henry Vereker, and Francis Johnson.

Ensigns Jas. Smith, Chas. Bowles, Henry Le Toller,* John B. Tresilian, Francis Barry, Thos. Boggie, Michael Carey, Fred. Irwin, Chas. O'Neill, and Chas. Watson.

Adjutant, Lieutenant Henry Brahan.

Quartermaster Thos. Wight.

Surgeon J. A. Bruff, Assistant-Surgeons John Glasco, Alex. Stephenson, and Walter Ward.

Paymaster Henry Cross.

Serjeant-Major Joseph Swinburne.

On the 7th the 2nd Battalion 83rd Regiment disembarked, the men, with one subaltern for each company, being quartered at the cavalry barracks, near the Calçada de Ajuda at Belem, while the remaining officers were billeted in private houses. On the 10th the battalion was ordered to march out to join the rest of the army, which was at this moment encamped in various villages to the north of Lisbon. In company with the two newly-arrived battalions of the Guards, the 29th, and the 1st Battalion of Detachments, the 2nd Battalion 83rd Regiment advanced to the neighbourhood of Mafra; it was quartered for two days in the Patriarch's half-ruinous summer palace at San Antonio de Tojal, and moved up on April 13th to the big village of Sobral, which was

* This officer's odd name was a constant source of error to regimental and other scribes. I find him down in pay books and other documents as Le Toller, La Toller, La Toler, Letoller, and once (in Bray's history of the 83rd, p. 10) as Lord Tulloch.

to be eighteen months later the central point of danger during the occupation of the famous lines of Torres Vedras.

On April 22nd Sir Arthur Wellesley landed at Lisbon and assumed command of the army. There was not a moment's delay in taking up the offensive; with one French army, under Soult, in possession of Oporto, and another, under Victor, threatening Central Portugal from the line of the Guadiana, it was necessary to strike at once, lest the two French forces, hitherto completely out of touch with each other, should combine for a concentric attack on Lisbon. Wellesley resolved to fall upon Soult, as the nearer enemy, and set the whole of his army, with the exception of a few regiments left on the Zezere to observe Victor, in march for Oporto.

In the organization of the newly-assembled host, the 2nd Battalion 83rd Regiment was brigaded with the 2nd Battalion 9th Regiment, under the command of Brigadier-General Alan Cameron, a gallant old Scottish officer, who had raised and long commanded the 79th Cameron Highlanders. In company with the 2nd Battalion 9th Regiment, who were to be associated with them for no more than one short campaign, the 2nd Battalion 83rd Regiment marched by the western high road from Lisbon to the north, passing through Roliça, Caldas, and Batalha. At the great abbey of the last-named place, we are told that the officers were most hospitably entertained by the monks—so much so, indeed, that the Brigadier found it necessary to write an order concerning over-long potations in the refectory after dinner. It was not often that such good quarters were to fall to the lot of the battalion.

From Batalha the route lay through Leiria to Coimbra, which was reached on May 2nd. Here the whole army was assembled for the blow against Soult. The battalion here had the honour of being inspected by Sir Arthur Wellesley, on whom it cast eyes for the first time. The diarist whose notes in the "Military Chronicle" give us our only glimpse at the 83rd from the inside during this year,* informs us that the inspection passed off with great success. "On the approach of the staff to Cameron's Brigade, I was flattered by General Sherbrooke's saying, 'The 83rd, Sir Arthur, is one of the finest bodies of men in the field.' 'Very handsome corps, indeed,' replied the Commander of the Forces."

In the Morning State of the Army taken at Coimbra on May 6th, just before the advance began, we find that the Regiment showed present on parade 39 officers and 833 serjeants and rank and file. It had already dropped behind the very appreciable number of 73 sick; all regiments found the air and food of Portugal trying, even in the balmy month of April, till they had become acclimatized. The 2nd Battalion 9th Regiment, which had been in Portugal since the time of Vimeiro, and had suffered from heavy marches in

*This diary only extends down to the end of the Oporto Campaign; it is scattered over three years of the "Military Chronicle," and is excellent for routes and description of the scenery, but singularly reticent about individuals.

the rain during January, had at this time no less than 227 sick to 562 present in the ranks. But this was the most sickly battalion in the army.

In the advance from Coimbra against Soult, Cameron's Brigade was at first attached to Sherbrooke's column, but was soon set aside to form, with Hill's Brigade, a flanking detachment, which was to advance along the sea-coast and turn the French line of defence along the River Vouga, while five other brigades and all the cavalry, under Wellesley himself, attacked it in front. On starting from Coimbra, a Portuguese battalion, the 2nd of the 10th Line, was attached to Cameron—an experiment which was not to last for long, for the combination was not found to work well. The regimental diarist did not think much of this Portuguese unit—" the men were little advanced in discipline and very short in stature." There were only ten officers present for 700 men—none of them but the lieutenant-colonel and major seemed to know their work, and the juniors were quite inefficient. But another arrangement here made was to be permanent—a rifle company of the 5th Battalion 60th Regiment was also attached to the brigade for skirmishing purposes. Wellesley was determined that he would always have a strong skirmishing line, and, to aid the two light companies of the regiments, gave the brigadier this proportion of trained marksmen from the famous German Jäger Battalion of the " Royal Americans." They were found invaluable, as Wellesley had prophesied in his General Order commending the riflemen to the special care of his brigadiers.*

1809

Hill's turning movement did not prove so decisive as had been hoped, for Franceschi, the general in command of Soult's advanced line, withdrew hastily without waiting to make serious resistance. The flanking column, transported by boats on the long lagoon between Aveiro and Ovar, landed in what had once been Franceschi's rear, to find that he was already out of reach (May 10th), and could only skirmish with some voltigeur companies near Ovar.

On the next day Hill was again sent out on a flanking march to turn the enemy's right, while Wellesley attacked him in front at the combat of Grijo. As the French once more gave back without any great trouble, Hill's and Cameron's battalions had no more than a fatiguing march, and were not engaged. That night the French went behind the Douro, after blowing up the great boat-bridge of Oporto.

In the splendid exploit of the next day, the " Surprise of Oporto," or " Passage of the Douro," the only British infantry seriously engaged was Hill's Brigade, which supplied the force which crossed in such a daring fashion between high cliffs and over a very swift river, to throw itself, boatload by boatload, into the midst of the criminally negligent French. It was Hill's three battalions which defended the Seminary, where they had taken post, when the enemy at last woke up to find themselves surprised. The brigade of

* See General Orders for May 2nd, 1809.

Cameron only crossed the Douro later in the day, and lower down, at Villa Nova de Gaya, and were with the Guards and the 29th when the lingering rearguard of the French was swept out of the city. They were not, however, seriously engaged, and the Regiment was not (like Hill's battalions) granted the battle honour of "Douro" to bear on its colours.*

Cameron's Brigade joined in the long and fatiguing pursuit of Soult's fugitive army over the mountains, in the drenching rain, which continued for several days after the Passage of the Douro. It marched near the head of the column, immediately behind the Guards, who were the leading infantry. On the 15th the 2nd Battalion 83rd Regiment stopped at the episcopal city of Braga, where it was welcomed with enthusiasm by the people. "The hospitality of the inhabitants might have proved our ruin," notes the regimental diarist—"hardly a sober soldier was to be found late that night," and if the French had turned to fight, the consequences might have been tiresome. But Soult was set on flight only. Late on the following day (May 16th) the British advanced guard at last came up with the rearguard of the retreating French, at Salamonde, just as dusk was closing in. The Guards' Brigade drove in the French containing force in a lively skirmish, and the 2nd Battalion 83rd Regiment and 2nd Battalion 9th Regiment came up just as the combat was ending, and had the satisfaction of seizing and eating much food which the enemy had left boiling in their camp-kettles all about the encampment from which they were dislodged.†

Marching next day (May 17th) the column reached Ruivaens, passing through several villages set on fire by the French, and noting many dead soldiers and horses lying by the roadside. A certain number of stragglers and footsore men were captured, but the main body of Soult's army had got a long start, and had reached the high lying town of Montalegre, near the Spanish frontier. Six miles short of that place Wellesley ordered the pursuing column to halt, on May 18th. The 2nd Battalion 83rd Regiment spent a most uncomfortable night; the staff officer, sent to bid them cease from the pursuit, gave them a direction for their encampment which was hopelessly vague, and the battalion, wandering among steep ravines, had to find what shelter it could in the waste, amid a desperate downpour of rain.

Wellesley had done by this time all the damage that he could to the enemy, and, as he wrote that day to Castlereagh, "it is obvious that if an army has

*Major Bray's "Regimental History of the 83rd" says that the light company of the battalion joined Hill in the Seminary, and under its commander, Captain T. Somerfield, distinguished itself in the defence, and lost 14 rank and file wounded. I can find no trace of this elsewhere; the diarist already quoted makes no mention of it, which he would surely have done, as he was keen for the credit of the battalion, and the official return of casualties shows no losses whatever for the 2nd Battalion 83rd Regiment.

† Bray's history says that the Regiment was up in time to join in the skirmish at Salamonde, and lost four wounded. The diarist, already so often quoted, makes no mention of this, nor do the casualty lists.

thrown away all its cannon, equipment, and baggage, and has abandoned the sick and wounded, who are entitled to its protection but impede its progress, it must be able to march by roads through which it cannot be followed by an army which has not made the same sacrifices."*

The chase of Soult was therefore abandoned, and the pursuing brigades turned back towards Oporto, marching by moderate stages, for the troops were exhausted by much travel in the incessant rain; food in this devastated region was almost unprocurable, and many men had worn out their shoes in the long and rapid advance from Coimbra. The sick were many, and before starting to accomplish the second half of his task—by treating Victor as he had treated Soult—Wellesley had to give his men a long halt at Abrantes, where the army rested from the 11th to the 27th of June. The 2nd Battalion 83rd Regiment was quartered for this period at Punhete, where the Zezere runs into the Tagus, a little below Abrantes. It had suffered appreciably from the heavy marching in pursuit of Soult, and had left a good many men in hospital at Oporto. Its sister battalion in the brigade, the 2nd Battalion 9th Regiment, had fared so much worse, that Wellesley declared it incapable of going on with the campaign, and sent it down to Lisbon, for removal to Gibraltar. Cameron was to receive instead the 1st Battalion 61st Regiment, a strong first battalion, which had just arrived from that same fortress; but it had only just reached Lisbon, and did not join the 2nd Battalion 83rd till July 20th. At the same time the army was organized into divisions for the first time—the Oporto campaign had been fought with no larger units than Brigades. Cameron's Brigade formed, from June 18th, the second of Sherbrooke's 1st Division, in which the others were Henry Campbell's (the two Guards' battalions), and Löw's and Langwerth's, composed of battalions of the King's German Legion. The Portuguese unit (2/10th Line), which had been serving with Cameron since June, ceased to belong to his brigade.

On June 27th the 2nd Battalion 83rd Regiment, still unaccompanied by the new sister battalion in its brigade, which had not yet got up from Lisbon, marched for Spain, attaching itself provisionally to Henry Campbell's battalions of the Guards, in whose wake it followed during the next fortnight. It was now about to go through the most bloody experience of its whole regimental history—the battle of Talavera. Wellington had made up his mind to march against Victor in conjunction with Cuesta and the Spanish army of Estremadura. He had yet to learn the danger of entering the field with a Spanish colleague, and his prospects at this moment looked bright. Victor had just been forced by famine to retire northwards, behind the Tagus, to the vicinity of Almaraz; his corps of 22,000 men was the only body of French troops between Wellesley and Madrid, where King Joseph had only some 6,000 or 8,000 men available as a central reserve. It was considered that Soult had been too much mauled in the Oporto retreat to be dangerous

*Wellington to Castlereagh, from Montalegre, May 8th.

for many a day. Of the other French corps that of Ney was believed to be still in Galicia; that of Sebastiani, in La Mancha, was being watched and contained by a superior Spanish force under Venegas; the only reinforcement that seemed able to join Victor was Mortier's Corps in Old Castile, whose exact position was unknown, but which was, at any rate, far off the present field of operations. It was hoped that Victor and King Joseph might be overwhelmed by the 22,000 British of Wellington and the 38,000 Spaniards, whom Cuesta was bringing up from Estremadura. Wellesley marched his troops from Abrantes eastwards in two parallel columns; the 2nd Battalion 83rd Regiment accompanied the rest of the 1st Division in the left-hand column, which moved by way of Cortiçada and Castello Branco, where the other column joined it. From thence the route lay to the Elga stream, the boundary of Spain, and to Coria, the first town in that kingdom, which was reached on July 8th. The next stage was Plasencia, where the battalion arrived on the 12th, and halted till the 17th, having hutted itself outside the walls. The march from Abrantes to Plasencia turned out far more exhausting than had been expected.

1809

The weather was very hot; the country on both sides of the frontier was desolate and thinly peopled; Junot's army had been almost destroyed by privation on these same roads eighteen months before. Wellesley, ignorant as yet of the resourcelessness of these forbidding regions, had not made the elaborate preparations for feeding his army by convoys from the rear, which he always adopted in his later campaigns. The troops were moving very lightly equipped, with little transport, and expecting to draw a great part of their food from the countryside. This proved an impossibility, and the army was suffering great privation, and losing many men by sunstroke, fever and dysentery, before it reached Plasencia, the first place where provisions were to be obtained. The 2nd Battalion 83rd Regiment, which had started from Abrantes with over 800 of all ranks, could only show 658 officers and men fit for service on July 15th, and had no less than 271 sick, of whom some few had been left at Abrantes, a good number at Castello Branco, but much the greatest proportion in the general hospital at Plasencia, from which the army continued its march on July 18th.

Though some provisions were obtained at Plasencia, the march, when resumed, proved almost more fatiguing in the boundless arid plains north of the Tagus than in the mountains which had been left behind, and the sick, largely sunstroke cases, became more numerous than ever. At Oropesa, on July 20th, Cameron's Brigade was at last completed by the arrival of the 1st Battalion 61st Regiment, which had been following by forced marches from Lisbon. This battalion was very strong, showing 900 of all ranks when it passed through Plasencia, so that it considerably outnumbered the 2nd Battalion 83rd Regiment.

On the same day (July 20th) the British fell in with Cuesta's Spaniards, who had moved up from Almaraz for the junction, and two days later both armies reached Talavera, from which Victor and his army corps hastily withdrew to the direction of Madrid.*

Here Wellesley halted, having again almost exhausted his provisions; the country about Talavera was incapable of supplying him with any more, as Victor had swept it clean during his sojourn in the district. It was for this reason that he refused to follow Cuesta in a further advance, which the Spanish General, nevertheless, undertook on his own account, little knowing the danger into which he was thrusting himself. Meanwhile, the British troops had to be placed upon half-rations on the 23rd, and remained on the same meagre fare for several days, convoys ordered up from Plasencia being vainly expected. The heat and semi-starvation were so debilitating to the men that on the day before the battle the 2nd Battalion 83rd Regiment, which had shown 658 officers and men fit for duty on the 15th, had now only 535 on parade. It discharged over 100 rank and file into the general hospital at Talavera.

Early on July 27th Cuesta's army came rushing back into Talavera in great disorder. It had run into the French, and found them in unexpected force, on the 25th; for not only had King Joseph joined Victor with the reserves from Madrid, but Sebastiani, with the 4th corps, had slipped away unmolested and undiscovered from in front of the Spanish army of La Mancha, and had brought 17,500 men to join the King, who had now 48,000 men collected, when the Allies had credited him with a possible 30,000. They had still a considerable superiority of numbers, the British (despite of all the sick sent into hospital at Plasencia and Talavera), having just over 20,000 sabres and bayonets, and the Spaniards, who had dropped some battalions to guard the mountain passes and the bridge of Almaraz, about 34,000. It was fortunate, however, for the Allies that their enemies' counsels were much perplexed by the rivalry of Victor and of Marshal Jourdan, who was acting as King Joseph's Chief of the Staff. They could agree in nothing, and Victor, though the junior of the two, frankly disregarded Jourdan's more cautious plans.

On the 27th July the Allies took up their position beside the town of Talavera, in none too good order, for the Spaniards had been much hunted, and Victor thrust in Mackenzie's Division, which Wellesley had left out in his front as a covering force, with great vigour and considerable loss. He even made an attempt after dark to seize the hill on the left flank of the

* The daily stages of the 2nd Battalion 83rd Regiment from Punhete to Talavera were: June 27th, Punhete to Abrantes; 29th, Abrantes to San Domingo; 29th, to Cortiçada; 30th, to Sarzedas by Monte Gordo; July 1st-2nd, Monte Carlo to Castello Branco; July 3rd-4th, to Salvatierra; 5th, to Zarza Mayor; 6th, halt at Zarza; 7th, to Montalagon, on the Alagon River; 8th, to Coria; 9th-12th, Coria to Plasencia; 12th-17th, halt at Plasencia, in huts; 18th, Plasencia to the Tietar River; 19th, to Centinello; 20th, to Oropesa; the 1st Battalion 61st Regiment joined; 21st-22nd, to Talavera; 23rd to 29th, halted at Talavera.

British position, which commanded the whole battlefield, and was only evicted from it with some difficulty by Stewart's Brigade of the 2nd Division. This fighting in the night does not concern the history of the 2nd Battalion 83rd Regiment, and need not be entered into, any more than the panic among the Spaniards, which irritated Wellesley at about the same time.

The part of the 2nd Battalion 83rd Regiment in the great battle of the following day requires more careful explanation. The allied army was drawn up on a front of three miles, reaching from Talavera town on the right to the Cerro de Medellin, the isolated hill, to which allusion has already been made, on the left. Cuesta's Spaniards held the town and its suburbs, with the adjacent olive groves, for a mile and a half, from the Tagus to a small redoubt which had been thrown up on a hillock near the end of the olives. From this redoubt, going northwards, the British took up the line, Campbell's Division on the right, Sherbrooke's in the centre, Hill's on the Cerro de Medellin, at the left end of the line, the one marked feature on the whole battlefield. Mackenzie's Division was in reserve, one brigade at the back of the Cerro de Medellin, the other behind Sherbrooke. The six regiments of cavalry were massed behind the centre. Cameron's two battalions, therefore, as the right-centre brigade in Sherbrooke's Division, were exactly in the middle of the British line of battle. The ground on which the 2nd Battalion 83rd Regiment and 1st Battalion 61st Regiment lay was almost level, the position being marked out only by the dry bed of the Portiña brook, which had shrunk in the dry summer weather to a chain of shallow stagnant pools. They were in fact, in the most exposed portion of the front, not covered by the redoubt and the woods like Mackenzie's Division on the right, nor with the advantage of the lofty hill, like Hill's Division, on the left. Wellesley seldom drew up his army on ground so destitute of cover or support.

With the first episode of the great battle of the 28th July, the 2nd Battalion 83rd Regiment had nothing to do. Victor, contrary to the desire of Jourdan, for a second time attacked, with one of his own divisions, the Cerro de Medellin, thinking thus to gain possession of the key to Wellesley's position. He suffered a bloody repulse at the hands of Hill; this partial engagement took place at about 6 o'clock in the morning. There followed several hours of quiet—a sort of informal armistice was established, and French and English alike went down to the Portiña, to fill their canteens from the muddy pools, for this was the only water upon the ground. It was not till the morning was far spent that the French were seen once more falling into their ranks; their leaders had determined to execute a general attack upon the whole British line. Jourdan and King Joseph had shrunk from the idea till news reached them that the Spanish army of La Mancha had at last discovered Sebastiani's absence, and was marching upon the almost unprotected Madrid. It was necessary to make an end of Wellesley and Cuesta promptly, or the capital would be seized behind the backs of the French army

Setting aside only a division of cavalry to observe and contain the Spaniards of Cuesta, the French Marshals determined to throw the whole of their infantry upon the front of Wellesley's line; if the British were broken, the Spaniards would not be very formidable.

About 1 o'clock the divisional batteries of Victor's and Sebastiani's Corps opened a tremendous fire upon the whole length of Wellesley's front. There were over sixty guns in action, and only thirty British and six Spanish pieces were so placed as to be able to reply. The superior fire of the French beat in the most deadly fashion upon Wellington's centre, where the Guards, with the 2nd Battalion 83rd Regiment and 2nd Battalion 61st Regiment had no shelter whatever. Colonel Gordon made his battalion lie down, but the losses were, nevertheless, very considerable. At about two o'clock the whole of the French infantry were seen to be advancing to the attack. We have no concern with what happened to right or left, but must only note the progress of events in the British centre. While Lapisse's Division of Victor's Corps came down upon the left of Sherbrooke's line, the two German brigades of Löw and Langwerth; the troops opposed to Cameron and the Guards were Sebastiani's own Division of the 4th Corps. The odds were appalling, for Lapisse and Sebastiani were each leading forward twelve battalions to attack four. The French divisions were both arrayed in double lines, the front one composed of six battalions, drawn up side by side, "in column of divisions," i.e., of double companies; while the second consisted of six others in the still heavier formation of columns of battalions. The force immediately opposite the 2nd Battalion 83rd Regiment and 1st Battalion 61st Regiment consisted of the three battalions of the 28th Line, of Rey's Brigade. Sweeping before them the screen of skirmishers, composed of the light companies and the Rifles of Cameron's Brigade, the French crossed the Portiña brook and came into close fire-contest with the two-deep line opposite them. All through the Peninsular War the one salient point that may be noted in the contests between the British line and the French "column of divisions" was, that in the duel of musketry the line was invincible. A battalion of 600 men in the British order could put every weapon in action with effect. The same number in column of double companies had only 130 men in its two front ranks able to fire. The seven files behind gave solidity to the mass, which might impose on a weak or demoralized enemy, but they could not use their muskets. A steady enemy like the British, who refused to be cowed, and fired low, invariably stopped the advancing column by shooting down its front ranks before it could approach near enough to use the bayonet. The odds of 600 balls received against 130 returned were too great.

1809

Cameron was quite aware of his advantage, and determined to use it to full effect. He kept his men lying down till the French were across the Portiña brook, and actually allowed them to approach within thirty yards before he opened fire. The first volley, so long husbanded, and delivered at such close quarters, was decisive; the Frenchmen went down in swathes, the columns reeled under the shock, and when the 2nd Battalion 83rd Regiment and 1st Battalion 61st Regiment advanced upon them, immediately after the murderous discharge, the whole of the three columns broke and gave way. The victorious brigade followed them for a short way beyond the Portiña brook, pelting the huddled masses with independent firing, and then halted at Cameron's order and re-formed their line.

Matters did not go so happily to right and left. The Guards and the German Legion, on each flank of the 2nd Battalion 83rd Regiment and 2nd Battalion 61st Regiment, had received the French as steadily, and had as quickly worsted them in the combat of musketry. They charged at the same time as Cameron's Brigade, but, unfortunately, they did not halt and re-form in the same style. Pressing on in straggling and disorderly lines, they went forward far into the rolling plain in front of them—the Guards, especially, seem to have driven straight forward, pushing the routed masses before them, till they had got more than a quarter of a mile beyond the Portiña. Here the three pursuing brigades were brought up, by running against the French second line, which had moved forward to support the first. Neither the Guards nor the Germans were in a condition to face a new fight with fresh troops, and between them was the broad gap where Cameron's battalions, halted far to the rear, were missing. The Germans, on the left, were taken in flank by artillery fire; the Guards, on the right, were attacked by two dragoon regiments, as well as by the unbroken infantry reserve in front of them. Neither could stand, and both fell back, in two separate disorderly masses, towards the position which they ought never to have quitted.

A desperate moment for Cameron's men followed; their routed comrades came running in upon their flanks, mixed with the pursuing French. The brigade stood for some time, keeping back what was in its front; but presently both its wings were turned, as the flying troops gave way on each side, and it was flung back with dreadful loss beyond the Portiña and on to its old position. It was with much difficulty that the main mass of the 2nd Battalion 83rd Regiment preserved itself by holding together in the tumult. Finally it was saved by taking shelter behind Wellington's last reserve—Mackenzie's Brigade, of the 3rd Division, which had been brought forward to stop the terrible gap in the centre of the line. While these three battalions (2nd Battalions 24th and 31st Regiments and 1st Battalion 45th Regiment) held back the advancing columns of Sebastiani in a desperate struggle, the Guards and Cameron's men rallied promptly behind them, but in terribly diminished numbers. The battle was saved, the French at last gave back, and the First Division resumed its old position. In other parts of the field fighting went on for some time, but in the centre it was at an end.

The losses of the 2nd Battalion 83rd Regiment had been frightful; out of 535 officers and men present it had more than 50 per cent. of casualties.

There were 4 officers and 38 men killed, 11 officers and 202 men wounded, and 28 men prisoners, a total loss of 283. All through the division the losses were similar; several other regiments lost in the same dreadful proportion, though the sister battalion in Cameron's Brigade (the 1st Battalion 61st Regiment) got off comparatively lightly, with 265 casualties out of 778 men present. The Colonel of the 2nd Battalion 83rd Regiment (Alex. Gordon) had been killed early in the fight, when the first charge was delivered. With him fell three lieutenants, Montgomery, Dahman and Flood. The wounded included two captains—Somerfield and Reynolds,—seven lieutenants—Abell, Johnson, Nicholson, Baldwin, Pyne, Boggie and Ferris,—and four ensigns—Le Toller, Barry, Carey and Irwin,—as also Lieutenant and Adjutant Brahan. The four ensigns were all hit carrying the colours, as was also Lieutenant Robert Pyne, who received a captain's commission in the 66th for saving one of them in the moment of the disaster. Serjeant-Major Swinburne was given an ensigncy for distinguished service, and took up the Adjutant's duty instead of the severely wounded Brahan.*

The gallant remnant of the battalion, not more than 300 strong, came under the command of Major Napper for the rest of the campaign. Its troubles were not yet at an end, for on the report that Soult was unexpectedly appearing in the rear of the British army, with three army corps, and threatening the line of retreat to Portugal, Wellington had to retreat hastily south of the Tagus, only five days after the victory. All the severely wounded officers and men, who could not bear transport, had to be left in the hospital at Talavera, to fall into the hands of the French. In this unhappy fashion Captains Reynolds† and Somerfield, Lieutenants Abell, Johnson, Nicholson, and Boggie, Ensign Le Toller, Adjutant Brahan, eight serjeants, four corporals, and seventy-two privates became prisoners of war. Nor was this all; Soult, in his advance, had occupied Plasencia, where the general hospital of the army had been left behind. All the sick who could not drag themselves away by a desperate effort fell into his hands, among them two serjeants, two corporals and thirty-nine men of the 2nd Battalion 83rd Regiment. Thus the battalion lost (including the twenty-eight rank and file captured in the battle) no less than eight officers and 155 men prisoners of war in this short campaign. In addition, a great number of the wounded from Talavera and the sick from Plasencia, who had started off as best they could on the near approach of the

* Joseph Swinburne is one of the only three Peninsular officers promoted from the ranks, so far as is known, who fought his way up from private to Lieutenant-Colonel. He served all through the Peninsular War, received the Peninsula medal, with ten clasps, and was 44 years as an officer with the Regiment, retiring at last as a Brevet-Lieutenant-Colonel, with Major's full pay, in 1853.

† The casualty returns show that Reynolds had lost a leg; the others were severely wounded. Of the officers hit, only Baldwin, Pyne, Ferris, Barry, Carey and Irwin were strong enough to get away from the hospital.

COLONEL GORDON'S GOLD TALAVERA MEDAL

Colonel Gordon was wounded in leading the successful charge of the 83rd Regt. at Talavera and was blown to pieces by a French shell as he was being carried off the field in a blanket — by four of his men who also were killed —

French, died by the way from weakness and privation. On October 25th, the general returns show that the battalion, which had 763 effective rank and file at Coimbra on May 2nd, had now only 367 present, with 280 sick and 47 detached. The slight rise in the effectives since July 28th was due to the return of a few wounded men to the ranks. The battalion would appear, therefore, to have lost between May and October about 235 men killed, dead of disease or privation, or prisoners of war, without taking account of officers. The number of the prisoners of war had been somewhat diminished, before the October morning state was drawn up, by the unexpected reappearance of Serjeant John Hyde, one corporal, and twelve privates, who, having been left sick or wounded in Talavera Hospital, succeeded in escaping from their guards, when they became more or less convalescent, and made their way over the mountains to join their battalion in the valley of the Guadiana. It would appear that about 100 men must have been dead, and about 135 in the hands of the French at the end of the campaign. In addition, there were the 280 sick, of whom the majority were never to join the ranks again, for those of them who had escaped from the Plasencia and Talavera hospitals on the approach of the French had been so debilitated by rough transport in carts, or even on foot, across mountain ranges, and with insufficient food, that many died and more were invalided home as unfit for further service.

The small effective remnant of the 2nd Battalion 83rd Regiment, under Major Napper, only remained with the army for a month after Talavera. They starved, with the rest of the 1st Division, on the high mountain positions south of the Tagus, to which Wellington had retired, from August 4th to August 20th, while headquarters were at Jaraicejo and Deleytosa. The privations were terrible; on the 5th and 6th the battalion received no bread at all; on the 4th only a half-ration; on the 7th a very little flour only. Meanwhile, they had to provide working parties to help the artillery, whose horses were half dead, to drag the guns up a series of steep mountain paths. No wonder that the note is added by the regimental diarist that "the troops are beginning to sink under their fatigues." On August 11th there was a halt near Deleytosa, and food was forthcoming again, but in quite insufficient quantities. Hence it was with intense relief that the division welcomed, on the 20th, the order to retire to Truxillo and march for the valley of the Guadiana. On the 23rd the battalion descended from the passes and reached the old town of Medellin, where it was once more in an inhabitated district with some resources.

On the 29th, when it had reached Talavera Real, near Badajoz, the 2nd Battalion 83rd Regiment, only 300 strong, received orders to leave the army and march for Lisbon, where it was to take up garrison duty, till it should once more reach a strength sufficient to enable it to operate in the field as a battalion. It was now the weakest, or the weakest but one, of all the units in the army, and Wellesley habitually in 1809-10 sent to the rear corps which had dwindled down to under 300 men. After 1811 he often united two of such skeleton units as a "provisional battalion," but this practice had not yet begun.

Before its departure the 2nd Battalion 83rd Regiment received a handsome testimonial in the Brigade Order, issued on August 29th, by General Cameron:—

"The death of Lieutenant-Colonel Gordon, who so gloriously fell at the head of his battalion while charging the enemy, on the 28th ult., leaves Brigadier-General Cameron the painful necessity of regretting the loss of a sincere friend and gallant officer, and his regiment that of a brave and worthy commander.

"The conduct of the 83rd in the arduous contest of Talavera merits the Brigadier-General's warmest thanks, and he hopes that Major Napper will justly appreciate the merits of those few that are left.

"The very weak state of the 83rd renders it necessary to send them to Lisbon. The Brigadier-General requests them to accept of his best acknowledgment for their uniform good conduct while under his command, and has (at the same time) to assure them that he shall be proud to have the 83rd again in his brigade when re-established in health and numbers."

Cameron's wish, however, was never to be fulfilled; the 2nd Battalion 83rd were forced to remain for nearly a year at the base, unable to recruit themselves to a strength sufficient to induce the Commander-in-Chief to call them up to the front. They were paying the penalty of being a "second battalion," and not one of the original foreign service units. The depôt at home had to feed, not one, but two battalions, and the senior unit, at the Cape, had a preferential call for drafts, as men fit for the front were collected. It would appear that the 2nd Battalion 83rd Regiment received no appreciable draft for more than a year after it had first sailed from Cork. When it had reached Lisbon, in October, its number of effective rank and file rose, as has been already stated, to 367, with 280 sick. The sick recovered slowly, or not at all. By February, 1811, the battalion had been restored to a full complement of officers, of whom it had 37, but had only 451 rank and file efficient, though the sick had gone down from 280 to 53. It is clear that only 84 men had returned to the ranks, and 143 had died or been invalided home. The deaths were terribly numerous; in the two months, December, 1809—January, 1810, thirty-five men died in hospital, and the number for October-November, 1809, was probably higher still.

1810

Receiving no appreciable draft, or perhaps none at all, from home, the battalion was unable all through the spring and summer of 1810 to raise a strength sufficient to induce Wellington to recall it to the front. In March it had 457 rank and file on parade; in May, 427; in June, 404 only, the summer heat having raised the number of sick to 73. Hence the 2nd Battalion 83rd Regiment had the mortification of seeing regiment after regiment land at Lisbon and march up country, while it was still tied down to wearisome garrison duty in the Portuguese capital.

It was not till it had been eleven months at the base that it would seem at last to have received a draft, though apparently a very small one.* On the first of September its total numbers had gone up to 530, including sick, and on September 12th it received orders to give up its quarters to the 2nd Battalion 88th, newly landed from Cadiz, and to prepare to move up to the front and join the 3rd Division. Ninety-two sick were left behind at Lisbon, and, after deducting a few men "on command," the battalion had only 420 effective rank and file when it joined the army. It was still one of the smallest units in Wellington's host.

The officers present with the 2nd Battalion 83rd Regiment at the moment of the commencement of its third Peninsular campaign were:—

Lieutenant-Colonel Richard Collins (who had just joined, from the 1st Battalion).

Major Henry William Carr (also a recent arrival; Major Napper had just gone home).

Captains Wm. Geddes, Samuel Hext, Chas. Laird, Geo. Mansergh, Hon. Henry Powys, Gilbert Elliott, and Robert Thompson.

Lieutenants Connell J. Baldwin, Wm. Cotter, Nicholas Colthurst, Thos. Elliott, Jones Ferris, Geo. Fry, Jas. Hingston, James Jackson, John Ramsay, Henry Vereker, Chas. Bowles, Thos. Gascoyne,† John B. Tresilian, Fred. Irwin, Chas. O'Neill, Geo. Mee, Charles Watson, Thos. F. Smith, Jas. Ormsby, and Henry Richardson.

Ensigns Francis Barry, Michael Carey, T. Broomfield, Robt. Bloxham, Henry Shepherd, Wm. Strangeways,† Joseph Swinburne, and R. Woodhouse.

Ensign Swinburne was acting as Adjutant; his predecessor, Brahan, had just been reported as having got away from his French prison, but had not rejoined the Regiment.

PART II.

SEPTEMBER, 1810, TO DECEMBER, 1811.

1810

ON being posted to the Second Brigade of the 3rd Division on September 12th, 1810, the battalion appears to have made good speed to get to the front, as it joined the field army about the 22nd of the month. It relieved the 2nd Battalion 58th Regiment, which was a depleted second battalion with only 268 effective rank and file, and 165 sick, and was sent

* If the effective rank and file and sick combined were 481 in June and 530 in September, the draft can only have been about forty men.

† Lieutenant Gascoyne and Ensign Strangeways were left sick at Lisbon.

back to take up the distasteful garrison duty at the base, from which the 2nd Battalion 83rd Regiment had just escaped.

The brigade in which the battalion took its place had as its senior unit the 2nd Battalion 5th Regiment, whose numbers were about the same as its own—some 418 effective rank and file—with one of those rifle companies of the 5th Battalion 60th Regiment, which Wellington had attached to his brigades in order to strengthen their skirmishing line against the French tirailleurs. The brigadier was Major-General Lightburne, an officer for whom Wellington had no great esteem, and whom he gladly sent home in the succeeding winter. The divisional commander was that thunderbolt of war, Thomas Picton, the most energetic and one of the most capable of all Wellington's lieutenants, a man whose fiery temper brought his troops to the front in every engagement, and earned for them the honourable name of the "Fighting Division."

Nearly all the rest of the Peninsular service of the 2nd Battalion 83rd Regiment was to be passed under this formidable veteran, whose faults and virtues made him sometimes the most hated and sometimes the most admired of all the generals at the front. At this time Picton had still his reputation to make—or rather to re-make, for he had come out to Portugal under a cloud. As Governor of Trinidad, a newly-seized Spanish colony, he had ruled over a motley and demoralized population with an iron hand, and had been put on his trial on his return to England for various acts of doubtful legality. The prosecution had been dropped, for Picton's conduct was found to have been no harsher than the situation demanded. But some of his doings had been widely and unfavourably discussed; most of all, his inhuman permission given to a local Spanish magistrate to apply torture to suspects over a matter of burglary—a permission legal according to the Spanish law, which still ran in Trinidad. A Mulatto girl of bad reputation, one Luisa Calderon, had been subjected to the torture of "picketing," with Picton's leave, and her confession had brought the burglars to light. Consequently the general's name was known all over the British Empire as that of the Governor who had tortured a woman. Wellington, nevertheless, applied for his services, on his excellent military reputation, and he had come out to command the 3rd Division a few months before the 2nd Battalion 83rd Regiment joined it.

Picton was a splendid fighter, as his Peninsular career was to show, but he was a rough, foul-mouthed soldier of the 18th century type, and his temper had been soured by the abuse that had been lavished upon him during his trial for his doings in Trinidad. He was liable to sudden and blasphemous outbursts of rage, and prone to quarrel with his equals (his colleague, Craufurd, of the Light Division, was his special detestation), and to hector his subordinates. Yet, according to his lights, he was just; he was no respecter of persons, often did his best to get obscure merit rewarded, always kept down jobbery, and was accessible and even considerate to good soldiers. A serjeant in one of his battalions sums him up as follows:—" He was strict sometimes, especially about plunder, always talking about how wrong it was to rob the poor peasantry, and he used to flog the men whenever they were found out; but where he flogged, many generals took life. Besides this, the men thought that he had their welfare at heart. Every soldier in the division knew that, if he had anything to complain of, 'Old Picton' would listen to his story, and right him if he could. On the whole, our fellows always thought him a *kind* general, in spite of his strong language." His officers liked him worse than his rank and file—it was they who had to bear the brunt of his rough and unparliamentary language,—but no one could fail to admire his quick military eye, his tenacious courage, and his clever management of troops. Many instances of his ability will appear while the annals of the 2nd Battalion 83rd Regiment are being narrated.

When Colonel Collins's battalion joined the 3rd Division the army was retiring towards the famous hillside of Busaco, where Wellington had resolved to offer battle to the oncoming host of Masséna, if the Marshal should be daring enough to attack him on that formidable defensive position—nine miles of steep slope, covered with heather, and with occasional outcrops of granite showing through its ribs. As the British army retired into its destined fighting ground, the 3rd Division found itself placed on the right centre of Wellington's line, with its first brigade (Mackinnon's) blocking one of the two passes which climb the long ridge, and its left brigade (Lightburne's) on higher slopes, trending up from the pass toward the loftiest summit of the position, where lay the 1st Division. The 2nd Battalion 83rd Regiment (Lightburne's left regiment) put in line on September 27th 39 officers, 25 serjeants, and about 420 rank and file, besides 16 drummers, etc.

On the foggy morning of the battle-day of Busaco it seemed at first as though the brigade of Lightburne was destined to be in the thickest of the fighting. Of the four great columns, each a division strong, which Masséna sent to storm the hillside, one (Merle's Division of the 2nd Corps) was started from the valley right in front of the 2nd Battalion 83rd Regiment. While the mist still lay deep on the slopes, the picquets of the brigade, which had been thrown out far down in front of the crest, found themselves pushed back by masses of tirailleurs, behind whom battalion-columns were dimly visible. A bickering fire went on for some time, and the light company lost an officer (Lieutenant Nicholas Colthurst) and four men wounded. But it soon became evident that the attack was not to be pushed at this point. The French skirmishers were edging away towards their left, and the columns behind them were crossing the hillside diagonally, and not mounting it directly. The stress of the attack was about to be delivered against Mackinnon's, not Lightburne's, Brigade, and when the fog rolled up it was to show Merle's Division furiously engaged with the 1st Battalions 45th and 88th Regiments and the 8th Portuguese, on the lower crest, more than half a mile to the right of the position of the 2nd Battalion 83rd Regiment. The struggle did not last for long; the French were presently hurled down the hill, and no more fighting occurred at any point near Lightburne's Brigade. If they had been wanted, they could have been brought down to support Mackinnon, but the crisis was over in this part of the field. The second attack, this time delivered by Foy's brigade of the 2nd Corps, was delivered at a point still further to the right of the 2nd Battalion 83rd Regiment, straight in front of the pass of St. Antonio de Cantaro. The other assault, that made by Ney's Corps upon Craufurd's Light Division and Pack's Portuguese, was more than a mile away, to the left, and out of sight, owing to the loftiness of the ridge on which the First Division was embattled. Thus, in strong contrast to its fate at Talavera —its last battle—the 2nd Battalion 83rd Regiment was never seriously attacked, and lost no men save the few hit in the skirmishing at the outbreak of the action.

Repulsed with the loss of 4,500 men in his attempt to turn Wellington out of his chosen position, Masséna fell back on manœuvring, and by a wide-sweeping flank march turned the left of the Allies. When aware of the movement of the enemy, the British Commander-in-Chief might have drawn back and fought again a few miles north of Coimbra. But being still much outnumbered, and having no good ground to defend across the line of Masséna's new route, Wellington refused to risk a second general action, and withdrew in a leisurely fashion toward the famous Lines of Torres Vedras, which he had planned out more than a year before. They were by now in a thoroughly defensible condition, and ready for occupation. The army retreated by short and easy stages toward them, only the cavalry of its rearguard seeing anything of the French, whose pursuit was not particularly vigorous. Masséna, quite unaware of the existence of the Lines, thought that he was driving the British into the sea by his irresistible forward movement, and that Wellington would in the end embark and abandon Lisbon.

1810

The 3rd Division formed the left-hand column of the three into which Wellington divided his army after leaving Coimbra, as it was destined to defend the left section of the Lines; it marched not far from the sea-coast by the route Alcobaça, Caldas, Vimeiro, and took up its position in and about the town of Torres Vedras. Thus it was the most westerly of all Wellington's units in the line from sea to sea. The retreat was absolutely unmolested by the French, who followed the other two columns, but not Picton's. It would have left no trace on the memory of the division if it had not been for the miserable rainy weather, and for the distressing sight of the masses of Portuguese peasantry, who, evacuating their houses by Wellington's orders, jammed the road in front of the division at many narrow defiles, and were continually dropping sick or weary aged people, women, and children in the path. The friars of the great convent of Alcobaça, as is recorded to their credit, threw open all their stores as they departed, and left a dinner prepared for the officers of the division in their immense refectory.

On reaching the Lines, the three brigades of the 3rd Division received orders to construct huts for themselves behind the River Zizamdre, the forts in front of their line being left to be occupied by Portuguese Militia; for Wellington did not lock up his field army in the entrenchments, but kept each division massed behind the section which was in its charge, ready to strike at the head of any French column which should endeavour to burst through the line of redoubts held by the garrison-troops in front. It is an entire mistake to suppose, as Continental critics have often done, that the British army was disseminated in small detachments, or tied up in the forts.

The western end of the Lines was never even approached by the French, who drew up in front of the central and eastern fronts alone. Not even a single battalion was sent beyond the Pass of Runa, which separates the head waters of the Zizamdre from those of the river of Alemquer. Hence Picton's men saw nothing of the tentative efforts of Junot's Corps near Sobral, or Reynier's Corps near Alhandra. Only the distant sound was heard of this skirmishing, which revealed to Masséna the impregnability of the Lines, which had brought him to a stand. Unable to advance, unwilling to retreat, the Marshal lingered in front of Wellington's positions from the 12th of October to the 14th of November, till the sufferings of his army, half-starved and exposed to continuous rain-storms, drove him to give back to the country about Santarem. His useless obstinacy cost him some 5,000 or 6,000 men, disabled by dysentery and rheumatism, of whom few ever returned to the ranks.

When it became clear, in the third week of October, that Masséna was now about to attack the Lines, the brigades of Picton's Division were allowed to spread themselves a little, and Lightburne's battalions moved out from their huts and got more comfortable quarters in the village of San Pedro de Cadeira, five miles to the west of Torres Vedras, where they spent nearly a month, not uncomfortably, as winter-campaigning goes; for the men had solid roofs over their heads again, and were regularly rationed. The brigade had now been completed to a strength of three battalions, having been joined in October by the 94th, a single-battalion Scottish regiment, from Cadiz, with a strength of some 600 bayonets. During the month of November it changed its commander, Lightburne going home, to no one's great regret, and being replaced by Major-General the Hon. Chas. Colville.

When Masséna finally evacuated his position in front of the Lines and drew back behind the Rio Mayor to the hilly country by Santarem, Picton's Division moved out to join in Wellington's cautious pursuit. When the French were found to have settled down on a definite line of defence, the division was placed opposite their extreme right, still maintaining its place as the last unit of the allied left wing. Colville's Brigade was placed at Alcoentre, a village some miles from the course of the Rio Mayor, on whose opposite bank the sentinels of Junot's Corps were visible. There being some possibility that the enemy might make an offensive movement, now that the allied army was no longer covered by the redoubts of the Torres Vedras lines, Picton ordered the fighting positions of his brigades to be strengthened with *abbatis* and breastworks, a task which kept the troops busy for some weeks. But the French made no serious movement, contenting themselves with a fruitless demonstration against Pack's Portuguese, the corps next to Picton's right, on January 19th, 1811, in which Junot received a ball in the face, which disabled him for some weeks. Colville's Brigade seem to have had a dull and uneventful time at Alcoentre, which was no comfortable place of sojourn, the houses having been wrecked by the French during their retreat, and left in a condition of complete dilapidation, while the countryside had been thoroughly plundered.

1811

From this point the history of the 2nd Battalion 83rd Regiment is brightened up by a light of personal reminiscence, such as has not been available since the summer campaign of 1809, an officer who kept a diary having once more joined the battalion. This was Ensign William Strangeways,* who had been in hospital at Lisbon all the autumn and winter, but rejoined his battalion at Alcoentre about February 10th, when his notes commence, and last for seven months. His diary shows the battalion settled down for a long stay, with a number of small messes organized, whose only drawback was the difficulty of obtaining regular subscriptions; for already the dreadful arrears of pay, familiar to every officer in the later years of the Peninsular War, had begun to accumulate, and the regimental paymaster could often make no statement on the 24th of each month, save that he had obtained nothing from the Paymaster-General; and when once a deficit had been established, each instalment, as it came in, only served to defray old expenses, and gave nothing for the future. It may be noted that Colonel Collins was no longer in command, having been moved into the Portuguese service, where he had been given a brigade; his succession fell to Major Henry William Carr. Brahan, the Adjutant, captured at Talavera, was at last back from French captivity, and had taken over the duties which Ensign Swinburne had so long discharged in his absence. Colville, the new Brigadier, is mentioned as somewhat of a martinet; " he gave the battalion the trouble of going through the manual and platoon more than it liked." From a diarist in another battalion of the brigade it may be gathered that he gained some unpopularity by his inability—or feigned inability—to understand an Irish or a Scottish brogue; he would order his aide-de-camp to interpret for him, " as he did not understand Gaelic." But he turned out a good leader when the

* His short (and much discoloured) file of notes were lent by his grandson, Mr. Leonard Strangeways, of Holland Road, Kensington. It is a pity that he missed the Busaco campaign by sickness, being left behind at Lisbon, in September, 1810.

brigade came under fire, and the men came to like him better than they had ever thought to do.

The strength of the battalion at Alcoentre on February 15th, 1811, shows no great variation from that of the previous autumn; the stay behind the Lines had evidently been a healthy time. The Morning State shows present one major commanding (Carr), five captains (Hext, Thompson, Noleken, Powys, G. Elliott), fifteen lieutenants, seven ensigns, twenty-seven serjeants, eighteen musicians, and 447 rank and file. There were eighty-four sick, no great number at the end of a hard winter.

On Monday, March 4th, the whole of the 3rd Division was reviewed by Lord Wellington, and showed (including its Portuguese brigade) nearly 6,000 men present. This review marked, as chance was to have it, the beginning of the campaign of 1811, for on the same evening the news reached Alcoentre that the French about Santarem seemed to be on the move, and were probably about to retreat.

This was true; the much-enduring army of Masséna was at last starved out, and was preparing to fall back towards the north before its last rations should be consumed. The Portuguese peasants who, a few days before, had informed an officer of the 83rd that the enemy was now living on horseflesh, and was even eating carrion like dogs and cats, had given perfectly correct information. On the 5th the 3rd Division received orders to be ready to advance next morning. Wellington did not start them off at once, because, though their position for falling on the flank of the retreating French was good, he wished to have his whole force in hand, ready to support his leading column, and the troops which would join Picton were still a march to the rear.

It was only at 11 o'clock on the morning of the 6th that the 83rd received their orders to push forward, along with the rest of Picton's column. They slept that night at the village of Rio Mayor, on the river of the same name, and found it thoroughly devastated. Next morning took them to Alcanhede, where the diarist of the 83rd, entering the church, was shocked to see the altar deliberately broken and the words "Vive, Napoléon, Empereur des Français," chalked above it, "as much as to say after driving away the superstitions of a superfluous religion, all that remains is the *Devil's name.*" Both on the 8th and the 9th of March there were reports that Ney, commanding the enemy's rearguard, had halted to fight. But they were untrue; the French were still marching hard, and the 3rd Division had not yet got near them.

1811

On the 9th the column crossed the steep watershed between Alcanhede and Porto de Moz, "the country the ugliest that I ever saw, mountainous, rocky, and, in fact, as disagreeable as Nature could frame it." In the once flourishing town of Porto de Moz the half-consumed bodies of nearly 200 peasants were to be seen in the sacristy of the Convent, which had been fired over them by the French when they took refuge in it. Here the division got into the main *chaussée* from Lisbon to Coimbra, and, marching faster along a good road, passed Leiria, which had been sacked and mostly burnt, and on the next day (March 11th) at last got in sight of the French rearguard, near Pombal.

Here the 83rd heard shots fired in anger for the first time since Busaco. Ney had at last halted, to gain time for Masséna to force the passage of the Mondego and capture Coimbra, where he intended to stay his retreat, if fortune favoured him. In the combat of Pombal, however, the Light Division had all the work to do; the 3rd was told off to execute a flanking movement, and found that the enemy had absconded before its arrival. "We did not get either rum, bread or meat this day," writes the 83rd diarist, "and it rained much before we got to quarters."

On the following afternoon, however, Colville's Brigade was seriously engaged, and the 83rd had its first casualty in the campaign of 1811. This was at the combat of Redinha, where Ney again made a front, to hold back the pursuers, at the head of two divisions of the 6th Corps. He was found holding the village of Redinha and the heights on each side of it, and with strong supports behind the Ancos River, on which the place is situated. Wellington, seeing the French in such force, waited till he had three divisions up, and then attacked in a crescent-shaped formation, intended to outflank Ney's line. Colville's Brigade, with the rest of the 3rd Division, was on the right, and had to deploy among woods and low rolling ground. "We struck off the main road and passed over many hills difficult of access. On one of them we formed close columns, and were advancing, when a round shot came and knocked over three of the 94th, one killed, another wounded, the third only stunned. We then retired a few yards, and moved off to our left, when we came in sight of a village (Redinha), and formed line to attack a French column beyond it. We were ordered to make a dash at the bridge, which lay upon the river, between us and the village; this we performed in superior style, nor did the enemy fire a shot at us until we were on the flat near the river. Here they saluted us with two or three more round shot, but (thank God) none of them took effect on us, though Lieutenant Clarke, of the 5th, was wounded. But, to conclude, we advanced and the French made off as quickly as possible. Night coming on, we took up our quarters in a wood, where we got rum and bread, but no meat. Our baggage came up and joined us."

In short, the turning movement of the 3rd Division compelled Ney to move off somewhat earlier than he had intended, but without any severe loss—only fourteen officers and 216 men. The British casualties, not much less, were twelve officers and 193 men, of whom fifty-eight belonged to the Light Division, which made the frontal attack, and fifty-one to the 3rd Division; the 2nd Battalion 83rd Regiment got off lightly, with one private wounded, though their neighbours in the brigade, the 94th, had two killed and fourteen wounded.

Meanwhile, Ney's rearguard actions at Pombal and Redinha had failed in their destined purpose of giving time for Masséna and his advanced troops to cross the Mondego, and occupy Coimbra. Opposed at its broken bridge by the Portuguese militia of Trant, Montbrun failed to force a passage, and thus left the French army in an uncomfortable position, for Wellington was pressing hard behind, and an impassable river was in their front. Masséna was forced to abandon his designs on Coimbra, and to make off by the only road left open to him, that to the east, toward Miranda de Corvo and the Alva River, which would ultimately bring him to the frontiers of Spain. He took it unwillingly, for it was rugged and through a barren land—but there was no choice.

Meanwhile, Ney was once more left as rearguard, to hold back the advance of Wellington's columns till the main body of the French had filed off to their right. This led to the combat of Condeixa (March 13th), in which the 3rd Division played a decisive, but a bloodless, part. Wellington used them to outflank Ney's left, sending them by a rugged track along the crest of the heights which flank the main road to the right. When he saw Picton's column approaching in the afternoon, the French Marshal hastily fired Condeixa and retreated eastward. The 3rd Division camped on top of the heights, where it had been advancing, with the blazing town below. If Picton had only known of it, there was another French column, which he might have cut off, on his left, in the lateral valley of Fonte Cuberta. But he never heard of it, and this body, Loison's Division, escaped in the dark over the hills. Masséna himself was present with it, and had some uncomfortable moments, while he imagined that the 3rd Division might be about to descend on him. The 83rd diarist describes the day as follows:—"The march was over mountains of rocks and stones almost inaccessible, the French still retreating. It was truly fatiguing, and we were glad to halt on a bleak mountain for the night; our baggage did not come up. All I know about the place is that we were just over the town of Condeixa, which the enemy left on fire. We heard firing in the night." This last firing was caused by Loison's Division bursting past the outlying picquets in its hurried march of escape.

On the 14th of March Picton's Division had to extricate itself from the hill-tops in order to descend on to the Miranda de Corvo road, where the Light Division was pushing Ney's retreat. It was a day of dense fog, and General Erskine, commanding the advanced guard, got himself into a dangerous position, by thrusting some companies of the 52nd and 95th right into the centre of a French Division, which Ney (unknown to him) had halted and deployed at the village of Casal Novo. Erskine was only saved from a severe check by the opportune descent of the 3rd Division on the left rear of the French, who had at once to abscond. "The enemy about this time were driving back our Light Troops, and the rifle balls flew thick about our ears, but (thanks be to God) only one corporal was hit, and he not badly. We advanced to a high hill, and immediately upon receiving a discharge from our light artillery, they made off in double quick time. Our cannon was carried up on mules, to my no small surprise. The French burn every house they come near."

On the following day (Friday, March 15th), fell the last of the many skirmishes in which the 3rd Division was engaged during Masséna's retreat. This was the surprise of Foz de Arouce in the late evening. The 83rd were marching all day through scenes of horror—"numbers of the enemy's waggons and carts half-burned, some of the enemy dead, and among them the unfortunate Portuguese, whom they had butchered, including women and children," as also a distressing group of 500 ham-strung horses, mules, and donkeys outside Miranda de Corvo, sprawling and screaming in the mud. At five o'clock the head of the column suddenly ran into Ney's troops bivouacking around the village of Foz de Arouce, with no proper picquets set, and no cavalry screen. The Light Division plunged into the right of the French encampment; the 3rd Division attacked the slopes on its left, where a brigade of Mermet's infantry opposed them. "The enemy will recollect the beating our light troops gave them this evening for a long while. Nothing that I have before heard came near the tremendous fire that was kept up till night fell. Our brigade formed line upon a hill in a wood, when the balls flew among us, but (thank God) no one was hurt. The coat of Sergeant Potter, who stood next me, was ripped up along the shoulder." The French, fighting confusedly, were driven into and over the Ceira River, where many of them were drowned, and a regimental eagle was fished out of the water. Strangeways is not quite right in saying that "*no one*" was hit in the 83rd—the official return shows one private wounded. The whole 3rd Division lost thirty-three men, a small loss for the amount of mischief that they did. The troops would have received no food that night if they had not captured the French camps, where a certain amount of not very appetising viands was found cooking round the fires.

From this combat of March 15th, onward, the 83rd were not engaged for nearly three weeks, though they were at the head of the marching columns with which Wellington was pushing Masséna out of position after position in Central Portugal. The French committed themselves to no more rearguard actions, but retired whenever pressed and outflanked, failing, even, to defend the strong line of the Alva River on March 19th, where a serious combat was expected. The march was through scenes of horror and devastation, for the enemy were living by plunder and murdering every peasant whom they could catch. The 83rd diarist gives distressing details of the sights that he passed, e.g., at Villa de Porco, on March 27th, he found the mutilated body of the parish priest, whose fingers the French had cut off, one by one, to make him disclose his little hoard of 800 dollars. When he gave in, and revealed his treasure, he was shot at once. "Opposite the church, which has been destroyed, the inhabitants are burying their neighbours, whom the enemy have murdered."

1811

In this long advance the British had quite out-marched their convoys, which had to toil up all the way from Lisbon, and Wellington had been forced to drop nearly half his army on the Alma River for some days, to allow provisions to come up. Only the cavalry and the 3rd and Light Divisions kept close to the heels of the French, living as best they could, by purchasing food from the peasantry off the immediate line of the French retreat, for the enemy was in such a hurry that his marauding extended no further than the villages on the road, and those on the flanking hills were generally intact. The 83rd diarist gives many details of his hand-to-mouth diet. Only half-rations were being issued, but a careful scouring of the hills generally produced a useful supplement. On March 21st he buys " a nice little cheese." On the 22nd, in a village two miles off the road, he finds some honey and a loaf, while a more fortunate messmate returns from a longer cast in the hills with a lamb, dried fish and onions. Next day no one can find anything, but on the 24th a farmer appears and sells the officers of the regiment fourteen loaves. Then come days of comparative starvation, till on the 28th, when the diarist is " absolutely faint from hunger," the first convoy from the rear struggles up, and everyone dines " early and heartily " on potatoes and ration-beef.

On the following day fell the capture of Guarda, when the 3rd Division, coming in unexpectedly over the hills upon the quarters of the French 6th Corps, expected to have a stiff combat, but saw, to their surprise, their adversaries make off in haste. The hard-fighting Ney was no longer in command, having been displaced by Masséna, and his successor, Loison, was only set on getting as quickly out of Portugal as possible. " We could see numbers of the enemy amongst the rocks upon the summit of the mountain, and all expected a bloody approach. We were ordered to form line and move forward, while the 5th, on our right, and the 88th, on our left, kept in close column.* In my life I never saw the Regiment march so well in line! Suffice it to say that upon our approach the enemy fled, and we passed through Guarda and halted beyond it, astonished to see the French fly from a position so fine without firing a shot. Their hurry was so great that they left many things of value behind, but I cannot help saying that they retreated in a very neat style. The beautiful Gothic cathedral of Guarda is destroyed, but in no part did the ruin appear so disgusting as in the organ, which they wantonly tore to pieces. There are libraries of books thrown about the streets! Nearly 200 prisoners taken this day. I bought a little bag of flour and got a goat killed for me."

The last stage of Masséna's retreat had now been reached; he was drawing near the Spanish frontier and his base; at the fortress of Ciudad Rodrigo. But the campaign was not to end without a sharp fight on April 2nd—the largest affair that had taken place since the French broke up from Santarem. Masséna

* The 2nd Battalion 88th Regiment, a battalion fresh from Lisbon, was with Colville's Brigade for a short time only, being soon drafted into the 1st Battalion 88th Regiment, in Mackinnon's Brigade.

had halted on the Upper Coa, with Reynier and the 2nd Corps on his left, at Sabugal, and Loison and the 6th Corps on his right. Wellington was now in force again, having received convoys sufficient to bring up to the front the divisions left on the Alva on March 17th. Seeing the French spread on a long line, with a gap between their wings, he resolved to encircle and crush the 2nd Corps. For this purpose the Light Division was to cross the Coa high up and turn Reynier's flank, while the 3rd and 5th forded the river lower down and tackled him in front. The details of the scheme went wrong, partly from a dense fog at dawn, and partly from the incompetence of Erskine, then commanding the Light Division. It resulted that the turning column, passing the river at the wrong spot, was ahead of its time, and became hotly engaged against triple numbers before the front attack was ready. By the brilliant fighting of Brigadier Beckwith and the 43rd, the enemy was held back longer than might have been thought possible. Meanwhile, the 3rd Division was hurried forward to strike before the Light Division should be overwhelmed. "We heard to our right a fire of artillery and small arms, not very hot. At length we reached the Coa, through which we waded nearly up to our middles. We then marched up a hill through a wood, the most difficult ground to pass that might be. Notwithstanding, we formed in divisions, then in grand divisions, and near the end of the wood in line. Then, in double-quick, we rushed out to attack the 'murderers of nations,' who were advancing both infantry and cavalry to drive back our Light Division. But, upon seeing our brigade, they ran down the hill like lightning, and formed a front against us, under cover of their artillery, with which they aimed a fire against our line. I had the honour to carry the King's Colour, and their first shot went over the staff of it. They threw many more, but all fell either behind or short of us. We were then ordered to retire a few paces, to allow our artillery to come up, which soon silenced the enemy's. Then we charged; the 5th Regiment got within three yards of the French, and poured in two destructive volleys; it was then intended to give them the bayonet, but they 'begged to be excused.' Where our Regiment halted there were many French lying about, killed and wounded, among them an officer, whose groans extracted pity from the most unfeeling. After the French had run off we encamped on the side of the hill near the village of Sabugal. God be praised, we had beat them right well! Our situation during the night could not have been worse—not a bit of bread, and pouring rain."

In elucidation of Strangeways's simple narrative, it may be explained that Colville's Brigade, coming up unseen through the wood, struck directly upon the flank of Reynier's reserve (the 17th and 70th regiments—six battalions), which the French general was just preparing to throw upon the already over-matched Light Division. The two hostile regiments made a gallant attempt to show front to flank, and to hold back Colville, but failed entirely, never having properly got into their new fighting order. Hence their loss was heavy, especially in the pursuit—400 men, including 100 prisoners—while Colville only lost absurdly little—not much over twenty men, of whom only two were in the ranks of the 2nd Battalion 83rd Regiment, which was in the centre of the brigade, while the right-hand regiment, the 2nd Battalion 5th Regiment, was leading the rough échelon in which the line came out of the wood.

After Sabugal the French withdrew behind Ciudad Rodrigo, apparently *hors de combat*, and unlikely to move again for many weeks. Wellington, having made provision for the blockade of Almeida, cantoned his army between the Coa and the Agueda, and went off for a short visit to Beresford's army in the Guadiana, which was at this time commencing the first unlucky siege of Badajoz. Thus, during his absence, the 2nd Battalion 83rd Regiment got a much-needed rest of nearly a month. For the greater part of this time it was placed in the village of Alamedilla, just across the Spanish frontier, while the rest of the brigade lay in neighbouring hamlets.

The diarist of the battalion "found the Spanish villages infinitely more clean and pleasant than the Portuguese, but though the people are civil enough, I really think they do not like us. They are as impudent as possible, cursed extortioners at selling. . . . Bread was five shillings a loaf to-day!"

The battalion remained at Alamedilla apparently from the 7th to the 24th of April. "On Good Friday, the 12th of the month, we had, thank God, Divine Service in the church of the village; it was performed by the Brigade Major, a plain form of prayer and thanksgiving offered by soldiers to their Maker for having protected them in the hour of danger." It is to be feared that the sight of their local place of worship requisitioned for a heretical service performed by a layman, must have made the villagers even less sympathetic to their military visitors than before. Probably the church was reconsecrated after the departure of the battalion!

On April 24th, the 2nd Battalion 83rd Regiment received orders to march forward to Fuentes de Oñoro, a large straggling village on a hillside above a brook. This was part of a general concentration toward the front, news having been received that Masséna was once more showing signs of life, and had brought several divisions forward to Ciudad Rodrigo. Fuentes was a well-provided place, but prices still ruled high—"sugar, 1s. 6d. a pound; a large loaf of bread, 5s.; but good milk in plenty, at 6d. a quart." Here the Regiment lay, in a state of great alertness and constantly turning out on false alarms that the French army was advancing, from April 24th to May 2nd. Sir Brent Spencer, left in command of the troops during Wellington's absence, was a nervous man, and not much trusted by his subordinates. "Lord Wellington is to be back from Badajoz to-morrow," writes the regimental diarist on the 27th, "of which I am extremely glad, and shall enter the field in twice a better humour than if Sir Brent commanded."

1811

On May 2nd, only four days after Wellington's reappearance, Masséna actually made his long-threatened advance, thinking himself bound in honour to attempt the relief of Almeida, which was now hard pressed. Having picked up Drouet's 9th Corps, and also two cavalry brigades of the Army of the North, under Marshal Bessiéres, he was in great strength, with about 48,000 men. Wellington, after deducting six battalions (five Portuguese and one British), left to watch Almeida, had about 37,000 men to hold him back. While the French were engaged in driving in Wellington's outlying cavalry and light troops from the line of the Azava, the 3rd Division was put under arms for the greater part of the day, and the baggage sent to the rear. At six in the evening the men were dismissed, but with orders for no one to take off his accoutrements or go far from his company alarm post, as the French might be up at any moment, and a general action was expected on the morrow.

Fuentes de Oñoro, where Colville's Brigade had been quartered for the last nine days, was destined to give its name to a battle, of which the most bitter part was to be fought in its streets, and the 83rd was now to see two days of hard fighting. Wellington's original position, taken up to cover the siege of Almeida, lay behind the line of the Dos Casas brook, from Fuentes, on the south, to Fort Conception, on the north; and in the first disposition of his forces the 3rd Division formed his extreme right wing, and was specially charged with the defence of the village, which formed an effective guard for the flank of the army, so long as the French should not extend themselves further to the south. It consisted of a mass of small houses, surrounded by stone-walled crofts and gardens, sloping up from the shallow Dos Casas to two bold, rocky mounds—one of them crowned by a chapel—at whose level began a broad plateau, on which the army was arrayed. North of Fuentes the brook lies in a deep ravine hard to pass, but at the village itself the ravine dies away, and the water is easily to be forded. Hence it was the obvious point for the enemy to attack, since no other section of Wellington's position was so accessible. But the village was eminently defensible and had been barricaded and loopholed. It was held by the twenty-eight light companies—English and Portuguese—of the 1st and 3rd Divisions, under Colonel Williams, of the 5th Battalion 60th Regiment—about 1,800 men. The 83rd, which on this day counted thirty-three officers and 427 men present, contributed its light company, under Captain Hext, to the defence of Fuentes, with a strength of three officers and about 50 men. The rest of the battalion was drawn up on the plateau above the village, ready to resist any attempt to turn it on the left, if the enemy should break over the brook in that direction. This operation was not tried by Masséna, so that only the light company was engaged this day, but that small section of the battalion was in the very thick of the fight.

On the first day of the battle, Masséna made no general attack on Wellington's position, but simply tried to beat in the allied right wing, by storming Fuentes, for which purpose he employed Ferey's Division of the 6th Corps, supported by that of Marchand. All through the afternoon there was bitter fighting up and down the streets of Fuentes; the Marshal put in battalion after battalion, till he held the whole division of Ferey engaged, and with accumulated force drove the defenders up to the top of the place, where they rallied around the chapel and the rocky mounds. Determined that the French should not keep the lower parts of the village, Wellington sent down into it three battalions from his reserves, the 1st Battalions 71st and 79th Regiments, and 2nd Battalion 24th Regiment, who, charging fiercely, drove the enemy back to the brook and recovered the whole of the lost houses. Masséna continued the strife by putting in some of Marchand's troops, but could make no further headway, and the combat stopped at nightfall.

The total loss of the British this day was 250, to which the 2nd Battalion 83rd Regiment's light company contributed nine wounded and three missing (probably captured by being trapped in a house, when the French stormed the lower part of the village). The much heavier casualties of the French were 652. The main body of the 2nd Battalion 83rd Regiment did not suffer at all, no attack having come near the point where they were placed. Strangeways narrates its fortunes as follows:—"After having been halted a long time, our Regiment was ordered out to support the light companies. We marched forward to a rocky hill above Fuentes de Oñoro, and here were posted as light troops, along with a brigade of Portuguese artillery, commanded by Major Arentschildt. His guns were well served, and dismounted one of the enemy's. I only have time to say that the French got and lost possession [of Fuentes]. It was delightful to see our light companies putting back their heavy columns. Late in the evening they made a desperate push to gain the village, but were charged by our light troops and driven back in 'double quick.' Thanks be to God, our little regiment did not suffer so much as might be expected. I am told the light company have in wounded and missing lost eleven men. I was with the King's Colour. We halted among the rocks for the night."

On May 4th, to the surprise of the British, no further attack was made upon them. The fact was, that Masséna did not like the look of Wellington's chosen position, and had resolved to turn it, rather than to attack it frontally for a second time. All this day his staff officers were casting about for tracks by which a wide encircling movement might be made, and at last reported that a long détour through woods and marshes might bring a column round Wellington's extreme right into the upper valley of the Dos Casas, without too much risk. Not quite unsuspicious of what might be in his adversary's mind, the British commander, this day, sent out all his cavalry to his right, and moved the 7th Division as an observing or detaining force to Pozo Bello, a village two miles south of Fuentes, which thus ceased to be his extreme right, and was soon to become his centre. He also made some changes in the composition of the garrison of Fuentes. Sending back the twenty-eight light companies, which held it on the 3rd, to their regiments, he replaced them by three whole battalions from the 1st Division—the 1st Battalions 71st and 79th and 2nd Battalion 24th Regiments.

On the early morning of the 5th the French appeared in great strength in the quarter where Wellington had suspected that they might be about to move. All through the night their columns had been carrying out a great turning movement to the south, and at dawn 3,000 horse, under Montbrun, drove in the English flank cavalry, while a column of two divisions of infantry rushed in upon Pozo Bello and evicted the 7th Division from it with heavy loss. There were more troops coming up in the rear, and the attack was so heavy that Wellington could only draw in his southern wing, and make a new line at right angles to his centre. Thus Fuentes de Oñoro, and the 3rd Division behind it, became the obtuse angle of an army drawn up *en potence* in a fresh position.

Masséna directed his turning columns to hold back somewhat, till his projecting point of Wellington's array should have been crushed. He launched against Fuentes three whole divisions, Ferey's, Conroux's, and Claparéde's successively. The attempt to carry the village became the main operation of the day, and remained undecided for many hours. Wellington fed the defenders with detachments from the plateau above, including all the light companies of the 3rd Division and many Portuguese Caçadores. At length much of the village was conquered by the French, and the 83rd, standing near above it, got engaged with numerous tirailleurs, pressing out of its flanks and working up the plateau. The battalion sent its colours to the rear, and extended itself along the slope in skirmishing order to keep them back. It inflicted and suffered some loss, one lieutenant (Jones Ferris) and six men killed, another lieutenant (Vereker, commanding No. 2 Company) and twenty-seven men wounded. But the French got no further forward in this direction, and at last the main attack, which had finally struggled up to the chapel on the rocks above the village, was cast back by a furious charge of the 74th and 1st Battalion 88th Regiments from Picton's reserve. The assailants recoiled beyond the river, and the day was over; for on the right Masséna refused to push his advance till Fuentes should have been carried, and gave it up when the village proved impregnable. There is no good account of this day's fighting in the diary of Strangeways, because he was detached, with the colours and their guard, to a position much behind the battalion's fighting ground, and near the 2nd Battalion 5th, 94th, and 2nd Battalion 88th Regiments, which Colville was holding back as an eventual support for the 2nd Battalion 83rd Regiment, if it were needed. He writes only: "The French made furious attacks upon our right and centre, which continued until evening. Poor Ferris was killed early. The enemy took the village again, but were driven out by our brave troops, with much loss. On our right they met with no better success. I am sorry to mention that Vereker was wounded—not (thanks be to God!) badly. I was with the colours in the rear, and was knocked down by the wind of a shell which burst just beside me. In the evening our Regiment was relieved and came back to the brigade. . . .

1811

I was with the colours in the rear of the 94th; their officers endeavoured to make my situation as agreeable as they could, especially Surgeons Enright and Ross."

On the 6th, 7th and 8th the British army was anxiously awaiting a fresh attack, and trenches and breastworks were being thrown up all along the line. The battalions lay out in battle order all day—baggage far to the rear, "no clean linen to be got, and servants, even, not available." But Masséna was really beaten, and was only deferring his retreat till he should have got a message into Almeida, to bid its Governor blow up the place, and cut his way through the blockading force if possible. On the night of the 9th-10th the whole French army was seen in retreat, and the 2nd Battalion 83rd Regiment were ordered to take up their old quarters in Fuentes de Oñoro. "This place, so bloodily disputed, is, of course, much altered, the houses having been dreadfully used by both parties. I got into my old quarters, but the people were, of course, flown: it is fortunate that the dead are almost entirely buried." On the 11th the baggage came back, and the officers got their clean linen and the small creature comforts of tea, sugar, etc., stored on their mules' packs.

The governor, Brennier, having blown up Almeida, on May 11th, and escaped with two-thirds of his men, owing to the mismanagement of Generals Campbell and Erskine, Wellington considered that he might turn his attention to another quarter. For Masséna's army had broken up and retired to distant cantonments, and the Marshal himself had just been superseded and sent back to France. Accordingly, the British general resolved to march with two divisions to reinforce Beresford before Badajoz, and to strengthen him against Soult, who was known to be moving for the relief of that fortress.

The divisions designated to go south with the Commander-in-Chief were the 3rd and 7th; so on May 14th the 2nd Battalion 83rd Regiment broke up from Fuentes de Oñoro, and, with the rest of Colville's Brigade, made a fatiguing march of 20 miles to Alfayates. This was the first stage in a most exhausting and interminable pilgrimage over the mountains of Beira.* The roads were bad, the weather very broken, thunderstorms alternating with spells of absolutely tropical heat. The country was very thinly peopled, and little was to be procured in the way of food to supplement the beef and biscuit of rations. The brigade did not halt at Castello Branco, the only considerable town on the route, but was quartered, night by night, in small and miserable villages. On May 20th the floating bridge of Villa Velha, on the Tagus, was reached, and after tiresome delays, caused by the defile being blocked by the baggage of Mackinnon's Brigade, which was marching in front, the 83rd got across the great river and toiled up hill to Niza, the first place in the Alemtejo.

* The stages, by Strangeways's diary, were: 14th May, Alfayates; 15th, Sabugal; 16th, Meimoa; 17th, Penamacor; 18th, Losa; 19th, Sarnedas (passing Castello Branco on the way); 20th, Villa Velha and Niza; 21st, Alpahao; 22nd, Portalegre; 24th, Arronches; 25th, Campo Mayor.

Straggling was now beginning—shoes were worn out, sunstroke was prevalent, and men, and even officers, began to be dropped behind in forlorn villages. A short halt on the 22nd-23rd in the large town of Portalegre, which had never seen the French, and was in a more intact condition than most Portuguese centres of population, was very acceptable to the regiment, and allowed many foot-sore men to come up from the rear. At this place the Regiment got its first news of the battle of Albuera, fought on the 16th, in which its old colonel, Collins, had lost a leg while leading on his Portuguese brigade in the successful advance at the end of the day. Wellington had hoped to arrive in time for the fight, at the head of his two marching divisions, but was somewhat consoled for his absence by the fact that Beresford had inflicted a decisive, if a very costly, check on Soult, so that the siege of Badajoz could proceed, without any immediate danger from the French army of Andalusia.

Leaving Beresford's much depleted troops to lie out as a "containing force," in front of the defeated Soult, Wellington undertook the siege with the 3rd and 7th Divisions, aided by some Spanish and Portuguese battalions. Colville's Brigade, marching from Portalegre on May 23rd, reached Arronches on the 24th, and the Portuguese frontier fortress of Campo Mayor, only twelve miles from Badajoz, on the 25th. Two days later Picton's Division crossed the Guadiana and invested the fortress on the south side, while Houston's Division (the 7th) took up its position on the nearer side of the river, and shut it in from the north.

The second British siege of Badajoz (May 27th—June 10th, 1811) was a mismanaged and disappointing business. The siege artillery, old Portuguese guns, drawn from the ramparts of Elvas, was inadequate in the number of pieces and ineffective from their antiquated make and irregular calibre. Moreover, the points of attack were ill-chosen, being the two strongest points of the place—the Castle on the south bank, Fort San Cristobal, on the north bank—both of which proved to be more formidable than Wellington's engineers had expected. Colville's Brigade was, like the rest of the 3rd Division, allotted to the attack of the Castle, and supplied men for trench work on alternate days.

The 83rd, after its very appreciable losses at Fuentes de Oñoro, and its fatiguing march across the mountains, was now very weak. It seems to have had before Badajoz about 404 of all ranks (return of June 1st), having at the time seventy-three sick and wounded left in hospital. From this modest strength it lost three men in the trenches on May 30th-31st, one on June 4th, one on June 6th, two on June 7th, and one on June 9th—losses which were moderate (as were all of those of the 3rd Division regiments) compared with the dreadful casualties on the San Cristobal side, where the 7th Division suffered terribly on the bare cliffs in face of the fort of that name. The attack on the Castle was never really pushed home, the artillery failing for many days to make a practicable breach. Indeed, the Castle was only just commencing to crumble, when, on June 10th, Wellington raised the siege, having received news that Marmont, the successor of Masséna, was bringing the whole army of Portugal to the assistance of Soult. There was no way of preventing their junction, and their united force was far too strong to be faced by the five Anglo-Portuguese divisions then in Estremadura; wherefore the Commander of the Forces withdrew behind the Guadiana, to a line between Elvas and Campo Mayor, to which he had directed the troops left behind in Beira (1st, 5th, 6th, and Light Divisions) to come up as quickly as possible.*

The retreat was made with ample margin of time to allow for safety; on the 17th Wellington was established in his new position; on the 20th the troops from Beira began to arrive; the last of them were up and in touch with the main army on the 23rd. Meanwhile Soult and Marmont had met at Merida, on June 18th, and entered Badajoz on the 20th. On the 22nd they executed a general cavalry reconnaissance of the whole of Wellington's line. The 3rd Division was now forming its right wing, near Campo Mayor, and was drawn up in olive woods close to the right of that place, expecting a serious attack to follow the advance of the great body of French horse which had swept in the allied cavalry vedettes with some loss. But no hostile infantry appeared: "We remained in our position till near 11, and were then ordered back to our quarters; so we went out expecting a bloody battle, and returned in perfect peace, thank God! Nothing is more common in war."

The next six days formed a real crisis in the Peninsular War. The French were collected in front of Wellington's line in great force; the two Marshals had over 60,000 men. Wellington's position was strong, and flanked by two fortresses, but it was very long, and he had only 54,000 men in hand, including the Portuguese. His cavalry, in particular, was little more than half that of the enemy. The temptation, therefore, to try a pitched battle was very enticing to Marmont and Soult; but after six days the elder Marshal, taking with him 15,000 men, marched off for Seville, which was being threatened by the Spaniards of Blake, whom Wellington had launched against his adversary's rear. This rendered it impossible for Marmont to attack, but he lay with 47,000 men round Badajoz till July had come. Wellington refused to take the offensive, and the armies faced each other across the Guadiana till, on July 15th, Marmont broke off and retired northward. Wellington followed his example on the 19th, and moved parallel with the Army of Portugal back to the north bank of the Tagus and the borders of Beira and Leon.

While the French and English faced each other across the Guadiana from the 22nd of June to the 15th of July, battle was for some time expected every morning. The 83rd diarist reports that his brigade took up its designated position each day from the 22nd to the 28th. "To fight, or not to fight? That is the question. At least it is ours every morning. I wish to God that

* Strangeways, the invaluable diarist of the 83rd, was sick at Elvas during the siege of Badajoz, and only rejoined the battalion on June 20th.

it was all well over, for I am getting tired of it." After a last alarm 1811 of a French advance—some exploring cavalry were seen—on the 28th, it was at last ascertained that a considerable part of the enemy had departed; but the battalion was still employed for some days more in making abattis and trenches along its front. The banks of the Caya and Guadiana are unhealthy, and the army began to suffer severely from fever—the 83rd less than many corps, though its sick went up from 73 to 129 in the month.

The orders to break up from Campo Mayor and march back to the Beira highlands were received with pleasure on July 18th. Everyone was tired of the hot and pestilential valley of the Caya. The 3rd Division was ordered to march northward by the same route that it had taken in May, and, passing Niza and the bridge of Villa Velha, had reached Castello Branco by the 24th. Here the brigade halted for a week, there being no such occasion to hurry as there had been in the downward march, but on July 31st orders came for a renewal of its northward progress. Moving by easy stages, it passed Penamacor and Sabugal, and came to a halt at Carpio in Leon, close to the gates of Ciudad Rodrigo, on August 10th. The regiments were cantoned in the villages around—Pastores, El Bodon, Robleda and others.

Wellington had now another project in hand. Foiled before Badajoz by the junction of Soult and Marmont, he was now about to try a blow at Ciudad Rodrigo, the other outlying fortress in the hands of the French; but he was destitute of a battering train, and while one was being organized for him at Villa da Ponte, behind Almeida, he did no more than observe Ciudad Rodrigo, with the 3rd and Light Divisions, and turn the Spanish guerilla bands loose to prevent any provisions entering the place.

Their operations were so far effective that Rodrigo began to suffer from severe privations long before the English siege-train was ready, and Marmont resolved to march to its relief not only with his own "Army of Portugal," but with large succours borrowed from his neighbour, Dorsenne, the commander of the "Army of the North," which occupied Old Castile. On September 20th no less than 60,000 French appeared in the neighbourhood of Rodrigo, and Wellington had to abandon the blockade of the fortress, and order a general concentration. Meanwhile, he was under the impression that Marmont was aiming at nothing more than throwing in a convoy for the relief of the fortress, and maintained the Light and 3rd Divisions in a forward position only a few miles from its gates. A further and sudden advance of the enemy he did not expect, and he was, therefore, caught in an uncomfortable position when, on September 25th, Marmont and Dorsenne sallied out, with 60,000 men, to drive in his advanced troops. He ordered, but too late, a general concentration to the rear on a position at Fuente Guinaldo, which he had selected as his fighting ground. The onset of the French drove the Light Division to a hasty and difficult retreat by a flank march; but the 3rd Division, which was nearest to the enemy, had to fight for its life, and was in serious danger of being cut up.

The 2nd Battalion 83rd Regiment had its share in the running fight, which is generally known as the combat of El Bodon. The regiment was along with the 94th and the 9th Portuguese, in the neighbourhood of Campillo, when the news came in that Montbrun, with four brigades of horse, had driven in the British cavalry screen, and had burst in among the scattered cantonments of the 3rd Division, of which the remainder of Colville's Brigade (the 2nd Battalion 5th and 77th Regiments) and Wallace's Brigade (1st Battalion 45th, 74th, and 1st Battalion 88th Regiments) were scattered between El Bodon and Pastores. The division had to unite as best it could, by a hasty concentric march, which only came to a safe end owing to the hard fighting of the 5th and 77th and Alten's cavalry brigade at El Bodon; their heroic stand enabled the flank detachments to come in, and by the afternoon Picton had his whole force united and on the march for Fuente Guinaldo.

His retreat, however, was dangerous: the column was beset by the whole of Montbrun's cavalry, which had horse artillery with it; the division had to retreat for many miles in a perfectly open country, shelled all the way, and with a charge apparently impending every moment. The 2nd Battalion 83rd Regiment, in the middle of the column, had to march under artillery fire for more than two hours, with no possibility of deploying or quickening the pace, which would have been fatal with so many French horse surging around. The battalion lost five men killed, fourteen wounded, and five left behind exhausted by the way. An eye-witness in a neighbouring corps gives an admirable description of the perilous march.*

"For six miles, across a perfect flat, without the slightest protection from any incident of ground, without artillery, and almost without cavalry, did the 3rd Division continue its march. During the whole time the French never quitted them; six guns were taking them in flank and rear, pouring in a shower of round shot, grape, and canister. It was a trying and pitiable situation for troops to be placed in, but it in no way shook their courage or confidence. General Picton conducted himself with his usual coolness. He rode on the left flank of the column, and repeatedly cautioned the different battalions to mind their quarter-distance and the 'tellings-off.' We had at last got close to the entrenched camp at Fuente Guinaldo, when Montbrun (impatient that we should escape from his grasp) ordered his troopers to bring up their right shoulders and incline towards our marching column. The movement was not exactly bringing his squadrons into line, but it was the next thing to it, and they were within half pistol-shot of us. Picton took off his hat, and, holding it over his eyes as a shade from the sun, looked sternly and anxiously at the French. The clatter of the horses and the clanking of the scabbards were so great, when the right half-squadron moved up, that many thought it the

* Grattan's "With the Connaught Rangers," pages 116, 117.

forerunner of a general charge; some mounted officer called out 'Had we not better form square?' 'No,' replied Picton, 'it is but a *ruse* to frighten us, and it *won't do.*' A moment later English cavalry reserves from Fuente Guinaldo came up and covered the marching column. Montbrun drew off, and in half an hour the 3rd Division were safe in the lines."

The losses of the 83rd in this critical hour, twenty-four in all, as has been mentioned above, were heavier than those of any other battalion in the division—even more than those of the 2nd Battalion 5th and 77th Regiments, which had been so heavily engaged in the morning. It is clear that they had more than their share of Montbrun's shells.

Though Wellington had now gathered three divisions at Fuente Guinaldo, and was joined by the Light Division next day, he was, by his own overlate orders for concentration, too weak to hold his ground when Marmont's and Dorsenne's infantry came up in full force: for his left wing, 15,000 strong, under Graham, was still a march away. Accordingly, he evacuated the Fuente Guinaldo position in the dark (September 26th) and fell back to another in front of Sabugal, between Aldea Velha and Rapoulla. Here his missing left wing joined him, and with 47,000 men, concentrated on very strong ground, he was ready to fight a defensive battle. But the French, after feeling his centre at the combat of Aldea da Ponte (September 27th), thought him too strong to be meddled with, and retired. So Marmont had achieved no more than the revictualling of Ciudad Rodrigo by his effort.

This was the last campaigning which the 83rd was destined to see in 1811. Wellington sent back the Light Division and cavalry to observe and blockade Rodrigo, but dispersed the rest of his army into winter cantonments. The 3rd Division had its headquarters at Fuente Guinaldo, but Colville's Brigade was dispersed in villages more to the right, and the 83rd was allotted that of Navas Frias, just across the Spanish frontier. In this high-lying place, on the lower slopes of the Sierra de Gata, it abode, in no great comfort, for nearly three months, badly sheltered from the snow, in cheerless houses, destitute of fire-places. Its strength dropped very low; on October 1st it showed only 26 officers, 19 serjeants, 15 drummers, and 287 privates present—the lowest figure since Talavera. This was not the result of the casualties of El Bodon, but of sickness—it had 186 men in hospital that day. During the month it received the first draft that had reached it since it left Lisbon—a small one, of fifty-five men. By the aid of this reinforcement, and by the rejoining of convalescents, the rank and file rose to 342 by December, when there were only 128 men in hospital. Major Carr went on sick leave during October, and the attenuated battalion was apparently commanded during this time of cantonments by Samuel Hext, the senior captain present. On December 22nd the Brigadier, under whom the Regiment had served so long, the Hon. Chas. Colville, was transferred to the command of the 4th Division, and Colonel Jas. Campbell, of the 94th, the senior officer with the brigade, became its interim chief.

SECOND BATTALION EIGHTY-THIRD REGIMENT.

It is unfortunate that the doings of the Regiment are no longer chronicled by Ensign William Strangeways, from whose diary so many details have been drawn during the first nine months of 1811. Invalided home in the autumn, he exchanged in 1812 into the 3rd Garrison Battalion, where he obtained a lieutenancy. His invaluable touches of actuality are not available for the oncoming sieges of Ciudad Rodrigo and Badajoz, in which the 2nd Battalion 83rd Regiment was to have so much bloody work when the New Year came round.

PART III., 1812.

1812

ON January 3rd, 1812, in a time of bitter weather, when the snow was falling and melting alternately, and the roads were in an abominable condition, Campbell's Brigade received orders to break up from its cantonments, concentrate, and march forward once more towards Ciudad Rodrigo. The order cannot have been entirely unexpected, since, on December 18th, all the regiments of the 3rd Division had been directed to set themselves to the task of making gabions and fascines for siege work, under the direction of military artificers: and an engineer officer had been sent to the headquarters of the division with cash, from which he was to pay two vintems (2½d.) for every fascine, and four vintems (5d.) for every gabion, which he passed as properly made.* This could only mean that a decisive move was in contemplation.

The change in the general aspect of affairs in Spain, which led to Wellington's mid-winter enterprise, was directly due to the action of the Emperor Napoleon. He had sent orders that Marmont was to detach three divisions of the Army of Portugal to the eastern side of the Peninsula, in order that they might aid Marshal Suchet in his attack on Valencia. Montbrun had marched at the head of these 15,000 men towards the coast of the Mediterranean, in obedience to an Imperial despatch received on December 11th: and in consequence of this weakening of the garrison of the province of New Castile, the remaining divisions of Marmont's army had to disperse themselves over a larger extent of territory, and in many cases to remove themselves further from Wellington's neighbourhood.

These movements were promptly reported to the British Commander-in-Chief by his Spanish correspondents, who secretly sent him intelligence from every quarter, and by January 1st he was fully informed of the fact that a large fraction of the Army of Portugal had gone off on a distant expedition. This was the opportunity for which he had long been waiting: when Marmont should be too weak to face him at short notice, he had always been intending

* Orders, Freneda, December 18th. Printed in Jones's "Sieges of the Peninsula," I., p. 99.

to strike at Rodrigo, and now the Emperor's orders had put the Marshal in such a condition that he could only assemble 30,000 men at the end of twenty days, even if he moved promptly. From such a force Wellington need fear nothing, till it was succoured by the Army of the North or by Soult from Andalusia—and these succours would take much time to collect.

His own preparations had long been made—the siege train was parked at Almeida, only twenty miles from Rodrigo, the gabions and fascines and platforms had long been ready and only needed to be moved up. But the venture would be a time-problem: the fortress must be taken in three weeks, or there would be danger from a relieving army. Hence all the movements had to be rapid and decisive. Four divisions were designated for the siege—the 1st, 3rd, 4th, and Light—together with Pack's Portuguese Brigade.

On January 4th the 83rd marched from its miserable winter quarters at Navas Frias, forded the Agueda, and on the 6th was in front of Ciudad Rodrigo. The brigade headquarters were at Ceradilla de Arroyo, a village a few miles from the suburbs of the fortress. The march was conducted in most inclement weather; on the 4th sleet was falling almost all day, and the 5th was hardly better. On the 8th the actual siege operations began with the storming by the Light Division of an outlying French work (Redoute Rénaud) on the hill called The Teson, from which Wellington intended to batter the town: and on the same night the trenches were opened and the battery-emplacements commenced. The siege operations were conducted by the different divisions on successive days, each relieving the other at intervals of twenty-four hours. The turns of the 3rd Division came on the 11th, 15th, and 19th. When not on duty, the battalions had to bivouac in the open, the camps being far away from the siege lines. There were no tents or huts of any description, and the ground was covered with snow, the only protection against the weather being the building of great fires. On working days the trenches were full of slush from the melting snow, in which the men had to labour, sometimes ankle-deep, sometimes almost knee-deep, under a furious fire of artillery from the fortress. Nevertheless, the advance went on rapidly; on the 14th twenty-two siege guns were in battery, from the new train, which was infinitely superior to that used at Badajoz in the preceding year. On the 15th a second parallel, only 150 yards from the walls, had been established, and a breach was beginning to be visible in the northern front; a second and smaller one to its left was established two days later. By the 19th they were both practicable, though the artillery fire of the defence was still unsubdued, and a furious response to the battering guns was kept up. Up to this time the 83rd had suffered in its trench work the moderate loss of three men killed and two officers (Lieutenant Henry Vereker and Ensign Joseph Mathews), and eight privates wounded.

On the night of the 19th, when Wellington determined to storm Rodrigo, the 83rd, with the exception of the light company, was set to guard the part of the second parallel from which the attack of the 3rd Division on the great breach was to be delivered, and to keep up an unintermittent fire upon the ramparts, under cover of which the storming party was to advance. The light company was detached to join in a small subsidiary attack by escalade, which was to be delivered by the 2nd Portuguese Caçadores, on an outwork in front of the Castle, which, being far from the breaches, Wellington suspected to be neglected by the defenders.

Since the 83rd did not form part of any of the storming columns, its losses were smaller than those of the other 3rd Division battalions, which took part in the actual assault. They only amounted to one rank and file killed and four wounded. The light company, under Captain the Hon. H. Powys, headed the Portuguese column, which operated below the Castle, and there distinguished itself, the work being escaladed in quick time. Picton specially thanked the light company for its services in divisional orders.

Thus Rodrigo fell, within eleven days from the night on which it was invested, and Wellington's time-problem had been most successfully solved. His adversary, Marmont, had only heard of the commencement of the siege on January 15th, though he was no further off than Valladolid. He at once ordered his army to concentrate; but Montbrun's three divisions were too far off—on January 16th he was before the gates of Alicante, on the Mediterranean coast. The rest of the Army of Portugal was gathered at Salamanca by the 25th, and some succours had been borrowed from Dorsenne—but meanwhile Ciudad Rodrigo had fallen six days before. Extremely disgusted at the evil news, the Marshal resolved that no more could be done for the present, and dispersed his troops once more into winter cantonments. A few days later Montbrun came back from the east, with his troops absolutely exhausted by a forced march; but his return did not inspire Marmont with any desire to undertake offensive operations. He gave his army a rest, thinking that Wellington would do the same.

This was far from the intention of the British general. The moment that the breaches of Rodrigo were repaired, and a Spanish garrison thrown into the place, he began to move his army very unostentatiously towards the south. His aim was to make, at Badajoz, the other great French advanced fortress, the same sort of swift, sharp stroke that had been so successful at Rodrigo. It would take Soult three weeks to collect a relieving army, and Marmont could not concentrate and come down to Soult's help in a less time. Part of the battering train was sent forward over the mountains of Beira early in February, and the infantry waited till it should be well on its way before starting.

The 3rd Division, after the fall of Rodrigo, was dispersed in scattered cantonments to the south of that place. The village of Martiago fell to the 2nd Battalion 83rd Regiment, who remained there from the end of January to the middle of February, when they were moved to Villar Mayor, a little more to the west and on the Portuguese side of the frontier, where Campbell's Brigade was concentrated in preparation for the approaching march to Estremadura. The battalion had now got back its Commanding Officer, Major Carr, who came back from sick leave just in time to see the fall of Ciudad Rodrigo, and to obtain a bar to his gold medal given for Fuentes de Oñoro. The battalion also received a draft of 115 men from home, the first really large reinforcement that it had obtained for two years—but second battalions were always badly served in this way. It therefore marched off on February 26th stronger than it had been at any time since it last left Lisbon, with 24 officers, 28 serjeants, 15 drummers, and 395 rank and file, even though it had 148 sick at the time. This total of 462 of all ranks, was the highest that it was to attain for many a day, for very heavy losses were about to be incurred in the spring and summer of 1812.

1812

The route taken by the battalion was the familiar one already twice traversed in 1811. It moved by Sabugal, Castello Branco, the boat-bridge of Villa Velha upon the Tagus, and Niza, till it reached Portalegre on March 5th. There it had three days of rest, but continued its march to Elvas on the 9th, and on the 16th crossed the Guadiana, and, joining with the rest of the division, took its place opposite the same Castle-front of Badajoz, where it had spent a fruitless fortnight in the May and June of the preceding year. The 4th and Light Divisions continued the line to the left, and enclosed the other sides of the fortress.

On the seventeenth of March the third and most famous British siege of Badajoz began, the French outposts being driven into the town, and a close reconnaissance of its whole circuit being made by Wellington himself and his chief engineer, Colonel Fletcher. The points of attack selected on this occasion were entirely different from those of 1811; the fort of San Cristobal and the north side of the place were entirely ignored—only a Portuguese brigade was set to watch them. Nor was any serious attempt made to attack the Castle—the front of attack was to the left of that very strong post, on the south-east section of the circuit. Wellington's intention was to begin by capturing the Picurina Fort, which lies on a low but commanding height three hundred yards outside the walls, and then to play upon the bastions behind it with batteries placed on a line of which the Picurina was to form the salient part, while the rest extended down hill into the flat ground facing the Castle. The 3rd Division started its work on the night of the 17th-18th, by throwing up a long line of trench on the slopes of the Cerro de San Miguel, less than 200 yards from the Picurina Fort. The night was dark and stormy; the enemy discovered nothing of what was going on, and at dawn the first parallel was three feet deep and many hundred yards long.

Daylight showed the French what had been done, and they commenced a furious fire against the trench, which cost the workers some casualties, on the 18th—but only one man from the 83rd was wounded. On the next day, however, losses were much heavier; appalled at the near approach of the British to his fort, the French Governor, Philippon, sent out a sortie of 1,800 men, who formed unobserved behind an outlying lunette, and rushed at the works headlong, almost before any movement had been spied. The working parties were driven out of the parallel, and their supports with them, but before they had fallen back fifty yards, they were rallied by their officers, and charged in to the lost works, dislodged their opponents ere they had done much harm, and chased them back towards the town with heavy loss. In this affray the 2nd Battalion 83rd Regiment lost one private killed, and two serjeants, one drummer and fourteen rank and file wounded.

From the 20th to the 26th of March the advance towards the Picurina Fort continued; for the first five days the weather was most inclement, and the work very toilsome, for the rain filled the trenches, and sometimes accumulated in such pools that a part of the line could not be occupied. The earth excavated became a mere slime, which would not pile up into parapets, and ran away in liquid streams out of the gabions into which it was cast. Only by lavish use of sand bags could anything above the ground level be built up. Nevertheless, with incessant toil, the batteries were completed, and on the 25th opened upon the Picurina Fort with some effect. The 83rd lost a few men every day in this muddy and laborious work—one on the 20th, two on the 21st, five on the 22nd, one on the 23rd, two on the 24th. On the 25th, though the fort seemed not irreparably damaged, Wellington ordered that it should be stormed, by 500 volunteers from the 3rd Division, under Brigadier-General Kempt (the successor of Brigadier Mackinnon, killed at Rodrigo). The 2nd Battalion 83rd Regiment supplied two officers and fifty men for the assault.

The details, as settled by Kempt, were that two flanking columns of 200 men each should run in upon the sides of the Picurina, and endeavour to enter in at the gorge, where its defences were supposed to be weakest. The third party of 100 men forming the centre column, was commanded by Captain Powys, of the 2nd Battalion 83rd Regiment light company, and contained the volunteers from his battalion. It was directed to charge at the salient angle of the fort, the only part of it that seemed to have been seriously damaged by the British batteries, and to endeavour to escalade it the moment that the two side attacks should have come into effect and distracted the enemy's attention.

The storm proved a desperate and costly business. The left-hand column was completely checked, with heavy loss. The right-hand one, composed of men from the 74th and 88th, finding the gorge too heavily defended by palisades to be practicable, swerved round to the flank of the fort, and strove to get in by the expedient of casting their ladders across the narrow ditch on to the *fraises* (projecting beams) of the parapet. Some men crossed, but they were shot down, and no lodgment had been made, when Kempt let loose the central column. Its charge proved decisive; the ditch of the salient was partly filled by débris, the ladders were planted in it, and, scrambling up by twos and threes, the men made good their footing on the summit. Captain Powys was the first man in; he was shot on the parapet, and would have been bayonetted if his serjeant, Thomas Hazlehurst, had not stood over him warding off thrusts with his pike. But the volunteers fought their way in past him, and entered the body of the fort, just as the right column succeeded in breaking through on their flank. Once within, the stormers made short work of the garrison, who were nearly all slain or taken. The 83rd lost Captain Powys mortally hurt, three men killed, and Ensign Hackett* and six men wounded. These casualties—eleven out of fifty-two officers and men engaged—were sufficiently heavy; but the volunteers in the flanking columns suffered in even greater proportion, for, out of the total of 22 officers and 500 men, who took part in this desperate storm, no less than 19 officers and 300 men were killed or hurt.

The Picurina and the knoll which it crowned being in his possession, Wellington was now able to proceed to the next stage of his projected attack on the city. New batteries were thrown up on the conquered ground, to play on the Trinidad and Santa Maria bastions, the front of the place which the Picurina commanded. At the same time, an attempt was made to sap forward from the second parallel toward the San Roque lunette, which covered the south-eastern gate of the place. This second part of the siege lasted from the 27th of March to the 6th of April, and was as toilsome and costly a business as the preliminary operations, for although the weather had much improved, work close under the guns of the fortress was even more deadly than at a distance. The 83rd lost in the trenches two killed and one wounded on March 27th, one wounded on the 28th, and so on till the actual night of the storm—that of the 6th-7th April.

By that day three large breaches had been made in the two bastions attacked and the curtain between them, and the fire of the besieged had been to a certain extent subdued. But the advanced trenches were still far from the walls, and the counterscarp was still intact, though the ditch was partially filled by the débris that had fallen into it. The Light and 4th Divisions alone were directed upon the breaches, a separate task having been set aside for the 3rd Division, at Picton's special request. This was no less a venture than the storming by escalade of the Castle, whose lofty walls were perfectly intact, since no attempt had been made to batter them. This old Moorish stronghold formed the highest corner of the city, and lay on the summit of a precipitous hill, at whose foot wound the sluggish but swollen Rivillas brook. To ascend the slope even in daylight is no easy matter, to drag heavy ladders up it in the dark looked well-nigh impossible; but Picton argued, and argued rightly, that for these very reasons the defence of such an apparently

* Ensign Isaac Hackett died of his wounds, like Powys.

impregnable section of the defences would probably be neglected to some extent by the French, who would concentrate all their attention on the three yawning breaches on the south front. For similar reasons Wellington resolved, at the last moment, that he would order another attempt at escalade to be made, on a separate point of the walls very remote from the breaches. Walker's Brigade of the 5th Division was told off to assail the bastion of San Vincente, near the river, at the north-west corner of the city.

It was intended that the three assaults should be simultaneous at 10 p.m., but the chances of war caused them to be delivered with some interval of time between. The Third Division first came under fire. It had been drawn up out of sight of the fortress in deep column, the stormers and ladder party at the head, then Kempt's Brigade (1st Battalion 45th, 74th and 1st Battalion 88th Regiments), next Champlemond's Portuguese Brigade (9th and 21st Line), and last Campbell's Brigade (1st Battalion 5th, 77th, 2nd Battalion 83rd, and 94th Regiments); but the place in the column that any regiment occupied was to make little difference in the danger that it incurred, or the losses that it suffered—every corps was fully engaged that night. The head of the column had moved up silently to the right end of the trench forming the first parallel, when it was discovered by the gunners in the Castle, who sent a salvo of shells in its direction. Seeing that he was discovered, Picton judged it of no use to wait till the appointed time for the assault, and a quarter of an hour before it was due (9.45 p.m.) gave his orders for the advance. The column sprang over the parapet of the trench, and crossed at a rapid pace the ground between the works and the foot of the Castle hill. The Rivillas brook was passed without much trouble, some men wading it, others using a broken mill-dam and bridge which spanned the stream. The palisade at the foot of the hill was torn down by many willing hands, and the stormers started up the slope, dragging their heavy ladders with them. The fire by this time was tremendous; the artillery of the castle was intact, and the column was now in close musketry range of the defenders. Soon, however, the leading brigade reached the narrow space at the head of the slope below the walls, and then began to plant the ladders as best it could. The first rush was fruitless—the defenders kept up a hot fire, rained live shells and heavy stones on the men massed around the ladders, and of the many who tried to mount few saw the parapet, and those few were shot ere they could set foot upon it. Picton had been disabled early; a spent shot struck him in the groin before he had mounted the slope, and kept him out of the fight for some time. Kempt, his successor in command, was also wounded not long after. But the efforts did not fail for a moment, and as more men straggled up the hill the front on which the escalade was being attempted extended every moment to right and left. Champlemond's Portuguese came up to the front in due course, and made a creditable effort to succeed where Kempt's Brigade had failed, but all to no effect; the casualty list grew, the time wore on, and no

1812

lodgment had been made upon the ramparts. At last, the rear brigade (Campbell's), which included the 2nd Battalion 83rd Regiment, mounted in its turn, and flung itself among the crowd. This third attempt was as fierce as those which had preceded it, and more successful. The defenders had used up most of their irregular missiles, and were reduced to their musketry alone; their numbers were somewhat thinned, and their strength was being exhausted in the attempt to keep down 4,000 desperate men, whose successive assaults never flagged for a moment. They got no reinforcements, for the fight at the breaches below, where the 4th and Light Divisions were hard at work, though prospering little, absorbed all the attention of the Governor Philippon. There, in his idea, lay the danger: if an attack was being made on the impregnable Castle, it was some mere demonstration, or at least bound to fail before the formidable and uninjured defences.

He was wrong: the 3rd Division achieved the impossible. The assault was raging all along the Castle wall, when at last at one point an entry was made. It is said that the first man to make good his footing on the ramparts was Major Ridge, of the 5th. Rearing again one of the oft-overthrown and blood-stained ladders against an embrasure which lacked a gun, he leaped into it, and did not meet his death, as so many had before him.* He was closely followed by an officer (Ensign Canch) and a few men of his own battalion, and cleared a small point of access for more. The sight of a foothold won shook the constancy of the defenders, and encouraged the stormers below. In an instant several other ladders were planted to right and left; up some of them went the men of the 2nd Battalion 83rd Regiment, who were keeping together and following their officers closely. It is said that Major Carr, Captain Hext, Lieutenants Broomfield and Swinburne were all among the earliest to mount the ramparts; the third-named was severely wounded. When once the assailants were pouring over the walls at several points, the chance of the defenders was over. It is a different thing to contend with adversaries straggling one by one up a ladder, and against men who have reached equal fighting ground and can use their superior numbers. The garrison of the Castle were soon overwhelmed, and for the most part bayonetted; few prisoners were made.

A rush was then made to throw open the Castle gates, in order that the division might descend into the town, and take the defenders of the breaches in the rear. But exit proved for some time impossible, for Philippon (who had intended to use the Castle as his last place of refuge) had walled up all the exits a few days before, save a postern, which proved hard to find. When at last it was discovered, and the men began to issue forth, they found the narrow way blocked by the last French reserve, a few companies whom Philippon had sent up to the Castle when the incredible news of its fall

* The 88th claim that Ensign McAlpin was up before Ridge, but was killed immediately. The general balance of evidence, however, seems in favour of Major Ridge.

STORMING OF BADAJOZ BY THE 83RD REGIMENT. 1812

From the drawing on the wall of the Patio at Government House, Gibraltar, by Capt. Marshman, 28th Regiment, 1870.

reached him. In the scuffle to break out, Major Ridge, the first man into the Castle, was killed—a lamentable loss.

Picton, who had come to himself after his wound, and had been carried up to the foot of the walls before the assault was over, sent up orders that the brigades were to re-form themselves before sallying forth, when Champlemond's Portuguese were to clear the ramparts to the right, the British brigades those to the left and towards the breaches. But presently directions were received from Lord Wellington that the division should keep itself close in the Castle, whose capture made the surrender of Badajoz a certainty, and not move out till dawn.

Meanwhile, however, the defence had fallen to pieces from another cause. Not only was the occupation of the Castle decisive, but another and equally successful blow had been given in another direction. Though the Light and 4th Divisions had completely failed at the breaches, with terrible loss, Walker's Brigade, of the 5th Division, had been triumphantly successful in the escalade of the San Vincente bastion. After a struggle on the ramparts, they poured into the streets of the city and arrived in rear of the defenders of the breaches, who thereupon were forced to throw down their arms and surrender. Thus, by a strange chance of war, the regular and formal main assault failed—but both the subsidiary attacks were completely successful.

The losses of the 2nd Battalion 83rd Regiment in the assault were very heavy—Captain Fry, one serjeant and twenty-one rank and file killed; Lieutenants Bowles, O'Neill, Barry and Broomfield, Ensigns FitzGibbon, Lane and Vavasour, three serjeants and thirty-six rank and file wounded—a total of sixty-nine casualties out of about four hundred of all ranks engaged. The total loss during the siege, including the trench work and the storm of the Picurina was three officers and thirty-seven men killed, and seven officers and eighty-one men wounded—a total of 128 out of 462 of all ranks shown as present on March 8th, in the last return made before the leaguer began. Thus the battalion lost nearly thirty per cent. of its strength in this bloody affair. When it marched away from Badajoz it mustered only seventeen officers and 274 men, leaving 244 sick and wounded in hospital.

Major Carr was mentioned in Wellington's despatch, and received another clasp for the commanding officer's gold medal already granted to him for Fuentes de Oñoro. Captain Hext was given a brevet-majority, and the name of Badajoz was placed on the regimental colours, to form a worthy pendant to that of Talavera.

Into the horrors of pillage, arson, and drunkenness, which filled the day after the storm of Badajoz, when the excited and exhausted men poured into the streets, and started an evil morning's work by draining the immense store of French brandy in the great depôt near the Cathedral, it is unnecessary to proceed. The tale is not one on which any historian of the British Army wishes to linger.

On April 8th it seemed not unlikely that Wellington might have to fight a battle to defend what he had just won; for Soult, hastening up from Andalusia with such forces as he could collect, was on his way to relieve Badajoz, and had reached Villa Franca, only thirty miles away, when he heard the news of its fall. Wellington hoped to fall upon him, but the Marshal, on learning that he was too late, retreated by forced marches, and was wise to do so, for his adversary, bringing up the siege troops to aid the covering army under Hill, might have attacked him with a great advantage of numbers.

The British army was not led in pursuit of Soult, but turned in another direction, to look for Marmont. That commander, on hearing of the investment of Badajoz, had not marched to join Soult, as in June, 1811, but thought to draw off Wellington by thrusting a raid deep into Central Portugal. On March 27th he had started from Salamanca with five divisions, and leaving Ciudad Rodrigo and Almeida alone, pushed straight before him, devastating the countryside as far as Guarda and Castello Branco. On April 13th he dispersed a great body of Portuguese militia near the former place, and spread terror all over the Beira. But already he was too late—Badajoz had fallen a week before, and Wellington (since Soult had flown) had his hands free. He determined to go in pursuit of the Marshal, having some hope of cutting off his retreat on Spain, since he had gone so deeply into Portuguese territory. On April 11th the British army began to move northward, by the old familiar route by Portalegre, Niza and the bridge of Villa Velha; but it seems that the 3rd Division, having suffered so heavily at Badajoz, was allowed a day or two extra of rest. It only left Badajoz on the 15th, but, marching by long stages, reached Alpedrinha, near Castello Branco, on the 20th, and Bemquerencia (south of Sabugal) on the 22nd. But on this day it was ascertained that Marmont, warned that a great force was on the march against him, had abandoned his raid and was in full retreat towards Salamanca. The pursuit was, therefore, given up, and, as Wellington did not intend to make his next move till he had revictualled Almeida and Ciudad Rodrigo, and had broken the easiest line of communication between Soult and Marmont, by destroying the bridge of Almaraz, the divisions in the Beira were sent into cantonments for a month. Those of the 2nd Battalion 83rd Regiment were at Aldea da Ponte, near Alfayates, where they abode for some weeks, and received a small draft of thirty-six men from England—a contribution quite acceptable, but altogether insufficient to fill up the gaps left in the ranks by the siege of Badajoz. On June 1st the battalion was at Sedavim, with a strength of 16 officers, 20 serjeants, 13 musicians, and 272 rank and file, with which modest force it marched out for the Salamanca campaign. There were still 272 sick and wounded in various base hospitals—mostly at Elvas.

1812

Having re-victualled Almeida and Rodrigo, and received the welcome news that Hill had broken the bridge of Almaraz and ruined the forts that protected it, Wellington concentrated his army and crossed the Agueda on June 13th for a blow at Marmont, whose divisions (as he was well aware) were widely scattered, and could not unite for many days in strength sufficient to allow him to give battle. Picton, thinking that his wound was sufficiently healed, rode for some days at the head of the 3rd Division; but he had miscalculated his strength, and was invalided again ere long. To replace him, Wellington nominated his brother-in-law, Major-General Edward Pakenham, who was to command Picton's veterans during the short and glorious campaign that followed.

The advance of the allies had not been altogether unforeseen by Marmont, but he was unable to concentrate his army before it was actually begun, on account of commissariat difficulties. Hence Wellington was able to drive in the French outlying troops with ease, and on June 17th he laid siege to the forts of Salamanca, where 800 men had been left in garrison to protect the important depôts which had been formed in that city. While the 5th and 6th Divisions conducted the leaguer (which lasted from the 18th to the 27th), the rest of the army, including the 3rd Division, was drawn up on a low line of heights named from the village of San Cristobal, to hold off any attempt of Marmont to bring succour. The Marshal appeared on the 21st, but with only 25,000 men, for his army was not fully collected, and he was obliged to be very cautious, lest he should be overwhelmed. On the next day more French divisions came up, but he still found himself inferior in numbers, and, having looked at Wellington's strong position, refused to attack it. After making a faint-hearted attempt to turn the allied right on the 24th, he drew back, to wait for further reinforcements, and had remained halted at a considerable distance from the San Cristobal heights for three days, when, on the 27th, the Salamanca forts were set on fire and stormed. Thus Wellington had his hands free, and no siege to cover; wherefore, lest the allied army should take the offensive before his expected reinforcements came up, Marmont retreated behind the Douro (June 28th). There was no serious fighting during the week, when the armies were in presence, and what little there was in the way of skirmishing and affairs of outposts, did not come in the direction of the 3rd Division.

Pursuing the retreating French, Wellington reached the line of the Douro on July 1st, and halted there, being as reluctant to attack Marmont in position behind an unfordable river with high banks, where there was no cover to conceal his approach, as the Marshal had been to assail the heights of San Cristobal a week before. The 3rd Division was quartered in front of the ford of Pollos, towards the allied left, from the 2nd to the 8th of July, waiting for orders to attempt a passage, which never came. On the last-named day the 2nd Battalion 83rd Regiment, which had picked up a few convalescents from the rear, showed twenty-two officers and 324 men present under arms.

The curious dead-lock along the Douro, which lasted from the 2nd to the 15th of July, was at last ended by Marmont's taking the offensive. He had by now gathered in the last outlying fractions of his own army, and was 7,000 or 8,000 men stronger than he had been on the first of the month. Moreover, he had ascertained that his colleagues would give him little help, for a vigorous naval diversion, under Sir Home Popham, had fallen on the coast of Cantabria and Biscay, and was drawing off the attention of the French " Army of the North," while the " Army of the Centre," round Madrid, was slow to assemble, and still slower to move. Wherefore the Marshal, on July 15th, concentrated his whole army by a night march, crossed the Douro near Tordesillas, and came rushing down in full force on Wellington's right wing.

The English general had to concentrate in order to meet him, and there followed six days of delicate and dangerous manœuvring, while Marmont continually turned the allied right by rapid marching, and Wellington, gradually getting his columns closer together, drew slowly back, offering to fight more than once on the defensive, but refusing to attack. He was thus gradually manœuvred out of thirty miles of ground, and had drawn back close to Salamanca, when Marmont at last made the mistake for which his adversary had for six days been waiting. On the morning of July 22nd the two armies, still moving parallel with each other, were south of the Tormes, near the two curious isolated hills, called the Arapiles, from which the French and Spaniards have christened the battle that followed.

Wellington's army was on this day disposed behind a low ridge of heights, with its back to Salamanca, and its centre protected by the northern Arapile. The 3rd Division, which formed its rearguard on the previous day, had only just crossed the Tormes River, and was in reserve behind the extreme right of the allied line, near the village of Aldea Tejada. Marmont, thinking that only another day of manœuvres was before him, and not fully making out his adversary's position, ordered the flank march, which had hitherto been so successful, to be continued. He intended to force Wellington to abandon Salamanca, by threatening to cut off his retreat on Ciudad Rodrigo. While making this move, his leading columns outmarched his centre, and his rear columns had not even reached the field. Seeing the dispersed state of the French, who were now marching past his position in apparent security and fearing nothing, Wellington suddenly struck at them.

The decisive blow was given by the 3rd Division, which, from its place in reserve, was rapidly transferred across the head of the French marching column, under cover of a convenient wood. When it should have intercepted Marmont's van and brought it to action, four other divisions, which formed the British right and centre, were to descend from the heights, behind which they were hidden, and fall upon the enemy in front of them. The manœuvre was carried out with perfect accuracy and success. Pakenham marched unseen, with his division in three parallel columns of brigades, Wallace's (1st

Battalion 45th, 74th, and 1st Battalion 88th Regiments) to the left, Champlemond's Portuguese in the centre, Campbell's (2nd Battalions 5th and 83rd, and 94th Regiment*) to the right. He was covered by the 14th Light Dragoons and D'Urban's Portuguese Horse. In this array he emerged from the woods, and found himself at the foot of a slight hill, which Thomiéres' Division, the leading French column, was commencing to descend. The manœuvre was quite unexpected by the enemy, who, in great disarray, tried to form line of battle, but was attacked before he was ready. Pakenham had only to order his three brigades to front to their left flank, which they did by each company throwing forward its right shoulder without stopping their march, and they were in order, in three successive lines, overlapping the narrow head of the long French column.

The attack uphill was completely successful, the leading battalions of the French being swept away by Wallace's Brigade, while the British and Portuguese Light Dragoons rode in upon their flank and rear. Thus commenced a most triumphant advance, which went on for nearly two hours. For Thomiéres' Division was forced back against Maucune's, the next in line, and this was already frontally engaged with Leith's Division from the British right centre, so that when taken in flank by Pakenham, it, too, crumpled up and gave way. The movement was continued all along the French line, each brigade in succession being outflanked by the 3rd Division and hurled against its next neighbour. The rout was made still more complete by a charge of Le Marchant's Heavy Dragoons, who, striking in diagonally between Leith's and Pakenham's troops, cut up many of the retreating battalions.

At last the whole French left had been broken up and driven in upon the centre, which had now been joined by the rear divisions of Marmont's army, which had only just come upon the field. The Marshal had been wounded early in the fray, but his successor, Clausel, made a desperate fight to secure at least an orderly retreat. But he was foiled, partly by a gallant frontal attack by the 6th Division, which Wellington brought up from his second line, but still more by the continued flank movement of Pakenham and the 3rd Division; for that gallant leader, always persisting in his original tactics of outflanking the left of each French force opposed to him, now deployed all his three brigades, and with this long line turned and drove in each new force that withstood him.

1812

Campbell's Brigade, including the 2nd Battalion 83rd Regiment, was now in front line, not in reserve as at the commencement of the action, and took part in the breaking down of the last French stand, made by Ferey's Division, from the reserve. In this final phase of the action, the battalion suffered considerably, losing two privates killed and Lieutenants

* The 77th, long associated with the 2nd Battalion 83rd Regiment, had suffered so heavily at Badajoz that they were sent back to Lisbon.

Gascoyne and Evans and thirty men wounded*, but its work was triumphantly completed, and the bulk of the French army, driven off its proper line of retreat and dispersed among woods, ought to have been captured or cut up. But Wellington had counted upon their route being blocked by a Spanish force holding the bridge and fords of Alba de Tormes, and this force had been withdrawn, without his knowledge or orders; whereupon the Army of Portugal got away, much mauled indeed, but not annihilated. It had lost 6,000 prisoners, some 7,000 killed and wounded, and so many stragglers, that a fortnight after the battle Clausel had little more than 23,000 men in hand, out of the 45,000 who had fought at Salamanca.

Wellington, finding that Clausel had got off, marched in pursuit of him as far as Valladolid, where he captured great French depôts, and, when the enemy continued to retreat northward, left a large detachment to observe him, but marched, with the rest of his army, on Madrid, from which he was resolved to drive out King Joseph. The 3rd Division accompanied the main column, which covered the distance from Cuellar to Madrid in seven days, between the 5th and the 12th of August. This march across the Guadarrama Passes was very trying, and the 2nd Battalion 83rd Regiment, already much exhausted by the rapid marches before the battle of Salamanca, sank very low in numbers at this time. Its commander, Lieutenant-Colonel Carr, and so many more were left behind in hospital at Salamanca or Valladolid, that it entered Madrid little over 200 strong—16 officers and 218 men only—having the dreadful number of 348 sick or wounded. Brevet-Major Hext was in temporary command.

At Madrid the battalion rested for more than two months—August 12th to October 30th. Its second day of sojourn in the Spanish capital was the most lively one. King Joseph had unwisely left a large garrison of 2,000 men in the Retiro Fort, just outside the city. Wellington ordered it to be attacked and stormed on the 14th. Campbell's Brigade was one of the units employed, and the 2nd Battalion 83rd lost two men wounded in capturing its outer *enceinte*. When this had fallen, the Governor, who had a short water supply and a discontented garrison, very tamely surrendered. Nearly 200 guns and vast stores of clothing and food became the conqueror's prize.

After a stay of a fortnight in Madrid, Wellington took off the 1st, 5th, and 7th Divisions to march against Clausel, in company with the troops which he had already left in Old Castile; but the 3rd Division was left in garrison at Madrid, probably in order that it might have time to recover from the depletion that its ranks had suffered at Badajoz and Salamanca. The Light Division also stayed behind. Their long stay in the Spanish capital would have been more pleasant if the troops had been better clothed and paid. They

* The battalion's old commander, Colonel Collins, was killed this day, at the head of the Portuguese brigade, of which he had now been long in charge. Though he had lost a leg at Albuera, he was back to the front again in 1812.

were in absolute rags from the toils of a campaign that had lasted since January, and their pay was six months in arrears—hardly a dollar was in any officer's pocket. Nevertheless there was a good deal of gaiety abroad. The Madrileños were extremely friendly, and hospitable, so far as the poverty of the times permitted. There were several great bull-fights, nominally in honour of the British army, and a good deal of dancing and theatre-going. An attempt to amuse the citizens in return, by amateur acting by a troupe of British officers, seems to have had no great success. The bombast of *Zanga* and the somewhat heavy humour of the *Mayor of Garratt* are said to have left the Castilians unimpressed. A good many 3rd Division spectators seem to have viewed a sight of a very different sort, of which two of them have left long descriptions.* This was the public execution, in the *Plaza Mayor*, of the celebrated priest, Diego Lopez, the trusted agent and spy of King Joseph Bonaparte. Captured with secret despatches hidden about him, he was condemned to die by the *garotte*. Hanging was a familiar sight enough to the Peninsular army, but the prompt neck-twisting efficiency of the Spanish machine seems to have impressed even those most accustomed to the sight of every form of sudden and horrible death.

The not unpleasant garrison duty in Madrid was brought to a sudden end as October reached its close. Wellington had failed in his ill-starred siege of Burgos, and while he lay before its castle the French had been gathering in from all quarters. The armies of the North and Portugal, concentrated not far from Burgos, outnumbered the main body of the Allies. On the other side, Soult, who had evacuated the provinces of Andalusia, and moved up to join King Joseph at Valencia, was also far stronger than the British force at Madrid, even though Hill had brought his corps from Estremadura to the neighbourhood of the Spanish capital. Overmatched on both the theatres of war, Wellington raised the leaguer of Burgos, and began to fall back toward Valladolid and Salamanca. At the same time he sent orders to Hill to pick up the two divisions at Madrid, and make a concentric retreat over the Guadarrama mountains to join him. The 3rd Division evacuated Madrid on the 30th of October, to the great grief of its people, many of whom followed them, weeping, for two miles from the walls, before saying adieu. Some, too much compromised by their acts during the British occupation, preferred to follow Hill rather than wait for the probable vengeance of the French.

The retreat of the 3rd Division over the mountains was uneventful—they did not form the rear of the marching column, and never saw Soult's advance cavalry. But the road by the Escurial, Espinal, Peñaranda and Alba de Tormes was bad and rough, and the weather had turned to rain. Though privations had not yet begun, the men were much tried, and stragglers commenced to drop to the rear. On the 7th of November Hill's column joined

* Grattan, of the 88th, and Donaldson, of the 94th.

the main army near Salamanca, while Soult and King Joseph had also completed their junction with the Armies of Portugal and the North. The full strength of both parties was now assembled on the same ground on which Wellington had won his great victory four months back. But the forces were no longer practically equal, as they had been in July; Wellington had about 70,000 men, including 12,000 Spaniards of the raw Galician and Estremaduran armies. Soult and King Joseph had nearly 100,000 men concentrated, from the four armies of Portugal, the Centre, the North, and Andalusia. Wellington could only dare to fight if the enemy made a frontal attack upon him in his chosen position, and gave him the advantage of the ground. He was for some time in hopes that this might be their intention, and kept his army assembled on the heights north and south-east of Salamanca, with a rearguard holding the line of the Tormes at Alba, and the main body concentrated behind it. The Third Division formed part of the right wing under Hill, and was encamped near the village of Calvariza Abaxo, without any shelter, the only comfort of the bivouac being large fires, which it was hard to keep up, for the rain was almost incessant, and wood was hard to procure on the downs, save by a long journey to the forest in front. Provisions, however, were still forthcoming in adequate quantities, and the men bore up well, in expectation of a battle near at hand.

Everything was changed on November 15th by the depressing news that the enemy, refusing to attack in front, was turning the allied position by a wide flank movement; he had brought his whole force south of the Tormes, and had moved it across that river by the fords above Alba. Presently he evicted Hill's rearguard from that town, and kept spreading more and more to the west, threatening the road to Ciudad Rodrigo. This march repeated Marmont's manœuvre of July 22nd, but it was made at a much greater distance from Wellington's front, and by a force double as great as that which Marmont had commanded. Since Wellington dare not attack such superior numbers, he had to move back towards his base on the Portuguese frontier.

Meanwhile the retreat could not have been made under more unhappy conditions. "The rain fell in torrents, almost without intermission. The roads could no longer so be called, they were become perfect quagmires; the small streams were rivers, and the rivers were scarcely fordable at any point. The men were obliged to carry their ammunition boxes strapped on their shoulders, to preserve them, while passing fords which in July were but ankle-deep."* "I never saw the troops in such bad humour,"† writes another 3rd Division diarist. The most trying thing of all was that the baggage and commissariat trains had been started off early, lest they should straggle and fall into the hands of the French. The troops never caught them up, for while retreating and making an occasional halt to beat off the pursuing enemy, they moved so slowly that the waggons got two days ahead of them. After the

* Grattan, of the 88th, p. 291. † Donaldson, of the 94th, p. 179.

16th no rations were issued, and the men had to live on raw beef, from the divisional herd of oxen, and on what they could glean from the countryside, which was mainly edible acorns from the great oak forests between Salamanca and Rodrigo. Many soldiers straggled after food, which they sought 1812 in villages far off the road; hundreds sank from fatigue and privation, and fell into the hands of the French cavalry. "It was piteous to see some who had dragged their limbs after them with determined spirit, until their strength failed, fall down among the mud, unable to proceed further; and as they were sure of being taken prisoners if they escaped being trampled down by the enemy's dragoons, the despairing farewell looks that the poor fellows gave us, when they saw the battalion passing on, would have pierced our hearts at any other time. But our feelings were steeled—we had no power to assist even when we felt the inclination to do so."*

The retreat lasted only from the 16th to the 20th of November, and did not cover more than fifty miles of ground, but the loss was heavy, and would have been still more considerable if the French had pursued with vigour. But the same vile weather and bad roads which hindered the British made it impossible for their enemies to move rapidly, and on the Huebra River the French halted, being as much without provisions as the retreating army, and unable to proceed further. Hence the last two days' march into Rodrigo was unmolested, and there the army found its baggage and received ample rations once more.

Soult now broke up the great mass of the French army, and dispersed it to recuperate, in winter quarters scattered from Toledo to Salamanca and Burgos—wherefore Wellington was able to do the same, and, all danger being over, dispersed his much-tried army into cantonments along and behind the Portuguese frontier. It was at this time that he issued the oft-cited and much-criticised objurgatory letter to the officers commanding divisions, brigades and regiments, in which he told them that "the troops had fallen off in discipline in the late campaign to a greater degree than any army in which he had ever served, or of which he had ever read. Yet they had met with no disaster, had suffered no privations which a trifling attention on the part of the officers could not have prevented, nor suffered any hardships, excepting those of being exposed to the inclemency of the weather at a time when it was most severe. Irregularities and outrages of every description had been committed with impunity, and losses sustained which ought never to have occurred." The truth would seem to be that Wellington underrated the trials to which the troops had been exposed—the man on horseback does not always realise the desperate state of the overloaded foot soldier, tramping along, knee-deep in mud, with sixty pounds on his back and no rations issued for four continuous days. But, undoubtedly, there had been much straggling,

* Donaldson, of the 94th, p. 181.

much marauding, and much indiscipline, and the rebuke was not wholly undeserved.

The 2nd Battalion 83rd Regiment, along with the rest of the 3rd Division, was sent back from the Portuguese frontier and quartered in villages some way behind it. Its cantonments were in and around Villar, those of the rest of Campbell's Brigade at Fonte Arcada and Vide. As was natural after such a retreat, its numbers were much depleted—the next figures in the battalion pay-book (there are none for December 1st) seem to show it with only sixteen officers and under 150 men present. It had started from Madrid on October 30th with little over 200 of all ranks. This is the lowest strength which it showed since the war began, even on the days immediately following Talavera. The loss was even heavier in proportion than that of other regiments in the 3rd Division. Its two British brigades, which had counted 3,780 rank and file on January 1st, 1812, had only 2,875 on January 1st, 1813; probably the December account, if it existed, would show even lower figures immediately after the retreat, before stragglers had come in and convalescents had begun to rejoin. There were no less than 2,253 sick in the division on January 1st, 1813; while 191 "missing" represent the losses in prisoners during the late retreat.

PART IV., 1813-14.

1813

THE campaign of 1813 was late in starting. That of the previous year had commenced as early as January 3rd, when the 2nd Battalion 83rd Regiment received its orders to march to the siege of Ciudad Rodrigo. It was not till the beginning of May that the operations of 1813 began. Meanwhile, the battalion lay for no less than four months in its winter quarters at Villar, slowly recovering from the fatigues of the late autumn, and filling up its depleted ranks with convalescents from the rear and a draft from England. The *personnel* of the officers was somewhat changed; Lieutenant-Colonel Carr still remained in command, but a new Major, G. T. Widdrington, had exchanged into the battalion from the 34th, so that Captain and Brevet-Major Samuel Hext, who had twice commanded the corps for months on end in 1812, was now only third in seniority. Lieutenant Joseph Swinburne was now not merely acting adjutant, but in formal charge of that post—his predecessor, Brahan, is found still with the battalion as a senior lieutenant in command of a company. Many new names crop up among the junior officers, and some old ones disappear, mainly by transference to the 1st Battalion on promotion. The serjeant-major is now Thos. Hazlehurst, the man who distinguished himself at the Picurina in March, 1812; William Duckett, who had held the post since Talavera, had been invalided home in the past autumn.

It may be worth noting, as a proof of the good discipline of the 2nd

PENINSULA MEDAL OF PTE. THOMAS HAZLEHURST,
83RD REGIMENT

*Now in the collection
of the Officers,
1st Bn. Royal Irish Rifles.*

Battalion 83rd Regiment during the last four years, that it had not down to this winter seen a single one of its officers or men sent before a general court-martial, to be tried for offences greater than could be dealt with by a battalion court-martial. Many very distinguished corps had seen ten or a dozen of their members so arraigned since Wellington's Peninsular command began, in April, 1809. Now, during the winter quarters of 1812-1813, we have at last two cases from the Regiment. Private Michael Foley was tried, along with four more men from other regiments, for having straggled from a convoy of sick during the retreat, for having plundered in the Spanish village of Cordovilla, and fired on the Alcalde of that place, and for having resisted arrest. He was sentenced to 500 lashes, getting off easily because the assault on the magistrate could not be proved for want of sufficient evidence. Two of his companions, a Dragoon and a Connaught Ranger, got 800 lashes apiece.* The second trial was that of Private William Marshall, for desertion, in January; he was sentenced to seven years' penal servitude in New South Wales.† There were to be three more grave cases of misconduct in the battalion before the year was out. Comparing this record of 1813 with the blank sheet of 1809-1812, it would, perhaps, be possible to infer that the class of recruit sent out in the latter years of the war was inferior to that which had been forthcoming earlier. Complaints of this sort are certainly to be found in other regiments of the Peninsular army: the colonel of a regiment which served beside the 2/83rd, wrote this winter that the officers at his depôt were sending him out men who were absolutely ruinous to his corps.‡

About the time that the battalion went into winter quarters it came once more under the command of its old chief, Colville, who had been absent since December, 1811, when (as will be remembered) he had been transferred to the temporary charge of the 4th Division. While holding that responsibility, he was wounded at Badajoz, and went home on sick leave. Returning to Lisbon in October, he joined the division in November, when the retreat was over. Thus the long period during which Colonel Campbell, of the 94th, had been acting as his deputy with the brigade came to an end. But, on January 26th, Pakenham, who had commanded the 3rd Division since Picton was disabled, was transferred to the charge of the 6th; so Colville, as the only major-general present, assumed charge of the whole division till Picton should return, and Campbell reverted once more to the position of acting-brigadier over a brigade somewhat stronger than it had been for some time, for it now included not only the 1st Battalion 5th, 2nd Battalion 83rd, and 94th Regiments, but also the 2nd Battalion 87th Regiment, a regiment which had come up from Cadiz and joined the 3rd Division after the campaign of 1812 ended. These new comrades, the "aiglers," of Barrosa, under Colonel Hugh Gough, were to stay

* Report in General Orders for April 3rd, 1813.

† Report in General Orders for April 10th, 1813.

‡ Life of Lord Gough, i., p. 183, Letter of December 9th, 1812.

with the 2nd Battalion 83rd Regiment for the rest of the war. There were no more changes in the brigade till Toulouse. The 2nd Battalion 83rd Regiment were now its left flank regiment—with the 94th next them; the 77th, which used to be the second senior battalion and, therefore, to take the left, had (as it will be remembered) been sent down to Lisbon in 1812.

In May Picton came back from England, where he had been undergoing a long course of Cheltenham waters, just in time for the campaign of 1813. Colville, therefore, gave up charge of the division, and reverted to the command of his old brigade, which was to see hot work under him before June was over.

The reasons for Wellington's long delay in opening the campaign of 1813 are not hard to find. The army was in a much more exhausted state than in the springs of 1811 and 1812, and needed a long rest. A number of new regiments had just been received from home, and Wellington did not wish to start on great operations till they had reached the front, and had been incorporated in the old brigades and divisions. But, most of all, was he delayed by his recently-given appointment as Commander-in-Chief of the Spanish Armies throughout the Peninsula. Hitherto he had arranged for the movements of his own Anglo-Portuguese host alone—to the Spanish generals he had been able to give no more than advice, which they generally disregarded. He had learnt to shape his campaigns in such a fashion that everything of real importance was to be executed by his own men alone, because he had no formal authority over any Spanish troops. The whole situation was changed when the Cortes at last made him their Commander-in-Chief, and gave him power to employ all their resources at his pleasure. For the campaign of 1813 he was, therefore, able to set in motion against the French, Spanish forces in far distant regions—Catalonia, Murcia, La Mancha, Andalusia, and he brought them all into his scheme of operations. But since most of these auxiliary armies were ill-equipped and destitute of magazines, and many were far remote from the future theatre of operations, it took a long time to get them prepared and set them in motion.

The Anglo-Portuguese could easily have moved in April, but in order to secure the advantage of the diversions which the Spaniards were to carry out, Wellington had to wait till May. About the fourth of that month the 3rd Division, still under Colville, for Picton had not quite reached the front, received orders to concentrate and to march northward, crossing the Douro into the province of Tras-os-Montes. This was new ground to the troops, who had never, since the war began, crossed its borders. The northerly movement was part of a broad scheme of offensive operations differing greatly from any that Wellington had yet carried out. The French were, as usual, scattered widely in cantonments, in order to be able to live on the country. Their divisions were disseminated from Zamora and Salamanca, in the west, to Pampeluna, in the east; from Madrid and Toledo, in the south, as far as

Bilbao, on the shores of the Bay of Biscay. The Anglo-Portuguese army, secretly concentrated far behind the Portuguese frontier, was to turn the right wing of the enemy with its main body, and appear on the north bank of the Douro; while a subsidiary force, under Hill, showed itself on the front of the French, on the old familiar ground in front of Salamanca; and Spanish detachments executed demonstrations on many points, with the object of distracting the attention of Jourdan and King Joseph, who were now in command since Soult had been drawn away by the Emperor to the great war in Germany.

No less than six divisions were devoted to the great turning movement through the Tras-os-Montes, of which the 3rd was one. The whole mass was placed under the charge of Sir Thomas Graham, the victor of Barrosa. It moved in three columns, of which the 3rd—now once more under Picton himself—formed the central one; it crossed the Douro on May 18th, and was at Vimioso, near the Spanish frontier, on May 20th; but having to wait for other troops, which had farther to march, only passed it on the 24th, and by the 28th was at Losilla and Carbajales, on the Esla River, to which neighbourhood the other columns had converged. The march from the Lower Douro to the Esla was a very inspiring one—the weather was fine, the mountainous country particularly beautiful, the commissariat was working well, and the men were confident, seeing great forces put in motion, and soon perceiving that the French were taken by surprise. Their cavalry retired almost without resistance, their infantry hurried off the moment that the Allies drew near.

Wellington, indeed, had caught the French unprepared—their divisions were scattered all over Central Spain, and could not unite till many days were over. Such measures as they had taken to resist Wellington had been formed on the hypothesis that he would advance on Salamanca, or possibly on Madrid. That he would strike for Zamora, Toro, and Palencia, north of the Douro, hardly occurred to King Joseph or Jourdan.

After the Esla was crossed on May 30th, and the enemy's right wing was found to be completely turned, and retreating with all speed, the march forward became even more exciting than before. The pace which the army kept up was astonishing, but, as an officer present remarked in his diary, " the men will do *anything* on an advance." After the Esla had been passed, the plains of Leon, the *Tierra de Campos*, had been reached, roads were fair, and the weather, fortunately, cool for a Spanish May, though there were occasional thunderstorms. The 3rd Division now formed part of the centre column of the advancing army, and was under Wellington's own eye. It never had to fight until the Ebro was reached, for the left-hand column (1st and 5th Divisions), under Graham, still kept turning every position which the enemy endeavoured to take up; the initial advantage of having outflanked the enemy was never lost for a moment. The French were increasing in numbers daily, as more of the outlying troops who had evacuated Madrid

1813

and New Castile kept coming up. But they were still not strong enough to face Wellington, who had his 70,000 Anglo-Portuguese all in hand, not to speak of the Spanish corps of Galicia and Estremadura. Perpetually outflanked, they had to abandon one after another the lines of all the rivers of Old Castile, the Rioseco, Pisuerga, Carrion, and Arlanzon.

"From the time of our crossing the Esla," writes an officer riding in front of the column, "we have been marching through one continued cornfield. The land is of the richest quality, and produces the finest crops with the least labour possible. It is generally wheat, with a fair proportion of barley, and now and then a crop of vetches or clover. The horses fed on green barley nearly the whole march, and got fat. The army trampled down twenty yards of corn on each side of the road (forty in all) where each column passed—in many places much more, from the baggage going on the side of the column, and so spreading further into the wheat. But the peasants must not mind their corn if we get the enemy out of their country!"

On June 6th Colville's Brigade passed Palencia, which King Joseph had been forced to quit on the previous day. On the 12th they reached Burgos, where it was thought that the enemy might stand and fight. But still outflanked, and knowing that they were outnumbered, the French gave way, and blew up the castle, which had served them so well in the preceding autumn. They then hoped that they might defend the pass of Pancorbo, on the steep descent from the plateau of Old Castile to the valley of the Ebro; but it was not by this way that Wellington moved. He turned sharply to the north from Burgos, still going beyond the hostile right, and, moving through a difficult and thinly-peopled country quite unopposed, came down to the Ebro far higher up its course than the King had expected. The 3rd Division crossed the river at the bridge of San Martin on the 15th. The positions which the French had hoped to occupy were again completely turned, and they had to cross the Ebro in haste, lest Wellington might fall on their line of communication with France if they tarried longer.

Thus, in a fortnight since the crossing of the Esla, the enemy had been manœuvred out of the whole of the kingdoms of Leon and Old Castile, and had lost not only Madrid but all Spain south of the Ebro, without having been able to offer any effective resistance. His only chance remaining was to give battle the moment that he had collected all his forces, which, if they could but combine, were considerably greater than those of Wellington. But the extremely rapid advance of the Allies had made this impossible hitherto, and there were still missing two great bodies of troops, which had been employed in hunting the Spanish insurgents when Wellington began his forward march. Clausel, with three divisions, was only just coming up from the direction of Aragon; Foy was in Biscay, with a force nearly as large, scattered among the mountains in moveable columns and small garrisons.

If both these corps had been able to join King Joseph, the moment that he crossed the Ebro in his retreat, Wellington would have had to halt on the line of that river and fight the rest of the campaign on a defensive plan. But he was well aware that these 35,000 men had not yet come in, and he was determined to push the French main army hard, and to endeavour to dispose of it before it received its last reinforcements. Hence the scheme of his movements, after the Ebro was crossed, became a time-problem. He had to force Jourdan and King Joseph to fight without their succours, or to push them in such a direction that it would be impossible for them to get into touch with one or other of the great columns which were coming up to their aid. This he accomplished by the extraordinary marches which followed his passage of the Upper Ebro on June 15th.

An officer of Colville's Brigade* describes the rapid advance for five days after the Ebro was passed, as one of incessant fatigue, gladly borne, because everyone saw that by swiftness alone could the enemy be forced to fight within the necessary time. "We crossed the Ebro at San Martin on the 15th," he writes, "and have been making long and distressing marches ever since, through the boldest, most mountainous, and romantic country that I have ever beheld. The Spaniards *deserved* to lose their country for not having defended these passes of the Ebro and all the country to the north of it. We have lately been badly off for bread, but our General (Picton) has been indefatigable in his exertions. The Seventh Division have now been with us for some days. The men are getting on capitally—only eight sick after all this fatigue. The whole army are in high order and spirits. The feeling of confidence is general; I am sanguine that this will be the most brilliant campaign for the 'Grand Lord'† that he has ever attempted."

On the 20th Wellington had once more driven in the French right, had seized, with Graham's column, the high road to France, so as to cut off Joseph and Jourdan both from their natural line of retreat and from Foy's troops in Biscay, and lay on the heights overlooking the plain of Vittoria, the last level ground at the foot of the Pyrenees. His divisions, forming four separate columns, encircled the French army from north-west to south, leaving them no retreat save due east, towards Salvatierra and Pampeluna, over a bad country road. Thereupon the King and Jourdan determined at last to fight, though they were cut off from Foy, and Clausel was three marches away. Other detachments, however, had come in, and they had about 65,000 men of all arms. This was a force still inferior in numbers to Wellington, who had over 75,000 men present, though he was short of the 6th Division and one of his two brigades of Guards, which had been left behind. But it must be remembered that some 20,000 of the allied army consisted of Galician, Cantabrian, and Estremaduran Spaniards, of whom many regiments were

* Lieutenant-Colonel Hugh Gough, commanding the 2nd Battalion 87th Regiment.

† "El Gran Lord" was the term usually employed by the Spaniards for Wellington.

newly raised, some were little better than guerilla bands, and few had any good fighting record. The Anglo-Portuguese, the solid front of the army, were about 58,000 strong. The main reason why the French leaders determined to fight was that they had with them about the most bulky and cumbrous collection of convoys ever seen in war—all the stores and heavy artillery from Madrid and other lost base-depôts, and the carriages and waggons of a horde of 10,000 French and Spanish refugees of all ranks. King Joseph's own baggage, collected for his final evacuation of his realm, covered a quarter of a mile of road, and included such things as famous pictures from the royal galleries at Madrid, furniture, plate, and miscellaneous valuables of all sorts. Unexpectedly cut off from the high road to France by Graham's last flank movement, all this mass of impedimenta had no good track to take. The King, to all effects, gave battle to cover his trains; he could not hope to get them away if he continued his retreat without fighting.

The line of the French front was marked out by the River Zadorra, a sinuous stream with large loops, so that at many points the ground along it was completely commanded from the British side, while at others re-entering curves cut deep into the French position. The left was on high hills, but the centre and right on undulating ground; some miles back from the river, however, there were two low ridges, which gave two successive lines of defence—the first behind the village of Ariñez, the second behind that of Armentia. There were seven bridges over the Zadorra, and for reasons hard to fathom Jourdan had neglected to break several of these; moreover, there were many fairly practicable fords. The whole position was thirteen miles long—overmuch for an army of only 65,000 men.

Wellington's attack was concentric and enveloping. The two wings under Graham (the left) and Hill (the right) were to press in upon the French flanks, and when they were gaining ground the centre was to pass the river and attack all along the line. Hill got forward early, and fought with success against the French left on the hills; but Graham was held back for the whole day by the desperate fighting of General Reille and the Army of Portugal, along the upper Zadorra. If he had been able to break through, behind Vittoria, the enemy would have been not merely defeated, but encompassed and annihilated.

About noon Wellington, seeing that Graham was making no way, resolved to strike in with his centre, to support Hill's flanking move. The 3rd Division, along with the 7th under Lord Dalhousie, formed the left section of the centre. Picton was growing excited as the hours wore on without any order to advance being given, though heavy artillery fire was audible on both flanks. "He was riding backward and forward, his stick going with rapid strokes upon the mane of his cob, and looking in every direction for the arrival of an aide-de-camp from headquarters." At length one galloped up, who said that he was searching for the 7th Division, 1813 having orders for it to commence the passage of the Zadorra at the neighbouring bridge of Mendoza; but Lord Dalhousie's column was still somewhat to the rear, having had difficult ground to pass. Wherefore Picton, forgetting all ideas of strict military obedience, said with some heat: "You may tell Lord Wellington from me, sir, that the 3rd Division, under my command, shall, in less than ten minutes, attack that bridge and carry it, and other divisions may support if they choose." Having thus expressed his intention, he put himself at the head of the right brigade, and started down towards the water, encouraging the leading regiment, as his biographer relates, with shouts of "Come on, ye rascals! Come on, ye fighting villains." It is easy to understand why Wellington, with his rigid rules of discipline, and the fiery and boisterous Picton did not get on together.

The attack was delivered with extraordinary speed and vigour, the right brigade (Brisbane's: 1st Battalion 45th, 1st Battalion 88th, and 74th Regiments) storming the bridge of Mendoza, while the left (Colville's) crossed at a deep ford a little higher up stream. Power's Portuguese followed in support. The French outlying troops along the river were driven in at a brisk pace, and then Picton, with his two British brigades deployed in line, started to attack the hostile right-centre, which had drawn itself up along the heights of Ariñez, two miles from the bank of the Zadorra. For some time the 2nd Battalion 83rd Regiment, as the left battalion of the left brigade, held the extreme flank of the whole advancing line, and was in a rather dangerous position, as, owing to Picton's haste, the 7th Division was not yet up to the front, or able to prolong the line of the 3rd Division westward. Nor were the 4th and Light Divisions, on the other wing, ready to continue Brisbane's line eastward.

Nevertheless, the 3rd Division, without any aid, plunged at the hostile line of battle, and, driving in all before it, got a lodgment on the heights of Ariñez. But the French soon discovered that Picton was far ahead of his supports, and made a desperate attempt to drive him down, turning forty guns upon his front and striking with heavy columns at both his flanks. The whole of Picton's line was engaged for hour or more in the most desperate defensive fighting, before it was at last succoured by the arrival of the 4th, 7th and Light Divisions, who aligned themselves on his flanks and disengaged him from the enemy who had pressed him so hard.

This, however, was only the first episode of the fight in the British centre; the enemy had given back, but only to occupy a second position to the rear of the first ridge. From this he had to be dislodged by the general advance of the whole British line. The 3rd Division now formed the left-centre of this long deployed array, and Colville's Brigade had to storm the village of Larmenda (or La Hermandad), which the 83rd and 94th Regiments carried with heavy loss but complete success. Twenty-eight guns were captured in clearing this part of the French position. It was after this achievement that, according to the regimental tradition, General Colville was so pleased with the conduct of the 2nd Battalion 83rd Regiment that he directed the Adjutant, Swinburne, to take down the names of a great number of the rank and file who had been foremost in the charge, and sent them, next day, a guinea apiece.

The French made yet a third stand after they had been driven from their second position, but this was little more than a desperate attempt to cover the retreat of their trains, which failed wholly. In the third attack Picton put his Portuguese brigade into the front line, and they, in company with the 7th, 4th and Light Divisions, gave the final blow. But, though the enemy were completely routed, and were forced in great disorder on to the bad road to Salvatierra, which they had not intended to take, no great number of them were cut off and captured. Down to the last, Reille's divisions held back Graham behind the upper Zadorra, and the flanking column, which was to have intercepted the French retreat, never came on to its destined ground. Nevertheless, the disaster to the enemy was one of almost unparalleled magnitude. Thrust aside on to a side path unsuitable for wheeled traffic, he lost all his artillery—about 151 guns—all his train, and the enormous and valuable convoys which contained all the wealth of Madrid. Some three or four thousand vehicles of all sorts fell into the hands of the victors, containing not only a well-filled military chest with some 5,000,000 dollars in gold and silver, but miscellaneous plunder of all sorts, some private, some public property, which constituted, in all probability, the biggest haul of booty that any army ever captured, since Alexander the Great took the camp of Darius—more than 2,000 years back. "The British camp that night was like a fair, and the dollars and doubloons were flying about in all directions." Some men are said to have pocketed as much as £1,000, and eight dollars (about 36s.) were offered for a gold guinea by those who were anxious to get their loot into portable shape.

The losses of the 2nd Battalion 83rd Regiment were naturally very heavy. They seem to have had in the field 20 officers, 24 serjeants, 15 musicians, and about 450 rank and file; of these one major (G. T. Widdrington), two lieutenants (Robert Bloxham and Thomas Lindsay) and eighteen men were killed, while one captain (J. Venables), two lieutenants (C. J. Baldwin and T. F. Smith), three serjeants and forty-seven rank and file were wounded—a total casualty list of seventy-four.* The Regiment was granted the name "Vittoria" to be borne on its colours and appointments; Lieutenant-Colonel Carr received a bar, to be worn on his already won gold cross; Volunteer Nugent (who had distinguished himself at the storm of La Hermandad) was given an ensigncy, and there was a considerable upward spurt of promotion among the ensigns and lieutenants.

* These figures and names differ from those in Bray's regimental history of the 83rd, but are those given in Wellington's detailed official returns, included with his despatches at the Record Office.

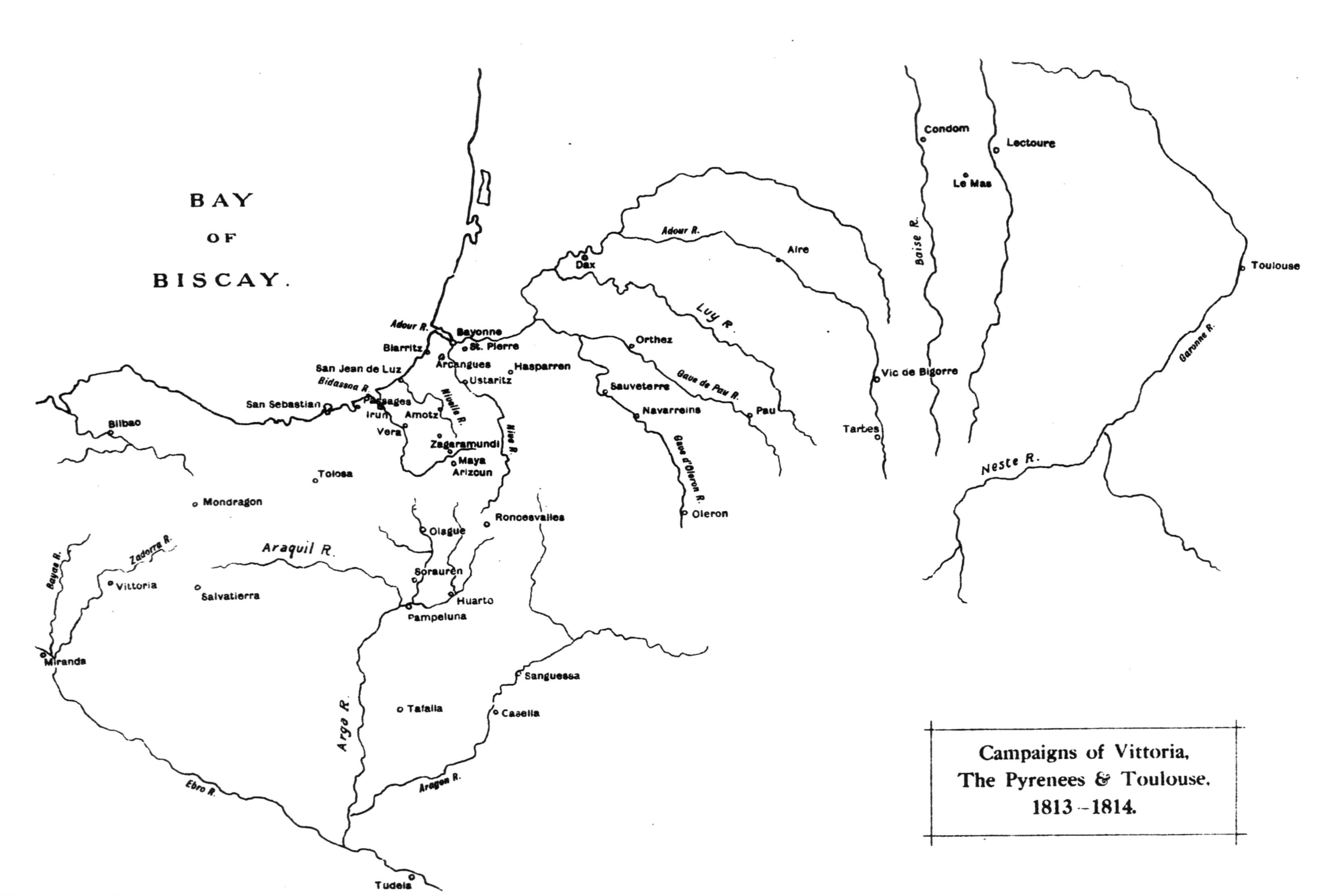

Campaigns of Vittoria,
The Pyrenees & Toulouse,
1813 – 1814.

SECOND BATTALION EIGHTY-THIRD REGIMENT.

On the day after the victory the 3rd Division marched, under Wellington's own command, with the main body of the army in pursuit of the flying French, while a large detachment, under Graham, took the other road, towards St. Sebastian and Irun, and drove Foy out of Biscay. The main column found that the enemy had passed Pampeluna and had escaped over the passes of the Pyrenees, so as to be out of danger. But Clausel's Corps, of 15,000 men, which was on the move for Vittoria and only two marches away upon the day of battle, was so close that Wellington made a dash at it, in the hopes of cutting off its retreat. This manœuvre cost the 3rd, 4th, 7th, and the Light Divisions some heavy and toilsome work, but to no effect, for fast though they marched, Clausel moved even more rapidly, and, covering nearly sixty miles in forty hours, escaped to Saragossa, and from thence crossed the Pyrenees.

Wellington then determined to lay siege to Pampeluna and St. Sebastian before crossing the French frontier, and placed his army in a position to cover these two leaguers, in a line parallel with the main chain of the Pyrenees. Picton and the 3rd Division were told off to the Pampeluna flank; they were for a few days blockading that fortress, till they were relieved, first by the 7th Division, then by Spanish troops, under Carlos de España and Henry O'Donnell, who came up to undertake the formal siege. On their arrival the 3rd Division took up its post at Olague and other villages upon the road from Pampeluna to the French frontier, in order to cover the operations of the Spaniards. Picton was thus placed as a reserve to a screen of troops, which was pushed farther forward towards the Pyrenean passes. It included Cole's 4th Division, the English brigade of Byng, and Morillo's Spaniards.

This precautionary disposition of the right wing of Wellington's army was soon to be justified. Marshal Soult, sent in haste from the Grand Army in Germany, superseded Jourdan and rallied the dispersed but still numerous fractions of the French army of Spain, with surprising rapidity. Only a month after Vittoria he had collected a disposable force of over 80,000 men, besides garrisons, and dared to attempt the relief of Pampeluna. He had the enormous advantage of being able to strike at given points of Wellington's long defensive line with his whole strength, while the allied army, guarding seventy miles of rugged frontier, cut up by passes and ravines, could only assemble in full force three or four days after the alarm should have been given. Soult left the western end of Wellington's front alone, and struck at two separate sections of its eastern half—those including the two passes of Maya and Roncesvalles. From each of these defiles good roads run downward towards Pampeluna, and the Marshal hoped to force them by a vigorous stroke, and to unite his two columns before the gates of the beleaguered fortress long before Wellington should have been able to bring up his outlying divisions to cover the siege.

The two simultaneous attacks on the passes were delivered on July 25th, and both succeeded, after heavy fighting. It is only that on Roncesvalles which concerns the annalist of the 2nd Battalion 83rd Regiment, for this alone fell within the section of front where Picton was in charge. The first attack fell on Byng and Morillo, at the actual pass and its neighbourhood. When they had lost it, Cole and the 4th Division came up, in time to bring off the defeated troops, but not in sufficient force to hold back Soult, who had 35,000 men, under Clausel and Reille, concentrated at this point. Cole, who had but 11,000 was compelled to fall farther back, till, at Zubiri, he was joined by Picton, who had hastily brought up the 3rd Division, 4,300 strong, from Olague, to reinforce him, and, as senior to Cole, took over charge of the whole corps, destined to " contain " Soult's main body for as long as possible, till Wellington should have brought up his reserves.

1813

But Picton was not long in ascertaining that the force in his front was at least double as great as his own, and, notwithstanding the strength of some of the defiles along the road, refused to commit himself to any serious fighting, evacuating several successive positions as soon as the French began to push him hard. (July 26th-27th.) On the evening of the second day, however, he turned to bay on a line of heights only a few miles north of Pampeluna, where he picked up 7,000 Spaniards from the blockading corps, and three brigades of British cavalry. In the line of battle which he now formed, the 3rd Division occupied the right wing, and lay along the lower hills trending back from the loftier heights of Oricain, on which the 4th Division was placed. Picton imagined that his own troops would be given the brunt of the fighting, since they lay on the more accessible ground; but Soult resolved to strike at the loftier and more commanding part of the position, and, sending one division, under Foy, to demonstrate against Picton, moved the bulk of his army more to his right, about the village of Sauroren, from which the battle that followed has been named.

Meanwhile, Wellington himself arrived on the scene, and took over the command from Picton. The shout of joy which greeted his arrival ran all along the line, and so impressed Soult, who easily interpreted its meaning, and soon caught sight of the British commander and his staff on the opposite heights, that he deferred his attack till the next morning, when all his rear divisions would have come up. The first battle of Sauroren, therefore, took place on July 28th; the 3rd Division had little or no share in it, since Soult confined all his attacks to the British left wing, and in front of the right there was only insignificant skirmishing. On the heights of Oricain, however, there was bitter fighting for the whole morning, which only ceased when the 6th Division, on whose arrival Wellington rested all his hopes, came up at noon and shattered the French attack by a sudden charge.

Soult's bolt was now shot; he had failed to beat the Allies while the superiority of numbers was still in his favour, and now Wellington's reinforcements were coming up from all directions. The Marshal should have retreated at once, but delayed his departure for a day, and on the early morning of the 30th Wellington suddenly took the offensive with his centre and left, stormed the village of Sauroren and the heights behind it, and drove the bulk of Soult's army in disorder in a north-westerly direction. The 3rd Division had, to Picton's great discontent, as small a share in this victory as in the fighting of the 28th. The French in front of him retreated hastily without offering resistance, and though he advanced and turned Soult's left wing, the movement did not bring him into contact with the enemy, who was already making off in retreat.

Hence it came to pass that the 2nd Battalion 83rd Regiment, like the rest of the 3rd Division, took no appreciable part in the "Battles of the Pyrenees." Between the 26th and the 30th July it only lost eight privates, wounded—and these apparently not in the main engagement about Sauroren on the 28th-30th, as the official casualty list of those days shows no figures for the battalion. Nor did Picton's men get any great share in the combats which attended the retreat of Soult: on the 31st of July and the 1st-2nd of August the 7th and 6th Divisions were heading the pursuit, and getting the pick of the fighting. Picton, sent off to chase the French left, under Foy, which fled separately over the pass of Roncesvalles, had a long stern chase, and never caught his quarry. Hence the name "Pyrenees" does not appear on the colours of the 2nd Battalion 83rd Regiment, for, by the usage of those days, grants of honorary distinction were only made to regiments which had been seriously engaged.

Soult's offensive had been completely foiled, and he had lost 15,000 men before he reached the French frontier. But Wellington did not follow the beaten enemy into his own country; he settled down to finish the interrupted sieges of San Sebastian and Pampeluna. The 3rd Division was once more put in the covering force, but no longer in the right-hand section of it, but in the middle. All through August it lay in the valley of the Bastan, in the allied centre, along the upper Bidassoa; the quarters of the 2nd Battalion 83rd Regiment appear to have been in the village of Arizcun, from which several "returns" are dated. At this time it lost its old brigadier: Colville was once more transferred to the charge of the 6th Division, whose chief (Pack) had been wounded at Sauroren. None of the senior officers of battalions in the brigade was given the interim command, which was entrusted by Wellington to Colonel Keane, of the 60th. On September 8th Picton went home on sick leave, and Colville came back, but to command the whole division, not his old brigade, which remained in Keane's hands all the autumn and winter.

San Sebastian fell on September 9th, but Pampeluna not till October was ending, and it was not till the latter fortress surrendered that Wellington committed himself to the invasion of France. All through August and September the 2nd Battalion 83rd Regiment rested in quarters that were pleasant enough till the early cold of the Pyrenean autumn began to make itself felt. The pass of Maya was close in front, from which those who chose to mount to the crest had wonderful views. "I thought the scenery of Roncesvalles splendid, but this is infinitely superior," writes an officer of Keane's Brigade. "From our camp we can see twenty or thirty leagues into France, studded with towns and villages, with a most extensive view of the Bay of Biscay. We look over on to the French camps, in which it appears that they have very few men."* It was while lying at Arizcun that the 83rd saw the first military executions which took place in its ranks during the whole war. Oddly enough, they came on two successive days. On August 30th Private Thomas Timmons was shot for attempted desertion to the enemy; he was apprehended by the Portuguese outposts of the division while making towards a French picquet. The court-martial found no extenuating circumstances, and he was executed in the presence of the whole regiment. On the very next day, August 31st, Private James Ryan was given the same sentence; his case was worse, in that he was actually captured acting as a guide to a French party during a reconnaissance. He blamed the tyrannical conduct of his captain and pay-serjeant as having driven him to the desperate act—but nothing could excuse actual taking service with the enemy, and he was duly shot.†

At the beginning of October, though Pampeluna had not actually fallen, Wellington made a slight advance of a few miles, which was to give him a foothold in France when he should be pleased to make a definite move forward. This was the forcing of the lines of the Bidassoa on October 7th, when the left wing of the allied army, suddenly and unexpectedly crossing the frontier river, by fords only practicable at low tide, swept the enemy out of a long line of redoubts, which he had created during the weeks of waiting. The 3rd Division was not among the corps which forced the passage of the estuary, being further to the right and well inland, but it made a corresponding forward movement over the pass of Maya, and drove the French outposts across the border. A diarist in the brigade of Keane notes that "while we were engaged in this business, there fell a tremendous shower of hailstones, some of them measuring five inches in circumference. Many men were hurt severely.‡ The division established itself in the old position of the enemy, and encamped for about a month on bleak hillsides above the French village of Zagaramundi, a few miles north of the border. A week or so after this move real winter cold began, and the troops found themselves very miserable in their tents, which gave very insufficient protection. This was the first campaign during which the whole army had been furnished with tents, and

* Letter of Gough, of the 87th. (See his life, p. 116 of vol i.).

† See General Orders for August 28th and 29th, 1813, and note on Ryan's case in the narrative of Donaldson, of the 94th, p. 210.

‡ Donaldson, p. 212.

regularly employed them; in the early years the alternative had been between billets in villages and bivouacking in the open. The weather was very severe —frosts at night and rain during the day. On more than one occasion a north-west gale from the Bay of Biscay brought down half the tents in the encampment in its first tremendous sweep.

Ten days after Pampeluna fell, Wellington commenced his long deferred invasion of France. He had no reason for further delay; not only were his communications clear behind him, but the news which kept coming in from Germany proved that Napoleon was losing ground, and had no chance of obtaining a peace from the Allied Sovereigns and sending reinforcements to Soult. Indeed, he was beginning to requisition troops from the Army of Spain to fill the depleted ranks of the "Grand Army." There was now no great danger, political or strategical, in pushing forward into France.

On November 10th the army of Wellington delivered a general attack on the lines to which Soult had withdrawn after the Bidassoa had been passed. They extended from the sea to the foot of the mountains on both sides of the River Nivelle, and every peak and ridge had been strongly fortified with abattis, redoubts, and trenches. But by endeavouring to hold a front of twenty miles by a passive defence, Soult had condemned himself to disperse his forces in a thin line, and risked the danger of finding parts of it attacked in overpowering force, while others were left almost entirely alone. This, indeed, was Wellington's plan. On the two wings of the French position he made little more than demonstrations, while he attacked three separate sections of the centre with masses of troops, which broke down all resistance.

1813

Though all the three British attacks this day were completely successful, that delivered by the 3rd Division—with Colville in command, and Keane leading the second brigade—was the most decisive of all. It was launched against the French left centre, where the Nivelle River cut diagonally through the hostile lines, and the bridge of Amotz, by the village of the same name, constituted the only good line of communication between the wings of Soult's army. While Hill, on the one side, and Alten, on the other, were driving in the French divisions on the hills, the 3rd Division, advancing across the one piece of fairly low ground in the whole line, struck at the entrenchments about the bridge-head of Amotz and carried them in the first rush, completely breaking up Conroux's French Division, the troops holding this part of Soult's position. Thus, the enemy's centre was pierced, and the success here gained was decisive, for Colville, turning against the line of redoubts to the left of the bridge, took them one after the other by assaults from the flank and rear, thus enabling the other British divisions of the centre section of the battle to drive in the enemy opposed to them with ease, for few troops, even if entrenched, can resist the combination of a frontal and a flank attack on their positions.

At the end of the day the 3rd Division crossed the Nivelle, and took post on its farther side, covering the passage of the remainder of Wellington's centre. But the day was by this time too far spent to allow of pursuit; the French had lost 4,500 men, including 1,200 prisoners, captured in various redoubts held too long, and 51 guns. The Allies had only 2,600 casualties. The 2nd Battalion 83rd had naturally suffered considerable loss in carrying the entrenchments at and above Amotz. It had seven privates killed, and Lieutenants F. M. Barry, H. Wyatt, and C. Watson, Ensign F. Burgess, one serjeant and thirty-five men wounded. Considering the importance of the work that it did, a casualty list of forty-six was not over heavy. The name "Nivelle" was very properly added, by Royal authority, to the names blazoned on its colours.*

For the next two months the 3rd Division was not destined to be heavily engaged, though many of the other fractions of the army had desperate fighting during the series of combats in front of the fortress of Bayonne, which are collectively known in official records as the battles of the Nive. Colville's troops now formed the reserve of the British centre, and were repeatedly moved from right to left and left to right, as one part or another of Wellington's line required support. But it was never necessary to bring them into the front of the fight. For the greater part of the mid-winter campaign they lay at Ustaritz or Hasparen, occasionally drawn out for an emergency, but soon sent back when it was over. In the combats connected with the crossing of the Nive, on December 9th, they were in support of the 6th Division, and not heavily engaged. On the 10th, 11th, and 12th, when Soult attacked Wellington's left, about Biarritz and Arcangues, the 3rd Division was brought up from Ustaritz, but its services were not required. On the 13th, when the Marshal, swerving to the other flank, attacked Hill, and was repulsed by him at the desperate battle of St. Pierre, Colville was once more sent across the Nive to aid Hill, but arrived late, to find him already victorious. It was not till the new year of 1814 was many weeks old that the 3rd Division was again to find itself in the forefront of the battle.

1814

By this time Picton was back from his sick leave to England; he had resumed command of his old troops late in December. His *locum tenens*, Colville, did not return to his former brigade, which remained under Keane, but was transferred to the temporary charge of the 5th Division. The officers present with the 2nd Battalion 83rd Regiment at the new year of 1814—the last of the Peninsular service of the battalion—appear to

* The last of the few general courts-martial held on men of the 2nd Battalion 83rd Regiment falls in connection with the battle of Nivelle. Private William Woodcock was shot "for leaving his colours during the action to plunder a house, and for resisting a serjeant of the Cavalry Staff Corps (military police), who was forcing him to quit the house, by levelling his firelock at him before he was disarmed." (General Orders, 1813, p. 391.)

have been:—Lieutenant-Colonel Carr, Major Blaquière (transferred from the 1st Battalion after Widdrington was killed at Vittoria), Captains Hext, R. Thompson, Gilbert Elliott, J. Venables, E. Gapper, Lieutenants J. Hingston, T. Gascoyne, C. Watson, J. Emslie, C. O'Neill, J. Evans, M. Carey, F. M. Barry, A. Lane, Ensigns J. Parnall, W. O'Neill, P. Nugent, T. Young, W. Maxwell, and E. Nihill. Lieutenant Swinburne was still acting as adjutant. The surgeon was Robert Sandford, his assistant surgeon A. Tonnerre. The paymaster was Henry Cross.

For the greater part of January, and the early days of February, the battalion was in winter quarters at Hasparen, "where the weather was bad, and the men much harassed, marching a distance of two or three miles every morning to the alarm-post before daylight, and remaining till the sun appeared." The villagers all talked Basque only, which seemed harsh and discordant to the ears of those accustomed to Spanish or Portuguese. All intercourse with them had to be by signs or dumb-show, French being of as little use as either of the Peninsular languages. Their round, flat *berets* reminded Scots of the headgear of their own Lowland peasantry.

About the middle of February Wellington drew his divisions out of their cantonments, and started on the last of his long series of campaigns against Soult. His object was to drive the Marshal from his hold upon the fortified lines around Bayonne, and to compel him to quit the neighbourhood of that stronghold, by turning his left flank. If Soult should refuse to move, he would run the danger of being shut up in the fortress with his whole army. If, as was more likely, he shifted himself inland to face the turning movement, he would lose the support which Bayonne gave him, and would have to face superior forces in the open field; for the French army was now much below Wellington's numbers—not only had the attrition of constant fighting borne heavily upon it, but on January 30th two divisions had, by the Emperor's peremptory orders, been drawn off to join the Grand Army, in Eastern France.

On February 14th Hill marched, with two divisions, eastward, to encircle Soult's flank by a long turning movement. Some days later Picton moved up on Hill's left, and three other divisions followed. Soult immediately transferred the greater part of his forces from the neighbourhood of Bayonne to face this advance, intending to defend in succession the lines of the various tributaries of the Adour, which lay across the route of the Allies. His main stand was to be made on the stream called the Gave d'Oleron, and he had fifteen thousand men concentrated in and near Sauveterre, the principal crossing on that water. Picton's orders were to demonstrate against this force, while the rest of the turning columns passed the river higher up. The main strategic purpose was achieved, for the Gave was successfully crossed with little opposition above Sauveterre; but Picton's demonstration was made somewhat costly by his venturesomeness. While his guns were cannonading the bridge of Sauveterre, he sent the four light companies of Keane's Brigade and a few Hussars across a deep and difficult ford a mile lower down. This was a rash act, for the movement drew the immediate attention of Villatte, the French general in command on this point, who sent a whole regiment against the 300 men of the 5th, 2nd Battalion 83rd, 2nd Battalion 87th, and 94th Regiments, who had passed the stream. They had taken up a position along a stone wall above the river bank, and prepared to defend themselves, but were attacked both in front and flank so vigorously that the line was broken at once. "The French drove in our light troops so precipitately that, in retreating down the narrow lane which led back to the ford, the men were wedged in so closely that they could not move. A number struck away to the right and attempted to swim the river, but were carried away by the current, and many of them drowned. Of those who reached the ford many were wounded in the river, and, losing their footing, sank to rise no more, among whom was an officer of the 94th. The French had come down close on to them, and few would have escaped being killed or taken had not a brigade of our guns been brought down to the bank of the river, and, by a heavy fire of grape, covered the retreat."* The loss of the four light companies was sixty killed and wounded and thirty prisoners—nearly one man in three. The share of the 2nd Battalion 83rd Regiment was two privates killed, two wounded, and seven captured, out of under fifty men who crossed the river.†

1814

This petty disaster had no importance, since Wellington's main column had successfully passed the Gave far above Sauveterre, and Soult had at once to withdraw all his scattered divisions, and to concentrate at Orthez, behind the Adour, two marches to the north. Wellington had thus forced him to drop his connection with Bayonne, and to retire towards the interior of France. And upon the 26th of February the forces left in front of Bayonne, under Hope, cast the celebrated bridge of ships across the mouth of the Adour, and completely invested the fortress, where Soult had left a very strong garrison—no less than 13,000 men—which reduced his field army to about 38,000 of all arms.

With this force he offered battle at Orthez, on February 27th, to Wellington, whose offensive wing, or main body, was some 45,000 strong, nearly all English and Portuguese, for the bulk of the Spaniards had been left among the 28,000 men with which Hope was undertaking the siege of Bayonne. Soult's position at Orthez extended from the town of that name, on the Gave de Pau River, along the heights for some three miles to its north. Wellington ranged opposite him five divisions, under Beresford and Picton, while two others, under Hill, were to turn the French left flank by crossing

* Donaldson, of the 94th, p. 223.

† These official figures of Wellington again differ from those given in Bray's regimental history of the 83rd, which are five killed and twelve wounded or taken—seventeen casualties instead of eleven.

the Gave up-stream, behind the town. The place of Picton's Division in the line was a very important one—the first troops to arrive on the ground, by crossing, at dawn, the difficult ford of Berencz, they had to cover the marching up of the Light and 6th Divisions from the rear, and then to take their place as the right-hand section of the main attack on the hills just north of Orthez.

This was not the place where Wellington intended to deliver the decisive blow—but it was necessary to keep the French left-centre hotly engaged, lest it should send troops to more important points of the battle-front. Picton fought an uphill offensive battle against the French divisions of Foy and d'Armagnac for three hours, with Brisbane's Brigade on his right, and Keane's, with Power's Portuguese, on his left. The 6th Division was in reserve behind him. The operations, which were intended to be decisive, by Beresford, on the left, and Hill, on the right, hung fire; the former was long checked by the French at St. Boës; the latter failed for some time to establish a passage over the Gave. Hence, Picton had to fight hard, and for some time gained little or no ground, and suffered severe loss. It was not till late in the day that the French defence suddenly crumpled up—the 52nd, at the head of the Light Division, having pierced a weak spot in Soult's centre, while at the same time Hill, having at last succeeded in crossing the river, appeared on his flank. The 3rd Division then took part in a simultaneous charge, delivered by the whole line, and drove the exhausted enemy in front of them over the hills. Soult's retreat, regular at first, soon degenerated into something not far from a rout, and he lost many prisoners and still more stragglers, before he got safely behind the Luy de Bearn, the next river in his rear.

The loss of the French was 4,000 men, among whom were 1,300 prisoners; in this total the stragglers are not included—they numbered at least 3,000. The Allies lost 2,300 only, of whom no less than 825 belonged to the 3rd Division, which suffered more than any other unit in Wellington's army. The 2nd Battalion 83rd Regiment had one serjeant and ten rank and file killed, and no less than nine officers (out of about twenty-two present) wounded. There were the Lieutenant-Colonel (Carr), Major Blaquiére, Captains Elliott and Venables, Lieutenants Baldwin, Watson and Lane, Swinburne (the Adjutant), and Ensign Nugent; also one serjeant, one drummer, and twenty-eight rank and file—a total casualty list of fifty. The number of officers hit is out of all proportion to that of men, as sometimes happens in attacks vigorously led against an enemy settled down in a good position. Colonel Carr got one more clasp for his gold cross, and Captain and Brevet-Major Hext (who succeeded him in command) a clasp for the gold medal that he had earned when once before in interim charge of the battalion. Another officer of the Regiment also earned the medal—this was Captain Gilbert Elliott, who was this day in command of a temporary unit—the massed light companies of the division, which Picton had used as a screen during the early part of his movement. The name of Orthez was granted as an honour for the regimental colours, to balance the recently-acquired "Nivelle."

This was the last general action in which the battalion was to be engaged during its service under Wellington, but not the end of its fighting, as it was to take part in one more sharp combat. But there was a short delay before the last stage of the campaign was reached. Soult had retired from Orthez, not in the direction of Bordeaux, as Wellington had expected, but eastward, towards Toulouse. His move in this direction had been intended to force his enemy to follow him, and to keep him from over-running the inland, by detaining him near the foothills of the Pyrenees. But Wellington, seeing Bordeaux uncovered, and knowing the Marshal to be much reduced in strength and confidence by his late defeat, thought himself able enough to make a side-stroke at the capital of South-Western France. He detached Beresford, with two divisions—some 12,000 men—to attack Bordeaux, where there was only a trifling garrison of conscripts, and where the governing classes were known to be secret foes of Napoleon and attached to the Bourbons. The enterprise was completely successful, the city was occupied without an effort, and the Mayor, Lynch, hoisted the white flag, proclaimed Louis XVIII., and organized a sort of provisional government in the King's name.

The moral effect of this development was notable, and Napoleon's hold on Southern France was severely shaken. But meanwhile Wellington, by making this large detachment, had lost his numerical superiority over Soult, and, therefore, halted for some days, till he should have brought up from the rear two Spanish divisions and received several isolated battalions of British reinforcements, which were marching up from the coast. He also ordered Beresford, when the fall of Bordeaux was known, to leave only one division there, and to come back to the main army with the rest of his column. On March 18th-26th the advance against Soult once more began, and led, on the 20th, to the combat of Vic de Bigorre, near Tarbes, the last engagement in which the 2nd Battalion 83rd Regiment took a serious part during the war.

In this fight the 3rd Division formed the frontal striking force, with which Wellington drove in Soult's rearguard, while the Light Division was turning its right by a flank march. The French opposed to Picton (the division of General Paris) made a long running fight. "We advanced up the main road to the attack till the enemy's artillery, which commanded it, forced us to strike into the fields to right and left, where we drove them from one field to another, while they contested each hedge and ditch, giving us a volley as we came up, and then retreating to the next fence. Having driven the enemy from the vineyards, we chased them into the town, retreating through which they killed a number of our men from various places where they had got under cover."* The whole army was assembled that night in front of Vic de Bigorre and Rabestens, the French retiring in the night upon Tarbes.

* Donaldson, of the 94th, pp. 227-8.

In this long skirmish against an enemy who always gave way when pushed, the Allies lost 250 men, mostly in the 3rd Division. The 2nd Battalion 83rd Regiment, which was in the front line and busily engaged had the appreciable loss of six privates killed, and two officers (Lieutenants Hingston and Lane), two serjeants and thirteen men wounded.

From March 20th to April 10th Wellington was driving Soult steadily before him across the rolling country of Languedoc, not without many small rearguard affairs, in which (as it chanced) the 3rd Division had no considerable share. On the day upon which the Allies arrived in front of Toulouse Napoleon had already abdicated, though no one in either army was aware of it, and the battle was unnecessary. Soult had prepared Toulouse for defence by erecting a line of redoubts and entrenchments upon the only side where it was not well protected by its ancient walls and the waters of the Garonne and the canal of Languedoc. The fortifications with which he had covered himself easily compensated for the difference between the numbers of the two armies—Wellington had about 50,000 men up, Soult about 38,000.

In the battle Picton, like Hill, was given a purely secondary part, that of detaining in front of him by demonstrations the French divisions which manned the north-western defences of Toulouse, along the canal of Languedoc. The offensive blow was struck on his left, by Beresford, with the 4th and 6th Divisions, and by the Spaniards of Freyre. It is well known how Beresford, after a desperate struggle, carried the line of redoubts in front of him and rendered Soult's position untenable, though the Spanish attack on his right had failed. But the 3rd Division had not the fortune to be used on the decisive point; it skirmished all day across the waters of the canal with the French voltigeurs, on the opposite side, and would have suffered no appreciable loss if Picton had not overstepped his orders, and ordered Brisbane's Brigade to try to capture the bridge-head over the canal at the crossing called Pont des Jumeaux. This attempt to storm a strong field work without the aid of heavy artillery fire, and without even ladders, failed, costing the right brigade of the division over 300 casualties, and Brisbane, the officer in command, was wounded. But Keane's Brigade, on the left, was not engaged in this unhappy venture, did no more than skirmish with the French in its front, as Wellington had intended, and suffered hardly any loss. That of the 2nd Battalion 83rd Regiment was only one man wounded.

In the distribution of battle-honours the 2nd Battalion 83rd Regiment was allowed to put the name " Toulouse " on its colours, though so slightly engaged. If any regular principle had been observed it should, arguing from this grant, have been also allowed to claim the honour, " Pyrenees "; for it was quite as heavily engaged during Picton's retreat towards Pampeluna in July, 1813, as at Toulouse on March 10th, 1814—indeed, it there lost eight men, as against one in the last battle of the Peninsular War.

Soult evacuated Toulouse on the night of March 11th, and Wellington was starting in pursuit when the news of the fall of Paris on March 31st, and the subsequent abdication of Napoleon, arrived, and put an end to the war. When it was known that Soult and his army had given in their adhesion to the new French Government, and changed the tri-coloured for the white cockade, Wellington gave orders for the breaking up of the army into cantonments and its leisurely transference to Bordeaux, where it was not wanted till ships should be available for its transport to the British Isles. The 3rd Division was quartered for the end of April in villages near the central Garonne, Keane's Brigade and the 2nd Battalion 83rd Regiment lying at Le Mas, near Lectoure. From thence they marched, in May, by slow stages, towards the coast, parting at Condom from their old comrades, the Portuguese of Power's Brigade (9th and 21st Line), who had served beside them most faithfully and honourably from Busaco onward. The separation was bitterly felt on both sides.

" On the morning that we marched the Portuguese, ranged along the street, saluted us as we passed, and their hearty " *Vivas!* " and exclamations of regret evinced that they really felt moved. But a scene of a more affecting nature took place when the Portuguese and Spanish women had to say adieu to our men, to whom they had attached themselves during the miserable state of their countries. Those of them who were duly married, with the Colonel's leave, followed the Regiment; but the generality of them were *not* married, though the steady affection and patient endurance of hardship which they exhibited would have done credit to a more legal tie. Being here (at Condom) ordered to return to their own land, with the Portuguese troops, and strict orders being given to prevent any of them from proceeding further, the scene which followed was most distressing. The poor creatures ran about concealing themselves, in the vain hope of being allowed to remain, but it was all to no purpose—although they were willing to sacrifice country and relations to follow us, the sacrifice could not be accepted."* Lord Wellington so far relented in the end as to permit that " a limited number of those who have proved themselves useful and regular, might accompany soldiers to whom they are attached, with a view to being ultimately married "—but this only covered a small percentage of the whole.

On the march from Toulouse to Bordeaux the battalion was rejoined by Major Blaquiére, who took command, having recovered from his Orthez wound. Colonel Carr was still disabled, and did not again see in France the men whom he had led through so many battles. At the end of May the 2nd Battalion 83rd Regiment were encamped at Blanquefort, some five miles from Bordeaux, where a great many regiments were waiting for transports. The quarters were comfortable, the weather fine, and officers and men were allowed to visit the city. Here there came to join the regiment the small remnant of the Talavera prisoners, who had neither died, escaped, nor been exchanged since their

* Donaldson, p. 232.

capture, and had been vegetating at Verdun Arras, or Briançon. They included one captain (Somerfield),* two lieutenants (apparently Boggie and Le Toller), Assistant-Surgeon Glasco, and fifteen rank and file. The whole of the troops at Blanquefort were finally reviewed by Wellington and the 2nd Battalion 83rd Regiment marched a few miles down the river to Pauillac, and there embarked for England on June 1st, five years and eighty-four days after their first landing at Lisbon on April 7th, 1809.

Of the forty-two combatant officers who had entered on their Peninsular service on the last-named day, eleven returned with the Regiment, including the three lately-recovered Talavera prisoners. These were Major Hext, Captain (and Brevet-Major) Gilbert Elliott, Captain Somerfield, Lieutenants Brahan, Hingston, Chas. O'Neill, Carey, Le Toller, Boggie, Barry and Watson. It may be noted, as a sample of the vagaries of promotion in the reign of George III., that all of them (putting aside the Talavera prisoners) had risen just one step in the Service since they landed in the Peninsula, save the unfortunate Hingston, who, a lieutenant since April 1st, 1808, was still a lieutenant on January 1st, 1815, and only thirteenth in the regimental list, though he had served out the whole war and had been twice wounded. Brahan was also still a lieutenant, but, as will be remembered, he had been taken at Talavera, and only rejoined the battalion after nearly two years' captivity.

Of the other twenty-four officers who landed at Lisbon in April, 1809, six had been killed—Colonel Gordon, Captain Powys, Lieutenants Montgomery, Dahman, Flood, and Ferris. One (Captain Reynolds) had died a prisoner in France. Eleven—Major Napper, Captains Geddes, Creagh, Noleken, Lieutenants Abell, Baldwin, Vereker, Johnson, Irwin, Jas. Smith, and Bowles—were still in the Regiment, but serving with the 1st Battalion, at the Cape, or with the Depôt. The remaining five had exchanged into other corps, or sold out from the Army.

Colonel Carr, in the general distribution of honours at the end of the war, was made a Knight of the Bath, and Major Samuel Hext a companion of the same order. The Colonel stood very high among Peninsular officers in service done; his gold cross and three bars represented seven actions, in each of which he had commanded his unit—Fuentes de Oñoro, Ciudad Rodrigo, Badajoz, Salamanca, Vittoria, Nivelle, and Orthez.† Majors Hext and Gilbert Elliott each received the gold medal given to officers who had led a

* Captain Reynolds had died a prisoner; Lieutenants Nicholson and Brahan had either escaped or been exchanged. The latter rejoined the Regiment in 1811. The former was serving in the York Chasseurs by 1813.

† The only officers who got as many (or more) bars to their cross as Sir Henry Carr were Wellington himself, Beresford, Hill, Pack, Cole, Dickson (the artillery commander), Picton; of the Staff, De Lancey, D'Urban, Hardinge, Colin Campbell, Geo. Murray, Lord Fitzroy Somerset; of regimental commanders, Barnard, of the 95th, Berkeley, of the 35th, Greville, of the 38th, Sutton, of the 23rd, French, of the 94th.

unit in a general action. Hext had two bars, for Orthez and Toulouse; Elliott had the plain medal for Orthez.

It may be worth remarking that of all the second battalions of regiments which had their first battalions serving abroad already, and which joined Wellington's army in 1809, the 2nd Battalion 83rd Regiment was the only one which fought out the whole war as a complete unit. Of the seven others, two (the 2nd Battalions 7th and 48th Regiments) were drafted into their first battalions when the latter came out to Portugal later in the war. Four were cut down to half-battalions in 1811-12, and ended as forming parts of "provisional regiments" (the 2nd Battalions 24th, 31st, 53rd, and 66th Regiments). One (the 87th) was absent at Cadiz from February, 1810, to the end of 1812, and did not see the central campaigns of the war under Wellington, though it rejoined in time for Vittoria. It was a unique distinction for the 2nd Battalion 83rd Regiment that, though recruited with difficulty, owing to the perpetual necessity for sending drafts to the sister unit, at the Cape, it preserved its battalion organization throughout the whole war, alone among the second battalions present with the main army from 1809 to 1814. The honours on its colours show that it was engaged in every general action or siege conducted under Wellington's own leadership in those six years, save two, the Pyrenees and the Nive. The roll is a magnificent one—Talavera, Busaco, Fuentes de Oñoro, Ciudad Rodrigo, Badajoz, Salamanca, Vittoria, Nivelle, Orthez, and Toulouse.

EIGHTY-THIRD REGIMENT (1817-1857).

CHAPTER X.

EIGHTY-THIRD REGIMENT (1817-1857).

Strength of the 83rd Regiment on arrival in Ceylon—Losses of the 83rd in the Kandian campaign—Draft arrived from England via New South Wales—Peninsula battle honours conferred on the Regiment—General Order of General Officer Commanding in Ceylon regarding 83rd Regiment—" Toulouse " and " Busaco " battle honours conferred on 83rd Regiment—Embarked for England—Farewell order of General Officer Commanding in Ceylon—Arrival at Gosport—Moved to Glasgow—Proceeded to Ireland—Garrison duty in Ireland—Cholera—Embarked for Halifax, Nova Scotia—General Hastings Fraser appointed Honorary Colonel—Battle honour of " Cape of Good Hope " conferred on the Regiment—Regiment moved to Quebec—Canadian Rebellion—Description of country and inhabitants—Combat at Saint Denis and Saint Charles—Capture of St. Eustache—Regiment moved to Upper Canada—Filibusters driven out of Fighting Island and Pelée Island—Raiders landed at Prescott—Lieut. Johnston, 83rd Regiment, attempted to turn them out and was killed—83rd Regiment stormed entrenchments at Prescott—Establishment of the Regiment changed—Customs of the Officers' Mess—83rd embarked for England—Garrison duty in detachments in England—Moved to Ireland—More detachments ordered to Bombay—Moved to Karachi—Lieut.-Colonel Swinburne's services—Regiment ordered to Deesa—Services of Colonel Law.

1817

THE insurrection of the Kandian chiefs had broken out some few weeks before the 83rd Regiment arrived in Ceylon. All the details of the cause of this rebellion are stated in the chapter recording the experiences of the Eighty-Sixth in the same Island.

The Eighty-Third Regiment was exceptionally strong in officers and men on its arrival in Ceylon. The commissioned officers numbered 46, the serjeants, 48, whilst 22 drummers and 969 privates completed the handsome total of 1,085. This fine body of troops was, naturally, at once made use of, and was marched into the disaffected provinces, where it continued until the rebellion was finally quelled in November, 1818.

1818

The losses inflicted on it by the enemy were very small—only a total of twelve killed and wounded—but Lieutenants Crutwell and Smith, Ensign MacNab, three serjeants, and 112 rank and file died of disease during that time, whilst in the following year it suffered the further loss of Lieutenant Cox, four serjeants, one drummer and eighty-six privates, all from the effects of disease.

No event of any importance marked this desultory campaign, which really amounted to frequent skirmishes with an unseen enemy in the bush, very much like a West African campaign at present, with no pronounceable names even to the villages.

A draft of 89 men arrived from England on the 23rd of September, under command of Captain Campbell, whilst a further small draft, in charge of Lieutenant Mee, numbering twenty men, joined on the 7th of January, 1820, having travelled by the unusual route of New South Wales and Calcutta.

1820 In January the following letter was received from the Deputy-Adjutant-General, Horse Guards:—

"Horse Guards,
"11th August, 1819.

"SIR,—I have the honour to acquaint you that the Prince Regent, in the name and on behalf of His Majesty, has been pleased to approve of the 83rd Regiment being permitted to bear on its Colours and Appointments, in addition to any other Badges or Devices which may have heretofore been granted to the Regiment, the words:—

"'Talavera,' 'Fuentes d'Oñor,' 'Ciudad Rodrigo,' 'Badajoz,' 'Salamanca,' 'Vittoria,' 'Nivelle,' 'Orthes,' in commemoration of the distinguished Services of the Regiment at the Battle of Talavera on the 27th, 28th, and 29th July, 1809, at Fuentes d'Onor in the month of May, 1811, at Ciudad Rodrigo in the month of January, 1812, at Badajoz on the 16th of March, 1812, at Salamanca on the 22nd of July, 1812, at Vittoria on the 21st of June, 1813, at Nivelle on the 10th of November, 1813, and at Orthes on the 27th of February, 1814.

"I have the honour to be, Sir,
"Your most obedient humble servant,
"(Sd.) JOHN MACDONALD,
"D.A.G.

"The Officer Commanding,
"83rd Regiment,
"Ceylon."

1821 Major-General Sir Edward Barnes, K.C.B., Commanding the Forces in Ceylon inspected the Regiment on the 12th of March, 1821, and published the undermentioned order, which is given in full, with the intention of shewing the points in which the Regiment excelled:—

"Headquarters,
"Colombo, 13th March, 1821.

"GENERAL ORDERS.

"No. 1.—The Major-General Commanding had the highest satisfaction yesterday in inspecting the 83rd Regiment.

"No. 2.—The uniform, equipment, and general good appearance of the men was most striking.

"No. 3.—The regularity and precision of their movements, and correctness of their firing reflect the highest credit upon Lieutenant-Colonel Cother and the Regiment at large, but what has been most satisfactory to the Major-General with regard to this Regiment is their steady and orderly conduct in quarters."

General Sir Robert Brownrigg, K.G., C.B., had inspected the Regiment in 1819, and published an even more complimentary order, adding that the Regiment "excelled every other corps in Ceylon," and that this would be reported to the Prince Regent.

In June, 1821, 140 men "volunteered" from the 73rd Regiment, which was embarking for England, all electing to join the 83rd. On the 22nd of February the headquarters of the Regiment, consisting of eight officers and 329 other ranks, embarked at Colombo, and sailed for Trincomalee, arrived there on the 8th of March, and remained there until the following year, when six small parties were sent by various vessels back to Colombo, the last party arriving at that station on the 6th of November.

1823

The Regiment had received orders earlier in this year to return to England, but owing to the outbreak of the Burmese War, the authorities retained it at Colombo in case its services should be required. In the meantime it relieved the 45th Regiment at Kandy in the interior of Ceylon, and that corps proceeded to India.

1824

Another battle honour was now conferred on the Regiment by the following letter:—

"Horse Guards,

"26th January, 1827.

"SIR,—I have the honour to acquaint you that His Majesty has been pleased to approve of the 83rd Regiment bearing upon its colours and appointments, in addition to any other Badges or Devices which may have heretofore been granted to the Regiment, the word "Toulouse," in commemoration of the distinguished Services of the Regiment in the attack of the position covering Toulouse on the 10th April, 1814.

"I have, etc., etc.,

"(Sd.) F. C. TORRENS,

"A.G.

"The Officer Commanding,

"83rd Regiment,

"Ceylon."

Whilst following on this letter, notification that yet another battle honour was conferred on the Regiment, was made known through the "London Gazette." The extract is given below:—

"War Office,

"4th June, 1827.

"His Majesty has been pleased to approve of the undermentioned Regiments bearing on their Colours and Appointments the words stated against them, in addition to any other Badges or Devices which may have heretofore been granted to them: 83rd Regiment, 'Busaco,' in commemoration of the distinguished conduct of the Regiment at the Battle of Busaco on the 29th September, 1810."

From 1825 to 1828 the 83rd remained in Kandy. In October of the latter year it returned to Colombo and embarked in December on two transports, named respectively, "Amitz" and "Arab." The first vessel sailed, with the headquarters, under Lieutenant-Colonel Cother, C.B., on the 4th of December, whilst the remainder of the Regiment followed in the "Arab" a few days later; the destination of the whole was England.

During the services of the Regiment in Ceylon—a period of eleven years—it sustained a loss by death of seventeen commissioned officers and 491 other ranks.

1828

The Officer Commanding Troops in Ceylon was still Sir Edward Barnes, K.C.B., now promoted, however, to the rank of lieutenant-general. He attended the embarkation of the 83rd, and afterwards issued the following General Order:—

"Headquarters,

"Colombo, 4th December, 1828.

"GENERAL ORDERS.

"No. 1.—The Lieutenant-General Commanding cannot allow the 83rd Regiment to quit his Command without placing on record this testimony of his highest approbation of their conduct during a period of eleven years' service in this Island.

"No. 2.—The 83rd Regiment brought with it in 1817 a reputation arising from a long series of most distinguished service which was not easily to be tarnished, but no event, however trifling, has occurred to produce such an effect; on the contrary, year after year has rolled on in the most uniform and steady line of good conduct, adding lustre to the former brilliant career of this gallant corps.

"No. 3.—The Lieutenant-General desires to assure Lieutenant-Colonel Cother that he will carry with him his highest respect and esteem, and begs that he will convey to the Regiment at large the strongest assurance of his approbation and regard for their future welfare.

"No. 4.—There is one Officer (Lieutenant-Colonel Kelly) to whom a particular mark of distinction is due. To his talents and exertions is greatly to be ascribed the subjugation of the Kandyan Kingdom, and his conduct afterwards in the Command of the interior was eminently conspicuous, and the Lieutenant-General regrets exceedingly that he is about to lose the advantages of his valuable services.

"(Sd.) F. B. GASCOYNE,

"D.A.A.G."

Thus, in the full sunlight of official praise for humdrum duty, well and honestly performed, the Regiment set sail for England, numbering all told some seventeen officers (of whom only one was an ensign) and 332 other ranks.

During its stay in this agreeable, though deadly Island, the Regiment had lost its Colonel, General James Balfour, by death, and the vacancy had been filled by the appointment of Lieutenant-General John Hodgson to succeed him.

1829 The two transports, bearing the 83rd, duly arrived at Portsmouth on the 16th of April and 18th of May. Here the troops were landed and were quartered in Forton Barracks, where they remained till August. On the 27th of that month the Regiment embarked on three transports, "Hope," "Amphitrite," and "William Harris," and set sail for Scotland, and, being landed at Leith, proceeded to march to Glasgow. Lieutenant-Colonel Cother, C.B., retired on the 3rd of December, 1829, and Major the Hon. Henry Dundas, M.P., assumed command of the 83rd.

1830 After an uneventful stay in Glasgow, the 83rd embarked in steamboats and proceeded to Belfast on the 16th of August. On arriving at that port it marched to Enniskillen, and furnished detachments at Omagh, Lifford, Sligo, and Ballyshannon. It left Enniskillen and proceeded to Castlebar on the 1st of November, 1831, again marching from that place for Limerick on the 23rd of October, 1832, arriving there on the 29th of the same month. Whilst at Castlebar the Regiment was attacked by cholera on the 26th of June, and during the epidemic it lost two officers (Assistant-Surgeon Watson, and Lieutenant Bowles) and ten rank and file. It was moved out of the barracks and encamped about a mile away from it, but after a month returned and occupied its original quarters.

1832 Leaving Castlebar on the 23rd of October, the 83rd marched to Limerick, where it furnished six detachments to various small towns. Whilst at Limerick Major Somerfield and Lieutenant The Hon. R. Clifford died. Leaving Limerick on the 11th of

1833 August, the Regiment marched to Dublin, arriving there on the 19th of that month. It remained quartered in that city until April,

1834 when, on the 5th of that month, part of it embarked on the "Innisfail," which ship took it to Cork and returned to Dublin, to take the remainder of the Regiment.

The 83rd had been ordered to embark at Cork for Halifax, Nova Scotia, and 31 officers, 30 serjeants, and 489 other ranks, duly embarked on the freightships, "Brunswick" and "Ruckero," on the 21st of April and 15th of May, arriving at their destination on the 26th of May and 20th of June respectively.

In the meantime, the depôt companies remained in Ireland and marched to Mullingar, under the command of Major Trydell. One man died of cholera in Halifax, and the disease was again combated by encamping the Regiment away from its barracks. It was kept under canvas until the 1st of October, and then returned to its quarters.

1835 The honorary colonelcy again became vacant, by the transfer of Lieutenant-General Hodgson to the 4th Foot, and the vacancy was filled by the appointment of Major-General Hastings Fraser, C.B.

Detachments were furnished by the 83rd to the Island of Cape Breton and Prince Edward's Island, whilst "Evangeline's Home"—the Annapolis Valley—was also garrisoned by a detachment from the Regiment, but all the detachments were withdrawn by the 20th of July.

1836 The following letter added yet another honour to the Regiment's already proud list:—

"Horse Guards,
"17th March, 1836.

"SIR,—I have the honour to acquaint you by the direction of the General Commanding-in-Chief that His Majesty has been pleased to permit the 83rd to bear on its Colours and Appointments, in addition to any other Badges or Devices heretofore granted to the Regiment, the words "Cape of Good Hope," in commemoration of the distinguished gallantry displayed by that Regiment at the capture of the Town and Garrison of the Cape of Good Hope on the 8th of January, 1806, when it formed part of the First Brigade employed on that occasion.

"I have, etc.,
"(Sd.) JOHN MACDONALD,
"A.G.

"The Officer Commanding,
"83rd Regiment,
"Halifax, Nova Scotia."

1837 On the 29th of June the 83rd embarked in Her Majesty's frigate, "Vestal," and on the sloop-of-war, "Champion," and sailed for Quebec, arriving there on the 12th and 13th of July, and occupied the Citadel Barracks.

At this time the present Quebec province was seething with unrest, and the disloyal party received countenance and much material support from the inhabitants of the United States, who lived upon the south side of the St. Lawrence River.

The population of Canada was then about 1,082,000, of which 432,000 belonged to Upper Canada and 650,000 to Lower Canada, which divisions correspond, roughly, with the present provinces of Ontario and Quebec. The populations differed materially, for, whilst the inhabitants of Upper Canada consisted very largely of English Royalists driven out of the United States, during the rebellion there, for their sympathy with England, the province of Lower Canada was peopled by the French-Canadians, who had been taken over with the ceded country in 1760, after the conquest of Quebec by General Wolfe's force.

Upper Canada gave no trouble, but it was not surprising that these French-Canadians should be disloyal. They had never been asked whether they would be transferred to English Sovereigns. Two points, however, were noteworthy. One, the extraordinary number of the French-Canadians who held out loyally for the British Crown, and the other the difficulty which the agitators experienced in obtaining any grievances to incite the population against the governing power. So far as can be ascertained, the demand was really made for self-government, but the rebels' movements were ill-considered and badly directed.

The country was principally inhabited along the St. Lawrence River and the lakes of Ontario and Erie, and this fact facilitated the movements of British troops by water to put the various outbreaks down. There were but few British troops in the country, amongst them being the 1st Battalion 24th, and 32nd and 66th Regiments, at Montreal, the 83rd Regiment at Quebec, whilst the 34th and 43rd Regiments were sent forward from New Brunswick, and the 85th Regiment from Nova Scotia via New Brunswick. It is recorded that such hospitality did all ranks receive from the inhabitants, both French and English, on their journey, that they openly questioned whom they could be going to fight, as all vied in their courtesy to the troops.

On the 1st of November two companies of the 83rd went to Three Rivers, and on the 10th ascended the St. Lawrence to Montreal.

Whilst they were here, and the remainder of the Regiment still at Quebec, the first act in the drama took place, this being the release from custody of two political agitators, MM. Demaray and Davignon, on the 18th of November, by a party of some 300 armed men, an officer and three men of the party guarding them being wounded.

This was quickly followed by two actions—one at Saint Denis, on the 23rd of November, the other at Saint Charles, on the 25th of November—by two distinct forces.

The first was a British reverse with the loss of six killed and ten wounded, and the leaving of a howitzer in the hands of the rebels as the force retreated. The other was a distinct victory, as the village was stormed, with a loss of three killed and eighteen wounded, whilst the enemy are supposed to have lost 152 killed, 300 wounded, and thirty prisoners.

These two villages are some distance south-east of Montreal. On the 1st of December Saint Denis was occupied, without fighting, by the British, and the howitzer, etc., recovered.

This ended the trouble at this place. On the 9th of December the remainder of the 83rd left Quebec and joined the other two companies at Montreal two days later, being all quartered in Quebec Gate Barracks.

In the meantime a party of rebels had established themselves at a village called St. Eustache, some distance to the north-west of Montreal.

General Sir John Colborne decided to march to this place. With wise forethought, he organised the expedition with all necessary trains for the conveyance of provisions and the materials for the repair of bridges, etc. Finally, he left Montreal with his force on the 13th of December. This force numbered some 2,000 men, and consisted of the 1st and 32nd Regiments, and the 83rd, under Lieutenant-Colonel Dundas, eight field guns, the Queen's Light Dragoons, Captain Walter Jones's Troop, the Montreal Volunteer Cavalry and Rifles, and Globensky's Volunteers.

Montreal stands on an island in the St. Lawrence. Crossing over from it to yet another island, called Laval, the brigade halted at St. Martin's for the night. It then pushed over the ice to the mainland, and, on the 14th, proceeded to attack Saint Eustache, having lost one of the tumbrils through the ice whilst crossing the river.

On approaching the town, the guns opened a sharp fire on the village (about 11.30 a.m.), whilst the infantry surrounded the town, so as to cut off the retreat of the enemy. Fortunately, many of the rebels who had been forced into rebellion by their leaders, had time to fly before the cordon was complete. The insurgents, who, though Roman Catholics, had treated the priests with great discourtesy, now made the Church their citadel. The guns were driven back by a heavy fire from the upper windows of the organ loft, and their shot could make no impression on the stout walls of the church. However, its door was riddled. Then the order to storm the village was given. The 1st Regiment led the storming column, and the houses near the church were taken. The church still held out. In some way one of the houses caught fire and the roof of the church also became ignited. Not wishing to incur heavy losses in storming the rebels' last stronghold, the troops were held back for a time, but by 2.20 p.m. the church was well alight and the roof fell in at 6 p.m., burning many misguided wretches in its fall. Chénier, the rebel leader, was shot whilst trying to escape; 108 prisoners were made, and the number of killed is stated variously from seventy up to some hundreds. Next day the troops marched to a village called St. Benoit, near by. This place was one of the very few who had a priest disloyal to the Government. He had wisely fled, and was interdicted by his own Church. The village offered no resistance, and the troops returned to Montreal on the 17th of December. Their losses were one killed and eight wounded.

This was no mean performance, when the great cold is considered—at this time of the year often 20° or 30° below zero.

So far the rebellion had been crushed, and Sir John Colborne, a most humane man, arranged that none of these misguided persons should lose their lives for their conduct. However, unfortunately, peace did not follow, for raiders came over constantly from the United States side of the border and established themselves in various places, hoping that the disaffected would again rise and assist them. These parties were inclined to loot what they required, and sometimes they murdered noted loyalists, so that the whole borderland was in a state of unrest. The United States authorities, in many cases, did not assert their authority and break up these bands. Again and again they openly embarked, amidst cheering crowds, to cross the river to raid Canada.

1838

The effect of all this was that the 83rd had again to take the field against disreputable foes, who were rightly styled "brigands." Two companies of the 83rd were sent west, into Upper Canada, in January. They travelled in sleighs some 600 miles and halted at St. Thomas, but one company was shortly afterwards sent on to Amherstberg, under Lieutenant Kelsall.

During this month an expedition of brigands from the United States had set sail in an armed schooner, and in passing Amherstberg had first neglected to answer the sentry's challenge, and when he had fired had returned the fire with a charge of grape-shot, discharged into the village of Amherstberg. On the 9th of January the schooner again came by Amherstberg, but this time the Canadian Militia, though ill-armed, were on the alert, and, fortunately for them, the vessel grounded, when they promptly rushed into the water (no pleasant place in Ontario in January) and proceeded to carry the vessel by boarding. The crew were taken, with a loss of nine killed and wounded and thirteen prisoners. A large quantity of arms and ammunition rewarded the Militia.

However, another raid was made in February by these filibusters, under the command of a "General" McLeod, said to have been a serjeant in the British Army, and afterwards a schoolmaster at Brockville. They seized Fighting Island, in the Detroit River. Lieutenant Kelsall, on the 25th of February, took his company of the 83rd, supported by a company of the 32nd Regiment, to turn them out. Crossing the river on the ice, they at once carried the island. The enemy fled, leaving a field piece and a large number of muskets, perfectly new, stamped as the property of the United States!

In March Pelée Island was raided by 500 men from over the States border.

The unfortunate inhabitants were made prisoners and their property was confiscated to the use of the invaders. Again Lieutenant Kelsall's company was called upon, and, strengthened by two companies of the 32nd Regiment, twenty volunteer troopers from Sandwich, and two 6-pounder field guns, they proceeded to the attack on Pelée Island. This island is in Lake Erie, some fifteen miles from the shore, between Leamington and Kingsville. With the British troops went a small party of Indians. Starting on the 2nd, at 5 o'clock, the party marched eighteen miles along the lake shore. Here they bivouacked in the snow. At 2 a.m., on the 3rd of March, the whole party proceeded to cross the ice to Pelée Island. They performed the march of twenty miles in splendid time and arrived at the island at daybreak. Two companies of the 32nd, with the cavalry, were placed in position to the south, to intercept any attempt at escape. The remainder of the force moved down on the raiders from the north. The latter at once fled and took to the bush. The troops followed, but the advance was slow, as the bush was very thick and the snow very deep. The invaders, seeing that the force to the south only amounted to some ninety men, assembled 300 men and advanced in line. The little containing force at once moved forward, receiving and returning a volley, and charged with the bayonet. The 300 invaders promptly fled back into the bush, put forty wounded men into sleighs, rushed to some boats, which were on an open tract of water on the lake, and so escaped, leaving one "colonel," one "major," and two "captains" and seven rank and file dead, whilst some prisoners were taken, several of whom were wounded. The British troops lost four killed and twenty-eight wounded.

Lieutenant Kelsall's company went to Kingston on the 17th of June, where the remainder of the Regiment had already arrived on the 6th of May.

To people nowadays it seems surprising that anyone should have been found foolish enough to take part in these raids into Canada. But it must be taken into consideration that in thirty years we had had two wars with the United States, and this had left an aftermath of ill-feeling; whilst every scamp who was wanted by the police, or those who had found it wisest to leave England if they had been connected with the industrial troubles in the 1820's had found a convenient haven of refuge in the States. Thus there were many in the place who had grudges to satisfy against England. There were others who were honest enthusiasts and believed that Republican government was better than Monarchical. These two classes found the funds. The officers were supplied by adventurers of all sorts, Poles, Irish, etc., whilst the rank and file were recruited from the population employed on the vessels, etc., of the large lakes and rivers and those dependent on them. As navigation closed, on account of the ice, these men had to look out for some winter occupation. Lumber camps, in many cases, claimed them, but joining a raiding force doing nothing but receiving pay, with an occasional descent on some unprotected settlement in Canada, was a pleasanter and more interesting life than many others. Thus, the forces were recruited.

In the autumn of the year 1838 a serious outbreak took place in Lower Canada, which was suppressed with tact and great firmness by General Sir John Colborne. A raid was made into Upper Canada, which raid brought the 83rd again on to the scene to suppress it.

On the 10th of November a steamer, called the "United States," set out from Oswego, on the United States side of Lake Ontario. It contained a large body of men, and steamed down the River St. Lawrence. On the way she took in tow some schooners filled with guns and munitions of war. This flotilla anchored just below Prescott, on the American side of the river.

Between 300 and 400 of the men then landed on the Canadian bank of the St. Lawrence, just below Prescott, bringing with them two field guns. They occupied a point of land on which stood a strongly-built stone mill. They threw up breastworks and utilized the mill and three or four stone houses to cover their rear.

A party of marines were on board H.M. Steamboat "Experiment," and they were reinforced by Lieutenant Johnson and forty-four men of the 83rd, from Kingston. Having arrived off Prescott, the "Experiment" found the "United States" steamboat trying to land reinforcements of some 500 men. Three times the attempt was made, and each time the vessel was compelled to return to the States side of the river. A schooner then tried to get past the "Experiment." She was also loaded with men, but she was equally unsuccessful. It was afterwards ascertained that the raiders lost heavily in their futile attempts. The 83rd and the marines then landed, and it was decided to try and clear the filibusters out of their entrenchments. Two columns were formed. The total force amounted to some 200 men. The 83rd led one party, under Lieutenant Johnson, the marines the other, under Lieutenant Parker. Some militia, etc., were also present, making up the number as stated to some 200 men. The 83rd was to take a circuitous route, whilst the marines followed the road by the river.

The attack at first was perfectly successful; the fortified position was carried. Some forty or fifty of the invaders retired into the houses and mill. Lieutenant Johnston called upon the 83rd to clear them out and led the way. He fell dead at the threshold of a house, with four of his men wounded beside him. The senior officer then stopped the attack, considering that many lives would be lost if it was continued unsupported by artillery. Whilst this engagement was going on the raiders made desperate efforts to reinforce their garrison by sending boatloads of men across the St. Lawrence, but they were steadily driven back by the British.

Besides the losses of the 83rd, one private of the marines was killed, and one lieutenant and twelve rank and file of that gallant corps were reported as wounded, whilst five of the militia were killed and twenty-seven wounded.

The enemy were reported to have lost forty killed and twenty-eight were made prisoners.

This action took place on the 13th of November.

Colonel Dundas arrived on the scene on the following day, with four companies of the 83rd and a " demi-brigade " of field guns. The attack, for some reason not stated, was deferred until the 16th of November, when the guns opened fire on the mill at a distance of 400 yards.

It being now late in the day, the Colonel ordered the 83rd to rush the houses, which was promptly done, and one was set on fire. The defenders of the mill at once hoisted a white flag and surrendered at discretion. The enemy lost thirty killed, sixteen wounded, and eighty-seven unwounded prisoners were also taken. The 83rd lost one man (Corporal Downes), who was killed while bravely storming a house, but the Canadian Militia, who took part in the assault, suffered very heavily, losing one officer killed and two wounded, whilst the returns further show that thirty-seven of the rank and file were killed and wounded. Having passed the night on the captured position, the 83rd returned next day to Kingston.

It is interesting to follow these raids a little further to see how they came to an abrupt conclusion.

On the 4th of December some 400 men crossed the river from Detroit to the Canadian bank, four miles from Sandwich. Having committed much havoc and killed several of the militia in fair fight, they proceeded towards Sandwich, murdering some of the settlers. They were now met by a Colonel Prince, with some 170 militia. This gentleman had been present at several raids, and he saw no reason that such atrocities should go on, and decided once and for all to stop this nonsense. He at once attacked the enemy, who, though in superior numbers, were routed; twenty-one were killed and four captured and brought before Colonel Prince. His despatch says:—" I ordered them to be shot, and it was done accordingly."

Following hard after these bandits, Colonel Prince captured twenty-six other prisoners, whom he reserved for trial by the regular courts. The remainder, who could, fled across the river; several, however, could not effect their escape and hid in the bush, where they were afterwards found, frozen to death.

Detroit was placarded with offers of rewards up to £200 for Colonel Prince, alive or dead. The House of Lords held an inquiry in England and admonished all concerned in the putting to death of the four prisoners. But this exhibition of force and determination had its effect, and the raids ceased; whilst the Attorney-General reported that the twenty-six other prisoners might have been put to death by their captors at the moment they were taken as outlaws.

So ended the disturbances in Canada in which the 83rd took part.

SHAKO PLATE OF 83RD REGIMENT, 1828-1844.

All gilt, the star and crown dead gilt, numerals burnished on a frosted back

1840 A letter received on the 2nd of February, sent from the Horse Guards on the 20th of November, 1839, fixed the establishment of the Regiment at—

Six service companies, consisting of	Four depôt companies, consisting of
7 staff-serjeants.	16 serjeants.
24 serjeants.	4 drummers.
10 drummers.	16 corporals.
24 corporals.	184 privates.
576 privates.	

Embarking in steamboats on the 20th and 21st of May, the 83rd proceeded to London on the River Thames in Upper Canada, sending a detachment to St. Thomas. The Corps arrived at its destination on the 29th and 30th of that month respectively.

1841 Lieutenant Wynniatt was accidentally drowned on the 14th of May whilst attempting to ford the River Thames on horseback.

It may be of interest here to note how customs have changed in small matters in the Service. An officer of the Regiment who was serving at that time states that all the unmarried officers dined at mess, at 7 o'clock in the evening, wearing uniform. The orderly officer in the 83rd did not wear his sword at mess at that time. This time-honoured custom was kept up in the 86th until 1891, when Colonel Wyndham, rather unnecessarily, abolished it. The orderly officer in that corps wore his sword until the officers were actually proceeding from the ante-room to the mess-room, when he obtained the permission of the senior officer present to remove it.

Breakfast was at this time always prepared by the soldier-servants in the officers' rooms, and luncheon generally consisted of a few biscuits and cheese, also served in quarters.

The St. Thomas detachment rejoined on the 17th of May.

1842 Half the 83rd marched from London to Toronto, leaving the former place on the 7th of July and arriving at Toronto on the 14th. The remainder followed on the 8th of July. Colonel the Hon. H. Dundas went on half-pay on the 2nd of August, being succeeded by Brevet-Lieutenant-Colonel B. Trydell.

1843 Leaving Toronto on the 22nd of May, three companies went to Three Rivers, whilst on the following day the remainder of the Regiment went to Quebec, arriving there on the 27th inst. All the companies rejoined by the 10th of June, and on the 16th of that month the 83rd embarked on two freight-ships, named "Countess" and "Jamaica"; both ships sailed next day, and dropped anchor together at Spithead on the 10th of July. The total strength brought home was 19 officers, 734 other ranks, and 70 women and 130 children, Lieutenant-Colonel Trydell being in command.

Landing at Gosport, the 83rd remained in Forton Barracks until the 17th of July, when it proceeded by detachments to Weedon by rail. From now for some years the Regiment appears to have been broken up into numerous detachments; a list of these stations is given in table form at the end of this chapter, as an example of how a regiment was dispersed previous to the Crimean days. Leaving Weedon on the 4th of August, the headquarters of the 83rd went to Northampton. Whilst here, the new percussion musket was served out to the Regiment.

1844 The headquarters of the Regiment moved to Weedon on the 8th of April by route march, and from there by rail to Leeds, on the 10th of October, remaining there until the following 13th of June, when it moved to Manchester. 1845 Re-united here, the 83rd went by rail from Manchester to Liverpool, on the 22nd of July, embarked on three steamboats for Dublin, arrived there on the 23rd inst., and 1846 headquarters went to Limerick, returning to Dublin from thence on the 3rd of October, 1846, being quartered in Richmond Barracks.

1847 Leaving Dublin on the 21st of May, the headquarters, with one company, proceeded to Kilkenny, remaining there until the 12th of September, 1848, when the headquarters moved to Fermoy. 1848 Its continuance there was of short duration, as on the 9th of January, 1849, it moved to Cork. Here the whole Regiment gathered by degrees. 1849 On the 11th of January the headquarters and two companies embarked for Bombay, on a ship of the same name, under the command of Lieutenant-Colonel B. Trydell. The "Bombay," sailing on the 17th of January, arrived at her destination on the 8th of May, and the troops landed and proceeded to Poona, reaching that place on the 18th of that month.

The remainder of the Regiment came out as follows:—Two companies, on the "China," leaving 6th February, arrived 27th May; two companies, on the "Mermaid," leaving 7th February, arrived 26th May; part of one company, on the "Earl of Balcarres," leaving 9th February, arrived 25th May; one company, on the "Ursula," leaving 15th March, arrived 25th June; one company, on the "Marian," leaving 15th March, arrived 23rd June; part of one company, on the "Lion's Hope," leaving 16th March, arrived 5th July. All these various detachments had reached Poona by the 14th of July.

1850 On the 2nd of November the Regiment sent the first detachment away, en route for Karachi. It marched to Panwele, where it "took steam conveyance" (presumably steamboats) to Bombay. Here it transhipped to the Honourable East India Company's steamer, "Bernice," on the 7th of November, and sailed for Karachi, where it arrived on the 14th of that month.

The remainder followed in three detachments, leaving Poona respectively on the 3rd, 4th, and 11th of November, arriving at Karachi on the 15th and

22nd of November and 1st of December. During its stay at Poona, the 83rd lost, by death, two officers and seventy-five of other ranks.

1851 On the 22nd of January Lieutenant-Colonel B. Trydell, being appointed to the command at Poona, handed over the Regiment to Lieutenant-Colonel Law.

1853 On the 8th of February, half the 83rd proceeded up the Indus, in river steamers, to Hyderabad, where they remained.

On the 22nd of May, Brevet-Lieutenant-Colonel Swinburne retired on full pay, after forty-four years' service in the 83rd Regiment. He served throughout the whole of the Peninsular campaign (the greater part of the time as Adjutant), and received the medal and ten clasps for Talavera, Busaco, Fuentes d'Oñor, Ciudad Rodrigo, Badajoz, Salamanca, Vittoria, Nivelle, Orthes, and Toulouse. He also took part in the suppression of the Kandian and Canadian Rebellions with the Regiment.

The headquarters, with half the Regiment, embarked on the Honourable East India Company's steam frigate, "Semiramis," on the 17th of December, under the command of Lieutenant Mainwaring, and proceeded via Gogo to Deesa. The cause of this officer being in command was, that all those senior to him were detained in Karachi on general court-martial duty. This party was halted at Gogo, awaiting orders, and was joined by Major Henry Lloyd, from England, in January. He took the headquarters, etc., on to

1854 Deesa, leaving Gogo on the 22nd of January, and arrived at Deesa on the 13th of February.

Major Lloyd left Deesa to go to Karachi on the 23rd of April, to take command of half the Regiment which had been brought there from Hyderabad. He, unfortunately, died from cholera en route to take over the command, on the 6th of May, at Cambay. Lieutenant-Colonel Law did not

1855 rejoin until the 4th of January, and on the 2nd of April the remainder of the 83rd moved from Karachi to Bombay.

During its stay in Scinde, i.e., up to April, 1855, the 83rd lost, by death, four officers and 152 other ranks. The companies at Bombay were embarked on the 22nd of December on board the "Adjaha," for

1856 Downes. There they disembarked and marched, via Surat, Baroda, and Ahmedabad, for Deesa. They arrived at their destination on the 25th of January.

1857 On the 25th of March Lieutenant Dickinson and ninety-nine men proceeded to Mount Abu. In May, Colonel W. H. Law retired from the Service with the rank of Major-General. He was also a Peninsular veteran, having been present at the battles of Nivelle and Nive. Lieutenant-Colonel Kelsall then took over the command of the Regiment.

During the same month the 83rd received orders to hold itself in readiness to act against the mutineers of the Bengal Native Army.

NOTE.—Where the 83rd was quartered from 1843 to 1849:—

Strength of Detachment.	Place it was sent to.	Date.	Rejoined Headquarters or went elsewhere.
Headquarters and 2 Companies.	Northampton ...	4th Aug., 1843	To Weedon, 8th April, 1844.
2 Companies ...	Birmingham ...	,,	Headquarters, 8th April, 1844.
1 Company	Wolverhampton	,,	Headquarters, 8th April, 1844.
1 ,, ...	Burslem	,,	Headquarters, 19th April, 1844.
1 ,, ...	Coventry	,,	Headquarters, 8th April, 1844.
1 ,, ...	Hanley	,,	Headquarters, 19th April, 1844.
2 Companies ...	Newcastle-under-Lyme.	,,	Headquarters, 18th April, 1844.
2 ,, ...	Weedon	6th Oct., 1843	To Headquarters, 14th October, 1843.
Headquarters and 3 Companies.	Leeds	10th Oct., 1844	
1 Company	York	8th Oct., 1844	Headquarters at Manchester, 11th June, 1845.
2 Companies ...	Bradford	,,	Headquarters at Manchester, 13th June, 1845.
1 Company	Sheffield	,,	Headquarters at Manchester, 11th June, 1845.
1 ,, ...	Keighley	10th Oct., 1844	Headquarters, 17th Mar., 1845.
1 ,, ...	Huddersfield ...	,,	Headquarters, 28th Mar., 1845.
1 ,, ...	Halifax	,,	Headquarters to Manchester, 11th June, 1845.
1 ,, ...	From Leeds to Hull.	17th Mar., 1845	Headquarters to Manchester, 11th June, 1845.
1 ,, ...	Scarboro' Castle	28th Mar., 1845	Headquarters to Manchester, date not stated.

Remaining at Manchester until 22nd July, 1845, went via Liverpool to Dublin, and went on detachment as follows:—

Strength of Detachment.	Place it was sent to.	Date.	Rejoined Headquarters or went elsewhere.
Headquarters and 2 Companies.	Castle Barracks, Limerick.	25th July, 1845	On detachment as below.
3 Companies ...	Strand Barracks, Limerick.	"	" "
3 "	Keith Barracks, Limerick.	"	" "
1 Company	Tipperary	"	To headquarters, 17th Dec., 1845. Relieved by another Company from Limerick.
1 "	Cahir	"	To headquarters, 19th Dec., 1845. Relieved by another Company from Limerick.
1 Company	Kilrush, furnishing 6 detachments to Forts on Shannon.	1st Dec., 1845	Headquarters, 14th May, 1846. Relieved by another Company.
1 "	Rathkeale	27th Jan., 1846	Headquarters at Dublin, 15th October, 1846.
1 "	Newcastle	26th Jan., 1846	Headquarters at Dublin, 20th October, 1846.
1 "	Castle Connell ...	16th May, 1846	Headquarters at Dublin, 10th October, 1846.
1 "	Croom	23rd May, 1846	Headquarters at Dublin, 17th October, 1846.
1 "	Ennis	17th June, 1846	Headquarters at Dublin, 19th October, 1846.
1 "	Clare Castle ...	24th June, 1846	Headquarters at Dublin, 17th October, 1846.

From the 20th October, 1846, to 13th May, 1847, the Regiment remained in Dublin, having sometimes four and sometimes two companies in Aldborough House Barracks, sometimes two companies in Island Bridge Barracks, and the remainder in Richmond Barracks.

Strength of Detachment.	Place it was sent to.	Date.	Rejoined Headquarters or went elsewhere.
Headquarters and 3 Companies.	Kilkenny	13th May, 1847	Fermoy, 12th September, 1848.
1 Company	Carlow	,,	Headquarters, 28th June, 1847.
1 ,, ...	Callan	,,	Fethard, 12th September, 1848.
1 ,, ...	Castlecomer ...	,,	Fethard, 13th September, 1848.
1 ,, ...	Carrick-on-Suir..	14th May, 1847	Fermoy, 30th September, 1848.
1 ,, ...	Bagnalstown ...	,,	To New Ross, 4th Sept., 1847.
1 ,, ...	Thomastown ...	,,	Date of movement unknown.
1 ,, ...	Wexford	16th May, 1847	To New Ross, 19th Aug., 1848.
1 Company	Enniscorthy ...	9th July, 1847	To Wexford, 14th April, 1848.
2 Companies ...	Graiguenamanagh	11th Aug., 1848	To Headquarters, 19th August, 1848.

The headquarters moved to Fermoy on 12th September. Detachments went out as follow:—

Headquarters and 4 Companies.	Fermoy	12th Sept., 1848	All to Cork in January, February, and March, 1849.
3 Companies ...	Fethard	14th Sept., 1848	
1 Company	Cahir	15th Sept., 1848	
1 ,, ...	Lismore	23rd Sept., 1848	
1 ,, ...	Clogheen	30th Oct., 1848	

The above is only the actual list of stations. The companies changed incessantly between each place. It really appears as if no company remained more than three months in a place purposely.

CHAPTER XI.

EIGHTY-SIXTH REGIMENT (1819-1857).

Eighty-Sixth quartered at Canterbury—Establishment reduced—Moved to Chichester—Ordered to Weedon—Sent to Chatham —Establishment still further reduced—Marched to Bristol—Sent to Ireland—Sent to various detachments—Further changes in Establishment—Drowning of Lieut. Close—Regiment ordered to the West Indies—Stations in the West Indies—Death of Major Richardson and Boy Brown—Eighty-Sixth sent to British Guiana—Murders occasioned by drink—Regiment returned to West Indies—Changes in Honorary Colonelcy—Eighty-Sixth returned to England—In detachment in garrisons in England—Sent to Dublin—Ordered to India—Receives volunteers from other regiments—Regiment embarked for India—Serjeant Lindtner—Quartered at Belgaum—Casualties of the 86th—Eighty-Sixth ordered to proceed on active service for the Sikh War—Account of the Sikh War—Cholera epidemic in the Eighty-Sixth—Quartered at Deesa—Ordered to Hyderabad—Right wing moved to Aden—Detachment sent to Gujarat owing to native unrest.

THE Eighty-Sixth had landed at Gravesend on the 23rd of October. 1819 They marched to Canterbury on the 25th of the same month, where they joined the depôt company. This company was commanded by Brevet-Lieutenant-Colonel Lanphier, who, it will be remembered, was in charge of the grenadier company at the taking of Bourbon.

Here Lieutenant-Colonel Fraser, C.B., received his well-earned promotion to Major-General, and Lieutenant-Colonel John Johnson assumed command of the Regiment. Two companies were sent, under Captain Creagh, to Brighton.

On the 10th of November, 1819, a War Office letter was received ordering the establishment of the Regiment to be reduced to ten companies, and directing this to be carried out from the 25th of November, 1819. (The regimental records state 1818, but from the context this must be a clerical error.) The letter further directs that all officers ordered to be made supernumeraries should immediately be released from regimental duty, but should remain on full pay until the 10th of January, 1820, after which date they would be placed on half-pay. The strength of the Regiment under this order stood as follows:—

Colonel	1	Ensigns	8
Lieutenant-Colonel	1	Paymaster	1
Majors	2	Adjutant	1
Captains	10	Quartermaster	1
Lieutenants	12	Surgeon	1

Assistant-Surgeon	1	Serjeants	20
Serjeant-Major	1	Corporals	30
Quartermaster-Serjeant ...	1	Drum-Major	1
Armourer-Serjeant	1	Drummers	21
Paymaster-Serjeant	1	Privates	620
Schoolmaster-Serjeant	1		
Colour-Serjeants	10	Total	746

It did not remain long at Canterbury, for, on the 17th of December, 1819, it marched for Chichester, occupying the hut barracks on the road from Chichester to Havant. The Regiment only remained here until the 9th of April, 1820, when it proceeded to Weedon, performing part of the journey on canal boats. It took until the 25th of April to reach Weedon. On the 2nd of August of that year it detached two companies to Northampton.

1820

Whilst it was stationed at Weedon thirty-two recruits were raised locally and 251 were recruited by parties sent to Ireland for that purpose.

On the 10th of February the Eighty-Sixth left Weedon and marched to Chatham, passing through London en route.

1821

On the 10th of August a detachment was sent under the command of Major Marston, to Harwich. Here they formed the escort to the remains of the late Queen Caroline, wife of George IV., which were being sent for burial to the Continent. On the 16th of this month a letter was received from the War Office still further reducing the establishment of the Regiment from the 25th of August. It now stood as follows:—

Colonel	1	Quartermaster-Serjeant ...	1
Lieutenant-Colonel	1	Paymaster-Serjeant	1
Majors	2	Armourer-Serjeant	1
Captains	8	Schoolmaster-Serjeant	1
Lieutenants	10	Colour-Serjeants	8
Ensigns	6	Serjeants	16
Paymaster	1	Corporals	24
Adjutant	1	Drum-Major	1
Quartermaster	1	Drummers	11
Surgeon	1	Privates	552
Assistant-Surgeon	1		
Serjeant-Major	1	Total	650

Of course, all these reductions were caused by the pressure brought to bear on the Ministry by a Parliament bent on reducing expenditure.

The long war with Napoleon had raised the National Debt to over £700,000,000, and the interest on this, though gladly paid whilst Napoleon was a danger, was now found to press heavily on peaceful citizens, who considered that a certain amount of relief might be obtained by reducing to a dangerously low level the numbers of that very army which, by its gallant fighting, had saved them and Europe from coming under the all-embracing sway of the great Corsican.

Major Marston retired on the 21st of October by the sale of his commission, after long and honourable service in the Regiment. He was succeeded as Major by that very good officer, Captain Michael Creagh.

Having received orders to embark for Ireland, the Eighty-Sixth marched from Chatham to Bristol. There they embarked on the 25th of October and landed at Waterford, in Ireland, on the following day. The Regiment was at once broken up into ten detachments. The headquarters remained at Waterford, but detachments were sent to Kilkenny, Wexford, Duncannon Fort, Clonmel, Dungarvan, Portlow, Mullingar, Ballinamull, and Newtown.

1822

On the 1st of April, 1822, the headquarters of the Eighty-Sixth marched from Waterford and reached Naas on the 8th. All the detachments rejoined on the march.

From Naas, further parties were detached to Drogheda, Finn, Kilcock, Edenderry, Ballinglass, Athy, Robertstown, and Wicklow.

The Eighty-Sixth was now ordered into Dublin, and leaving Naas on the 10th of July, occupied Richmond Barracks, bringing in all detachments.

The Regiment was ordered to Athlone in the following year, and, leaving Richmond Barracks, in two detachments, on the 12th and 13th of May respectively, it marched to its new garrison, duly arriving there on the 16th and 17th of that month. On the 3rd of October of the same year it moved to Armagh, arriving there on the 10th of October. It remained there until the 16th of March, 1824, when it moved to Newry, leaving a detachment at Armagh, and sending others to Aughnacloy, Clones, Cootehill, Dungannon, Downpatrick, Drogheda, and other places.

1824

1825

It left Newry again on the 21st of January, 1825, and arrived at Naas on the 26th of that month, where it was quartered for a few months.

Whilst stationed here a circular memorandum from the War Office was received, dated the 25th of April, to the effect that each regiment of infantry was to be increased by two companies, and that only eight of these would serve abroad and be maintained to their full strength by the remaining four companies, which were to act as depôt companies, under command of one of the field officers. This order was not carried into effect until the Regiment was about to embark for foreign service towards the end of 1826.

The actual strength is shewn below:—

SERVICE COMPANIES.

Field Officers	2
Captains	6
Lieutenants	8
Ensigns	4
Paymaster	1
Adjutant	1
Quartermaster	1
Surgeon	1
Assistant-Surgeon	1
Serjeant-Major	1
Quartermaster-Serjeant ...	1
Paymaster-Serjeant	1
Armourer-Serjeant	1
Schoolmaster-Serjeant	1
Hospital Serjeant	1
Colour-Serjeants	6
Serjeants	18
Drum-Major	1
Drummers and Fifers	9
Corporals	24
Privates	492

DEPOT COMPANIES.

Field Officer	1
Captains	4
Lieutenants	4
Ensigns	4
Assistant-Surgeon	1
Colour-Serjeants	4
Serjeants	8
Corporals	12
Drummers	4
Privates	212

One acting-adjutant to be selected from the Subalterns.

One acting-serjeant-major to be selected from the serjeants.

All detachments were now called in, and the Regiment marched in two detachments for Clonmel, leaving Naas on the 26th and 27th June. From Clonmel it furnished no less than twelve detachments to the neighbouring towns.

Whilst the Regiment was stationed here it lost a very fine young officer—Lieutenant Frederick Close. He was found drowned in the Suir River, together with a Miss Grubb. Nothing is known as to how they met their untimely fate, but the following was all that transpired:—Lieutenant Close was supposed by his brother officers to be in love with Miss Grubb, who was the daughter of some most respectable Quakers. He went out to meet her at nightfall, and Miss Grubb is believed to have fallen into the water, and Lieutenant Close was supposed to have been drowned in attempting to rescue her, but nothing is really known of the matter. The really curious part of the matter was that a boy named Stuart, who afterwards joined the Regiment and commanded it, dreamt that Lieutenant Close's hat was found at a certain point on the River Suir. He mentioned the dream, and two days afterwards the hat was found at that identical spot and the bodies near it.

On the 12th of March, 1826, the Eighty-Sixth marched to Buttevant. On leaving the Limerick district (Clonmel), Major-General Sir Charles Doyle published a most complimentary order, thanking the Regiment for its good behaviour.

1826 The Regiment furnished the usual detachments—in this case eleven in number. On the 21st of September orders were received for the Eighty-Sixth to be in readiness to embark for the West Indies, and, with reference to the memorandum of the 25th of April, 1825, to form service and depôt companies. Not wishing to proceed abroad, Lieutenant-Colonel Johnson retired on half-pay, and was succeeded in command by Lieutenant-Colonel Mallett, C.B., who was transferred from the 89th Foot.

On the 9th and 10th of October the service companies commenced their march for Cork in two divisions, for the purpose of embarking, leaving the four depôt companies behind them, under the command of Captain Stuart, as Major Baird, who should have been in command, was retiring from the Service.

The Regiment embarked in three transports. The "Waterloo" carried two companies and sailed on the 27th of October, the "Princess Royal," with two companies, sailing on the 1st of November, whilst the remaining two companies followed in the "Thetis," sailing on the 5th of the same month. The total numbers embarking were twenty-three commissioned officers and 562 non-commissioned officers and men. The regimental records note as a most unusual circumstance that, despite the fact that the Regiment was almost entirely Irish, and that it was proceeding to a most unhealthy climate, only *one* desertion took place during this embarkation.

The "Waterloo" reached Barbadoes on the 24th of November, and was ordered to Trinidad, to relieve the 9th Regiment, which was stationed there. The men disembarked from this transport at Trinidad on the 29th of November. The "Princess Royal" disembarked her troops at Trinidad on the 4th of December, whilst the "Thetis" landed her two companies at Tobago on the 18th of December.

At Trinidad the Regiment occupied Orange Grove Barracks until the completion of the new barracks at Saint James. Being relieved by the 1st Battalion of the 1st Regiment, the Eighty-Sixth sailed for Barbadoes on the 1st and 5th of January, 1828.

1829 During its short stay at these two islands the Regiment lost Ensign Selway and forty-three non-commissioned officers and men, and its total casualties at Barbadoes, where it remained until the 14th of January, 1830, were the Adjutant (Lieutenant Dolman) and fifty-nine non-commissioned officers and men—all from fever.

1830 On the 14th of January the orders arrived for the headquarters to go to Antigua, whilst the left wing sailed for Saint Kitts, furnishing a subaltern, a serjeant, and thirty rank and file to Montserrat. The left wing arrived at their destination on the 19th of February, whilst the right wing reached their goal on the 24th of February.

On the 2nd of December Major W. Richardson died, universally regretted. He was the last of three brothers who had all served as officers in this Regiment. The officers and men erected a monument to his memory in the Parish Church at Newry, County Down.

Colonel Mallett, C.B., having gone home on leave of absence previous to this, Captain James Creagh assumed command of the Regiment until the arrival of Captain McLean, from Saint Kitts.

Whilst at Antigua, one of the band boys, James Brown by name, died one evening whilst playing in the band. He played his kettle-drum in perfect time until he fell to the ground dead.

1831 In 1831, in Antigua, the slaves objected to the Sunday market being closed, and set fire to several plantations. A strong detachment of the Regiment was sent out on the 13th of June, and remained in garrison in St. John's until tranquillity was restored.

Lieutenant-Colonel Mallett, C.B., rejoined the Regiment from leave on the 1st of April, and proceeded, on the 11th of September, to Saint Lucia, to assume the civil government of that island. He, unfortunately, died there in December, and was succeeded, on the 24th of January, by Lieutenant-Colonel M. Creagh, who had been appointed from a half-pay lieutenant-colonelcy to the 4th Regiment, and had been removed to the 86th Regiment.

1832

In March the further distinction was granted to the Regiment of bearing the motto of "*Quis Separabit*," in addition to the Harp and Crown, already borne.

Subalterns' detachments were sent in the summer from St. Kitts to Tortola, and from Antigua to Barbadoes, to control the slave population, which was, at this time, very much excited throughout the West Indies.

On the 3rd of December the Colonel of the Regiment, General the Earl of Kilmorey, died, and was succeeded in the colonelcy by Major-General Lord Harris, K.C.H.

Colonel Stuart, in his "Reminiscences of a Soldier," mentions no less than six cases of murder committed by men of the Regiment during their stay in the West Indies. In all cases the crime was committed under the influence of drink. New rum was plentiful and cheap, and its consumption drove the European temporarily mad. In some cases the culprit shot his own best friend, and in others the first person whom he chanced to meet.

In one case, at least, the murderer was really mad, even without the assistance of the new rum. He was named Fenelon. He had been drunk when for duty. One of the men took his equipment and went on parade, answering to his name and performed the duty. Coming off parade Fenelon shot him dead. When on the scaffold Fenelon threw an orange at a friend in the ranks, called to the band to play "Patrick's Day," and proceeded to

Colonel Sir M Creagh

Colonel, 86th Regiment, 1832

dance a jig, and was actually hanged whilst dancing. There could be no question that such a person would now find himself in a lunatic asylum.

The Eighty-Sixth was then ordered to Demerara and Berbice, and, being relieved by the 36th Regiment, the headquarters and right wing set sail on the 5th of February in three men-of-war, H.M. ships " North Star," " Columbine," and " Arachne," and in an *Army* brigantine, called " The Duke of York."

The Regiment landed at Demerara and Berbice respectively on the 14th and 17th of that month. The left wing arrived on the 27th of February, having voyaged in the troopship " Athol."

On the 1st of March, Lieutenant-Colonel Sir Michael Creagh, K.H., arrived from England, and assumed command of the Regiment. He had recently been created a Knight of the Hanoverian Guelphic Order, and received the honour of a British knighthood. He brought with him the new Colours of the Regiment, presented to them by their late Colonel, Lord Kilmorey.

On the 1st of December nineteen volunteers joined from the 93rd Highlanders, who had been ordered home to England.

The climate was not the only unpleasant part of the stay in Demerara. The mosquitoes and sand flies were so vicious that the Government actually supplied large bundles of wet straw, which were placed outside the men's barrack-rooms at night and set on fire, to keep these pests away. The men slept as best they could in the midst of the smoke. Often the sand flies became so bad that even in the officers' mess wet straw had to be placed in iron pots under the chairs and set fire to, to smoke these active intruders away. Fort Cange, Berbice, appears to have been the worst place.

1835 On the 4th of December the colonel of the Regiment, Lord Harris, was transferred to the 73rd Regiment, and was succeeded by Major-General the Honourable R. C. Ponsonby.

The colonelcy of the Regiment again became vacant by the transfer, to the 1st Dragoons, of Major-General Ponsonby; he was succeeded, on the 31st of March, by Major-General Watson.

1836 On the 8th of May the Regiment was ordered to Barbadoes, and the part at Berbice was embarked on H.M.S. " Gannet," and taken to near Demerara, where they were transhipped to H.M.S. " Vestal," which already had the remainder of the Eighty-Sixth aboard. The transfer was carried out at sea, and the troops landed at Barbadoes on the 17th of May.

1837 On the 21st of March the Regiment embarked on the " Moira," transport, for England.

Lieutenant-General Whittingham, Commanding the Forces, having seen the embarkation, directed a General Order to be issued, calling attention to the orderly and soldier-like manner in which the embarkation was carried out in every detail. Thirteen officers and 424 rank and file were embarked, also seventeen women and forty-five children.

After a rapid voyage, the "Moira" anchored at the Cove of Cork on the 22nd of April, and on the 24th of that month orders were received to sail for Chatham.

During its ten years' tour in the West Indies the Regiment lost five officers and 299 non-commissioned officers and men by death. On the 3rd of May the depôt companies (four in number) joined the service companies at Chatham, bringing in fourteen officers and 260 non-commissioned officers and men. They had spent the ten years, whilst the Regiment was in the West Indies, in various parts of the United Kingdom. Up to 1830 they remained in very many places in Ireland; in that year the depôt companies embarked for Jersey from Cork, returned to Portsmouth in 1831, and in 1835 returned to Ireland.

On the 24th of May Major-General Sir Arthur Brooke, K.C.B., was appointed colonel of the Regiment, in succession to Major-General Watson, removed to the 14th Regiment.

General Lord Hill, Commander-in-Chief of the Army, inspected the Regiment on the 8th of August, and expressed his satisfaction with it. The regimental records state that it was believed to be the first time that any regiment had returned from a full tour in the West Indies in a perfect state of efficiency.

Certain troubles having arisen in the industrial centres in Lancashire, the Eighty-Sixth was ordered to Manchester, sending one company to the Isle of Man. They remained in or near Lancashire for some two years, being generally split up into detachments. For instance, on the 14th of April, 1838, there were two companies at Stockport, three at Burnley, three at Blackburn, one at Rochdale, and one in the Isle of Man.

1838 and 1839

On the 2nd of June the Regiment assembled at Liverpool and sailed to Belfast, arriving there next day and remaining there until the 22nd of September, when a portion of the Eighty-Sixth *sailed* for Dublin, followed on the 23rd of that month and on the 7th, 12th and 15th of October by the remainder of the Regiment. At first they were scattered in Richmond, Island Bridge and Beggars' Bush Barracks, but by the end of the year all were concentrated in Richmond Barracks.

1840

To the elections of this year the Regiment owes its first trip in a steamboat, for two companies were sent round by steamer to Cork on the 1st of July, and were kept very busy in preserving the peace of this city during the hotly-contested election there, whilst other companies were on the same peaceful errand at Templemore, Clonmel, etc., and their labours did not cease until the 26th of August.

1841

On the 19th of that month orders were received directing the Eighty-

SHAKO PLATE OF 86TH REGIMENT, 1828-1844.

Dead gilt star and crown, silver cut star with gilt labels "India" and "Bourbon" mounted on, other mounts gilt, the garter pierced, blue enamel under the garter and centre

Sixth to be in readiness to proceed to the Cape of Good Hope, and the four depôt and six service companies were duly separated, and the depôt companies were sent to Spike Island.

1842 On the 12th of March, however, the War Office changed its mind and directed the establishment to be made up to 1,000 rank and file, as the Regiment was to proceed to India. This alteration of destination altered the composition of the depôt companies, which was reduced to one company, whilst the Regiment took nine companies to India. This depôt company returned to the Regiment, and on the 7th of April still further orders were received directing the establishment to consist of forty-eight officers and 1,066 warrant officers, non-commissioned officers and men, exclusive of the depôt company, which consisted of one captain and three other officers, one colour-serjeant, five serjeants, five corporals, and one drummer, with no establishment of privates mentioned.

On the 19th of April a further letter was received stating that notwithstanding any orders to the contrary received from the Secretary of State for War, it was Lord Hill's intention that the Regiment was to be formed into ten service companies and one depôt company, and adding the information that steamboats would convey the Eighty-Sixth to the "River."

The ten service companies were formed on the 19th of April.

To bring up the Regiment to the required Indian establishment, volunteers were allowed to join from other regiments, and the following numbers joined from the under-mentioned regiments:—

2nd Battalion Royal Depôt, 40; 8th Foot, 50; 27th Foot Depôt, 48; 35th Foot Depôt, 73; 36th Foot Depôt, 69; 70th Foot Depôt, 11; total, 291, besides seven rejected as bad characters.

On the 21st, 22nd, and 25th of April, the Regiment embarked in three detachments in hired steamers and came to Liverpool. Here, as the regimental records state, the journey was continued by "land steam conveyance, via Weedon, to Gravesend." Each train passed the night at Weedon for some reason.

The Regiment embarked on five different ships, four of which—the "Inglis," "Margaret," "Rajasthan," and "Berkshire"—sailed on 2nd of May, whilst the "Eliza Stewart" brought the remainder of the Regiment on the 27th of May.

The regimental records note with pride that only twenty-three desertions occurred from the time the Regiment's destination was changed from the Cape of Good Hope to its embarkation for India, viz., from the 14th of March to the 2nd of May. Nowadays this would be looked upon as extraordinarily bad, but then it was regarded as exceptionally good.

The total strength of the Eighty-Sixth embarked appears to have been 33 officers, 1,052 non-commissioned officers and men, and 127 women and 152 children.

Several points appear to call for remark in looking over this embarkation. One was that though most of the troops were aboard by the 29th of April, they did not sail until the 2nd of May, presumably waiting for a favourable wind; whilst those who sailed in the "Eliza Stewart" embarked on the 20th of May, and did not get away until the 27th of May. Again, at the embarkation one lieutenant-colonel, one captain, and eight lieutenants were reported as absent without leave, of which number four lieutenants were reported as being absent without leave in India, whilst one lieutenant was absent without leave in New South Wales. The lieutenant-colonel returned in time to go out to India on the "Eliza Stewart." The probable explanation of all these officers being absent is that undoubtedly they really had leave, but owing to the slowness of posts and official delays, the necessary authority for their absence had not reached the Regiment by the time it sailed. Two officers received leave to proceed to India "overland."

The ship "Inglis" arrived at Bombay on the 30th of July, having cleared the Land's End, in England, on the 14th of May, thus making one of the quickest voyages on record.

Before the other ships had arrived cholera broke out on the 3rd of August. There were about thirty cases, of which at first only three were fatal. On the 16th of August Captain Bennett died of this disease whilst sitting on a court-martial, and the bandmaster, Serjeant Henry Lindtner, also died of it on the 26th of August. He is recorded as having served with the Regiment for sixteen years. Out of that time, however, he had only been enlisted for one year, permission having been given for his enlistment on the Regiment being ordered to India. In those days regiments obtained their bandmaster where the colonel thought fit—often from Germany. This bandmaster was a German, and for fifteen years of his service had been "hired" by the officers, to conduct their band.

By the 26th of September the last transport, the "Eliza Stewart" had arrived, and the Regiment was ordered to Belgaum, leaving in vessels from Bombay in detachments from the 6th of October, landing at Vingerla, and all arrived at Belgaum by the 5th of November, excepting 120 non-commissioned officers and men left sick in Bombay. Between the 1st of August and 31st of October 2 officers, 94 non-commissioned officers and men, 13 women, and 29 children on the strength of the Regiment died, about two-thirds of them from cholera.

1843

On the 1st of February, 1843, the Governor-General of India ordered the Regiment to be reduced from ten service companies to nine, as ordered by War Office originally and altered by Lord Hill, Commander-in-Chief in England.

On the 25th of August Lieutenant-General J. Maister was appointed Colonel of the Regiment, vice Lieutenant-General Sir A. Brooke, K.C.B., who had died.

1844 On the 1st of January the first portion of the Regiment left Belgaum for Karachi, arriving there on the 16th of January. Part of the journey was performed, of course, in ships; the remainder of the Regiment joined headquarters on the 2nd of March.

On the 11th of March the Eighty-Sixth commenced to march to Hyderabad. It is noted that on the 25th and 26th of May eleven non-commissioned officers and men died of *coup de soleil* (sunstroke).

On the 3rd of September the Regiment was back in Karachi, having left 1 officer and 127 men sick in Hyderabad, and having lost there by death 1 officer and 56 non-commissioned officers and men, and 12 women and 34 children.

1845 On the 1st of January the strength of the Regiment was forty-six officers and 1,043 non-commissioned officers and men. A constant stream of officers and men had had steadily to come out from England to keep the Regiment up to this strength.

On the 29th of December the Eighty-Sixth received orders to join the Army of the Lower Sutlej, ordered to assemble at Koree.

The Eighty-Sixth went on this campaign in the happy position of being fourteen over strength—that is, including drummers and corporals, they had 964 rank and file, instead of the 950 allowed by the establishment. It, however, took no active part in the campaign.

The Sikhs were the cause of the trouble on the Sutlej in December, 1845. It is a complicated story, but runs more or less as follows: While Runjeet Sing, the great ruler of the Sikhs, lived, peace was maintained, but when this ruler, known as the "Lion of the Punjaub," died in 1839, the prospects for peace were not so hopeful. His son succeeded him, all power being really in the hands of his son again—one Nonehat Sing. The latter was bitterly hostile to the British. He was murdered by a weight thrown upon him as he passed under a gateway. Shere Sing, supposed to be one of Runjeet Sing's sons, succeeded him.

He had a struggle to obtain possession of the throne, his opponent being Nonehat Sing's mother, the Ranee. When he did obtain possession, he took to drink, and was murdered, and the Ranee returned to power. The army now usurped the power of the State, and the only territory to plunder was British territory. Sir Arthur Hardinge assembled an army for the protection of the East India Company's possessions.

On the 13th of December the Sikh army crossed the Sutlej, and was assembling on the left bank in great force in British territory.

On the 18th of December, after a 22-mile march, 12,350 British troops attacked 30,000 Sikhs, with forty guns, at Moodkee, and swept them from the field, losing 872 killed and wounded.

On the 21st and 22nd of December 16,700 British troops, with 69 guns, attacked 50,000 Sikhs, with 108 heavy guns, in an entrenched position, at

Ferozeshah. The Sikhs, after a most gallant resistance, were thoroughly routed, the British losses being 2,415, and the Sikhs losing some 9,700.

The Sikhs, after this defeat, retired over the Sutlej, but the British did not at the moment follow them, as they were awaiting the arrival of a battering train from Meerut.

Plucking up heart, the Sikhs again came across the river on a bridge of boats to the British side, and a party was caught by Major-General Sir Harry Smith, G.C.B., with 10,000 British troops, at Aliwal, on the 26th of January, 1846. The Sikhs had 19,000 men and 68 guns, and held an entrenched camp, but the British troops drove them headlong into the river, though not without severe fighting.

The main body of Sikhs held their bridge, and Sir Hugh Gough, Commander-in-Chief in India, marched on the Sikh bridge-head. It was defended by over 30,000 Sikhs, and the fighting was of the hardest. Finally the entrenchments were carried, and as the Sutlej had conveniently risen some seven inches that morning, it was barely fordable, and numbers of the gallant foe were drowned; 2,383 were killed and wounded on the British side, whilst the Sikhs again lost about 10,000 men.

1846

On the 22nd of February, 1846, the British occupied the citadel of Lahore, the capital of the Sikh nation, and the war was over.

On the 27th of January, 1846, the whole Regiment was concentrated at Koree.

On the 10th of February Lieutenant-Colonel Derinzy, then commanding the Eighty-Sixth, was appointed to command the 3rd Infantry Brigade, and Lieutenant-Colonel Aplin took over command of the Regiment. By the 2nd of April the Regiment embarked for Karachi, arriving there on the 17th, having travelled 1,100 miles, and lost three men from disease, and two from drowning in the Indus.

On the 14th of June No. 1402, Private William Bradley, died of cholera, and during the next three days 300 men were in hospital with it. The records contain a most heart-rending account, compiled by the surgeon-in-charge, who states that a dark cloud and dust storm appeared to usher in this plague. It is sufficient to say that no officers suffered from the disease, but that 410 men were admitted to hospital, of whom no less than 240 died, whilst nineteen of the married women also suffered from cholera and fifteen of them died. The children escaped with only seven deaths. Curiously enough, the grenadier and light companies suffered most heavily, losing thirty-six and thirty-one men respectively. The heat was not great at the time, being only about 90°.

1847

In January, 1847, the Regiment left Karachi for Bombay. During their stay in Scinde from 1844 to the end of 1846, the total deaths amongst the non-commissioned officers and men were 391, whilst 106 children of the married families were buried there. The headquarters arrived at Bombay on the 3rd of February, and marched for Poona, arriving at the latter place on the 15th of that month, leaving half the battalion at Bombay, but on the 2nd of January, 1848, the whole were united at Deesa. In December, 1847, 122 men volunteered to the 86th Regiment from the 28th Foot, whilst 229 recruits arrived from England, bringing the Regiment to seventy-one above strength.

1850 After an uneventful tour of duty at Deesa, the Eighty-Sixth left there on the 2nd of November, and, by marching and by sailing in country craft, etc., moved, via Bombay, to Poona, where it arrived on the 10th of December. Previous to leaving Poona, the General Commanding there called attention to the good behaviour of the Corps, and particularly noted the extraordinary number of good conduct badges amongst the men.

1854 Leaving Poona on the 20th of January, by detachments, the right wing was assembled at Hyderabad by the 27th of March, and the left wing was at Karachi.

Poona had proved a healthy station, as only forty non-commissioned officers and men had died there from 1850 to January, 1854.

On the 29th of April Colonel Aplin was commanding the brigade at Hyderabad, whilst Lieutenant-Colonel Creagh assumed command of the Regiment at Karachi, every preparation being made for its being sent to the Crimea, if necessary.

Major-General Lord James Hay was appointed colonel of the Eighty-Sixth Regiment on the 8th of May, in the place of Major-General Roger Parke, appointed 26th of May, 1852, who had died.

On the 1st of September Lord Hill's long-cherished plan of having ten companies instead of nine in regiments serving in India was brought into effect. On the 28th of this month Colonel Aplin died at Karachi.

1855 On the 8th of May the right wing was sent to Aden, under the command of Lieutenant-Colonel Tudor, whilst the left wing left Karachi and went to Bombay, arriving there on the 20th of December.

1856 Ensign Dartnell joined from the depôt on the 11th of March. This officer afterwards became well-known from his good work in South Africa during the Boer War of 1899-1902, and was created K.C.B.

1857 On the 27th of May a detachment of 150 non-commissioned officers and men, under the command of Captain Jerome, was sent into Gujarat, as the natives there were showing signs of unrest, whilst the wing at Aden rejoined the headquarters at Bombay on the 7th of July, having been away for two years.

CHAPTER XII.

EIGHTY-THIRD REGIMENT (1857-1859).

Description of India in 1857—Effect of 100 years of British Rule—Numbers of Native and European Troops in India—Mutiny of Vellore—Difference between Madras and Bengal Native Armies—Annexation of Oude—Questionable dealings of Financial Authorities with the Military—Greased Cartridges and their effect on the Sepoys—Rising at Meerut—Delhi—Cawnpore—Lucknow—State of the country north of the Nurbadda River—Rajputana—Mutiny at Nusseerabad—Ajmere—Mutiny at Neemuch—Attempted Mutiny at Nusseerabad—Further trouble at Neemuch—Attack on Mount Abu—Jodhpur Army beaten by Jodhpur Legion—Repulse of British at Awah—Destruction of the Jodhpur Legion—Operations at Nimbharia—Jeerum—Blockade of Neemuch—Rajputana Field Force formed—Awah—Kotah—Combat of Sanganeer with Tantia Topi—Kotaria—Half-yearly Inspection by General Roberts—Combat at Sikar—Pursuit at Kushana—Capture and execution of Tantia Topi.

1857

In the beginning of 1857, peace reigned throughout India. The East India Company, familiarly known as "John Company," had fared on, since it was formed in A.D. 1600, through small beginnings, through many a disastrous day, until it had arrived at being the undisputed ruler of two-thirds of the Indian Peninsula and the "predominant partner" in alliance with the rulers of the remaining one-third of that huge tract of country. Well might its Directors in Leadenhall Street, London, consider that they were more firmly established in India than they had ever been before, and yet it was just when absolutely nothing appeared to threaten the Indian Dominions that the most appalling crisis that the Company had ever experienced arose and had to be dealt with.

In dealing with this crisis both the 83rd Regiment and the 86th Regiment took a part, and it will explain their action more clearly if the situation in India during 1857, 1858, and 1859 be studied first as a whole.

To quote a recent author* whose work and opinion will often be referred to in these chapters, the effects of one hundred years of British rule since the battle of Plassey, were as follow:—

He says:—"Few Englishmen care to learn how a handful of their countrymen established that rule and steadily widened the sphere of its operation; for they do not know that they are refusing to look upon a unique historical drama, full of picturesque incident. . . . gorgeous potentates, intriguing courtiers, subtle diplomatists, ambitious queens hatching plots in the recesses

* Rice Holmes's "History of the Indian Empire."

of their palaces, clan chieftains founding empires, daring upstarts forcing their way by craft and violence to the command of armies and the conquest of kingdoms, cunning priests inspiring awe alike in king and noble, soldier and statesman, merchant and artizan, while suddenly the strong figure of the 'White Man' appears in the midst, dominates all, evolves order out of chaos, bids the contending rulers hush their quarrels, and holds out hope to suffering millions. But though each successive page of the drama contains fresh revelations of the dauntless courage, the adventurous generalship, the farseeing statesmanship of the Englishman, it would have only a tragic interest if it did not bear witness to his righteousness of purpose. It had been with this purpose before him that he had given order, peace and justice to the country which he had found a scene of anarchy, intestine war, and injustice; that he had disabled the monster Famine—and looked forward to destroying it—that he had reclaimed vast tracts from the ravages of wild beasts, repressed crime, stimulated industry and developed commerce.

"Yet his rule had been no unmixed benefit. Sometimes the very energy of his benevolence had intensified the evil which his ignorance had wrought. At other times the faults of his character had led him astray. An eminent Frenchman has characterised his government as 'just but not amiable.' That terse criticism exposes its weak side; while the ruler had laboured for the material well-being of his subjects, he had too often failed to reach their hearts, and in his calm sense of superiority he had forgotten that his intrusive reforms might not always be appreciated.

"It was not that the natives resented the thoroughness with which he exemplified the maxim 'Everything for the people; nothing by the people.' They were accustomed to depend for their happiness upon the favour of their rulers; and they could appreciate the benefits of a strong and just rule. They might boast idly of their own superiority; but they were persuaded in their inmost hearts that the Europeans were their superiors. It was only necessary for the master race to assert its supremacy, and it would have secured a willing obedience. But, unhappily, while it had sometimes shrunk from avowing and righteously exercising the supremacy which it in fact possessed, it had too often provoked an unmerited distrust of its benevolence. Its land legislation had aroused the ill will of a class whom it was important to conciliate, and who complained that, having made use of their influence over the lower classes to conquer the country, it no longer cared to treat them with common civility. It had heedlessly thrown a host of native officials out of employment by filling up their places, after each new conquest, with men of its own choice. By occasional acts of indiscretion it had shaken the old confidence in its tolerance. It had once been hailed by the victims of tyrannical princes as their deliverer. But a new generation had arisen, who felt no gratitude for the deliverance of their fathers from a tyranny which they had never suffered, and who, moreover, saw in the traditional deliverers actual conquerors." Such was, in well thought out phrases, the state of men's minds in India as the result of one hundred years of British rule.

The hard facts of the case were that many men of influence, whose rapacity the British had checked, hated them. Their retainers joined them in their hatred from clan feeling, or from having their share of the spoils also taken away. Many protected rulers feared that the Company proposed to do away with their family's succession to their thrones under various specious pretexts. Men of daring found their careers of violence, which formerly led to wealth and chieftainship, rudely checked by the strong hand of British law, and deeply resented it, whilst the Hindu teachers feared that their people would grow faint in their adhesion to the faith, and they would, as a rule, gladly have seen the last of the English. Finally, the Sepoy, or native army of Bengal was none too loyal, and it was this fact, fanned no doubt by religious fanaticism, which gave rise to the outbreak known as the Indian Mutiny.

It was at Bombay that the first native troops, drilled in English fashion, were raised, and soon large bodies of these useful auxiliaries were formed all over Southern India. The defence of Arcot, even before the great victory of Plassey in A.D. 1757, was where the Indian soldier, under British leaders, first proved his worth. Gradually three strong armies grew up—the Madras, the Bombay, and the Bengal—and in 1857 the number of native mercenaries had risen to 233,000 men. To this total must be added the armies of independent princes and chiefs, not, however, including in those armies such forces as the British were supposed to keep up for the protection of their allies, and for which those allies paid a subsidy (such as the Nagpor Subsidiary Force), as these troops would be included in the grand total given above.

To balance these large numbers of native troops, the East India Company maintained 45,000 European soldiers. Some of these were regiments of the Regular Army, lent by the British Government, others were troops raised by the Company itself.

It is said, however, that the European troops were located throughout India on such false principles that their controlling power was seriously impaired. Be that as it may, it is very doubtful if the authorities, civil or military, had considered the chances of a mutiny of the Indian troops when they arranged the stations for European regiments, and certainly neither had ever thought that any mutiny amongst those troops would assume such vast proportions. Sir Charles Napier was one of the few generals who had predicted a great mutiny in our Indian Army. Mutinies had occurred before in the Native Army from the time of Clive downwards. Perhaps the most notable one was that which took place at Vellore. Though it has but little bearing on the Great Mutiny, a rather full account is given of this outbreak, to show how one good soldier promptly crushed the plot which might have been as dangerous to the British rule, if it had not been at once checked, as the one of 1857 was through not being dammed at once at its source. Vellore was a strong and large fort in Southern India, surrounded by formidable granite walls. In it resided the four sons of Tippoo Sahib. It will be remembered that Tippoo was killed at the storming of his capital in 1799. The garrison of Vellore consisted of four companies of the 69th Regiment, three companies of the 1st Madras Infantry, and the whole of the 23rd Madras Infantry. Tippoo's sons made a party amongst the Mohammedans of the Native Infantry, and at 2.30 o'clock on the morning of the 11th July, 1806, the native troops rose and proceeded to murder their white comrades. The 69th Regiment, 372 strong, fought until 8 a.m., though only provided with six rounds of ammunition each. The Sepoys turned two guns on them. Repeated attacks with the bayonet were made by the Europeans, of whom only 200 remained unhurt. All their officers having fallen, the charges were led by the two surgeons of the regiment. In the meantime Colonel Robert Gillespie was at Arcot, some sixteen miles away, with the 19th Light Dragoons and the 7th Madras Cavalry and some guns. A breathless messenger met Colonel Gillespie riding, and told him the news. Within a quarter of an hour a squadron of the 19th Light Dragoons and a troop of the 7th Madras Cavalry were thundering down the road to Vellore, being shortly followed by the remainder of the 19th, with the "Galloper" guns. Unable to enter the gates, as they were barred and were swept by the fire of the revolted troops, Colonel Gillespie was hauled up by a rope which he threw up to the European survivors on the ramparts; then, heading a bayonet charge of the men of the 69th Regiment on the mutineers, he held them in check for a moment whilst the "Galloper" guns blew in the gates. The cavalry charged into the fort, and 350 Madras mutineers were picked up afterwards dead in the fort, whilst the remainder were killed outside; 116 of the British were killed and 80 wounded. At Java, in 1811, Colonel Gillespie led the assault on the Dutch fortress with quite a small column, turned the large garrison out in a panic, led the cavalry after them on an artillery horse cut loose from a captured gun, brought up the infantry and again headed their attack when the flying Dutch took to country unsuitable for cavalry action, and alternately led the cavalry and infantry in repeated attacks until their Dutch Governor, General Janssens (who had met the 83rd Regiment at the Cape in 1806) was forced to surrender. General Gillespie, as he afterwards became, was shot through the heart in the Nepal War of 1814, when attacking the fort of Kalunga, and was buried at Meerut, where, within one mile of his grave, the first shots of the revolted troopers in 1857 were fired. It is probable that if he had been present this mutiny would also have ended in one day. Colonel Gillespie came from Comber, in County Down.

The armies of Bengal and Madras were, in one way, quite different to each other. The Madras Sepoy was almost always a married man whose wife and family lived in the lines of his regiment. The Bengal Sepoy was probably married, too, but his wife and family lived in his own village, so the Bengal Army was, on the whole, a bachelor army. Two points are plainly marked from this difference. The first was that the man from Bengal, if he mutinied, had given no hostages to fortune in the shape of families in his barracks, and all he had to do was to fly, if the mutiny failed, as fast as he could go to the hills or thick jungle, to avoid the avenging British. The second point was that the Madras Sepoy, having his family with him, made his regiment his home, and the greater part of his interests lay there; thus he was less liable to be moved by outside influences. On the contrary, the Bengal Sepoy had his wife in the native village from which he came; he went there frequently and heard from it frequently, and thus kept more in touch with the local sentiment and feeling of his own countryside. This sentiment and feeling being, unfortunately, of a hostile nature to the British, reacted on the Bengal Army, and whilst nearly the whole of that army mutinied, the Madras Army remained staunchly loyal to their British rulers.

Why there should have been bad feeling entertained towards the Company in Bengal is a very long story. It is sufficient to say that most of the Bengal Sepoys were recruited from the villages of Oude, the capital of which country was Lucknow. Here the King of Oude held his court. He was an ally of the British, but his territory was not administered by them. Still, by treaty he had to be supported against all comers by the arms of the Company. Thus the Resident at his Court found that the most awful tyrannies were practised on the unfortunate inhabitants of the country; when these people rose against their oppressors in petty revolts, the British power could be used to suppress their outbreaks, but not to right their wrongs. Seeing the unfairness of this, the Indian Government annexed Oude in 1855, and tried to reform these abuses. In doing so they offended the great landowners of the kingdom. These had originally started as tax farmers over certain villages, and had gradually acquired a vested right in these districts. The British Commissioners refused to recognise that these gentlemen had any such vested rights, and the result was that they became hostile to British rule. The country people, not realising the meaning of all this change, to which they were, on principle, averse, sided with their land denuded aristocracy.

So much for the feeling of the countryside in Oude; another question to be considered is whether the native Army had no grievance of its own. The term " just but not amiable," has been used to describe the British Government in India. It is, however, a question as to whether the term "Just" could be applied to the Military Administration of the East India Company. Readers of this work will remember how the mutiny of the officers of the Madras Army was caused by gross injustice in reducing those officers' allowances. Another case directly affecting the pay of the men was constantly arising as follows:— When troops were sent into non-British territory for warlike or other operations, extra pay was given to the Sepoy to pay for the increased cost of his provisions and other incidental expenses. So soon as the operations were successful, the usual outcry was made by the purseholders of the East India Company against the great expense of the armed forces. This cry was met by the authorities proclaiming the country to be British Territory, and then the wretched troops automatically lost their extra pay, though they still had to pay the same high prices for their provisions. Other instances could be cited as affecting the pay of the troops, but enough has been said to show that they had grievances, and the Military Financial Authorities were not the most likely people to view these troubles in a sympathetic spirit, judging by their decision in a well-known case, viz., when the 15th Bengal Infantry mutinied the adjutant, at the risk of his life, tried to recall them to their duties, and had his horse killed by some shots directed against himself. The Military Auditor-General, however, refused him compensation for the loss on the grounds that the regiment having mutinied he had no longer occasion to keep a charger, and therefore it was not necessary to replace it.

As things fell out, when the Mutiny did break forth the Bengal Army was quartered over a larger extent of country than usual, owing to the fact that the Bombay Army had furnished some regiments for the Persian War and the Bengal Army had taken over some of their garrisons in Rajputana, thus enlarging the probable area of disturbance. The ostensible cause of the great outbreak is generally considered to be the introduction of the greased cartridge. The new Enfield rifle required these cartridges, and the soldiers, when using them, were supposed to bite off the end with their teeth. What they were greased with is an open question, but the Sepoys had the idea firmly fixed in their minds that the British Government intended to issue these cartridges with a view to destroying their caste. This idea was sedulously cultivated not only by agitators who hoped to turn the Sepoys' discontent to account, but also by the lower caste men, who did not care for their higher class brethren, and taunted them with their imminent loss of caste.

As soon as the authorities realised the feelings of the men they very sensibly gave permission for the Sepoys to use what fat they liked for greasing the cartridges, but it was too late. The Bengal Sepoys, some disloyal, many ill-disciplined, but the majority genuinely frightened of horrors which their superstitions conjured up, were ripe for disorder. This aspect of the Mutiny is very plainly apparent to an unbiassed observer, and though no words are strong enough to convey the utter abhorrence of their treacherous and dastardly conduct in shooting down their own officers, and in murdering and ill-treating unfortunate European women and children, yet it cannot be denied that many evinced as great courage when fighting against the British even as they had on many occasions when battling on their side, and it appears that the mutinies were in many cases the result of their undefined fears. In no other way is it possible to account for some of the outbreaks. To take a typical case. On the 28th August, 1857, the 51st Native Infantry, which had already been disarmed, had their lines searched by the European authorities, and swords, muskets, pistols, and ammunition were found hidden under floors and in the roofs. The 51st Native Infantry, entirely driven by fear, for they must have known in their sane moments that to resist was certain death, seized the piled arms of a newly-raised irregular regiment, overpowering the regiment, and were immediately overpowered themselves. Sixty escaped. Seven hundred were buried in three deep trenches.

Minor mutinies occurred at various military stations on the 26th of February, on the 29th of March, and on the 3rd of May, but on the 10th of May the first really serious outbreak occurred in the evening at Meerut. The garrison there consisted of the 11th and 20th Bengal Infantry and the 3rd Bengal Cavalry, the European troops being "The Carabiniers," one battalion of the 60th Rifles, and some Horse and Field Artillery. Here the Native Cavalry mutinied and burned some bungalows and murdered their officers, assisted by the Native Infantry, and then fled unpursued to Delhi, whilst the unfortunate British troops, commanded by a general in his dotage, were refused permission to attack the mutineers. At Delhi the rebels were joined by the native battalions and artillery quartered there, and all proclaimed the last of the Mogul Kings, who lived in the palace there as a pensioner of the East India Company, as their king. The possession of Delhi placed the resources of that arsenal at the disposal of the rebels, though it had been partially destroyed by some brave Englishmen. On the 30th of May a British force marching on Delhi was opposed by the mutineers at the river Hindan. The mutineers were defeated and a British force was established on the ridge at Delhi, and held its own until the city was stormed on the 14th of September, with a loss on that day alone of 1,170 officers and men. Cawnpore was another place where the Mutiny broke out very fiercely. Here the Nana Sahib was a power in the land. He was the adopted son of Peshwa, chief ruler of the Mahrattas, and had just been refused the continuance of that father's pension by the Government. He was bitterly opposed to the British, though he concealed his feelings under an appearance of friendliness. His right hand man was named Tantia Topi. A copy of a Court-Martial in which the latter took a leading part will be found at the close of the next chapter. Situated some forty miles from Lucknow, it was garrisoned by four native regiments, who mutinied, and by a few European gunners and some of the 32nd Regiment, totalling 210 Europeans in all. On the 6th of June, these unfortunate people were attacked by the mutineers; on the 27th they capitulated, and all—men, women, and children—were basely murdered, excepting two British officers who fought their way through, and one lady.

Lucknow was garrisoned by the remainder of the 32nd Regiment and by several Native Infantry regiments and one Native Cavalry regiment. The Mutiny broke out here on the 30th of May, but several hundreds of the native soldiers remained faithful. The residence of the Commissioner, Sir Henry

Lawrence, was held, first under his own command, and then, when he was killed, by Sir John Inglis, the colonel of the 32nd Regiment, a Nova Scotian by birth. The fighting here was of the most desperate description; mining and countermining went on merrily, but on the 25th of September a small force fought its way into Lucknow, under the command of General Havelock. This column had several times essayed to come up from Cawnpore, which place they had seized, and had, as far as possible, avenged the awful massacre there. On the 17th of November, General Sir Colin Campbell, Commander-in-Chief in India, came up with a small column, and leaving a couple of thousand men, withdrew the women and children to a place of safety. He returned again with a much stronger force on the 9th of March, 1858, and cleared all the enemy out of Lucknow, after desperate street fighting, which continued up to the 14th of March. Such, in brief, is the history of the Mutiny, as popularly known, for when the Mutiny is mentioned the imagination always conjures up Delhi, Lucknow, and Cawnpore. But if that had been all the Mutiny, the military authorities would not have been very hard pressed to put it down. But there was much more than that. If the map of India is looked at, a river called the Nurbadda is seen to flow to the western side of India, falling into the sea at a point some 170 miles north of Bombay. Carrying that river's course on a straight line right across India, it can be roughly said that almost all the native troops north of that line either mutinied or had to be disarmed, and that many of the inhabitants were up in arms. The following extract gives a capital impression of the state of the country during this time:—" It is hard for a reader unacquainted with the characteristics of Indian society to picture to himself the headlong violence with which the floods of anarchy swept over the North-West Provinces when once mutiny had let them loose. Neither the Hindus nor the Mahometans generally regarded the English with any particular dislike; they acknowledged, notwithstanding all their grievances, the comparative justice and efficacy and the absolute benevolence of English rule; but they were too ignorant to perceive that it was their interest to support it; they knew nothing of the reserve force that was available to rescue it in case of danger; and, therefore, when the defection of the Sepoy Army seemed to threaten it with destruction, they naturally relapsed into the turbulent habits of their ancestors, and prepared to make their profit out of the new order of things. Bands of mutineers and hordes of escaped convicts roamed over the country and incited the villagers to turn upon the Feringhees; rajahs emerged from their seclusion, gathered their retainers around them, and proclaimed their resolve to establish their authority as vassals of the King of Delhi. Mobs of Mahometan fanatics unfurled their green flags and shouted for the revival of the supremacy of Islam. Rajputs and Jats renewed old feuds and fought with one another to the death. Swarms of gujars (professional robbers) starting up on every side and girding on their swords

1857

and bucklers and shouldering their matchlocks, robbed the mail carts, plundered peaceful villages and murdered the villagers. Mobs of budmashes set fire to tahsils and drove out the tahsildars (petty native revenue officers). The native police, who had generally been recruited from the dangerous classes, and whom interest, not loyalty, had hitherto kept on the side of authority, felt that there was nothing to be gained by endeavouring to prop up a doomed government, and threw in their lot with the evildoers. Dispossessed landowners, clutching at the opportunity for which they had long waited, gathered their old tenants together, hunted out the purse-proud upstarts who had bought up their estates, and triumphantly re-established themselves in their ancestral homes. Insolvent debtors mobbed and slaughtered without pity the effeminate baniyas (money lenders) whose extortion they would have punished long ago, but for their dread of the strong arm of the law. Suttee, and other barbarous customs which benevolent rulers had abolished, were re-established. The mass of the people enjoyed the excitement and the freedom of the time. How disastrous was the collapse of authority will be understood from the fact that public works, except those undertaken for military purposes, absolutely ceased. Civil justice could only be administered in a few isolated and favoured spots."

In this welter of violence, here and there in places where European troops were quartered tranquillity was maintained. Yet some of the peoples came to the aid of the paramount power. The Sikhs of the Punjab convinced that we would win came forward in thousands to assist, and the Chief Commissioner of the Province, John Lawrence, brother of Sir Henry Lawrence, who was killed at Lucknow, sent many regiments to Delhi and elsewhere. The ruler of Nepal offered to send some of the famous Gurkha soldiers to assist. The Governor-General of India, Lord Canning, weakly first accepted the offer of a thousand men, then declined them, and shortly afterwards asked for 3,000. These brave little men, in marching forward and backward and again forward through the belt of pestilential jungle which stretched along the base of their hills, suffered grievously from sickness.

To the north of the Nurbadda River, already mentioned, however, lay a country called Rajputana. This country includes some twenty native states, as well as the British District of Ajmere-Merwara. It is of irregular shape, touching on Sind and the Punjab and a number of the states of Central India, the British District of Ajmere being in the centre. The Aravalli Mountains intersect the country from north-east to south-west, the heights of Mount Abu lying at the south-western extremity of the range. The tract which stretches from Sind on the west to the Punjab near Delhi on the north-east is mainly a sandy desert, comparatively fertile, however, towards its eastern extremity, and characterised by sand hills of varying height and length sparsely clothed with vegetation. But the south and south-eastern divisions of Rajputana is more fertile, being well watered by the drainage of the Vindhya

Mountains, the Chambal, the Banas and the Parbati Rivers. This region is characterised by wooded hills and valleys, fertile plains, and rich cities. The people are mainly Hindu, but there is also a large Mohammedan population. The Rajputs are the ruling race, but there are numbers of Brahmans, Jats and others, as well as such aboriginal races as Bhils, Minas, and Mers. On the decline of the Mogul Empire and the rise of the Mahrattas, the States of Rajputana came under the domination of that enterprising people, and were long subject to extortion and desolation. The Mahrattas being checked by the British, the plundered territories were restored to their Rajput owners and Ajmere was ceded to the Company. When the Mutiny broke out there was not a European soldier in the whole country, the garrison consisting of two batteries (one horse and one field), two cavalry regiments, and eight battalions or contingents of infantry—all natives. The nearest European troops were at Deesa, 230 miles from Nusseerabad. The main strength of the British at Deesa was the 83rd Regiment. The first intelligence of the outbreak at Meerut reached Colonel Lawrence, an elder brother of Sir Henry Lawrence, on the 19th of May. He was styled Agent to the Governor-General in Rajputana, and at that time was residing at Mount Abu. He at once sent to Deesa for a light field force, to consist of three Horse Artillery guns (Europeans), one squadron of Native Cavalry, 250 European Infantry, and 200 Native Infantry. The European Infantry were Nos. 5, 6, and 7, and the Light Companies of the 83rd Regiment, consisting of:—

1 Field officer (Major Steele).
2 Captains.
6 Subalterns.
276 Other ranks.

285 Total.

Part of the force marched on the 23rd of May, but Major Steele's party marched on the 26th of May. Nusseerabad was reached on the 12th of June, and was reckoned to have been 237 miles distant by the way the detachment marched. It speaks well for the spirit of the men that, despite its being the hottest time of the year, there had not been a single casualty. Previous to the arrival of the 83rd Regiment, however, the Nusseerabad garrison had mutinied. It consisted of a Native Field Battery, the 15th and 30th Bengal Infantry, and the 1st Bombay Cavalry. The authorities had taken stringent precautions against an outbreak by night only, so the Bengal Infantry naturally seized the opportunity of mutinying by day. The guns were first secured by the Native Infantry about four o'clock in the afternoon. The 1st Bombay Cavalry was ordered to charge the infantry and guns. When within a few yards of the guns the cavalrymen turned back, leaving their officers to ride home. Two of the officers were killed and two badly wounded. As the cavalry refused to charge again, the few Europeans, including women and children, had to go to Beawar that night, 37 miles away, the Bombay Cavalry going with them. After destroying the cantonments, the mutineers marched for Delhi. They were pursued by a thousand of the Jodhpur troops, commanded by British officers, but their officers could not get these troops to attack the mutineers, so they escaped unmolested. The 1st Bombay Cavalry afterwards performed good service at the capture of Gwalior, in 1858.

Sixteen miles from Nusseerabad stood Ajmere. It was garrisoned by the Light Company of the 15th Bengal Infantry. Ajmere was a populous native city. It contained an arsenal so large that it could supply all the troops in Rajputana. It was capable of furnishing a siege train of great strength, and also held an immense amount of treasure, yet its fort had been allowed by the British to fall into such a state of disrepair that it was said a gun fired on the bastions would have brought it all down. All this was guarded by a native company believed to be disloyal. The Grenadier Company of the 15th Bengal Infantry was sent to relieve the Light Company, as it was thought more dependable. The Light Company at first refused to let the Grenadiers into the fort. Afterwards the Grenadiers were admitted, but, being not above suspicion, one hundred Mers were sent by a forced march of 37 miles to Ajmere and the Sepoys were marched to Nusseerabad. On the arrival of the detachment of the 83rd Regiment at Nusseerabad, No. 7 Company was hurried to Ajmere. Nos. 1, 2 and 3 Companies of the 83rd Regiment left Deesa on the 17th of June, and joined the left wing on the 10th of July.

Another outbreak had occurred before this at a cantonment called Neemuch. This was situated 150 miles south of Nusseerabad. Its garrison consisted of the 4th Troop 1st Brigade Bengal Native Horse Artillery, a wing 1st Bengal Cavalry, 72nd Bengal Infantry, 7th Infantry Regiment Gwalior Contingent. These troops mutinied on the night of the 3rd of June. The officers made strenuous efforts to recall the men to their duty, but were fired upon and driven away, and within eleven hours the whole force had left the station, carrying away all the plunder they could and burning down most of the cantonment before they left. The following day the mutineers tried and executed some of their number on the charge of helping their British officers to escape with their lives. They then marched on towards Nusseerabad. Finding, however, that it was held by the 83rd left wing, they moved via Agra towards Delhi. They were joined by various parties of mutineers and were severely defeated by the British near Agra and near Delhi. To hold Neemuch, which had been re-occupied by the British, with a few native levies, Captain Read was detached from Nusseerabad with Nos. 5 and 6 Companies on the 9th of July. His force, all told, consisted of two officers, 106 non-commissioned officers and men. It arrived at Neemuch on the 18th of July without any casualties.

On the 24th of July an order was received augmenting the strength of the Regiment to a total of 1,339 all ranks, of which ten companies were to be abroad and two at home, with the respectable total of 46 officers.

On the 9th of August there was an outbreak at Ajmere jail, when about 50 prisoners escaped, after cutting down the policeman at the gate. Colonel Lawrence at once started in pursuit, but the civil police forestalled him and almost immediately brought in twenty-five of the runaways, having cut down fifteen in the encounter.

A commotion was caused at Nusseerabad on the 10th of August by a trooper of the 1st Bombay Cavalry galloping down the front of the lines of his regiment calling upon them to rise. Few, however, accepted his invitation, so he proceeded to the lines of the 12th Bombay Infantry and did the same there. The infantry passively sympathised with him, did not seize him, and refused to give him up, or to proceed to the rendezvous at the guns. Brigadier Macan called the guns out, and the trooper fired at him. He was at once mortally wounded by an officer of the artillery. The detachment of the 83rd Regiment turned out and the men of the 12th Bombay Infantry were disarmed. Five of the ringleaders were hanged on the 25th of August. This regiment afterwards fought well at Gwalior.

At Neemuch on the 12th of August the Colonel of the 2nd Bombay Cavalry, who was then in command there, received information that his one squadron proposed to mutiny that night. He called out the 100 men of the 83rd Regiment. They were able, despite the darkness of the night, to seize some of the ringleaders, but some escaped. Private Chambers was shot in the melee—it is stated by accident—and two other privates of the Regiment were wounded, as was also Lieutenant Blair, of the Bombay Cavalry.

So far no brilliant success had graced the arms of the mutineers in Rajputana. They now, however, determined to destroy a small party of the British, and the detachment of weakly soldiers of the 83rd Regiment stationed at Mount Abu was selected as the object of their next attack.

Mount Abu, the hill-station of Rajputana (3,930 feet) lies some forty miles north-east of Deesa. It was in August garrisoned by some sixty men of the Jodhpur Legion, and there were there also some thirty-five sick and convalescent men of the 83rd Regiment. There were also a few civil officers and the families of officers who were serving in the plains. The Jodhpur Legion was a particularly smart corps, maintained by the State of Jodhpur under the terms of a treaty with the Indian Government, and had several British officers in charge of it. The men were generally recruited from the same sources as the Bengal Sepoys. This latter point should have raised doubts in the minds of their officers, but apparently it did not. On the 19th of August a company of the Jodhpur Legion arrived at Anadra at the foot of the pass leading up to Mount Abu. A troop of the Jodhpur Legion Cavalry was distributed in small parties between Abu and Deesa to protect the roads. Captain Hall, commanding the whole legion, was at Anadra on the afternoon of the 20th, looking after the comfort of the detachment there, and returned that day to Abu. On the way he met a native serjeant of the detachment at Abu, named Gozan Singh, who said he was going to Anadra to see his comrades. It was afterwards discovered that he had been deputed to arrange the attack on the 83rd Regiment for the following morning. The morning of the 21st of August was thick with mist. Under the cover of this fog the detachment of the legion left at Anadra crept up the hill. It was almost as dark as night, so they were able to approach the barracks unperceived by the sentry. Creeping up to the windows, they poured in a heavy fire. The men of the 83rd Regiment were absolutely taken by surprise, but seizing their arms they returned the fire and closed with the enemy. The latter fled, leaving one dead man behind them, but the Europeans had escaped unscathed. Whilst the main body of the enemy had been employed in this futile attack on the barracks, another party of miscreants, under a native captain, had gone to Captain Hall's house to murder him. This party extended in line in front of his house and proceeded to fire volleys by word of command through the doors and windows. The inmates fled out at the back, whilst their assailants were attacked by a corporal and four men of the 83rd Regiment, who were quartered in a school close by. Captain Hall, assisted by a Dr. Young, led these five men straight upon the Sepoys' lines or barracks and drove the whole down the hill after a scuffle, in which a European civilian was severely wounded. Considering it unwise to pursue with only thirty-five men, the whole of the Europeans were concentrated at the school, which was instantly put into a state of defence.

The discomforted rebels retired towards the headquarters of the legion—Erinpura, fifty miles north of Abu. Here the remainder of the legion joined them excepting a portion doing duty at Nusseerabad. After burning and plundering Erinpura, they marched for Awah, whose chief, a vassal of Jodhpur, had been in rebellion against his liege lord for some time. After some hesitation this chief accepted their services. The native army of Jodhpur now marched against the rebel legion. They encamped close to one another, and by treachery the picquets of the Jodhpur army were withdrawn. The legion attacked the army on the 8th of September before dawn, and the army fled, save its commander, Anar Sing, who, surrounded by a group of brave companions, fought to the last, until he and all who were with him were killed.

Returning to Awah with the guns, treasure, and spoils of the Jodhpur army, the legion then prepared for future eventualities by strengthening the old fortifications of the town and fort. Awah was on the high road between Deesa and Nusseerabad, so Colonel Lawrence, who was now a brigadier (general) decided to clear the rebels out of the place. On the 30th of August three officers and 119 other ranks of the 83rd Regiment had been brought from Nusseerabad to Ajmere to help keep the peace during a great Mohammedan festival there. So, taking this party, thirty-one men from the companies already in Ajmere Fort, and calling up another captain and fifty-three men from Nusseerabad, Brigadier Lawrence proceeded by the mountain pass of Burr to reconnoitre Awah. To this party of the 83rd Regiment was added five field guns, two $3\frac{1}{2}$-inch mortars, 200 1st Bombay Cavalry, 200 Merwara Native Infantry, and 40 of the 12th Bombay Infantry. The force was detained by heavy rain for some days, but arrived at Awah on the 18th of September. This place was surrounded by a high wall and thick jungle which prevented the guns from opening fire on the town from a distance. The rebels fired briskly from their batteries inside and outside the town, and after a three hours skirmish General Lawrence withdrew with a loss of one officer and two men killed and three wounded. Two of the five guns were disabled. Out of the casualty list, three of the wounded men belonged to the 83rd Regiment. The force retired to a village three miles distant, where it remained in hopes that the rebels would attack it. Of course, the mutineers were not so obliging, so the whole force marched back to Ajmere, where it arrived on the 28th of September. There is no doubt that Brigadier Lawrence made a great mistake in these operations. They should either not have been undertaken or should have been carried through at any cost. As it was, it was noised abroad that the rebels had won a crushing victory at Awah.

The Jodhpur Legion remained at Awah until the 10th of October, when they marched towards Delhi. A British force was sent out from that city to meet them, under Brigadier Gerrard. The legion took up a strong position at Narnoul and was at once attacked by the British, who utterly defeated and dispersed them. Brigadier Gerrard was killed in the combat. The portion of the legion at Nusseerabad with the 83rd Regiment was promptly disarmed but remained there doing duty. They behaved so well that their arms were restored to them in March, 1858.

On the night of the 18th of September the detachment at Neemuch, under command of Captain Read, together with two guns and a party of native troops, marched to take the walled village of Nimbharia, sixteen miles from Neemuch, as it had been occupied by some mutineers from Tonk, with three guns. The whole business was greatly mismanaged, for the head man of the village had come out to surrender and with his adherents had thrown down their arms, when Colonel Jackson, who commanded the whole party, ordered the troops to drive them into the town with the bayonet, which was done and the gates were promptly shut. The guns could not breach the walls, which could not be escaladed. However, on the morning of the 20th the enemy were found to have evacuated the village, which was then occupied. The 83rd Regiment had Assistant-Surgeon Miles wounded and Lance-Corporal Thomas Young was killed, whilst the native troops had sixteen casualties. The prize-money is noted in the Regimental Records as having amounted to sixteen shillings per man for this capture. The force withdrew to Neemuch that night, leaving Ensign Chamley and thirty-two men to hold the village, but they were withdrawn next day.

In October the Mandiswar insurgents seized a fortified village named Jeerum, which also possessed an old fort. This place was situated some ten miles from Neemuch. On the morning of the 23rd of October a party started from Neemuch to turn the rebels out of Jeerum. Captain Tucker, of the 2nd Bombay Cavalry, was in command, and he had six officers and 120 of his own men, two officers and fifty men of the 83rd Regiment, and some of the 12th Bombay Infantry. Captain Tucker opened fire with guns, and then sent the infantry to the attack; but the enemy sallied out in great numbers, and drove the infantry back, killing Captain Read and wounding two privates of the 83rd Regiment and captured a mortar. Captain Tucker at once charged with the cavalry, and the infantry re-captured the mortar and the rebels were driven into the village, Captain Tucker being killed in the charge. The place was too strongly fortified to be taken by assault, but the enemy fled during the night and the village was occupied next morning, and its fortifications were blown up. Five other officers were wounded and one native soldier was killed.

On the 8th of November the Mandiswar rebels advanced to attack the garrison of Neemuch. Neemuch was suffering from divided command, as there was great friction between Captain Showers and Captain Simpson. The latter was in command of the troops, and retained the command though severely wounded. Captain Showers was political agent in Meywar. However, on the 8th of November Captain Showers moved out to attack the rebels with part of the force, and after a skirmish was ordered to retire by Captain Simpson. Next day the garrison of Neemuch was besieged in a small fortified square in which was situated the magazine. The enemy were rather more than 4,500 strong, with three guns. They were, however, short of ammunition for the latter, so the siege principally consisted of a heavy fire from matchlocks from an entrenchment which the garrison had constructed round the square, but which was found to be too large to be held by their small numbers. An hour before dawn on the 21st of November the rebels attempted an escalade but were signally repulsed, and left their ladders and a green flag on the ground; one of the ladders was mounted on four wheels. Once during the siege a Fakir, or holy man, with a mirror fixed on his breast, walked round the fort under fire, having stated to his dupes that if he succeeded in completing the circle the fort would fall into their hands. He passed unharmed round a large part of the circuit, but was then shot down, and as the native soldiers inside the fort were becoming uneasy at his supposed magical powers, a Christian bandsman went out and brought in the mirror and the holy man's head. Hearing of a force coming up to the relief of the little garrison, the rebels decamped during the night of the 21st of November. The regiment suffered no losses during this blockade, though two officers and six men of the small native force were wounded.

EIGHTY-THIRD REGIMENT (1857-1859).

On the 26th of October the headquarters of the regiment, which had up to then remained at Deesa, marched via Mount Abu to Nusseerabad, where it arrived on the 28th of November, whilst two companies, under Major Austen, relieved the Neemuch detachment.

Reinforcements now began to enter Rajputana from the Bombay Presidency as a field force was being assembled at Nusseerabad, and it was decided to settle with Awah once and for all. A column of about 1,800 men, with fourteen guns and howitzers, went to this place on the 18th of January. Two companies of the Regiment, under Lieut.-Colonel Heatley, formed part of this force. Awah was invested, and after five days' siege operations, during which there was incessant firing on both sides, a breach was pronounced practicable, and it was ordered to be assaulted next morning. During the night a fearful storm raged, and the noise and darkness were so great that sentries only a few paces apart could neither hear nor see one another. Under cover of this storm the enemy evacuated the place, and though it was completely surrounded by the cavalry with a view of stopping all means of escape only one of the Lancer picquets heard the fugitives and was able to cut up eighteen and take seven prisoners. The cavalry sent in pursuit next morning brought back 124 prisoners; twenty-four of these were ascertained to be Sepoys who had mutinied, and they were promptly shot. In the town were found thirteen guns and three tons of powder, etc. The defences were found to be very strong, consisting of a double line, the inner one being of strong masonry, whilst the outer one was an earthwork; both were well loopholed. After blowing up the keep, bastions, and masonry works, the force marched to Nusseerabad. The Regiment had only two men wounded in this operation.

1858 The campaign throughout the whole of India now began to assume a different aspect. Up to this time, about the beginning of 1858, the chief difficulty had been for the British to protect their lives from the revolted troops and other insurgents, whilst they strove to become the assailants only in a few isolated places, as, for instance, the attack on Delhi; and elsewhere kept their own cantonments, so to speak, clear of the enemy. Now, however, troops had been brought in from Mauritius, South Africa, England, the China Expedition, and even from Burma. The British now, therefore, were in a position to attack the mutineers in various places, break up their rebel bands, bring the treacherous rulers to justice, and generally to bring the country into an orderly state. Sir Colin Campbell was arranging for his second advance to Lucknow, which city it will be remembered was finally cleared on the 14th of March, and the Military Authorities also arranged that three columns should operate in Central India and Rajputana. One under General Sir Hugh Rose would sweep the country from Mhow to Kalpi, re-capturing the town and fortress of Jhansi. The movements of this column will be dealt with in the next chapter, as the 86th Regiment formed part of its force. Another force, under General Whitlock, was to operate from Jubbulpore to Banda. The movements of this column were successful though dilatory, whilst a Bombay column, under a Major-General Roberts, operated in Rajputana. The 83rd Regiment formed part of this column, and its next move, after the destruction of Awah, was the capture of Kotah. This was a city some hundred miles south-east of Nusseerabad ruled by a Rajput Rajah. It had been famed from the days of Zalim Shah as a secure emporium for treasure, opium, and merchandise. The Rajah kept up a contingent under British officers. This was sent to Agra when the Mutiny first broke out. Taken out to fight one day, the infantry were separated from the artillery, and fearing that the British had made some base scheme to destroy them they went over to the mutineers. Their Rajah held manfully by the British, but a force of rebels occupied the greater part of Kotah, and plundered its miserable inhabitants. General Roberts marched two strong brigades against it, which left Nusseerabad on the 10th and 11th of March respectively. Three companies of the 83rd marched with the 1st Brigade, and the headquarters of the Regiment with the 2nd Brigade. The force is stated in the regimental records to have been 4,500 strong, including many native troops, and also the 72nd, 83rd, and 95th Regiments, whilst the 8th Hussars joined outside Kotah. There were also eighteen heavy guns, beside some field batteries. The official records give the numbers of the force as 5,500.

On the morning of the 22nd of March General Roberts arrived opposite Kotah with his 1st Brigade and encamped to the westward on the left bank of the river Chambal. The 2nd Brigade with the siege train arrived the same evening. Three batteries were at once erected and opened fire on the rebel portion of the city on the morning of the 24th of March. On the 26th Lieutenant-Colonel Heatley took 200 men of the 83rd Regiment with a company of native infantry across the river to hold the portion of the town which was held by their ally, the Rajah, as the rebels had assaulted it on the 25th and 26th with 6,000 men, and had attempted to carry the palace by escalade. The enemy repeatedly attempted to storm this part of the town, despite the arrival of Colonel Heatley's party, and in the next three days he lost three killed and five wounded of his 200 Europeans. More batteries were built by the British, and on the morning of the 30th, instead of the slow fire of two shots apiece from each gun every hour, a very heavy rapid fire was opened from every gun. At 11.30 a.m. the guns suddenly ceased fire and three columns moved to the assault, whilst one remained in reserve. Passing through the Kitanpal Gate the columns moved so as to take the guns of the enemy, which were bombarding the Rajah's Palace, from the rear. Major Steele, with 250 men of the 83rd Regiment, led the centre column. Many of the enemy were shot or bayoneted, many escaped by letting themselves down by ropes over the walls, whilst others threw themselves over and were dashed to pieces. Some occupied houses and fought desperately. The reserve column, with Lieutenant-Colonel Heatley and 250 men of the 83rd Regiment, now came forward, and, sweeping round the opposite quarter of the city until it joined hands with the other columns, completed the enemy's discomfiture. The Regiment only lost one man killed and six wounded on this occasion. By 2.30 p.m. all resistance was over; 400 rebels were killed and numbers of prisoners taken. The total British loss was fourteen killed and forty-six wounded. The rebels escaped unpursued, despite the strong force of cavalry with the British. There were two gates by which they could retreat. The cavalry was sent to watch what was believed to be the only ford seven miles down the river. The enemy thought it desirable not to come in contact with the 8th Hussars, and as there was a thick jungle between the two gates stretching away for miles, keeping that between them and the cavalry, they were able to effect their escape untouched towards Gwalior. They were followed sixty miles; their track was marked by dead bodies and broken-down carts, but though seven field guns were found abandoned, not a single mutineer was overtaken. In fact, the cavalry pursuit was begun thirty-six hours after the rebels had gone. Kotah was abandoned to plunder by the troops for five days, but the articles collected by the prize committee were of inferior value and hardly worth the miseries they cost the poorer classes to whom they mostly belonged. Fifty-six guns were captured; at the ends of the streets were found so-called "infernal machines," consisting of forty matchlock barrels fixed on frames movable on wheels. This native substitute for a machine gun reflects credit on the inventor, but had fortunately caused but little damage. On the 18th of April the besieging force was broken up and the 83rd Regiment marched back to Nusseerabad, arriving there on the 29th of March.

Enfield rifles were now served out to the men. It is also noted in the Regimental Records that the waist-belt with a small pouch, worn in front, had been supplied about a year previously, to take the place of the cross-belt.

On the 13th of April Major Steele succeeded Lieutenant-Colonel Kelsall in command of the Regiment, the latter having retired on full pay. Colonel Steele died suddenly seven days after he retired from the Army.

Tantia Topi has been mentioned before as the right-hand man of Nana Sahib at Cawnpore, and he had done much fighting with the British since that massacre. Forced out of Gwalior by Sir Hugh Rose's column, he first moved in a north-western direction, but being headed by a force sent from Agra, he moved south-west and crossed the Chambal River with a force of 10,000 men and marched for Jeypore, one of the some twenty states of Rajputana. The Rajah of this state had always kept on the best terms with the British, so General Roberts hurried off with a force of 2,500 men to protect him; 600 of the 83rd Regiment marched with him, leaving Nusseerabad on the 28th of June. On the 3rd of July this force arrived within one day's march of the city of Jeypore. Tantia Topi changed his direction and marched off south to Tonk, another native state. The Rajah shut himself up in his fortified palace and left a portion of his army outside to operate against the rebels.

This part of the army fraternised with the mutineers and made over their guns to them. General Roberts followed the enemy by forced marches to Tonk, and Tantia Topi fled in a south-easterly direction, pursued by part of General Roberts's force. The general himself kept about 900 men with himself, including the 83rd Regiment. The weather was extremely hot, and six men of the Regiment died by sunstroke between the 7th and 10th of July. Tantia Topi was hard pressed by the detached force and changed his direction west, so that General Roberts, who had been steadily advancing by forced marches, came up with the enemy on the evening of the 8th of August, near the village of Sanganeer (Bhilwara in map). The numbers of the enemy were about 8,000, of whom two-thirds were cavalry. General Roberts attacked at once with his 900 men. The 83rd Regiment moved forward under a well-sustained fire from the artillery of the mutineers, and here for the first time used their new Enfield rifles with most satisfactory results. The fire of the rebels was directed too high, so not a man of the British was touched, whilst the rebels lost sixty in killed and wounded. They fled in the gathering darkness, and the 83rd Regiment, which had outmarched its baggage, having come thirty miles that day, lay down where they were for the night. The force then pursued. On two occasions they managed to rush and cut to pieces the rebel picquets, but the main body slipped away untouched on each occasion. On the 14th of August, about 7 a.m., the enemy was seen moving in heavy masses of horse and foot on the further bank of the Bunnass River, their march being covered by their artillery, which was posted on the crest of a rising ground close to the village Kotaria (south-west of Bhilwara).

General Roberts had been reinforced since his last skirmish, but the 83rd Regiment still led the advance, marching in "double column of sub-divisions" from its centre. The enemy's guns opened with round shot and grape, and were instantly vigorously answered by the British Horse Artillery, the fire on both sides being extremely heavy. The 83rd Regiment formed line as it waded knee-deep across the river and took four guns. The cavalry charged and great numbers of the enemy were killed. The artillery and cavalry followed for seven miles, then the gun horses could proceed no further, so the cavalry went forward alone a further eight miles, only giving up the pursuit when 100 of the enemy's infantry rallied with a strong body of cavalry in a village. There were only 180 British cavalry left, and they could not take the village. Coming back, many more fugitives were killed, whilst the infantry had cleared the nearer copses and villages. Over 1,000 dead rebels were seen on the field, and the British lost three killed and nineteen wounded, out of which the Regiment only lost one man wounded. Several elephants and camels laden with stores were captured on this occasion. The enemy now fled over the Chambal River, and the 83rd Regiment returned by easy marches to Nusseerabad, which place it reached on the 29th of August.

This lull in warlike operations was seized upon by General Roberts to make a half-yearly inspection of the Regiment, the first it had undergone since April, 1857. However, it was a very thorough one, for it lasted a whole week. The General then issued the following order:—

"Camp Nusseerabad.

"11th October, 1858.

"DIVISIONAL ORDERS BY MAJOR GENERAL ROBERTS,

"Commanding Rajputana Field Force.

"No. 1. Having completed the half-yearly inspection of H.M.'s 83rd Regt., the Major General would desire to notice his satisfaction. He has seen but little of the Regiment in the Garrison, but perhaps it has seldom happened that an Inspecting Officer has had such an opportunity of noticing the good conduct of a regiment on field service such as has occurred under the Major General's immediate command, and he would assure Lieutenant Colonel Steele and the officers and men of H.M.'s 83rd Regiment of his high appreciation of their merits, and that there is no regiment in the service he would rather have under his command on active service in the field than H.M.'s 83rd Regiment.

"By Order,

"ALEX. CARNEGY, Captain,

"Assistant Adjutant General,

"Rajputana Field Force."

On the 29th of September the Governor-General in Council granted a donation of six months full batta to all officers and men who had served with the Rajputana Field Force.

The Mutiny was now nearly crushed out, though Tantia Topi still fled before his many pursuers, but he was far away from Nusseerabad, where the 83rd Regiment was quartered, so Lieutenant-Colonel Steele proceeded on leave to England in December, and Lieutenant-Colonel Austen took over command.

Tantia Topi's day was not yet over, and, hard pressed, he made again for Rajputana, so a force was organised, under the command of Brigadier (General) Honner, to move towards Neemuch to meet the mutineers. Five hundred of the 83rd Regiment marched with this force from Nusseerabad on the 17th of December, but on the 23rd 300 men were sent back to Nusseerabad, arriving there on the 27th of that month, whilst the other 200 men, under command of Lieutenant-Colonel Heatley, went on with Brigadier Honner's column.

1859

Early in the month of January news came to Nusseerabad that the enemy were moving again on Tonk, so Lieutenant-Colonel Holmes took out a column on the 8th of January, of which 300 men of the 83rd Regiment, under Lieutenant-Colonel Austen, formed part. On the 10th of January he heard that the rebels were in camp at Aligarh, Rampura, twenty-four miles distant, threatening Tonk. Next day Colonel Holmes marched through Tonk to Bambaur, thus effectually covering the former place. At Tonk he heard that the rebels were marching towards Jeypore, so he followed, hoping to intercept them by keeping slightly to the east. Having marched twenty miles by 3 p.m., Colonel Holmes found that the enemy were well ahead and that a column commanded by Brigadier Showers, which was also in pursuit, had encamped ten miles ahead of him. Marching on, therefore, at 9 p.m. he passed through Brigadier Showers's camp, and at daybreak had marched forty-four miles in twenty-four hours. Halting for five hours he marched again. Still pressing on by forced marches he reaped his reward on the 21st of January, for on the 20th, at the end of his day's march, he heard that the enemy were camped at Sikar, twenty-eight miles to the west. Moving out at 6 p.m. the force reached the vicinity of the rebel camp at 4.15 in the morning, having marched upwards of 290 miles in thirteen days without a single halt, and having marched 54 miles in the last twenty-four hours.

The surprise was complete. The mutineers had no intimation of the approach of the pursuing column until their outlying picquets were driven in. Indescribable confusion followed. The rebel horsemen galloped off in every direction without attempting to make a stand and numbers without even saddling their horses. The British infantry at once advanced in line to the attack, whilst the cavalry galloped forward with four guns. The artillery quickly got into action, but as they only had the bright moonlight to work by little execution was done. Then the cavalry charged and the rout was complete. Over fifty of the enemy were killed and fifty-one were brought in as prisoners. A greater number might have been killed, but all those who had thrown away their arms were spared, and they seized the opportunity of slinking away from the rebel force. The three rebel leaders—Tantia Topi, Rao Sahib, and Feroz Shah—were pursued for some hours by the cavalry, but they managed to elude their pursuers. The day after the surprise at Sikar, Feroz Shah, with his followers, left Tantia Topi, and the next day the latter quarrelled with Rao Sahib. This leader, with some 3,000 followers, left Tantia Topi, joined Feroz Shah, and moved towards the west of Ajmere, where he arrived at a place called Kushana on the 10th of February. Brigadier Honner had marched over the greater part of Rajputana since he had left Nusseerabad with 200 of the 83rd Regiment on the 18th of December. His column had marched over 800 miles, including 130 miles in the last four days, and he arrived at Kushana at 5 o'clock on the 10th of February too. When he was within eight miles of that place the rebels heard of his approach, and they instantly broke up their camp and fled in two large bodies, one of which moved to the south-west, under Feroz Shah, and the other to the south-east, under Rao Sahib. Within three miles of Kushana the force was formed up; 146 of the 8th Hussars on the right, then 105 of the

1st Bombay Lancers; some irregular Sikh Horse then came on the left; but thrown forward in echelon were 133 of the 83rd Regiment, under Lieutenant-Colonel Heatley's command, with Lieutenants Marsh, Onslow and Huyshe, of the Regiment, with fifty-seven of the 12th Bombay Infantry all mounted on camels in line in front. In this formation this strange combination of races advanced towards Kushana at a sharp canter, the camels keeping their places in excellent order. When they came close to Kushana, seeing that the enemy had made off to the British left, they changed front in that direction, and increased the pace to a gallop. After going about two miles the camels with the infantry were sent back to hold the rebels' camp, whilst the cavalry dashed on for eight or ten miles farther. Aided by a bright moon, they cut up many of the insurgents and returned to their camp at 10 p.m. Including all killed by the cavalry and infantry, 226 dead rebels were collected, whilst the British reported the loss of two killed, two wounded, and five of the Sikh cavalry missing.

After the defeat at Sikar the rebels dispersed, as already related, and 600 of Tantia Topi's party surrendered. Tantia Topi himself, with three or four followers, went to the jungles of Paron; here he met with Man Singh, an under chief of Sindhia, who, having quarrelled with his own lord, had waged war against him. He had carefully notified the British Government that he had nothing to do with the rebels, but the general had directed him to keep the peace, and as he would not obey had moved against him, the 86th Regiment being employed in these operations. Paron was in what had been Man Singh's territory. He had no wish to have Tantia Topi, with all his base crimes unsettled for, as guest or ally, and he promptly borrowed a party of the 9th Bombay Infantry, arrested Tantia Topi, and handed him over to the British on the 7th of April, and he was tried and duly hanged on the 18th of April at Sipri. He behaved very bravely at his court-martial and on the scaffold, where he put his head into the noose quite cheerfully. After the defeat at Kushana, Rao Sahib and Feroz Shah took to the jungles, and they and their followers were killed in detachments up till the 15th of April. Then Feroz Shah fled by himself, disguised as a pilgrim, and was never caught, whilst Rao Sahib wandered about until the year 1862, when he was caught in the hills to the north of the Punjab, and was duly hanged.

After their harassing campaign the 83rd Regiment returned to Nusseerabad and to Ajmere.

Note.—The author is greatly indebted to "The Revolt in Central India," compiled in the Intelligence Branch Division of the Chief of the Staff, Army Headquarters, India.

CHAPTER XIII.

EIGHTY-SIXTH REGIMENT (1857-1859).

Bombay at the beginning of the Mutiny—Right Wing brought from Aden—Detachments sent to Mhow, Poona, and Belgaum—Native troops disarmed at Ratnagiri—Belgaum—Description of Mutiny at Indore—Siege of Dhar—Combat of Goraria near Mandiswar—Central India Field Force formed—General Sir Hugh Rose—Punishment of Bhopal Mutineers—Siege of Chanderi—Remainder of 86th Regiment joined force at Chanderi—Storming of Chanderi—Siege of Jhansi—Previous massacre at Jhansi, description of the Fortifications, etc.—Tantia Topi's Advance—Defeat of Tantia Topi at the Betwa—Storming of Jhansi—Retribution Hill—Pursuit of the Ranée of Jhansi—Affairs in Northern India after fall of Jhansi—March on Kalpi—Combat of Kunch—Operations about Kalpi—Mutineers retreat from Kalpi—Sir Hugh Rose's order to the Central India Field Force—Combat of Mutineers with Gwalior Army—Sir Hugh Rose's march on Gwalior—Operations about Morar—Concentration of troops on South-East side of Gwalior—Death of Ranée of Jhansi—Gwalior City taken—Fortress of Gwalior taken—Combat at Alipur—Operations against Man Singh—Siege of Paori—Combat at Ranod—Surprise of Mutineers at Sarpur—86th Regiment stationed at Morar—Officers mentioned in despatches—Complimentary Orders—Strength and losses of 86th Regiment in 1858—Copy of Tantia Topi's Court-Martial.

1857

THE general history of the Indian Mutiny has been sketched in the last chapter in connection with the services of the 83rd Regiment during that Mutiny, so it will be unnecessary to repeat it now whilst narrating the history of the 86th Regiment during that same eventful period. The beginning of the Mutiny is well defined by the outbreak of the 3rd Bengal Cavalry on the 10th of May, 1857; nominally it closed on the 1st of November, 1858, when the proclamation was read at every station in India, which stated that the Queen had assumed the government of India, and in which proclamation she offered pardon to all rebels who had not directly taken part in the murder of Europeans. The real close of the Mutiny is harder to define. Probably the defeat of Feroz Shah's force on the 15th of April, 1859, would be nearer the right date for the final cessation of armed resistance by mutineers. At the beginning of this period the 86th Regiment had its right wing at Aden and its left wing at Bombay, and a party of five officers and 150 men had been sent under Captain Jerome from the latter place to Gujarat, a province to the north of Bombay.

The Government of Bombay were in a difficult position. They had three native regiments in the city and 400 European soldiers to guard the place. The native regiments, or some of them, did conspire to mutiny and seize the city. The plot was discovered and two of the conspirators were executed and six were transported for life.

EIGHTY-SIXTH REGIMENT (1857-1859).

The first troops to hand were the right wing of the 86th Regiment, which had been stationed for two years at Aden, and was now hastily brought back from that desolate station and rejoined at Bombay, under the command of Major Keane, on the 8th of July.

A detachment of the Regiment, consisting of eight officers and 251 men, marched from Bombay for Mhow, arriving at the latter place on the 2nd of August.* They suffered a good deal on this march, owing to the incessant rains. On several occasions, when halting for the night, they found the whole country flooded. No tents could be pitched, and the troops had to bivouac as best they could in the rain. The consequence was that there was a good deal of illness and seven men died of cholera.

A second detachment of the 86th Regiment was then ordered to proceed to Poona, as an outbreak was anticipated in that station. They experienced the same trying weather on the march, but only lost one man, who also died from cholera. This party was commanded by Captain Darby, and left Bombay on the 14th of July. It consisted of 200 men.

The headquarters of the Regiment, with the remaining three companies, were ordered on the 8th of August to proceed to Belgaum. Embarking on a vessel, named the "Victoria," they steamed down the west coast of India. The "Victoria" was a very small-sized steamer, and when all were on board an officer present wrote: "We were packed like pigs on an Irish boat. It was impossible to go below, and on deck it rained as if the windows of heaven had been opened to pour down their waters upon us. I should think every man on board was seasick, in which 'pleasant' condition we continued for two days." This cargo of misery stopped on the 10th at a town called Ratnagiri. The 27th Bombay Infantry had a wing in this town, and the other wing was at Belgaum. The half-regiment at Ratnagiri had become disaffected and had killed two of their European officers. A number of intercepted letters proved the existence of an organised Mohammedan conspiracy for a general rising throughout the south Mahratta country. Major Stuart, who was in command of the 86th Regiment, at once disembarked eight officers and 150 men of his Regiment, and, taking fifty sailors belonging to the East India Company, with two guns, proceeded to disarm the mutinous troops. This was effected without trouble, and leaving the sailors to garrison the place, assisted by fifty men of the 2nd Bombay European Regiment, he re-embarked on the 13th of August and steamed on to Goa. The 86th Regiment had been stationed in this place from 1806 to 1809, and no British regiment had been there since that time. The force transhipped on the 15th of August into small boats and proceeded up the river some eighteen miles and stopped in a monastery for a couple of days' rest. On board the steamer they had been so closely packed that no man had been able to lie down, and

* Other records state that the 86th Regiment reached Mhow on the 6th of August, but the Regimental Records state the 2nd of August.

it was during this halt at the monastery that the men had the first opportunity to do so since leaving Bombay. Again entering the small boats, they pushed on up the river to Assinwood. Here the force disembarked and marched eighty-one miles to Belgaum. This march was accomplished in four days. The road was, if possible, worse than those followed by the Mhow and Poona detachments. No wheeled transport could traverse the country on account of the heavy rain. All the baggage had to be carried either by pack animals or by coolies. One man (Private Young) died of cholera, and all the kits and bedding of both officers and men were absolutely destroyed by the constant wet. Great attention was paid to the men's comfort, and their daily ration was increased for the march by an extra pound of beef and four drams of arrack (native spirit) per man, with the most satisfactory results.

The country in this part of the Bombay Presidency was ripe for revolt. Their chiefs were disloyal, as their practice of "adopting" sons had been suppressed. Curiously enough, only one amongst all of them had a son to succeed to his estates. Some of them had been deprived of their ancestral acres on various pretexts, whilst their people had been allowed to retain their arms and were quite prepared to use them. There had been an outbreak in this district in 1844, and the inhabitants had been charged with the cost of suppressing the rebellion, so the Government had many enemies. Three native regiments garrisoned the district, but the only regiment with which the 86th Regiment had any dealings was the one at Belgaum, which, as will be remembered, had a wing at Ratnagiri. The wing at Belgaum had been disarmed at the same time as the wing at Ratnagiri, and on the 24th of August the ringleader of the disaffection was blown from the muzzle of a gun in the presence of a full parade of the whole garrison. Major Stuart has left a record of this awful execution. He writes: "As I had only about one hundred men, I did not wait for orders, but before going on parade I quietly ordered them to load. As there were only detachments of the 2nd Europeans and artillery besides my own men, I was determined I would not be caught napping in case any disturbance should take place at the execution.

"It was indeed a fearful sight. The square was formed on three sides, the fourth being occupied by the Artillery with the field pieces, which was about to blow the poor wretch into eternity. I was directed to send an officer and twenty men with the gun to parade close behind it. I must confess that I felt a shiver of horror when I beheld the doomed man approach, a splendid looking fellow, the perfect cut of a Hindoo high-caste soldier. He stepped as firmly and resolutely as if on parade, not a shake or shiver of his limbs; not a trace of emotion on his countenance, denoting the slightest fear of the frightful fate he was about to encounter. He did not appear to be more than twenty-five years of age. He placed himself composedly before the gun to which he was then fastened. Although perfectly aware that every moment he might expect the word 'Fire,' that would blow him into a thousand pieces, his face never altered, but a slight sneer might be traced upon his upper lip. It was a moment of horror to all, and when the word 'Fire' was given it was almost a relief. We heard a dull 'thud,' a Scotch word more expressive than any English one I could give, and after a second or two the remains of the Hindoo soldier was falling to the ground like large hail stones, and particles of bone and muscle struck my officer and men who were stationed behind the gun.

"There was a dead silence for a moment, and the word from General Lyster came, and what a relief it was, 'March home your regiments.'"

Another regiment, the 28th Bombay Infantry from the same district, is stated in the Regimental Records to have been "daily expected to mutinize the station," so they were sent to Aden to put them out of the way of temptation.

Captain Weaver died on the 13th of September at Mhow, and the fortunes of that detachment can now be followed.

Mhow was the cantonment which was supposed to watch Indore, the capital of the territories of Holkar, one of the five great Mahratta chiefs. It was only a few miles from Indore, but instead of having a strong garrison of European troops, as it should have had to watch over Indore and other native states, all swarming with national and contingent troops, it had but one Bengal battery, of which the gunners alone were Europeans, the drivers being natives, whilst it also had a native infantry regiment and a wing of a native cavalry regiment as a garrison. Holkar may, or may not, have been loyal to British rule, but as soon as the news of rebel successes arrived his troops rose and attacked the Residency on the 1st of July. The native troops at Mhow also mutinied, and the whole having burnt the bungalows pushed northwards towards Gwalior, whilst the country generally lapsed into anarchy. Major-General Woodburn had marched from Poona on the 8th of June for Mhow. He had five troops of the 14th Light Dragoons, two batteries, and the 25th Bombay Infantry. This force arrived at Mhow on the 2nd of August with their artillery horses so exhausted that the Bengal battery horses had to be sent for to bring the guns in. Here the detachment of the 86th Regiment joined them.

Partly because of the impassable state of the country from unusually heavy rains, and also owing to the impossibility of striking a decisive blow at the scattered forces of the enemy, General Woodburn remained inactive at Mhow for some time. In the meantime a body of revolted troops, reinforced by levies and vagabonds from Afghans downwards, assembled at Mandiswar, 120 miles from Indore, a large and important town on a tributary of the Chambal River. This force amounted to 15,000 men, with eighteen guns, and was under the command of Feroz Shah, a prince connected with the Imperial House at Delhi. This leader plundered two towns and then marched on Dhar, which place was twenty miles north-west of Mhow. The chief at Dhar was a minor, so his council had to decide what was to be done. Influenced, no doubt, by fear, they made common cause with Feroz Shah and admitted him on the 31st of August and mounted his guns in the Rajah's

Palace. It was consequently decided to march on that place as soon as the monsoon had cleared off, and a siege train was organised for the purpose. An advance guard proceeded in the direction of Dhar, and awaited the arrival of the main body, which latter marched on the 20th of October in two parties, and arriving before Dhar on the 22nd by two different routes, invested the fort simultaneously on two sides.

A cavalry reconnaissance was fired upon from the native town, and as the column under Major Keane, of the 86th Regiment, was the first to arrive, it was directed against the north face of the fort and immediately engaged the enemy. The country hereabouts was dotted with lakes and trees and to the south between the fort and the other column, commanded by Brigadier Stuart, stretched a wide expanse of water, whilst the western side was a series of hills, rocks and ravines.

On the approach of Brigadier Stuart's force the rebels advanced to the attack, covering the movement with the fire of four brass guns, posted on a hill south of the fort. The British column formed line to meet them, and the artillery, opening on the fort and guns, soon disabled one of the latter. The guns were then charged and captured by a troop of 14th Dragoons and were immediately turned on the enemy, being served by the Sepoys of the 25th Bombay Infantry. The two advancing columns had now driven the enemy from their position, but as they continued to threaten the right (east) of Brigadier Stuart's column, the cavalry again charged and routed and dispersed the enemy in all directions. The Dragoons then made a complete circuit of the fort and drove the enemy into it, where they were invested as well as the small numbers of their opponents would permit. The troops occupied particularly a ridge of hills commanding the fort. This work was built of red granite and stood on a slight eminence detached from the town. The stone walls were about thirty feet high, having towers at intervals. The siege train not having arrived, the bombardment had to be delayed until next day, and the force was assembled in a basin amongst the hills, leaving strong posts to watch the fort. A small party of the beleaguered garrison broke out on the 24th, but was pursued by a detachment, which came up with them at the village of Chiklia. The village was burnt, some of the enemy were killed, and a number of elephants were captured. On the 25th, a sand-bag battery was constructed 2,000 yards south of the fort and fire was opened. On the night of the 29th the breaching guns were in position close to the walls and 24-pound shots were fired at the entrance. The firing continued at night, as there was bright moonlight. On the 30th the enemy displayed a white flag and asked for terms, but were informed that nothing less than unconditional surrender would be accepted.

The breach was found to be practicable on the night of the 31st and the assault was ordered for 10 o'clock that night, but two serjeants, who were reconnoitring the breach prior to the assault, found that the enemy had wisely evacuated the fort under the cover of darkness. The cavalry pursued the fugitives in the direction of Mandiswar, but fear lent the mutineers wings and few were overtaken. The British loss during these operations only amounted to a few killed and wounded. A great quantity of money was found in the fort, so the Regimental Records state, but though considered as "prize," it was not then distributed. Whether the 86th Regiment ever obtained it is not stated, but it is to be hoped that they fared better than Lieutenant Dartnell. At the storming of Jhansi a large amount of money and precious stones were captured and were understood to be "prize." Lieutenant Dartnell then (1858) refused £250 for his share of the prize-money, and in 1869 received £26 as his correct share from the Government.

A reinforcement joined the column which was now commanded by Brigadier Stuart, and the force pushed on from Dhar on the 6th of November in pursuit of the enemy, who was known to have retired on Mandiswar. The country people appeared to have had enough of the rebel rule and were very pleased to see the British returning. The Nawab of Jaora, in particular, most loyally supplied the column constantly with provisions. Without his assistance it is doubtful if the advance could have been continued. Two fugitive British officers came into the camp on this march from Mehidpur. Their Malwa Contingent Infantry and Artillery had been attacked by the mutineers from Mandiswar. The infantry would not stand the charge of the Afghan adventurers with this force and all the guns were captured. Some of the infantry joined the rebels, who set fire to the Malwa Hospital and burnt alive about forty sick and wounded soldiers. The Malwa cavalry had mutinied some time before, near Neemuch, murdering their two British officers, but the two officers who managed to escape to Brigadier Stuart's column were escorted by thirty-five faithful troopers. The rebels carried off the guns and as much ammunition as they wanted, both for the siege of Neemuch, where, it will be remembered, that some of the 83rd Regiment formed part of the garrison. On receipt of this news a force of 338 native cavalry was at once despatched to Mehidpur. They found the cantonments empty and pressed on, recovering four guns and spoils at every mile, until they came in sight of 500 men with two guns in a strong position near the village. The British force formed into two parties and each charged from a flank, forming line as they galloped forward. The enemy fought with great resolution to the last and 100 were killed and seventy-four made prisoners. This action took place on the 12th of November. By the 18th of November the whole of the seventy-four prisoners had been tried by court-martial and were shot at 5 p.m. on that date. Brigadier Stuart crossed the Chambal River on the 19th unopposed and arrived in the vicinity of Mandiswar on the 21st of November. The Regimental Records of the 86th Regiment briefly state that "a severe action was fought (on 21st November) which lasted till the evening of the 24th of the same month, when the rebel army were completely routed and put to flight. Loss of the detachment, 86th Regiment, was one serjeant and two privates killed and nine men wounded. The loss of the enemy was severe. The town was burnt and thousands of dead and wounded left on the field."

To go more into detail, Mandiswar is a large straggling town full of trees, situated on the river Sohna. It at this time had a fort, which was occupied by a strong force of insurgents.

The British camped at 9 a.m. on the 21st in a position covered to the front by some rising ground, flanked on the left by a little village and gardens, beyond which were several groups of trees and yet another village surrounded by more gardens. At about three o'clock in the afternoon the enemy moved forward in force, threatening the British flanks and centre, and advanced steadily, with banners flying. The British column was summoned to arms, the troops fell in, the guns opened fire on a village occupied by the rebels, whilst the enemy on the British right were suddenly charged by a picquet of the 14th Light Dragoons, supported by some native cavalry, and were driven back with heavy loss, including 100 killed. In the meantime, out of the village which the guns were bombarding strolled a man dressed in white, carrying a green flag. He sauntered leisurely along the whole line of the British guns while shot and shell whizzed past him. Having apparently satisfied himself with his reconnoitring promenade, he coolly turned and walked back to the village. Just as he regained it a round shot struck him, and dashed him to pieces.

Next morning, in order to intervene between the party of rebels besieging the 83rd Regiment at Neemuch and the main body of mutineers at Mandiswar, Brigadier Stuart crossed the Bahri Ford 1,400 yards south-west of the town and established himself on the Neemuch Road, with his flanks well protected by two branches of the river Sohna. The cavalry now ascertained that some of the enemy's baggage had been seen leaving the village of Goraria, some four miles up the Neemuch road, so they at once galloped up that road and came upon the enemy about six miles from camp. Here a village named Piplia was found to be strongly held by the enemy with infantry, and many standards were displayed there. After killing some 200 of the mutineers in the open country the remainder fled into this village, and as, of course, the cavalry could not carry it by storm without infantry they returned to camp.

Concluding that the infantry seen at Piplia was the advanced guard of the rebel force from Neemuch, Brigadier Stuart moved forward at 8 a.m. on the 23rd of November, crossed the northern branch of the river Sohna, and found the enemy in great force to the north, where they occupied a strong position with their right on the village of Goraria. Their centre and guns were stationed on a hill where the gunners obtained shelter from a large mud hut, whilst their left was on a ridge and was covered by a nullah screened by lines of date trees. In front of the left were standing crops of jowari (millet which grows to a height of eight feet). Some ruined houses behind the guns sheltered a reserve of infantry and cavalry.

Covered by a cloud of skirmishers, the British line advanced. The hostile infantry, with green banners flying, moved forward through the intervening millet fields to meet them, whilst their guns at the same time opened fire, and the dense masses of their main body appeared like dark clouds on the near horizon, lit up by the flash of arms and accompanied by the rumbling thunder of distant drums.

Brigadier Stuart halted his lines. His guns first opened fire at 900 yards distance on the mutineers and then moved to the right front and enfiladed the enemy's line. Thirty of the 14th Light Dragoons at once charged the rebels' guns and took them, but were forced to retire by the musketry fire from the huts. The infantry then moved forward and opened a musketry fire on the enemy's guns, and they were again charged by a squadron of the Dragoons and again taken, whilst the gunners were caught and sabred. The rebels then commenced to fly into the village of Goraria, and the native cavalry, led by their British officers, burst upon them, killed many of them and hurried the rest into the village. The village was promptly shelled and strong infantry picquets were posted round it.

In the meantime the mutineers had issued from Mandiswar and had attacked the rear guard and baggage. The enemy numbered some 2,000 in this part of the field, so two guns and some cavalry moved to the rear. The rebels had advanced boldly to within 800 yards of the rear guard, when the British guns opened fire and the cavalry at once charged the mutineers. Many were killed, but the remainder rallied and beat off the cavalry, holding on desperately to a gravel pit. However, on the approach of the British infantry they retired hurriedly into Mandiswar.

Having thus satisfactorily cleared his rear, the Brigadier made arrangements to storm Goraria next morning at his leisure.

Having shelled the village for three hours, the infantry were then sent forward.

The following account is by an officer who was present:—

" After a shower of shot and shell to clear the walls and houses of their sharpshooters, the Royal County Downs (86th), the 25th Regiment, Nizam's Infantry, and Madras Sappers and Miners, led by Majors Keane and Robertson and by the officers commanding the two remaining corps, rushed across the range under fire over the mud walls, and, amid the burning houses, began to shoot and bayonet the mutineers, who had themselves fired some of the houses and poured a deadly matchlock fire from the eaves into our redcoats as they dashed along the streets.

" From mid-day until evening the bloody work went on, the County Downs despising all means but the bayonet. Occasionally a son of the sister isle would be seen all covered with sweat and dust, his face blackened with powder and smoke, leading tenderly outside the walls a woman or child. . . . All the rebels who ventured to rush from the burning village were sabred by the cavalry. Hand to hand fights were going on in the patches of sugar cane near the village, and about 200 Walayatis* came out *en masse* under a flag of truce and surrendered as prisoners. At evening some few only remained in strongly-built houses at the upper end of the village. Captain Robertson was taking a nine-pounder, to blow open the doors, as the infantry bugles rang out 'The Assembly,' and a more thirsty and powder-besmeared body of soldiers could not exist than came forth after the day's work. Cavalry picquets were again thrown out round the ruins in which this handful of desperate men remained, and the brigade encamped. The next day not a living soul remained in Goraria."

This stern example had an immediate effect. The rebels evacuated Mandiswar and dispersed in all directions. A loyal chief attacked them and killed 80 of them. The main body then broke up their standards and said that their gods had deserted them.

Brigadier Stuart then tried to communicate with Neemuch, and found that the rebels besieging that place, terrified by the destruction of Goraria, had also fled.

Leaving a garrison at Mandiswar, the column marched for Indore, arriving there on the 14th of December, and then Holkar's troops were at once disarmed without opposition. On the 15th the troops encamped in the grounds of the Residency which the mutineers had burnt in the rising of the 1st of July.

The windows of the church had gone, the bell was torn down, and the furniture had been removed. Over the altar rail some poor hunted European had written "They have thrown down thine altars and slain thy prophets with the sword, and they seek my life to take it away." The blood of the murdered Europeans was still fresh about the jail, and a woman's scalp with long, fair hair was found near by.

Justice was, however, meted out with a firm hand, some of the mutineers being hanged and others blown from guns.

After this the force returned to Mhow.

Brigadier Stuart especially mentioned Major Keane, of the 86th Regiment, in his despatches.

Over the gateway at Mandiswar was found the head of Captain Tucker, of the Bombay Cavalry, who had been killed whilst leading a gallant charge at Neemuch. The heads of two rebel chiefs replaced it before the column marched from Indore.

About the end of 1857, as already mentioned, the Indian Government was at last in a position to cope with the mutineers on more even terms. Up to then fighting had generally amounted to various groups of Europeans holding their own desperately, able only to keep themselves and their wives and families from being massacred. Now, however, Lucknow had been relieved for the

* Walayatis literally translated means—strangers. They were generally Afghans.

second time, and all the women and children had been carried away by Sir Colin Campbell's force. Delhi had been recaptured some three months before the close of the year, and columns were being organised to march through the disaffected country. Leaving Sir Colin Campbell to deal with Oude and other provinces on that side of India, it is only necessary in this chapter to follow the fortunes of the 86th Regiment as they marched with General Sir Hugh Rose on his very celebrated campaign through Central India, whilst a Madras force, under Whitlock, was to operate to the east of Sir Hugh Rose's column, starting from Jubbulpore.

The great difficulty with all this part of the country arose from the fact that it was divided up into a number of small states, all of whom maintained strong bands of armed retainers, and many had subsidised forces kept up for them by treaty with the Indian Government. It was not a generation ago since all had been immersed in fighting, and their warlike deeds were still fresh in men's minds, as many still living had taken part in them. Thus when it seemed as if British power was to crumble as many another more ferocious, and therefore to the native mind, stronger power, had crumbled before this time, there were many adventurers ready to try their fortune at founding a petty kingdom. Armed men were forthcoming in numbers from the armies of these native states. The country lent itself to their operations, as it was studded all over with forts and jungles. Into this country Sir Hugh Rose was ordered to proceed, and his force was to be two brigades. Truly a small force to fight such a numerous enemy, but a good cause and stout hearts carried his force through ten times their own number of enemies from victory to victory.

The 1st Brigade, under Brigadier C. S. Stuart, Bombay Infantry, was to assemble at Mhow, where he already was with part of the 86th Regiment. His force was as follows:—

One squadron 14th Light Dragoons.
One troop 3rd Bombay Light Cavalry.
86th Regiment—three companies.*
25th Bombay Infantry.
Two batteries European Artillery.
A detachment of Royal Engineers.
A detachment of the Hyderabad contingent.

The other or 2nd Brigade assembled at Sehore, some 110 miles to the north-east of Mhow. It was commanded by Brigadier Stuart, of the 14th Light Dragoons, and was much the same strength as the 1st Brigade, having the 14th Light Dragoons and the 3rd Bombay European Regiment as the white troops.

Major-General Sir Hugh Rose, K.C.B., had thirty-eight years' service in the Army to his credit; he had fought in Syria in 1840, and again in the

* The remainder of the 86th Regiment joined the column on the 16th of March during the siege of Chanderi.

Crimean War, and had besides this dabbled in diplomacy, so he was a capable man and well suited for a campaign of the sort in front of him. The first step in the campaign was to put matters right with the Bhopal contingent. Bhopal was near Sehore. When Holkar's subsidised troops mutinied on the 1st of July the Bhopal contingent was guarding the British Residency. Holkar's gunners opened fire on the Bhopal lines. Only five of the Bhopal Cavalry would follow their leader, Colonel Travers, in a charge on the guns; the Bhopal gunners worked their guns loyally, whilst the Bhopal Infantry refused to act, and threatened their officers.

The chiefs of Bhopal were generally disloyal, whilst their Begum or Queen held firmly to the British. It was as well, therefore, to put matters right in Bhopal first.

The mutineers were therefore caught and tried; 195 were executed, 274 dismissed the service, and 228, who had proved faithful, were retained.

1858

The 2nd Brigade marched off on the 6th of January, 1858. On the 29th it captured Rahatgarh, a fortress of great strength, very strongly held by all sorts of miscreants. The headman had instituted a custom of cutting off the noses and hands of all those who were suspected of being well disposed towards the British, no proof being required; whilst before the rebels an effigy of the head of a decapitated European female was borne as a rallying point. A great many rebels of note were caught here and hanged, and the fighting was severe. The same brigade fought a well-contested combat on the 31st of January at Barodia; the enemy lost 500 killed, whilst the British loss was twenty-three killed and wounded. This action relieved the fort at Saugor, where 173 men, sixty-three women, and 130 children had been beleaguered by hosts of rebels since the 29th of June. A neighbouring fort, Rehli, had been most gallantly held throughout by some Sepoys of the loyal 31st Bengal Infantry.

The 1st Brigade was ordered to march from Mhow on the 6th of February. It marched up the Grand Trunk road as far as Goona, and then moved off the road heading north-east towards the town and fort of Chanderi, where the rebels, driven away by the 2nd Brigade, were assembling, the fort itself being an arsenal, where guns and powder were manufactured. On the 1st of March Brigadier Stuart had sent his Hyderabad detachment to the east to help the 2nd Brigade to force some mountain passes on their way to Chanderi. On the 6th of March the 1st Brigade encamped at a small village six miles from Chanderi. This fortress was of great strength, and was especially notable for a desperate defence by the Rajputs in 1528, when the whole garrison perished to a man attacking the besieging force.

A scouting party of cavalry sent out from the 1st Brigade had its advance marked by bonfires lit at every half-mile by the enemy and were driven back by a heavy musketry fire from the jungle which surrounded the fortress. Next morning Brigadier Stuart sent the infantry forward and a brisk fire was immediately opened upon them, but with the 86th Regiment on the right, and the 25th Bombay Infantry prolonging the line to the left, the hill-sides were soon cleared of the enemy, whilst the barricades on the road, erected by the rebels, were removed by the Engineers. When the troops arrived at the level ground beyond the jungle large numbers of the enemy were seen in the ruined temple and summer-houses in front. The artillery opened with round shot and shell, driving the mutineers into another tract of jungle beyond, from which they were again driven still nearer the fort and took shelter behind a temple called Fatiabad and behind a wall of recent construction, which extended across the valley from hill to hill. The wall was loopholed and furnished with bastions twelve or fourteen feet in height and several feet in thickness. The field pieces made no impression on it, and if well held it would have prevented any advance on the fort or town in that direction. But the infantry rushed on and the Royal County Down Regiment, with Lieutenant Lewis leading, rushed the wall, jumped into the enclosure, and drove out the enemy, who fled into the fort and town, which were about half a mile distant. The wall was destroyed and some native troops were placed to hold that ground whilst the remainder of the brigade skirted round a range of hills commanding the fort and pitched their camp. On the 7th of March the troops proceeded to clear the ridge of the enemy; it was of sandstone, thickly wooded, about 100 feet in height, with a passage cut through the solid rock called the Kati Gati. The fort was visible through this tunnel, over which was an inscription stating that a king of Delhi had caused to be made the lofty gate of Gumti, and presumably had cut this tunnel likewise.

A few of the enemy were killed here, and at a small village at the foot of the ridge.

The fort could be clearly seen from this ridge. It was built on an isolated hill about the same height as the ridge, and was only separated from it by a jungle-clad ravine some half a mile wide. At first sight it looked impregnable, but at one spot a ridge of rock ran across the valley forming a sort of road. Across this was evidently the vulnerable part, but the fortifications had been strengthened at the spot where the ridge joined the fort hill by means of two towers and a bastion of solid masonry. This curtain was chosen for breaching.

With great difficulty some siege guns and mortars were got up the ridge and fire was opened on the palace, which was a prominent feature, but the enemy's guns made good practice and could not be silenced. It now became necessary to make a road along the crest of the ridge in order to get the heavy guns into position. The men were much exposed, the trees being the only protection, and much work had to be done at night.

On the 10th of March the artillery and engineers, with the aid of elephants, dragged up the 24-pounder guns. The cavalry were employed in reconnoitring daily, but as the ground was unsuited to their action the results were small.

That night the enemy made a sortie and recaptured the wall, which had been stormed on the first day by the 86th Regiment and given into the charge of the native troops. The locality was retaken next morning and handed over to the care of the 25th Bombay Infantry.

An officer who was present wrote the following short description of the siege:—

"The breaching commenced, the range being a very short one and point blank. As the battery was slightly over the eminence [i.e., on the forward slope] anyone to approach it had to run the gauntlet of the enemy's fire. It was evident that Chanderi had not been so disturbed for many a year. Most of the trees were of a flowering description and covered with gorgeous blossoms, while flights of parrots screamed among them, monkeys chatted at the soldiery, an occasional panther was turned out of his lair, and wild duck wheeled overhead. But the sun was fierce and hot, and it was a very thirsty tour of duty in all the batteries, which were five in number. The breaching battery being nearest the fort, was the object of the enemy's special attention, and they kept up an incessant fire upon it, both from their cannon and small arms. One individual who possessed a European rifle and had learnt to use it, caused much annoyance and many wounds, and the bullocks bringing up ammunition afforded him excellent marks. They appeared to have an unlimited number of guns and wall-pieces extending completely round the fortifications. Our shells fell thick and fast into the fort and did them much damage, but as it was so large they had plenty of space and shelter to escape from them, and an underground passage down the rock into the town close beneath, whereby they got both provisions and water and occasionally stole the baggage animals which had left camp to graze, on one occasion killed some camp followers when foraging, and on the night of the 13th they ascended a hill overlooking our camp and fired a regular volley into it but did us no damage."

On the night of the 15th a party of thirty native cavalry arrived with despatches from Sir Hugh Rose, and on the 16th of March the remaining seven companies of the 86th Regiment, under Major Stuart, marched into camp, and the same day the breach was reported practicable.

Previous to describing the assault on Chanderi, an account of the concentration of the remainder of the 86th Regiment will be given.

In December, 1857, the stations of the various detachments of the Royal County Down Regiment were as follow:—

Three companies in Mhow, under Major Keane.

Two companies in Surat, in the Gujarat country, under Captain Jerome.

Three companies in Belgaum, under Major Stuart.

Two companies in Poona, under Captain Darby.

The total strength in India of the Regiment on the last day of the month was: 31 officers and 1,009 other ranks, whilst the large number of thirteen officers were shown as "on leave," but this included the colonel, who was being retired, and all those sent home by the Invaliding Board.

In December Captain Darby's detachment was relieved at Poona by a detachment of the 56th Regiment, which had arrived from Mauritius, and was sent to strengthen the two companies under Captain Jerome, which were never in one place for a week, but were constantly on the move in Gujarat, endeavouring (as the records state) "to keep the troops of the different native princes from rebelling."

On the 11th of November a draft had arrived on the "Genghis Khan," Lieutenant Hamilton being in command. The draft numbered ninety-eight men, whilst Lieutenant Coates and his soldier servant hurried out on the "Sedgemoor," arriving on the 28th of December at Bombay.

During 1857 the Regiment had lost 51 men from death, not to mention 7 women and 28 children, whilst 86 men had been invalided home to England, 10 had been sent home on the expiration of their term of service, and 3 had been transferred to other corps.

On the 12th of January the three companies at Belgaum marched for Vingerla, arriving there on the 19th of January. They embarked on the steamship "Berenice" and proceeded to Bombay, reached that place on the 23rd, and encamped on the esplanade.

On the 27th of January Lieutenant Lewis and forty-three other ranks proceeded by rail from Bombay to the terminus of the line only a few miles out and then got into a bullock train, which took them on to Mhow, at which place they arrived on the 5th of February. They were immediately followed by other detachments of the Regiment in the following order:—

28th of January—Lieutenant Coates and 43 other ranks.

29th of January.—Lieutenant Coran and 43 other ranks.

30th of January.—Brevet-Major Mayers and 43 other ranks.

31st of January.—Major Stuart and 43 other ranks.

1st of February.—Lieutenant Ord and 43 other ranks.

3rd of February.—A lieutenant (apparently an attached officer) and 20 other ranks.

4th of February.—Lieutenant and Adjutant Cochrane, Paymaster Heatley, Surgeon Stack, and 17 other ranks.

This last party arrived at Mhow on the 12th of February. Here the four companies from Gujarat were awaiting them, and the whole force started to join the 1st Brigade at Chanderi.

The Regimental Records give the following short account of the proceedings:—

"On the morning of the 15th of March the 'divisions' marched 17 miles, and on the same evening started again, at the urgent request of Brigadier Stuart (each man taking in his haversack a half-ration of bread and meat), for Chanderi, distance 30 miles, where it arrived on the morning of the

16th of March, thus accomplishing a distance of 47 miles in 24 hours; was present at the storming and capture of the stronghold fort of Chanderi, 17th of March; had 3 men killed, 2 officers and 19 men wounded."

A slightly more detailed account might be of interest.

On this very long march Major Stuart relates that the 86th Regiment lost its way in a native town in the dark. The guides had run away, and the order was finally given to lie down until daybreak. When daylight did appear the Regiment managed to find their way out and continued their route. As they marched they distinctly heard the report of the guns attacking Chanderi. The sound had a most magical effect upon the men; some, who were almost "beaten down" with this terrific march, seemed to derive new life and energy from it, and as the artillery reverberated through the morning air he thought the last seven miles of the march were the quickest on record. His horse had to break into a regular amble to keep up with the light-hearted soldiers. When they arrived at the encampment the whole force turned out to meet them and the cheering that greeted them was immense.

Two orders for storming were issued, one was made out on the 11th of March, before the arrival of the seven companies of the 86th Regiment, and reads as follows:—

"*Advance Post.*

30 Rank and File H.M. 86th Regt., under an Officer.
70 Rank and File H.M. 25th Regt., N.I., under an Officer. The senior to command the whole.

"*Column of Assault.*

50 Rank and File 21st Comp. R.E., under an Officer, half carrying ladders, half implements to burst doors, etc.
25th Regiment N.I.
Detachment H.M. 86th Regiment.

"*Reserve.*

Remainder of 21st Company R.E., under an officer.
2nd Company Bombay S. & Miners.
70 Rank and file 25th Regt. N.I., under an Officer.

"The Senior Officer to command the whole. Regiments to form under further instructions, which will be issued on the Plateau near the Breaching Battery, in column of sections right in front. The camp to remain standing, and it will remain in charge of Major Gall, H.M. 14th L.D., under whose orders the Field Officer and Subaltern of the Day will act. All guards and picquets to remain at their posts, all available men of

H.M. 14th L. Dragoons, 3rd Regt. C.H.C., and Detachments 3rd E. Regiment to form up under Major Gall's instructions.

"Further orders will be issued as to Field Artillery.

"By Order,

"(Signed) I. C. COLEY, CAPT., M.B."

The early arrival of the remainder of the 86th Regiment altered the composition of the "Advance Post," as it was strengthened by twenty Europeans, and the native infantry in it were reduced by a like number. Also it was decided to make a false attack at a point opposite the Kati Gati, where the rock could be climbed and where the wall was not very high. Part of the cavalry remained in camp, which was struck after all, whilst some of the cavalry had been sent away on the 12th to watch a fort, also held by the enemy in the neighbourhood. It was expected if the assault on Chanderi was successful that many of the mutineers would make for this other fort for refuge, and this is what actually happened. The fugitives were met by the cavalry, who killed numbers, whilst the enemy in this fort fled during the night.

At 3 a.m. on the 17th of March the attacking parties formed up, and after a short and rapid fire from the breaching batteries the storming parties both rushed towards the fort. The enemy opened fire with their guns, but fortunately they were aimed too high and the discharge did no damage. Major Stuart writes: "Daylight not having appeared, it was perfect guess-work in the dark, and we seemed to be climbing a steep precipice, stumbling and tumbling at every step over the broken ground." Both parties effected an entrance. At the breach a trench was found cut in the rock fourteen feet wide and deep, but that was crossed on the scaling ladders. Each gun was carried in rapid succession, much fighting going on with the bayonet. Many of the rebels driven back against the wall were dashed over it down the precipice into the ravine below. A magazine exploded and blew up seven men of the Royal County Down Regiment. The poor victims were a most piteous sight, as their clothing was either on fire or burnt off completely, whilst their bodies were shockingly burnt, and all died from their injuries. Here Private Roger Mathews saved Lieutenant Lewis's life by bayoneting a native who was in the act of killing that officer. Having cleared the fort, the stormers dashed on into the palace and into the buildings beyond, which were soon taken, as the enemy fled into the town and through it to the secure shelter of the jungle.

A hundred of the enemy were found dead in the fort, killed in the assault, besides those who had fallen over the walls or had been killed in the pursuit, whilst, according to the returns, three officers only of the British force were killed and wounded, besides the Regiment's already stated loss of three men killed and two officers and nineteen men wounded.

The 1st Brigade marched from Chanderi on the 21st of March and arrived at Jhansi on the 25th of March. Sir Hugh Rose had arrived there on the 21st of March with the 2nd Brigade, reinforced by the British cavalry from the 1st Brigade, which had caught 100 armed robbers entering the town to assist in the defence and killed them all.

The city of Jhansi was about four and a half miles in circumference, and was surrounded by a fortified and massive wall from six to twelve feet thick, and from eighteen to thirty feet high, with numerous flanking bastions which mounted guns, and also were loopholed with a banquette for infantry.

The fort was in the south-western portion of the city, and was of great strength. It stood on an elevated rock rising out of the plains. It was built of most massive masonry, and commanded the whole city and surrounding country. It was difficult to breach, as its solid walls were built of granite and varied in thickness from sixteen to twenty feet. It was surrounded by elaborate outworks of the same solid description. In some places there were five tiers of musketry. Guns had been placed on the highest towers to command the surrounding country.

The city walls met on the southern face of the fort in a high mound, on which were five guns, which mound was protected by a ditch faced with very solid masonry, the ditch being twelve feet deep and fifteen feet broad.

Outside the walls the city was girt with wood excepting some parts of the east and south fronts. On the former side was a picturesque lake and water palace, whilst to the south between the British camp and the city were the remains of the cantonments. Here the unfortunate European inhabitants had been butchered on the 8th of June, 1857. The history of that sad event was briefly as follows:—In 1854 the last ruler of Jhansi died. His widow wished to adopt a son to succeed to the kingdom, but the British Government, thinking it better that all such vacant thrones should not be filled, as the people were often misgoverned and oppressed by their native rulers, refused to allow such an adoption. They also gave her £6,000 a year as an allowance, and meanly called upon her to pay her husband's debts out of this amount. The Ranée said nothing, but brooded over the bitter resentment which she harboured against the British. The Indian Government had foolishly kept no European troops in this exceedingly strong fort. The Resident equally foolishly had made no preparations to protect the lives of the Europeans by provisioning the fort. When the native soldiers mutinied on the 6th of June all their officers except one were murdered. He was wounded. The civilian European population retired into the fort, and repulsed the mutineers, who had promptly attacked it. Having no provisions they unwisely surrendered on the 8th of June on the promise of the Ranée to send them to another English garrison. Having got them out of the fort, the Ranée remarked, "I have no concern with the English swine," and the unfortunates were cruelly slaughtered, men, women and children being tied to trees and then cut to pieces with tulwars, and their bodies were left exposed for three days where they were murdered, and then were thrown into gravel pits hard by and covered over. Such was the incident that the 86th Regiment and the remainder of Sir Hugh Rose's force had come to write the sequel of in letters of blood.

No plans existed of the city or fort, so Sir Hugh spent the 21st of March under fire examining the whole place, and also drove in to the city many outlying parties of rebels. On the 23rd the whole place was thoroughly invested by seven camps of cavalry, with guns in some places. The whole city was watched by infantry outposts as well as by cavalry vedettes. All roads, etc., from the city were obstructed by trenches and abatis. The natives very naturally had a great belief in the strength of Jhansi, and every night parties of rebels tried to break in to assist the defence. One party from Kalpi was captured bringing in a convoy of rockets.

The 1st Brigade, when it arrived on the 25th of March, encamped about two miles from the 2nd Brigade, at a place about a mile from the fort.

The difficulties of the attack on the Jhansi fort now became apparent. The only portion of the fort which could be breached, set on its rocky hill, as it was, and protected by the town on almost all sides, was its south front.

The south front was, however, flanked by the fortified city wall, and by the mound which has before been described.

It seemed, therefore, that it would be best to capture the mound, as not only did it enfilade two walls of the city, but it also commanded the whole southern quarter of the city, and likewise the palace. If these were all taken approaches could easily be constructed against the fort.

An excellent position for a breaching battery was a rocky ridge 640 yards from the mound. Two different attacks were decided upon and everyone instantly was hard at work filling sand-bags, making gabions and heating shot.

The garrison of Jhansi has been variously estimated. The lowest calculation places it at about 13,000 troops, besides the inhabitants. They had about forty guns and many well-trained gunners. There was hardly a house in Jhansi which did not contain British plunder, and unpleasant questions might be asked, so that though the inhabitants had first thought of leaving the city but had been prevented by the Ranée, they now prepared for a strenuous defence to keep the avenger out. The batteries opened on the 28th of March and cleared the enemy from the mound that day. The battery nearest the 86th Regiment's camp consisted of three 18-pounders, two 10-inch mortars, two 8-inch mortars, and one 8-inch howitzer.

The 10-inch mortars created great havoc in the fort, and a powder magazine blew up at the third shell.

A breach was made in the wall by the 30th of March, but the enemy retrenched the breach with a double row of palisades, filled in between with earth. The British gunners replied to that with red-hot shot, which set fire to the wooden palisades, whilst marksmen were posted in various places to fire on every loophole and embrasure, and they caused great loss to the defenders.

From the shelling alone the enemy afterwards acknowledged to have lost about seventy men a day killed. On the 30th ammunition began to run short, and arrangements were at once made to assault the breach and to escalade the walls in various places; but that morning Sir Hugh Rose received warning that Tantia Topi was coming at the request of the Ranée to raise the siege of Jhansi and was at that moment only three miles away from the Betwa River, which was some seven miles from the camp to the east. With him were 22,000 men and twenty-eight guns. The general marched with the 2nd Brigade at 9 p.m. to crush the enemy as he crossed the two fords, but after arrival there he decided that it would be better to draw him on, so he retired to the camp again and was followed on the evening of the 31st by Tantia Topi.

A great shout of joy went up from Jhansi when he appeared and took up a position in order of battle in rear of the 2nd Brigade's camp after sunset.

The British silently filed out of their camp, leaving a strong force to guard it and to prosecute the siege, and formed up in two lines opposite to Tantia Topi's force half a mile only behind the camp of the 2nd Brigade. The total strength of the two lines was 1,182 officers and men. A detachment of the 86th Regiment was in the second line.

This line was drawn up in contiguous columns at quarter distance. A weak troop of the 14th Light Dragoons was on the right and the Hyderabad Cavalry on the left flank; in the centre came 208 officers and men of the 86th Regiment, two field batteries and a detachment of the 25th Bombay Infantry. The British lines were not in position until long after dark, and the opposing forces slept opposite each other in their ranks. The mutineers called out repeatedly during the night that they were very numerous and the English very weak, and that they would finish the Feringhees off easily next morning. In the meantime Tantia Topi, who really was something of a general, had sent another force by another road to break the investment line round Jhansi. News of this came to Sir Hugh Rose just after midnight, brought by some native cavalry, whom he had detailed to watch the road.

The 2nd line was at once sent off under Brigadier Stuart and marched for Bangaon, eight miles from Jhansi, close to the Betwa River, with a view of taking the force in flank.

Sir Hugh had intended to attack at daybreak, but the enemy forestalled him by sending forward clouds of skirmishers, followed by heavy masses of infantry and cavalry. As the enemy bore rather to their right, the British conformed cautiously, as they anticipated it might be a ruse, and, as they expected, a strong body of horse appeared on the right of the British line to charge them in flank. The horse artillery now advanced and enfiladed the enemy's line, whilst the cavalry charged so soon as the rebels were at all shaken, half leading on the enemy's right and half, headed by the general himself, on the enemy's left. Forming a line four or five deep and throwing back their flanks until they each rested on a rocky knoll, the rebels met the charge most bravely with a rolling fire of musketry. Breaking through this dense line, the cavalry, wheeling right and left, charged down the line, breaking it up, and a rout ensued. But it was a rout with many a rally, for the bravest of the mutineers formed themselves into groups and fought to the last.

A second line of rebels under Tantia Topi himself was drawn up in rear at a distance of three miles. This force retired in splendid style, covered by skirmishers, but the British drove the latter in and attacked the main body, which withdrew, fighting, burning the jungle as they passed. The cavalry and guns galloped after them and broke up every attempt at a rally.

The whole now rolled across the river in one mass, the enemy's guns playing on the pursuers from the farther bank of the Betwa. The British infantry followed hard after the cavalry, being called up when any position was occupied by the enemy which the cavalry could not carry. Quantities of stores, eighteen guns, and enormous spoils of all sorts, were taken on the far bank of the river.

Brigadier Stuart, having found no enemy at Bangaon, had marched back towards the sound of battle. His cavalry cut up some fugitives, and then the force came on a party of the enemy 2,000 strong holding a village called Kushabir, with a certain number of guns to help them. Brigadier Stuart advanced with his infantry in skirmishing order, his cavalry on either flank and his guns on the main road until within about 600 yards of the enemy's position. The artillery then opened fire, and after a few rounds the infantry were ordered to storm the village. The 86th Regiment and the native infantry at once dashed forward and carried the village at the point of the bayonet, capturing all the enemy's six guns. Falling back from there, the hostile troops rallied at a second village, but were instantly ejected by a charge of the 86th Regiment.

The enemy then fell back in splendid order, pursued by the 1st Brigade, though 250 of the rebels were killed in this encounter alone. The troops had now been under arms for thirty-six hours, and besides killing 1,500 of Tantia Topi's army had taken all their stores and guns. So Sir Hugh Rose ordered the force back to their camp.

The losses of the British were surprisingly small in this combat or series of combats, being only nineteen killed and sixty-two wounded.

The Adjutant, Lieutenant Cochrane, had three horses shot under him, and was mentioned in despatches, as was also Lieutenant-Colonel Lowth.

Tantia Topi's discomforted army fled to Kalpi, and the siege of Jhansi was pressed with redoubled vigour. The besieged, whilst Tantia Topi's relieving force was trying to effect an entrance, had contented themselves by redoubling their artillery fire, and had poured forth volleys of musketry which appeared to presage a sortie, but the defeat of their brethren had chilled their ardour.

Tantia Topi's intervention had only delayed the assault on Jhansi by forty-eight hours.

The storming of that city was ordered to take place on the morning of the 3rd of April.

Detailed orders were given, so Sir Hugh states in his despatches, to the officers commanding the various units. Such detailed orders are not, however, shown in his despatches or in other records, so far as can be traced. But the actual arrangements for the assault are well known.

The brigade was to attack in two places. The main attack was made on the breach in the wall round the mound. It was headed by the 86th Regiment, and Lieutenant Jerome,* who won the V.C. later in the day, led the stormers.

At the same time 350 men of the 86th Regiment and 25th Bombay Infantry were directed to attempt an escalade. They were commanded by Major Stuart, of the Regiment, and the point chosen for the attempt was known as the Rocket Tower.

The 2nd Brigade was ordered to attempt to climb the walls of the town elsewhere. An officer not belonging to the 86th Regiment who was present states that he was surprised to be called at 2 a.m. on the 3rd of April to "Assault immediately." So evidently the secret of the design to storm Jhansi that morning had been well kept.

The troops fell into their respective places silently, but there was no lack of enthusiasm. On the 8th of June of the year before thirty men, sixteen women, and twenty little children had been brutally murdered hard by the spot where they stood, and the only crime of all these innocent victims was that their faces were white.

Inside the city in front the murderers stood defiantly.

Truly, the Royal County Down Regiment could ask for no better fortune than to be allowed to combat such people, and read them a much-needed lesson.

The column under Lieutenant-Colonel Lowth, of the 86th Regiment, forced the breach with comparative ease. Sir Hugh Rose mentions in his report that it was considered, previous to the commencement of the bombardment, that the point chosen for breaching was the easiest point of entrance. It may have been, but it was difficult enough, for at some distance in rear of the breach was found another high wall fifteen to twenty feet thick, with a deep tank again in rear of it cut in the solid rock. However, over all this the Grenadier Company of the 86th Regiment scrambled somehow, and the breach and the mound were won.†

In the meantime, Major Stuart led his 350 men valiantly forward to assault the rocket tower. The Light Company of the 86th Regiment went first, then

* The son of an old officer of the 86th Regiment, born in the Regiment. His brother also served in the Regiment and commanded it.

STORMING OF JHANSI BY THE 86TH REGIMENT, 1858

By kind permission of the artist
Miss Marjory Watherston

came 100 men of the 25th Bombay Infantry, and then two reserves of seventy-five men from each corps. Ensign Fowler, with a few skirmishers, were slightly in advance. This party got within 350 yards of the wall just before daybreak. The wall to be assaulted was about twenty-three feet high. So soon as the signal was given, the column moved rapidly forward until within 150 yards of the town, when Major Stuart called out, "Now, lads, for an Irish yell." The officer before quoted notes that "Such a yell was given as might have frightened the devil." A terrible fire was then opened on the party, and as they arrived close to the wall stinkpots, rockets, and red-hot balls came down on them in showers, and a good many casualties took place. However, the ladders were planted, and up them rushed Lieutenant Dartnell, followed by Ensign Fowler, Ensign Sewell (all 86th Regiment), Lieutenant Webber (Royal Engineers), and Major Stuart, of the 86th Regiment, with all his men at his heels. Lieutenant Dartnell was greeted by four sword cuts. Ensign Fowler shot a couple of Lieutenant Dartnell's opponents and saved his life. Then a hard struggle commenced to obtain a footing on the wall. Gradually, however, a space was cleared, and a party from the breach taking the enemy in flank, the rebels retired into fortified houses and other works, from which they kept up a heavy fire. They were followed by the infuriated Irishmen, who proceeded to clear the houses in their front in a most business manner. Suddenly this party came under a severe fire from the fort, under which they were obliged to retire, as they were unable to get into it, and lost in a very short space three officers killed—Captain Darby, Ensign Sewell, and Lieutenant Halroyd (an attached officer)—and many men severely wounded. Ensign Sewell was very badly hit, but Lieutenant Jerome and Serjeant Byrne carried him off at the risk of their lives. The whole party only retired to the shelter of the nearest houses, and took counsel as to the next thing to be done. Here Surgeon Stack, of the 86th Regiment, was killed. He was engaged in putting a wounded man into a dooley when Major Stuart said to him, "For God's sake take care; the fire is terrific." Surgeon Stack replied, "Never fear," and the next moment he fell dead, shot through the heart.

Here also occurred a very gallant action. Privates White and McCann, of the 86th Regiment, had just been shot dead in this place, and the fire from the fort was growing, if possible, more terrific, when out of a house came a poor little native boy about five years old. He was immediately severely wounded through the shoulder by a shot from the fort. Private Kavanagh, of the 86th

† There was a story current amongst the men of the Regiment when the author joined the Service that one officer leading fell down, and was immediately attacked by a crowd of natives with tulwars.—Another officer jumped in to save him and also fell. Serjeant Byrne—or, as he then was, Private Byrne—who won the Victoria Cross later on in the day, jumped in at once, and also coming down struggled to his knees and kept the natives back by swinging his rifle, held by the muzzle, at about the height of their knees, and that some of the men of the Grenadier Company came up and finished the matter. The story is uncorroborated, but probably true.

Regiment, rushed at once with a shout into the street, and, taking the poor wounded child in his arms, carried him to a place of safety. As he was doing so a gingal ball struck the wall within two inches of his head. The poor little wounded boy was kept and nursed all through the campaign by the gallant Irishman.

Every morning he carried him to the doctor to have his wound attended to, and when he recovered he was dressed in a little suit of European clothes, and he became the pet of the company to which Private Kavanagh belonged.

At the end of the Mutiny, Private Kavanagh brought him down to Bombay, and on embarking for England handed him over to the Roman Catholic Asylum in Bombay.

In the meantime Lieutenant-Colonel Lowth and his party, after carrying the breach, sent assistance to the attackers on the left and right, and pushed on with the main body into the palace.

Sir Hugh Rose, in his despatches, notes that " the 86th, on the road to the palace from the mound, sustained many casualties from their left flank, being exposed as they passed through an open space to a flanking musketry fire from an outworks of the fort, and from houses, and from the palace itself to their front," and he caused a covered way to be made to the palace through the houses.

" The skirmishers of the 86th [to quote Sir Hugh's despatches again] penetrated gallantly into the palace; the few men who still held it made an obstinate resistance, setting fire to trains of gunpowder, from which several of the 86th received fatal injuries."

In the meantime it had not fared so well with the attack made by the 2nd Brigade. The 3rd Bombay European Regiment led the assault there and were met by what is described by one who went through it as " a sheet of fire out of which burst a storm of bullets, round shot, and rockets, destined for our annihilation." The Royal Engineers, who carried the ladders, began to fall fast, and the party sheltered for a moment in some ruins to regain their breath and then dashed once more for the wall. The hail of bullets, rockets, huge stones and every possible description of missile, if possible, grew worse. Three ladders were placed against the wall, and up the three ladders ran three brave young officers, followed hard by their equally brave men. Lieutenants Dick, Meiklejohn and Fox arrived on the ramparts, and their three ladders broke behind them under the weight of men. Left standing alone on the wall, Lieutenant Meiklejohn closed with his assailants and was cut to pieces. Lieutenant Dick, pierced by shot and bayonets, fell dying from the wall, whilst Lieutenant Fox was shot through the neck. Some of the men were now hurried round to the breach to effect an entrance there. Other ladders were obtained and placed against the wall, and the following statement was made to the Author by Colonel Sir Edward Charles Ross, C.S.I., late of the Bombay Army, who died in 1912, who was present at the storming of Jhansi, and was

SERJEANT BYRNE AND HIS VICTORIA CROSS
JHANSI, 1858

Victoria Cross in possession of the

afterwards Resident in the Persian Gulf: "That he climbed the ladder which was set up after these three ladders were broken down; that several men or officers went up before him and all were knocked off the ladder by blows from tulwars as they reached the top; that he went up expecting the same fate, and as the first native with a tulwar struck at him he saw him bayoneted by a soldier of the 86th, who came running along the wall with several other soldiers belonging to the Royal County Down Regiment. As he clambered on to the wall the soldiers 'bounded'* at great numbers of rebels and engaged them with the bayonet."

The 2nd Brigade now poured over the wall, and the general strove to clear the enemy from a space in the city, from which he could operate, first against the remainder of the city and clear it of the enemy, and then against the fort.

So, whilst the 86th Regiment, which was now all together storming the palace, held the apex and one side of the triangle, the palace being the apex; the 2nd Brigade held the eastern side of the triangle, which ran north-east from the palace to one of the city gates, the base of the triangle being the walls which had just been stormed. The fighting in the palace was very strenuous. The mounted bodyguard of the Ranée, drunk with opium, defended the palace stables desperately, and these were enfiladed by the fire from the fort. They fired from behind the doors, through the windows, and through the loopholes, with matchlocks and pistols, and as the British approached some charged down on them with their tulwars; some retired into houses and fought there in the semi-darkness to the last. One body remained in a room off the stables which were on fire, and then rushed out with their clothes in flames and hacked at their assailants with their swords whilst they guarded their heads with their shields. Finally, the palace was captured but not without severe loss. In the palace was taken a silken Union Jack, which the Viceroy had given the grandfather of the late ruler of Jhansi, with permission to have it carried before him as a reward for his fidelity: a privilege granted to no other Indian prince. The 86th Regiment asked permission to hoist this flag on the palace at once, which was granted. Numerous incidents marked the desperate feelings which animated the defenders. A retainer of the Ranée tried to blow himself and his wife up. Failing in the attempt, he cut her to pieces and then killed himself, whilst two Afghans threw a woman who was with them into a well and then jumped into it themselves.

The palace won, the triangle in British possession remained to be carried.

The officer before quoted writes: "I went with 100 men to clear a part of the town. This house-fighting was no joke, but we killed upwards of 200 of the enemy. All are full of the praise of the 86th, and they richly deserve it, for no men could have behaved better. They have lost 1 officer and 12 men killed and 6 officers and 60 men wounded, all but 7 severely and most severely, and I fear that the deaths will be numerous."

The Regimental Records, however, state that the loss on the 3rd of April

* The actual word used.

was Surgeon Stack and 7 men killed, 5 officers and 58 men wounded. The wounds were so severe that many men died afterwards from them and most of the others had to be sent to England to be discharged as incurable.

Whilst all this was going on in the town Sir Hugh Rose received a report about 3 p.m. from the commander of one of the seven strong cavalry posts outside the city that a large body of the enemy flying from the town had tried to force his party back, but that he had driven some 400 of them on to a high and rocky hill some 600 yards to the west of the fort, and that he had surrounded this hill with cavalry, and awaited reinforcements before he stormed it. The general ordered out from the camps all the camp guards and any reserves he could lay his hands on; the rebels were surrounded and shelled; the infantry then went up and cleared the hill with the bayonet. When they saw that escape was impossible some of the rebels lay down on their powder flasks and blew themselves up. The British loss here was one officer killed and about twelve men killed and wounded. One of the very few who escaped from the hill was the Ranée's father. He was wounded, but was captured some days afterwards and duly hanged. The hill was afterwards known as "Retribution Hill."

Sir Hugh Rose states in his despatches that after clearing the quarter of the town in British possession of the enemy, he had intended to attack the remainder of the city, but Brigadier Stuart dissuaded him, as the men were too exhausted for further operations that day. At sunset the news came in that a force was again advancing to relieve the city and the tired troops had to fall in again and were marched out to where they had fought the action of the Betwa. Fortunately it was only a false alarm, troops from Tehree having been mistaken for the enemy.

Next day the remainder of the city was occupied, and on the following morning the enemy were found to have evacuated the fort. The officer before quoted states: "To our great delight, on the morning of the 5th we found the enemy had bolted from the fort, for had they not done so we would not have got in for 10 years. It is the strongest place I ever saw."

Major Stuart, writing on the 10th of April to Lieutenant Dartnell's father, says: "The enemy bolted from the fort next day, luckily for us, as we would have never got in; it is a place of immense strength."

The reason of this *débâcle* was that the Ranée had fled from the fort on the night of the 4th-5th of April, accompanied by a bodyguard of twenty-five mounted men and 300 Afghans. Unfortunately for them they ran into one of the cavalry picquets, but the mounted men and the Ranée managed to escape and fled towards Bandiri. The cavalry went in pursuit of the Ranée, whilst the 86th Regiment took possession of the fort.

Tantia Topi sent a body of horse to bring off the Ranée, and twenty-one miles from Jhansi, in sight of Bandiri, the pursuing cavalry caught sight of the rebel horsemen. These separated into different parties to give the Ranée a better chance of escape. The Ranée's party was pursued by an officer with some Indian cavalry; he caught sight of the Ranée mounted on a grey horse, but was detained by a party of Rohillas and Bengal Irregular Cavalry, of whom he managed to kill forty. Pressing on again, he was disabled by a severe wound; so the Ranée escaped for a time. The woods, the gardens, and the roads round the town were strewn with the corpses of the fugitive rebels. Sir Hugh, in his despatches, says: "It will give some idea of the destruction of insurgents which ensued when a party of the 14th Dragoons alone killed 200 in one patrol.

"The rebels, who are chiefly Afghans and Pathans, generally sold their lives as dearly as they could, fighting to the last with their usual dexterity and firmness. A band of forty of these desperadoes barricaded themselves in a spacious house with courtyard, vaults, etc. Before they were aware of its strength it was attacked by a detachment of Hyderabad Infantry, under Captain Hare, with the loss of Captain Sinclair, of whose conduct it is my duty again to make honourable mention. Reinforcements and several pieces of siege artillery were brought up by Major Orr, who commanded the attack against the house, but even when it was breached and knocked to pieces the rebels continued to resist in the ruined passages and vaults. They were all destroyed, but not without several casualties on our side. Major Orr expresses his obligation to Lieutenant Lewis and Ensign Fowler, of Her Majesty's 86th Regiment, the first being severely wounded, who led the men."

Besides these two officers seven men of the Regiment were wounded at this house. Speaking of the storming of the breach, Sir Hugh Rose says in his despatches: "Colonel Louth [Lowth], commanding Her Majesty's 86th Regiment, acted with cool judgment, and I witnessed with lively pleasure the devotion and gallantry of his regiment."

He further reports that 5,000 rebels have been killed, though, he adds, "the native accounts make the loss of the rebels amount to more than 5,000."

So ended a famous siege, in which the troops were on duty for seventeen days and nights without taking off their clothes, the horses being always saddled and bridled, and in which, opposed to the strongest fortress in Central India, with a garrison of more than twice the strength of the besieging force, they killed 5,000 of that garrison, only snatching a hasty twenty-hours' respite to go and fight and defeat a relieving army of four times the strength of the total British force.

The first order issued by Sir Hugh Rose after the capture of the city was that the destitute women and children should be fed from the prize grain.

The total British losses were 38 killed and 181 wounded, of whom 22 subsequently died.*

* The 86th Regiment recently asked that the name of Jhansi might be recorded amongst their battle honours, but the request was refused on the grounds that no doubt it had been considered at the time, and there must have been some good reason then for not awarding it.

It would be well to consider how matters stood in the rest of Northern India at the moment of the capture of Jhansi by Sir Hugh Rose. That city was stormed on the 3rd of April; on the 30th of March the Rajputana Field Force, in which was the 83rd Regiment, had captured the town of Kotah and had driven out a great host of rebels who were now at large in the country. Again the late garrison of Chanderi, who had got away fairly intact, had made a raid on the road to Goona, whilst Sir Colin Campbell had retaken Lucknow and had driven a force of rebels, estimated at 100,000 armed men, out of that city. On the much-used road from Jhansi to Cawnpore where it crossed the Jumna River, stood the arsenal of Kalpi. It was full of warlike stores and ammunition, and besides this so long as it was held by the rebels it prevented the free communication of the British armies in the West with those in East India, and gave them a central position to move against either British force at their pleasure. The orders Sir Hugh Rose received were that he was to take this fortress, so, leaving a small garrison in Jhansi, Sir Hugh marched with the 1st Brigade up the Cawnpore road, leaving Jhansi at midnight on the 25th-26th of April, being followed by the 2nd Brigade two days later.

The hot weather had now set in, and the troops suffered severely in consequence. The country they marched through was flat and without vegetation. The dust was several inches thick on the roads, and the heat during the day obliged the troops to march at night. Many of the soldiers who slept during the time the sun shone, never woke again and were found dead. The farther the column marched the scarcer became the water, which was now only found in small round wells at a very great depth, whilst it was lukewarm and often had a brackish flavour. A fight occurred on the 2nd of May at a small mud fort, called Lohari. It was noticeable because the enemy used a burning cloth full of loose powder thrown upon the stormers' heads, and also because of the garrison, some 100 strong, who were mostly 12th Bengal Infantry, not one escaped alive.

About fifty miles up the Cawnpore road is the town of Kunch, an unwalled town, but difficult to attack, as it is surrounded by woods, gardens and temples, with high walls round them. Here the Ranée of Jhansi and Tantia Topi had assembled an army. These two worthies had fled to Kalpi from Jhansi, travelling the distance, about 100 miles, in one day. Their army consisted of the remains of the Jhansi garrison, some Bengal regiments, the levies of rebel Rajahs, and of the Gwalior Contingent Cavalry from Kotah. They had brought this force forward to Kunch, at which place they had thrown up entrenchments. As the heat was so excessive, Sir Hugh Rose decided not to undertake a siege, so he turned the defences by marching round the north-west of Kunch, starting at 10 a.m. on the 6th of May. This march threatened the enemy's retreat to Kalpi, and besides, the rebels had not fortified that side of the town.

The 1st Brigade had marched fourteen miles to get into position, and when they arrived opposite to the town the order was given for a meal to be cooked, and to keep the enemy occupied two 18-pounders and an 8-inch howitzer shelled them, whilst the Horse Artillery were also able to effectively shell the mutineers. Sir Hugh Rose determined to drive the rebels out of the woods, gardens, and temples on this side, and then to storm the town, including a dilapidated mud fort, on which the mutineers' red flag was flying, which fort was opposite to the 1st Brigade. Once in possession of this portion of the town, the enemy's line would be cut in two, and those in the entrenchments on the right to effect their retreat would be forced to retire over the plain outside the town, where they could be caught and cut up by the cavalry.

The operation was effected by throwing forward the left wing of the 86th Regiment, under Major Stuart, and the whole of the 25th Bombay Infantry in skirmishing order, the 86th Regiment's skirmishers being on the left, both flanks being supported by some field guns and cavalry, whilst the remainder of the 86th Regiment, under Lieutenant-Colonel Lowth, formed a second line in reserve.

The 25th Bombay Infantry were able to clear the ground in their front and seize some guns, whilst Major Stuart made a circuit to his left, took all the obstacles in front of him, and then swinging slightly to his right advanced, despite artillery and musketry fire, through the whole northern part of the town, and stormed the fort. The enemy was now streaming out of the town in excellent order, but intent on getting away. The 2nd Brigade, to the right of the 1st Brigade, was in full cry after them, with a strong detached force of native cavalry, principally the Hyderabad contingent, working still further to their right: but the British infantry were completely exhausted, and the cavalry and artillery alone were able to continue the pursuit.

The enemy commenced their retreat across the plain with resolution and intelligence. The line of skirmishers fought well to protect the retreat of the main body. When charged they threw aside their muskets and fought desperately with their swords. Full of fight, they threw back the extreme right of their skirmishers so as to enfilade the line of pursuit, but the cavalry charged and cut off this enfilading line and annihilated them. The greater part of the rebel skirmishers were now killed, and the remainder driven in whilst the rebels' guns were captured, so the main body of mutineers began to lose their nerve, and crowded into the road to Kalpi a helpless column of runaways. The horses of the British cavalry and artillery were so fatigued by the heat that they were reduced to a walk, and the guns were only able to rake the flying column with round shot and shell, but could not approach sufficiently near to it to use grape, whilst the cavalry had to content themselves by cutting up the numerous stragglers who dropped out of the enemy's ranks unable to keep up from the same reason.

Seven miles from Kunch a wood was approached, and here the rebels broke up and fled, and the British force crawled back to Kunch, having been sixteen hours marching and fighting. Their losses amounted to one officer and seven men killed, two officers and forty-three men wounded, whilst the cases of sunstroke amounted to two officers and forty-three men, three of the 86th Regiment died from this cause, and twelve of the 71st Regiment, which had joined the force at Jhansi. The rebels lost about 550 killed. The thermometer registered 115° at Kunch and 118° at Kalpi in the shade, but burst in an officer's tent at 130° whilst on the way to Gwalior.

The rebels quarrelled as they fled home to Kalpi. The infantry accused the cavalry very justly of want of courage; the Afghans also were charged with not having shown their usual stern bravery at Kunch. It was suspected that their heavy punishment at Jhansi had shaken their *moral*, though they naïvely explained that they had hurried from the field feeling it to be their duty to escort the Ranée of Jhansi to a place of safety, whilst all abused Tantia Topi as a coward for having fled early in the day. So Kalpi was deserted by almost all the mutineers, and would have fallen easily if the rebel Nawab of Banda had not unexpectedly arrived with a large force of good cavalry mutineers and with some guns and infantry. The Ranée of Jhansi, as usual, was indefatigable, so the mutineers were re-collected and plucked up spirit to defend Kalpi. It was wretched as a fortification, but its position was good. In front there were eighty-four temples, with walls round them, built of most solid masonry. In front of the temples the rebels had formed some good entrenchments. Behind the temples, which were some three miles from Kalpi, was a network of ravines; behind that came the town itself, then a further line of ravines, and, finally, the fort, which stood on a steep and lofty rock springing from the right bank of the Jumna. At 2 a.m. on the 9th of May Sir Hugh Rose marched with his 1st Brigade on Kalpi from Kunch. He intended to get into communication with a Lieutenant-Colonel Maxwell, of the Bengal Army, who was co-operating with a column on the left bank of the Jumna. This column was bringing up ammunition for the Central India Force, as a great deal of their supply had been used up before Chanderi and Jhansi. He had written to Colonel Maxwell to say that he would meet him on the banks of the Jumna a few miles below Kalpi on the 14th of May. This letter was sent by a spy, and was either placed in the sole of his shoe or in a quill in his mouth, but in any case it never reached its destination.

On account of the want of water the whole force could not be concentrated against the defences before Kalpi, so the General moved off the road with the 1st Brigade to the village of Golauli about five miles below Kalpi, and then he intended to move up the Jumna, resting his right flank on that river, whilst his right flank was to be covered from any unwelcome attentions of the enemy across that stream by the column under Colonel Maxwell, which was to move up in line with him.

The 2nd Brigade was to march up the main road and demonstrate in front of Kalpi, but it lost its way, and followed the General to Golauli. It was a long and trying march as the sun was very hot. All the staff of the 2nd Brigade were on the sick list, including the Brigadier, whilst Sir Hugh Rose had had five slight sunstrokes up to this time.

On the morning of the 16th the two brigades were duly assembled beside the Jumna, and communication was established with Colonel Maxwell, who was some thirty miles away on the other side of the stream. Two pontoons were launched on the Jumna. These had been brought up with great difficulty from Poona for the purpose of communication and of bringing the ammunition across the river. The rebels were found to have destroyed or taken to Kalpi all the country boats.

The enemy had attacked the rear guard on the 15th, but had been beaten off though 200 out of 400 men of the 25th Bombay Infantry had fallen out on account of the heat of the day; whilst on the 17th of May an attack was threatened against the front of the 1st Brigade as it faced towards Kalpi.

From now on matters became rather embarrassing for Sir Hugh Rose. He had arranged to carry Kalpi by two different attacks. One was to be conducted by the 1st Brigade on Kalpi straight to their front, but a village named Rayar, between the 1st Brigade and Kalpi, was held by the enemy with a force and battery, which battery commanded the only pass through the ravines, by which guns could approach Kalpi. Therefore Colonel Maxwell's column on the left bank of the Jumna was to shell Rayar, and also the fort at Kalpi, with the object of blowing up the powder magazines. The second attack was to have been made by the 2nd Brigade well on the enemy's right, but here the British were foiled as the wells ran dry, and the 2nd Brigade had to be brought nearer to the Jumna, for the purpose of obtaining water. Colonel Maxwell, whose principal strength consisted in 458 Sikh Police and 578 of H.M. 88th Regiment, sent two companies of the latter over to reinforce Sir Hugh Rose and also 124 of the Sikh Police, for sickness was increasing enormously in the ranks of the Central India Field Force.

About this time their general wrote to the Commander-in-Chief that " they fell in their ranks struck down by the sun and exhausted by fatigue, but they would not increase the anxieties of their General or belie their devotion by a complaint. They were often too ill to march, but their devotion made them fight; it is almost superfluous to add that troops animated by so high a sense of duty were sober, orderly, and most respectful to their Officers."

On the 20th of May the enemy advanced through the ravines and attacked the British right flank. Reinforcing the picquets with four companies of the 86th Regiment, under Major Stuart, and some few other troops, the general simply directed them to hold their ground, which they did, driving off the enemy and losing in doing so four officers and forty men in killed and wounded.

Obtaining information from his spies that he was to be attacked in force on the 22nd of May, Sir Hugh Rose made the following dispositions:—

His front ran from the Jumna facing Kalpi until it came near his left flank, when that was thrown back to guard that flank, as there the ground was suitable for horse and guns. The picquets on the right were composed of the 86th Regiment and some of the 3rd Bombay European Regiment, whilst two companies of the Connaught Rangers and four companies of the 25th Bombay Native Infantry formed the camp guard.

At 8 a.m. on the 22nd the enemy advanced in force from Kalpi. Making a great show on their right with 1,400 cavalry and several battalions of infantry, and with horse artillery, they very quietly pushed a strong force up the ravines in front of the 86th Regiment. Though Sir Hugh Rose could not see them, he suspected their design and drew no men from his own right, but he made his cavalry retire on his own left, drawing the enemy on until he was able to open on them with great effect with his siege guns. He then sent a company of the 3rd Bombay European Infantry to the ravines in front and found the enemy there as he suspected, and a general engagement at once began along the whole front. The rebels rose out of their lairs and came forward in thick chains of skirmishers, firing heavily, followed by large supports and columns in mass at a distance. An unexpected danger now arose. The ammunition for the Enfield rifles was of two sorts, one good the other bad. With the heat it was found that the bad could not be used, as the men could not force the bullets down the barrels. Most of the good had been used up, and Sir Hugh Rose, hearing the firing slacken on his right, rode over and found the rebels, mad with opium, pressing recklessly forward, whilst no one could reply to their fire. Brigadier Stuart came up to Major Stuart, of the 86th Regiment, and drawing his sword said to him: "Stuart, we have nothing to do but die like Scotchmen." Fortunately the general had a company of the Camel Corps with him, composed of the Rifle Brigade and 80th Regiment, and he at once led them against the enemy, who had now reached the guns, where the gunners were defending their pieces with their swords. One volley of the enemy's killed or wounded every horse of the general's staff, but he led the Camel Corps straight on the mutineers, who after a struggle were driven back, losing very heavily, whilst the artillery plied them with grape. The Camel Corps had to be recalled at once as large numbers of them fell in their ranks with sunstroke. Lieutenant-Colonel Lowth, with some of the 86th Regiment, in the meantime moved forward through the ravines on the right, and by a skilful manœuvre cut off and surrounded a considerable body of mutineers who had advanced too far. Part were killed on the bank of the Jumna and the rest were driven into the river, where they were shot or drowned. In the meantime the 25th Bombay Native Infantry had been driven in, but, rallied by the 21st Company Royal

Engineers, they charged superior numbers with great pluck and drove the enemy in front of them.

The whole line now advanced, came up with the retreating rebels, dashed at them, and bayoneted large numbers. The left of the British line swung forward and took the enemy rather in flank, and the whole broke up and fled, pursued by the horse artillery and cavalry. Exhaustion and bad ground stopped the pursuit, in which numbers of the enemy were killed and the village of Rayar was cleared of the mutineers. Rightly judging that the rebels would be downhearted after their repulse, the general determined to capture Kalpi on the 23rd of May, and long before dawn his force was marching on that town in two columns. That on the right, which threaded its way through the ravines, contained the 86th Regiment, still with 520 rank and file on parade.

The enemy had re-occupied Rayar, and the 86th Regiment were met with a volley, but at once charged with the bayonet and cleared the village. Kalpi Fort and town were both found to be deserted. In the advance through the ravines a panther and two hares were put up and promptly killed by the men.

Great numbers of the rebels were seen retiring across the plain to the south-west and were charged by all the cavalry and were destroyed in great numbers. The Ranée of Jhansi had had the unpleasant experience of having one of the British mortar shells burst in her room, killing two of her attendants and she had fled betimes.

All the rebels' guns were captured and very extensive stores were found in the town.

The British casualties were twenty-four killed and forty-three wounded, besides the losses from sunstroke.

The Regimental Records do not mention the loss of the 86th Regiment on those two days, but remark: " On these two days the troops suffered as severely as at Kunch from the effects of the sun and the scarcity of water. For seven miles round Kalpi scarped and rocky ravines, almost impervious to air, rendered the heat unsupportable, and the losses from *coup de soleil* were very numerous."

Two small columns of cavalry and guns, with a few native infantry, were sent out; one down the Jhansi road, and the other down the Jalaun road, which runs south-west from Kalpi. On the 31st of May a rebel fort was captured at daybreak nineteen miles from Orai, on the Jhansi road, a fight occurred outside the fort and 130 rebels were killed and thirty-five taken prisoners. It was thought with the crushing defeat of the mutineers on the 22nd of May that the campaign in Central India was over, and General Sir Hugh Rose issued the following congratulatory order to the troops:—

" Soldiers, you have marched more than a thousand miles and taken more than a hundred guns; you have forced your way through mountain passes and intricate jungles and over rivers. You have captured the strongest forts and beaten the enemy, no matter what the odds, wherever you have met him.

You have restored extensive districts to the Government, and peace and order now where before for a twelvemonth were tyranny and rebellion. You have done all this and you have never had a check. I thank you with all my sincerity for your bravery, your devotion and your discipline.

"When you first marched I told you that you, as British soldiers, had more than enough courage for the work that was before you, but that courage without discipline was of no avail, and I expected you to let discipline be your watchword. You have attended to my orders. In hardships, in temptation and danger you have never left your ranks. You have fought against the strong, and you have protected the weak and defenceless, of foes as well as of friends.

"I have seen you in the ardour of combat preserve and place children out of harm's way. This is the discipline of Christian Soldiers, and it is what has brought you triumphant from the shores of Western India to the waters of the Jumna, and establishes without doubt that you will find no place before which the glory of your arms can be dimmed."

Sir Hugh Rose then commenced his preparations to take up command of a division at Poona. However, much to his surprise, he found that he had to take the field again. The Ranée of Jhansi, desperate and daring, had decided to march on Gwalior; so instead of the rebels leaving Central India by crossing the Jumna and going eastward into Oude, they were found to be going westward towards Gwalior.

A wing of the 86th Regiment was sent to reinforce the column on the Jalaun road, and now news came in which created a sensation throughout India, only equalled by that caused by the first mutinies.

On the 30th of May Tantia Topi and the Ranée of Jhansi, at the head of a force of 7,000 infantry, 4,000 cavalry, and twelve guns, arrived near Gwalior. At daybreak next morning Sindhia, the Mahratta ruler of Gwalior, marched out with 8,000 men and eight guns, to within two miles of the rebels, and taking up a position awaited an attack.

The mutineers advanced at 7 a.m., carried the guns by a charge of 2,000 cavalry. Simultaneously the whole of Sindhia's army, with the exception of his bodyguard, went over to the enemy. The bodyguard made a brilliant defence, and Sindhia arrived safely at Agra, whilst the rebels entered Gwalior and took the treasury and Sindhia's jewels, said to be of fabulous value. The garrison of the fortress, considered to be one of the strongest in India, had, after a mock resistance, opened its gates, and sixty guns and a fine arsenal, well stocked with warlike stores, fell into the hands of the rebels.

In short, the mutineers, who had fled from Kalpi a disorganised mob, were now fitted out with money, guns, ammunition, and Sindhia's army as their allies.

Gwalior was a city of 170,000 inhabitants, flourishing and prosperous. Sindhia's territories were of great extent, and the British communications ran for hundreds of miles through his country, and his rebel troops were the best drilled of all the native levies of India. The hot weather was at its height, and the rainy season coming on would be very bad for the British military operations.

Still, these operations had to be forced through. Sir Hugh Rose received information of the fall of Gwalior on the 3rd of June, and, leaving a garrison at Kalpi, he started on the 6th of June marching on Gwalior, making forced marches by night to avoid the sun.

Brigadier Stuart, with the 86th Regiment, was already well on his way to Gwalior, and on the 11th of June Sir Hugh Rose caught him up at the Sind River with the cheering intelligence that a strong Bengal force, with a siege train, was to reinforce him, and that a brigade from the Rajputana Field Force, under Brigadier Smith, was to co-operate. The Hyderabad Contingent were on their way home, but hearing of this new affair, like the brave soldiers they were, at once marched on Gwalior.

Sir Hugh Rose decided to invest Gwalior as much as its great extent would allow and then to attack its weakest side, the investing troops cutting off the escape of the rebels. The weakest point was the southern side, which was commanded by some hills, which hills were out of range of the fort.

Gwalior had been the scene of a mutiny on the 12th of June. The Gwalior Contingent had then risen and killed four of their officers. The Lieutenant-Governor had previously refused to allow the ladies to be sent in to Agra, where he was living in safety, and these unhappy ladies had to see or hear of their husbands being killed. The mutineers partially burnt the cantonments of Gwalior, which were at a place called Morar, some five miles east of that city. Sindhia's Government had been at much trouble and expense to repair the cantonments for future use, and the general now wished to occupy them so that he could have shelter for his men and his parks and hospital. Brigadier Smith invested the south-east side; the Bengal force, under Colonel Riddell, was ordered to guard the western and northern sides, whilst the Central India Field Force marched in from the east.

On the 16th of June Sir Hugh Rose arrived within five miles of the Morar cantonments. His Intelligence Department informed him that his prompt advance was having an effect on Sindhia's rebel forces, who were deserting the enemy in great numbers and rejoining their own Maharajah, so he sent these deserters orders to move down the old Bombay road to intercept fugitives. He then reconnoitred the Morar and found it occupied by strong forces of the enemy, but as the buildings looked inviting and promised shelter from the sun for his men, he determined to turn the enemy out before they could burn them, and accordingly formed his force into two lines for an attack. The 1st Brigade, under Brigadier Stuart, was in front, and the 2nd Brigade, or so much of it as was not garrisoning Kalpi, was kept in support.

Both lines advanced, the field and siege guns in the centre of the first line, the 86th Regiment on the right, and the 25th Bombay Infantry on the left, with the 14th Light Dragoons on each flank. The ground on the right and in front of the Morar was full of ravines.

The general's plan was to mask the dangerous ground to his left, toward which the enemy evidently wished to draw him, to outflank the enemy's left, double it up and cut off their retreat from the road over the bridge in rear of the cantonments leading to Gwalior.

The first line advanced in line across the plain, dressing by their centre with the regularity of a parade movement. The enemy retired from their positions in front of Morar into the cantonments.

The British line, led by Sindhia's agent, moved to the right to get on to a road to outflank the enemy, but the guide missed the road, and on getting into some broken ground a masked battery opened upon them. The siege guns dealt with this battery and some others, all of which they took obliquely in flank. The cavalry was now sent from the right flank to the left, as they were required there, whilst the 86th Regiment, in skirmishing order and firing as they advanced, took by storm, under the cannonade of the enemy's right battery, all the Morar cantonments in front of the advancing line, the rebels' guns retiring at a gallop. The Regiment now swept the whole cantonment and occupied it. The British batteries galloped past the right of the 86th Regiment and again opened fire on the rebels in flank and the mutineers withdrew their batteries so soon as they saw their left compromised by the successful advance of the 86th Regiment.

The rebels had occupied some advanced ravines as trenches, but some troops were now placed to enfilade them, and every man in the ravines was killed after a desperate resistance. In one ravine alone lay dead seventy of Sindhia's faithless army wearing English accoutrements, with "1st Brigade Infantry" on their breastplates. A general retreat began after this defeat at the ravines, and the 14th Light Dragoons caught the enemy on the plain and slaughtered great numbers of them.

The rebels had intended to make a stand at Morar, and had commenced storing supplies there. They were surprised by the rapid British march from Kalpi and had not been able to remove the supplies, and were driven out before they could burn the buildings.

On the 17th of June Brigadier Smith moved on Gwalior from the south with his brigade of the Rajputana Field Force. He had a very hard fight to effect his advance, and during the combat a dashing cavalry charge was made by a squadron of the 8th Hussars. In this charge the Ranée of Jhansi, dressed as a cavalry soldier, was struck by a bullet and later cut down by a hussar. By her death the rebels lost their bravest and best military leader. Brigadier Smith encamped on the field at Kote Ki Serai, from which place he sent to Sir Hugh Rose asking for reinforcements. Fortunately the troops left to garrison Kalpi had been relieved and arrived on the morning of the 18th at Morar; so, sending a reinforcement of all arms to Brigadier Smith, the general moved out himself that afternoon from Morar, leaving his Kalpi garrison to keep up the investment on that side.

With the general marched the 86th Regiment, two troops of the 14th Light Dragoons, and a weak wing of the 71st Highland Light Infantry, with some native troops. This force moved south-west on Kote Ki Serai. The distance was twenty miles and large numbers suffered from sunstroke; 100 of the 86th Regiment were prostrated by this malady, but were brought on in dhoolies. The whole of these men insisted on coming out next day with their companies to take part in the fighting.

Brigadier Smith's column was now being shelled in their bivouac by the enemy. He was in possession of some hills on his right front, from which he had driven the rebels; then came a pass which led through the hills for two miles on the way to Gwalior. He was also in possession of the road through this pass. South-east of the road and parallel to it ran a deep, dry canal, cut largely in solid rock; the enemy held the hills to his left and in front of him far back. It was clear that the mutineers must be driven from both positions. General Sir Hugh Rose's force bivouacked on the banks of the Morar River. He directed the Madras Sappers and Miners to make a bridge across the canal some way to the left rear of his position, the bridge to be ready by sunset. He determined to cross over this bridge with a force of all arms during the night, get on the south road to Gwalior through the hills above mentioned, place himself between Gwalior and the enemy's two positions, and to fall on them a little before daybreak, when the 86th and 95th Regiments, supported by the rest of Brigadier Smith's brigade, were, concealed by the ravines, to attack their front and turn their left. Large bodies of troops now came out from Gwalior, and the general received information that an attack was intended. Brigadier Smith's left was the selected place for attack, as being the weakest point.

The Central India Field Force was in none too good fettle after their last night's harassing march and a bad bivouac on the rocks, and the bridge across the canal was not ready. On the other hand, the position in the narrow pass was so false that it became necessary to free it from the risk of a serious attack, and this was done by changing the defensive for the offensive. He therefore directed Brigadier Stuart to take the 86th Regiment, supported by the 25th Bombay Infantry, and to cross the canal somehow; crown the heights on the other side of it, and attack the rebels on their left flank. He also ordered the 95th Regiment to skirmish with the enemy to facilitate the attack of the 86th Regiment. The rest of the force was disposed in support of these three regiments and for the protection of the camp in rear.

The 86th Regiment, under Colonel Lowth, crossed the canal steadily and ascended the heights. Here they took the enemy in flank and forced them hastily to retire towards their battery. The skirmishers of the 86th Regiment, with their usual ardour, pressed the rebel infantry so hard that they did not make a stand even under their guns, but retreated across the entrenchments, in rear of which the guns were in position. The County Down Regiment dashed into the entrenchment with a cheer and took the guns, which proved to be three excellent English 9-pounders.

The 86th Regiment, leaving a party with the guns, passed on after the enemy's cavalry and infantry, who had retreated, part towards Gwalior and part towards the hills to the south.

The guns were at once manned by parties of the 86th and 95th Regiments, which had been previously instructed in gunnery, and they opened fire on the hostile cavalry and infantry in the plain below, making excellent practice at 1,000 yards range.

Four rebel batteries now opened a heavy fire on the advancing 86th Regiment.

Some Bombay Infantry made a dashing attack and captured a couple of hills with five pieces of artillery.

The city of Gwalior lay at the bottom of the ridge just seized by the Regiment. To the right was the handsome palace of the Phul Bagh, with its gardens, and the old city, surmounted by the fort, the latter remarkable for its ancient architecture with lines of extensive fortifications round the high and precipitous rock of Gwalior. To the left lay the Lashkar or new city, with its spacious houses half hidden by trees. The slopes descended gradually towards Gwalior; the lowest one commanded the grand parade of the Lashkar, which was almost out of fire of the fort, and afforded an entrance into the city.

The general now felt convinced that he could take Gwalior before sunset. He determined to make a general advance against all the positions which the enemy occupied for the defence of the city, and then to take the Lashkar by assault. Throwing out cavalry on either flank, he moved his force forward; the 86th Regiment was on the left, and in advance of the other troops; the 95th Regiment was on the right. The left wing of the County Down Regiment, who had pursued across a deep ravine to the range of hills to the south the body of the enemy who had retreated to the left, had returned and rested their left on a hamlet situated on the crest of the range which commanded Gwalior. The rebels immediately brought two 18-pounders to bear on the hamlet from the city and sent round shot into this hamlet with great precision.

The skirmishers of the 86th Regiment dashed down the hill towards the barracks and advanced against some houses held by the enemy in the trees at the foot of the hill. The 95th Regiment detached a company to support them. The 95th Regiment kept up a heavy covering fire from their Enfield rifles and soon silenced the 18-pounders. Still more mutineers, however, kept pouring out of Gwalior to join in the fight.

The 86th Regiment were now ordered to halt and cover the left, whilst

MAP OF GWALIOR AND ITS ENVIRONS.

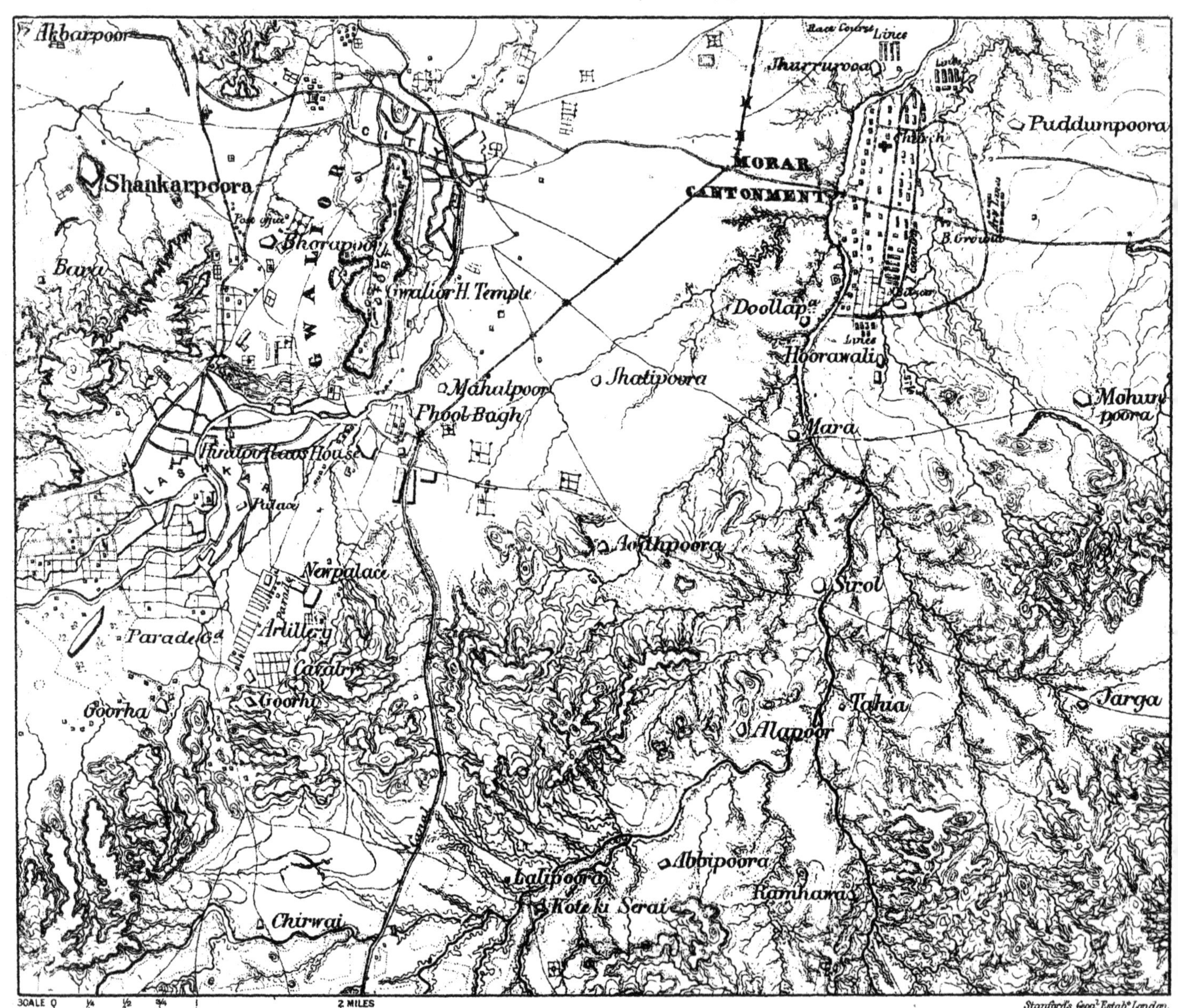

Reproduced by kind permission of Messrs. MacMillan & Co. Ltd., publishers of "History of the Indian Mutiny," by T. Rice Holmes

the attack was made on the right. Infantry, artillery and cavalry all pressed forward; the infantry cleared the houses whilst the cavalry cut down those in the open, the guns firing on every party of mutineers. The fort of Gwalior joined in, but the British advance was masked from its guns. By sunset the whole of the Lashkar was in British hands, including the palace. Though the rebels held the old city, many fled. Brigadier Smith pursued them far into the night, killing many and taking many guns.

The fort of Gwalior still remained to be carried, and its capture reads like some story of ancient romance.

It stands on a rock four miles round. This rock is from 200 to 300 feet high, and where it is not a precipice by nature has had the face scarped. The rampart conforms to the edge of the precipice all round, and the only entrance to it is by steps running up the side of the rock, defended on the side next the city by a wall and bastion and further guarded by seven stone gateways at certain distances from each other.

The area inside the fort is full of noble buildings, reservoirs of water, wells, and cultivated ground.

The morning after the combat the enemy again fired from the fort on the troops. Lieutenant Rose, of the 25th Bombay Infantry, took some of his men with another officer, a few of Sindhia's Police to show the way, and a local blacksmith to break the locks of the gates.

He burst in the gates one after another, closed with the rebels, and killed every man of the garrison in a desperate hand to hand struggle, in which he lost his own life. He had already been twice mentioned in despatches, and had originally escaped from the mutiny at Neemuch.

The British loss in this series of combats was only twenty-one killed and forty-four wounded, of whom five subsequently died.

Sindhia arrived on the 20th of January, and was escorted to his palace with great pomp by the British cavalry.

Tantia Topi was now rather nonplussed where to go. He was unable to move to Goona as it was occupied by a British force, and he was unable to go north on account of a column which came out from Agra under Brigadier Showers. Colonel Riddell's force should have kept him back in Gwalior itself, but its march had been delayed at a ford, and Tantia Topi slipped out with 12,000 men and twenty-two guns, and Brigadier Napier moved in pursuit at one and a half hour's notice on the morning of the 20th of June with 700 cavalry and guns.

Having marched twenty-five miles he halted for the night, and marching next morning at 4 a.m. found the enemy strongly posted at Alipur. Taking them in flank with artillery, the cavalry charged them in front, captured their twenty-two guns and chased them for six miles, and then returned and cleared the neighbouring villages, *all* of which were held by the enemy. They were first shelled, and then carried sword in hand by the dismounted cavalry. The

Regimental Records of the 86th Regiment state that: "With the fall of Gwalior operations on a large scale were at an end, but detachments of the regiment were constantly in the district maintaining order or engaged with flying bodies of the rebels."

The chief of these minor operations appeared to have been undertaken against Man Singh, a rajah under Sindhia. Sindhia would not let him succeed his father in the territory of Narwar, which lies forty-four miles south of Gwalior, and Man Singh, after sending in to warn the British that his quarrel was with Sindhia and not with them, took up arms and seized the fort of Paori. He had a force of 4,000 men, of whom 700 were rebel Sepoys. Brigadier Smith replied that he was responsible for the peace of the country, and that private wars could not be allowed, and moved down with some troops, including part of the 86th Regiment, and finding the fort too strong for field guns sent to General Napier, who brought up a small siege train, and the place was shelled on the 20th of August, but on the night of the 21st Man Singh and his followers slipped away into the jungle surrounding the fort, as the perimeter was so large that it could not be invested by the small besieging force. Man Singh lost forty killed and wounded, the British three wounded.

There was no reason that Man Singh should not have held out much longer, as the fortress he was in was very strong. It was a mile and a half round; one side rested on an impracticable precipice, flanked by ravines and jungles, and backed by a deep torrent and a forest extending for miles. In front of the remaining side was open ground, with tanks and marshes. The main gateway was very strong, having three gates and numerous lofty bastions to flank them. The "jungle" gateway had two gates opening on a ravine and jungle, a wicket led down by the one side of the precipice, and there was also a path down the other side by which men could escape in single file. The walls, though ancient, were generally ten feet thick, and from twenty-five to fifty feet high, built of massive stone, and well protected by the precipice, by deep tanks, and by a wet ditch except at two or three points.

Probably Man Singh, who was rather a sensible personage, felt it was foolish to contend with the British. Having elected to flee, he had to be pursued, and ninety-three men of the 86th Regiment and 118 of the 95th Regiment were mounted on elephants and camels, and attended by some native troops marched from Paori on the 27th of August. On the 29th they came up with a small party of Man Singh's infantry, some of whom they killed and some they captured, whilst the rest fled away into the jungle.

Still pushing on by forced marches, with just over 500 men all told, the column halted on the night of the 3rd of September, started at 2 a.m., and marched until 10 a.m.; marched again at 5 p.m., bivouacked when it became dark, marched again at midnight, and on the morning of the 5th of September arrived just before daybreak within a short distance of the village of Bijapur on the Chupet River. The village lay between the British and the enemy, who were encamped on the bank of the river. At daybreak the cavalry were sent round the hamlet by the right to attack the rebels on their left flank, whilst the infantry marched through the village straight upon the mutineers. The latter were posted on the high ground, and had 800 infantry, with 150 cavalry. For a short time the insurgents stood their ground, but the European troops closed with the bayonet, and they then broke up, crossing the river and taking refuge in ravines. The cavalry of the British force was very active, 450 out of the 950 rebels present on the field were killed, and they were found to be all mutineers, principally of the Gwalior contingent or of the Bengal Infantry.

The British lost one officer and five men killed, and four officers and fourteen men wounded.

The column then returned to Goona. Their marching had been a fine performance, as it was over black cotton soil in the height of the rains.

Again a portion of the 86th Regiment was called out taking part in some operations against a leader called Feroz Shah.

This personage had suffered several checks, but as the British troops were after more important forces he had not been harried before in the way he now found himself hunted.

Sir Robert Napier moved from Gwalior in pursuit of his band, and after a very long chase arrived at a town called Ranod on the 17th of December. General Napier's force was reduced to 275 cavalry and thirty-eight of the 71st Regiment mounted on camels. The people of Ranod were greatly excited, as the rebels were being guided by a revolted landowner of the neighbourhood, who evidently had a grudge against the town, and had arranged with the rebels to plunder it. The mutineers had not the slightest idea that any force was near them and came on in a mass, intent only on murder and plunder. The 14th Dragoons led the charge on the enemy, and the pursuit was kept up for eight miles; 450 dead bodies of rebels were counted, and the British loss was one killed and fifteen wounded.

Not in the least upset by this defeat, the enemy attacked a party of Bombay Cavalry and drove them into Goona, from which place on the 22nd a force some 370 strong came out under the command of Captain Rice, of which force the 86th Regiment furnished fifty-five men. No information could be obtained from the country people, but fortunately a man was met who had been robbed of his horses and beaten by the rebels. He reported that 2,000 of them were encamped in a deep glen, about eleven miles away, through thick jungle, at a village called Sarpur.

Leaving the camp standing, Captain Rice marched with part of his force at 6 p.m., and being favoured by bright moonlight reached the enemy at 11 p.m. with about 200 men. The picquet was shot and the camp stormed at once, but the enemy fled so rapidly that few were killed, but all the camp, many arms, and over 100 horses were taken.

Sir R. Napier was very pleased with this exploit, and especially brought to notice the officers and troops of this column, " for this very dashing and difficult enterprise, which has taught the enemy to distrust the security even of the deepest jungle, that have so often favoured their escape."

The column, after spending the night in the captured camp and sending their cavalry in pursuit of the enemy next day, marched back to Goona, whilst Feroz Shah fled on with his band and joined Tantia Topi.

Feroz Shah was the rebel leader who made the longest resistance, his force being finally broken up by Brigadier Showers with a force from Goona on the 15th of April, 1859. He fled disguised as a pilgrim, made his way to Kurbela, where he lived many years afterwards.

Tantia Topi's court-martial is appended to this chapter.

The 86th Regiment was now concentrated at the Morar cantonment, near Gwalior. The fatigue and exposure of the campaign had crippled and prostrated a good portion of the Regiment, and now all unnecessary drills were put a stop to, to let the men recover themselves. The duties were heavy enough all the same, as one-third of the Regiment was on guard every night.

The discipline was very good at this time, and there was not a single case of flogging during the whole year. The principal amusements of the men are noted in the Regimental Records at this time as being " Handball, Cricket, and 'Long Bullets.' " The latter was a favourite amusement of the Dragoons, who crushed out the Vellore Mutiny, under Colonel Gillespie, in 1806. This Irish game has now entirely died out in the Regiment and in the greater part of Ireland.

It was in this campaign that the 86th Regiment captured the drum which is still used in the regimental band on important occasions.

Three curious cases are mentioned in the Regimental Records of bad wounds received at Jhansi:—

Lieutenant Lewis was dangerously wounded by a sword cut which extended from the right shoulder across the back to the left hip, which presented a frightful spectacle. To the astonishment of all, Lieutenant Lewis recovered from his wounds. Ensign Dartnell received four sword cuts and a bullet wound. On the left arm and hand he received three sword cuts, one being seven inches long, which divided the muscles above the elbow to the bone, also a sword cut on the right leg, and lost three fingers.*

Lieutenant Sewell was severely wounded by a musket ball, which lodged in his *left* thigh; some days afterwards the ball was extracted from his *right* thigh. He also finally recovered.

The following officers of the 86th Regiment were mentioned by Sir Hugh Rose in his despatches:—

Lieutenant-Colonel Lowth (special mention), for gallantly and ably commanding his Regiment throughout the campaign. He took by storm the

*** This gallant Officer's death was cabled out to India 13th of August, 1913.**

heights on the left of Gwalior and the guns in the enemy's entrenchments (Companion of the Bath, Brevet of Colonel).

Lieutenant Brockman, mentioned for ably serving the captured guns on the heights of Gwalior.

Lieutenant-Colonel Stuart, twice thanked by Sir Hugh Rose for services before Jhansi in the trenches. Led the only successful escalade at the storming of Jhansi (specially mentioned and Brevet-Lieutenant-Colonelcy). Led the storming party at Kunch (specially mentioned). Commanded the advanced picquets before Kalpi on the 20th of May, 1858 (specially mentioned).

Lieutenant Jerome, severely wounded, mentioned for most gallantly leading his company against the enemy on the 20th of May, 1858, and for services at Jhansi (Brevet-Majority and Victoria Cross).

Captain Lepper, mentioned for skilfully directing the different companies of skirmishers against the attacks of the enemy on the 20th of May, 1858 (Brevet-Majority).

Captain Lepper and Assistant-Surgeon Barry, mentioned for having left their beds, being in the sick report, to join their Regiment in action on the 22nd of May, 1858.

Ensign Keane, mentioned for having on the 2nd of May rendered good service in the field.

Only seven distinguished conduct medals were awarded to the infantry during the Indian Mutiny. Two of these seven medals were awarded to the 86th Regiment, the recipients being Serjeant-Major Alleyne Wolfe, who slew three Sepoys in single combat at the storming of Jhansi, and Quartermaster-Serjeant William McNeill.

Many complimentary orders were issued to the Regiment, from the thanks of the Viceroy in Council downwards.

It is perhaps only necessary to notice the short order issued by the Peninsula veteran, Lord Clyde, now Commander-in-Chief in India, who, as Sir Colin Campbell, had fought in the Crimea and stormed Lucknow.

1859 Copy of a General Order by his Excellency the Commander-in-Chief, dated Headquarters, 26th of March, 1859:—

"Her Majesty's 86th Foot are now returning to the Bombay Presidency to embark for England after nearly seventeen years' absence from home, and the Commander-in-Chief cannot permit this distinguished corps to depart without the earnest expression of his appreciation of the service rendered by it to the State.

"The exemplary conduct in quarters of this Regiment throughout its career in India has been as conspicuous as its services in the field have been distinguished during the late eventful period."

And that issued by the General in Command of the Bombay Presidency about the time that the Regiment was leaving his command:—

"With feelings of gratitude and satisfaction Sir Henry Somerset remembers how the 86th Regiment served in Malwa under Brigadier C. S. Stuart, C.B., of the Bombay Army.

"The Regiment, owing to the pressure of the times, had to be collected from different stations and pushed to the front as rapidly as detachments could be moved. Present at the siege of Dhar, engaged during the three successive days at Mandiswar, eminently successful in the assault of Chanderi, this noble Regiment joined the column, under Sir Hugh Rose, before Jhansi, where it was foremost in the breaches in the capture of that stronghold.

"The time has come when this war-worn Regiment can look forward to the peace it has so nobly and so brilliantly fought for, and the rest it so richly merits.

"To Colonel R. H. Lowth, C.B., and to the officers, non-commissioned officers and men under his command, Sir Henry Somerset bids an affectionate farewell and God speed."

The duty state of the 1st of January, 1859, shows as follows:—

	Officers.	Other ranks.
Present fit for duty	18	591
Sick	1	217
Detached or on leave, etc.	16	280
Totals	35	1088

Statement of losses by death, invaliding, etc., for the year 1858 (officers not included):—

Killed	13
Died of wounds	40
Sunstroke and natural causes	38
Invalided	37
Total	128

MEDALS OF QR.-MR. SERJT. WILLIAM MCNEILL, 86TH ROYAL COUNTY DOWN REGT.

Only seven Medals for Distinguished Conduct in the field were given to the whole Army for the Mutiny. Q.-M.-S. McNeill's medals are now in the Collection of the Officers, 2nd Bn. Royal Irish Rifles

COURT-MARTIAL PROCEEDINGS OF TANTIA TOPEE.

Letters Giving Court-Martial Proceedings to the 2nd Battalion Royal Irish Rifles.

23, Brookfield Road,
Bedford Park,
29th March, 1905.

My Dear Stokes,

I am writing to you on a subject that might interest the old County Down Regiment. My mother married, when a widow, the late Colonel Thomas S. P. Field, R.A. On his death I found the Proceedings of a General Court Martial which tried Tantia Topee, there were two copies, one of which I have, and if you think the 86th Regiment, or their Depot, would care for it I will give it to you, the other copy I lent to a friend who never returned it.

Colonel Field told me owing to state of War, the proceedings were not sent to Headquarters and he retained them.

As you know the 86th were in Central India, they might like the document. My old regiment was not in India at the time. If you would like the document yourself, you can have it with pleasure, as I expect you know the 86th as well as the 83rd.

Yours Ever,

T. King.

23, Brookfield Road,
Bedford Park, W.

My Dear Stokes,

Enclosed is the Court Martial. I can certify to Col. Field's signature, and he told me all about it. I hope it will be of use to the 86th Regt. I am resting after all my work.

Yours Ever,

T. King.

PROCEEDINGS
OF
A GENERAL COURT-MARTIAL
ASSEMBLED AT SIPREE ON THE FIFTEENTH DAY OF APRIL, ONE THOUSAND EIGHT HUNDRED AND FIFTY-NINE.

At a European General Court Martial assembled at Sipree, on the fifteenth day of April one thousand eight hundred and fifty-nine. By order of Major Meade Commanding at Sipree, under the provisions of Act No. XIV of 1857.

PRESENT:

President.

Captain Baugh, 9th Regiment Bombay Native Infantry.

Members.

Captain Pierce, 24th Regiment Bombay Native Infantry.
Captain Webster, 3rd Bengal European Regiment.
Lieut. Orchard, 3rd Bengal European Infantry.

Officiating Judge Advocate General.

Captain T. S. P. Field, Royal Artillery.

Interpreter to the Court.

Lieut. Gibbon, Meade's Horse.

Tantia Topee a resident at Bithoor in the District of Cawnpore in the Territory of British India, and in the service of the late ex-Peshwa Bajee Rao a pensioner of the British Government, appears a Prisoner before the Court.

The orders for the assembly of the Court, appointing the President, Officiating Judge Advocate, and Interpreter, are produced and read.

The names of the President and Members of the Court Martial are read over in the hearing of the Prisoner and he is asked:—

Question: Do you object to be tried by the President or any of the Officers sitting as members of your Court Martial?

Answer: No.

The President and Members and Officiating Judge Advocate are duly sworn.

Lieut. Gibbon of Meade's Horse is sworn as Interpreter to the Court.

COURT-MARTIAL PROCEEDINGS OF TANTIA TOPEE.

Tantia Topee a Resident at Bithoor in the District of Cawnpore in the Territory of British India, and in the service of the late ex-Peshwa Bajee Rao a pensioner of the British Government, charged under Act XIV of 1857.

Charge.—With having been in rebellion and having waged war against the British Government, between June 1857 and December 1858, and having subsequently been leader of, and present with, the Rebel Army which fought against the British Force under Major General Sir Hugh Rose, K.C.B., near Jhansi, on or about the 1st April 1858, and also one of the leaders and present with the Rebel Army, which having attacked and defeated Maharaja Scindia near Gwalior, on or about 1st June 1858, occupied Gwalior, and subsequently fought at or near Gwalior, against the British Force under the same Major General Sir Hugh Rose, K.C.B., between 14th and 21st June 1858.

Sd. R. MEADE, Major,
Commanding Field Force.

Camp Sipree, 13th April 1859.

Question: How say you Tantia Topee, are you Guilty or Not Guilty?
Answer: Not Guilty.

1st Witness on the Prosecution, Meerhamed Alli, is called into Court and makes solemn affirmation.

Question: Do you know the Prisoner?
Answer: I do.
Question: Who is the Prisoner now before the Court.
Answer: Tantia Topee.
Question: Where used Tantia Topee to reside?
Answer: At Bithoor.
Question: In what District is Bithoor?
Answer: In the District of Cawnpore, I believe.
Question: In whose service was Tantia Topee?
Answer: In the service of the Nana.
Question: At what period do you allude to?
Answer: During the Mutiny.
Question: Where did you first see Tantia Topee?
Answer: At Parone.
Question: What was he doing there?
Answer: He was in the jungle with between 50 and 100 rebel Sowars.
Question: Did you see him anywhere else?
Answer: Yes. I saw him at Nehurghim with a large rebel Army.
Question: Was Tantia Topee in Command of the Rebel Army?

Answer: Tantia Topee and Rao Sahib were the Commanders.

Question: Where had Tantia Topee come from?

Answer: He came from the south, from the direction of the Nerbudda.

Question: What had he been doing there?

Answer: He had been fighting with the British Force and was now retreating.

Question: Where did Tantia Topee leave the Rebel Army?

Answer: Near the Chumbul, where a British Force was sent against him; and then went into the Parone Jungle as I have before stated.

Question: Where was Tantia Topee going to about eight days ago, when you were sent after him?

Answer: He was going to join the Rao's Army near Seronja.

Question: Where did you meet Tantia Topee about eight days ago?

Answer: I was sent by Raja Maun Singh to Tantia Topee to call him, which I did. I met Tantia Topee near Nahurghur.

Question: When was Tantia Topee captured?

Answer: He was taken the night I was sent to him, by a party of Sepoys who were sent out under Maun Singh's orders by Major Meade from Mahodra.

Cross examined by the prisoner.

Question: How do you know I was in the service of the Nana?

Answer: I have heard from everybody, but I do not know whether you were the Nana's Servant before the Mutiny.

Question: When were you in the jungle that you happen to know I was there with 50 or 100 rebel Sowars?

Answer: I have been there for the last eight or ten years, in the service of Raja Maun Singh in the same jungle.

Question: How can you, having made solemn affirmation, state that you have been in Maun Singh's service for the last eight or ten years, when I know you have been a Sepoy Sowar in the 5th Irregular Cavalry (Bengal).

Answer: Maun Singh can testify to my having been in his service, and the officers of the late 5th Cavalry might be appealed to.

Question: When I had been eight days at Parone, and when I met you with another man, in whose service were you?

Answer: I was a Sowar in Maun Singh's service.

Question: When did you see me at Nahurghur, with a large Rebel Army you have mentioned?

Answer: About three or four months ago.

Question: What time either in the day or night did you see me at Nahurghur?

Answer: About Mid-day.

The Witness retires.

COURT-MARTIAL PROCEEDINGS OF TANTIA TOPEE.

2nd Witness on the Prosecution, Binack Dainooder, Naib Sonbah of Sipree, makes solemn affirmation.

Question: Do you know where Tantia Topee resided before the Mutiny?
Answer: I have heard he resided at Bithoor.
Question: In what District is Bithoor?
Answer: In the District of Cawnpore.
Question: Is Cawnpore in British Territory?
Answer: It is.
Question: Do you know in whose service Tantia Topee was before the Mutiny?
Answer: In the service of the Nana.
Question: Do you know that Bajee Rao was a pensioner of the British Government?
Answer: I do.
Question: Do you know that Tantia Topee was for many years a resident at Bithoor?
Answer: Yes. I have before said so.
Question: Do you know whether Tantia Topee was in the service of Bajee Rao?
Answer: I do not know.
Question: Do you know the prisoner?
Answer: I do not.

The prisoner declines cross-examining, and the Court has no questions to put.

The Witness retires.

3rd Witness for Prosecution, Delaivar Khan, makes solemn affirmation.

Question: Do you know the prisoner?
Answer: I do. I know him to be Tantia Topee.
Question: In whose service are you?
Answer: In the service of Maun Singh.
Question: Where did you first see Tantia Topee?
Answer: I saw him at Jhansi.
Question: What was he doing there?
Answer: He was in Command of the Rebel Army.
Question: Do you know whether this Force attacked a British Force near Jhansi?
Answer: I do.
Question: In what month was this fight?
Answer: It was about a year ago.
Question: What was the British Force doing at Jhansi?
Answer: Attacking the Ranee of Jhansi who had rebelled against the

British Government, and who had been reinforced by a rebel Force from Calpee.

Question: Were you present in the fight near Jhansi, and who commanded the Rebel Force?

Answer: I was present and I know that the Ranee of Jhansi commanded in the Fort, and the prisoner Tantia Topee the Forces outside the Town and Fort of Jhansi.

Question: When did you see Tantia Topee again and where?

Answer: I saw him in the Parone Jungle about three months ago.

Question: Where had Tantia Topee come from at that time?

Answer: From Marwar.

Question: Who were with him?

Answer: He had about 150 Cavalry and Infantry with him, and two Native Cavalry Officers.

Question: From what Force did he separate?

Answer: He left the Rao Sahib who had with him about seven or eight thousand Infantry and Cavalry.

Question: Do you know where Tantia Topee was going to about eight days ago?

Answer: A messenger came from the Rao Sahib with a letter inviting him to re-join him. Tantia Topee was on his way to do so.

Question: Where was the Rao Sahib at this time?

Answer: In the jungle near Seronji.

Question: Were you present when Tantia Topee was apprehended?

Answer: I was.

Question: Do you know where Tantia Topee resided before the Mutiny?

Answer: I heard he resided at Bithoor.

Question: Do you know that Bithoor is in the District of Cawnpoor, and in the British Territory?

Answer: I have never been there but I know it.

Question: Do you know that Tantia Topee was in the service of the ex-Peshwa Bajee Rao at Bithoor?

Answer: I have heard so from everybody.

Question: Do you know whether Bajee Rao was a pensioner of the British Government?

Answer: I do.

Cross examined by the prisoner.

Question: Did you see me anywhere else?

Answer: Yes. At several places.

Question: You say a letter came from the Rao Sahib inviting me to join him. Did you see this letter or only hear of it?

Answer: I saw the man who brought the letter who told me the contents of it.

The Witness retires.

4th Witness for the Prosecution, Eda, makes solemn affirmation.

Question: In whose service are you?
Answer: I am a Sowar in Maun Singh's service.
Question: Where did you first see Tantia Topee, and is he the Prisoner now before the Court?
Answer: I first saw him at Jhansi, and the Prisoner now before the Court I recognize as Tantia Topee.
Question: What was Tantia Topee doing at Jhansi?
Answer: He was fighting.
Question: Against whom was he fighting?
Answer: He was fighting against a British Force.
Question: Against whom was the British Force fighting?
Answer: Against the Ranee of Jhansi.
Question: Who was in Command of the Rebel Army which fought against the British Force outside the Town and Fort of Jhansi.
Answer: The Prisoner Tantia Topee.
Question: Where did you see Tantia Topee a short time ago?
Answer: He was with me about three or four months ago at Parone.
Question: Was anybody with him at the time you mention?
Answer: Ajeit Singh was with him.
Question: Where had Tantia Topee come from?
Answer: He came from the direction of the Jewpore Territory, where he had been fighting and had been defeated.

The Prisoner declines cross examining and the Court has no questions to put.

The Witness retires.

5th Witness for the Prosecution, Rumzan Khan, makes solemn affirmation.

Question: In whose service are you?
Answer: I am a Jemadar of the Tehsaldanee of Sipree.
Question: Where were you in June, 1858?
Answer: I was here in June, 1858, but was sent by the Sonbah's orders to Gwalior.
Question: Do you know the prisoner?
Answer: I know the prisoner. He is Tantia Topee.
Question: Where did you first see Tantia Topee?
Answer: I saw him near Maharaja Scindia's palace at Gwalior, at about four o'clock in the afternoon.

Question: What was Tantia Topee doing at the time you saw him at Gwalior?

Answer: He was addressing the people and saying they were not to fear, that as they had hitherto been commanded by Maharaja Scindia, so they would now be by him and he would take care of them.

Question: After Scindia's defeat and flight to Agra, who commanded the Rebel Army at Gwalior?

Answer: Tantia Topee did.

Question: In whose name did Tantia Topee administer affairs at Gwalior?

Answer: In the name of the Peshwa.

The prisoner declines cross-examining, and the Court has no questions to put.

The Witness retires.

6th Witness for the Prosecution, Sheik Rumzan, makes solemn affirmation.

Question: Do you know the prisoner?

Answer: I know the prisoner Tantia Topee.

Question: Where were you in September 1857, and in whose service were you?

Answer: I was in Maharaja Scindia's service at a Chowkee called Eteinda in the Zillah of Bhind.

Question: Did you see Tantia Topee at that time and what was he doing?

Answer: He had the whole of the Gwalior Contingent who had mutinied, with him, and was on his way to Bithoor.

Question: Where did you see him again?

Answer: I saw him at the same place after his flight from Bithoor, on his road to Gwalior, accompanied by Rao Sahib, Ranee of Jhansi, the Bandah Nawab, and from fifteen to twenty thousand rebels.

The prisoner declines cross-examining, and the Court has no questions to put.

The Witness retires.

7th Witness for the Prosecution, Khorshially, makes solemn affirmation.

Question: Do you know the prisoner?

Answer: I do.

Question: What is his name?

Answer: Tantia Topee.

Question: Where did you first see him?

Answer: I saw the prisoner at Morar.

Question: What was prisoner doing there?

Answer: He came to induce the Contingent to join him, which they did and became rebels.

Question: Was Tantia Topee in Command of the Rebel Army?

Answer: He was.

Question: How long ago is it since you were at Gwalior?

Answer: About ten months ago.

Question: Did Tantia Topee manage affairs at Gwalior at that time?

Answer: He did.

The prisoner declines cross-examining and the Court has no questions to put.

The Witness retires.

8th Witness for the Prosecution, Sheik Abdoola Aziz, makes solemn affirmation.

Question: In whose service are you?

Answer: I am a Jemadar in the 9th Regiment of Native Infantry.

Question: Where did you first see the prisoner?

Answer: I saw him near a river at Parone.

Question: What were you doing there?

Answer: I was ordered to go there by Major Meade, with a party.

Question: Did the prisoner give himself up, or was he taken by force.

Answer: He was taken by force.

Question: Had he any arms with him?

Answer: Yes. He had.

Question: Is the prisoner now before the Court the man you captured?

Answer: He is.

The prisoner declines cross-examining, and the Court has no questions to put.

The Witness retires.

The Deputy Judge Advocate here hands into Court a document in the Vernacular and makes the following statement:—

I now lay before the Court in original a statement of confession which has been dictated by the prisoner Tantia Topee of his own free will since his capture. I shall proceed to prove the document before its being read and translated to the Court.

Lieut. Gibbon, Interpreter to the Court, is examined on his former oath.

Question: Do you know anything of the document I now shew you, and can you identify the signatures to the declaration at its feet?

Answer: I was present yesterday morning when the contents of this document were read over to the Prisoner Tantia Topee by Moonshee Junga Persad before Major Meade. The Prisoner stated that the document was in all respects perfectly correct, and that it had been dictated by him of his own free will and accord. I identify the signature at the foot of the declaration as that of Major Meade Commanding the Field Force which captured the Prisoner. I have made a careful translation of this document which is appended to it.

Junga Persad Moonshee, called into Court and makes solemn affirmation.

Question: Do you recognise the document I now shew you, and the signatures to it?

Answer: Yes. I wrote it myself. It is the statement of confession dictated by the Prisoner Tantia Topee, taken down from his own lips. There are two signatures of the Prisoner which I here point out, those of myself and Naib Kumasdar Pirbhloo Sali as witnesses, and that of Major Meade below the declaration at the foot.

Question: Was this statement made by the Prisoner of his own free will?

Answer: Yes. He was repeatedly asked if such were the case.

The Document having been duly proved, is now read to the Court in the Vernacular and the translation made by the Interpreter marked A and attached to the proceedings.

Question of the Court to the Prisoner: Have you any remark to make on this document?

Answer: None.

Major Meade is called into Court and being duly sworn is questioned.

Question: Is Bithoor in the British Territory?

Answer: It is.

Question: Was Bajee Rao the ex-Peshwa, a pensioner of the British Government?

Answer: Yes. He was for many years.

The Witness retires.

The prosecution is closed and the Prisoner being put on defence states:—

I only obeyed in all things that I did, my master's orders that is to say the Nana Sahib's orders up to the Capture of Calpee, and afterwards those of the Rao Sahib. I have nothing further to state except that I have had nothing to do with the murder of any European men, women or children, neither have I at any time given orders for any to be hanged.

The Prisoner declines recalling any witnesses.

The Court is closed and proceed to deliberate on its finding.

Finding: The Court from the evidence before it, finds the Prisoner Tantia Topee a resident at Bithoor in the District of Cawnpoor, in the territory of British India, Guilty of the charge preferred against him.

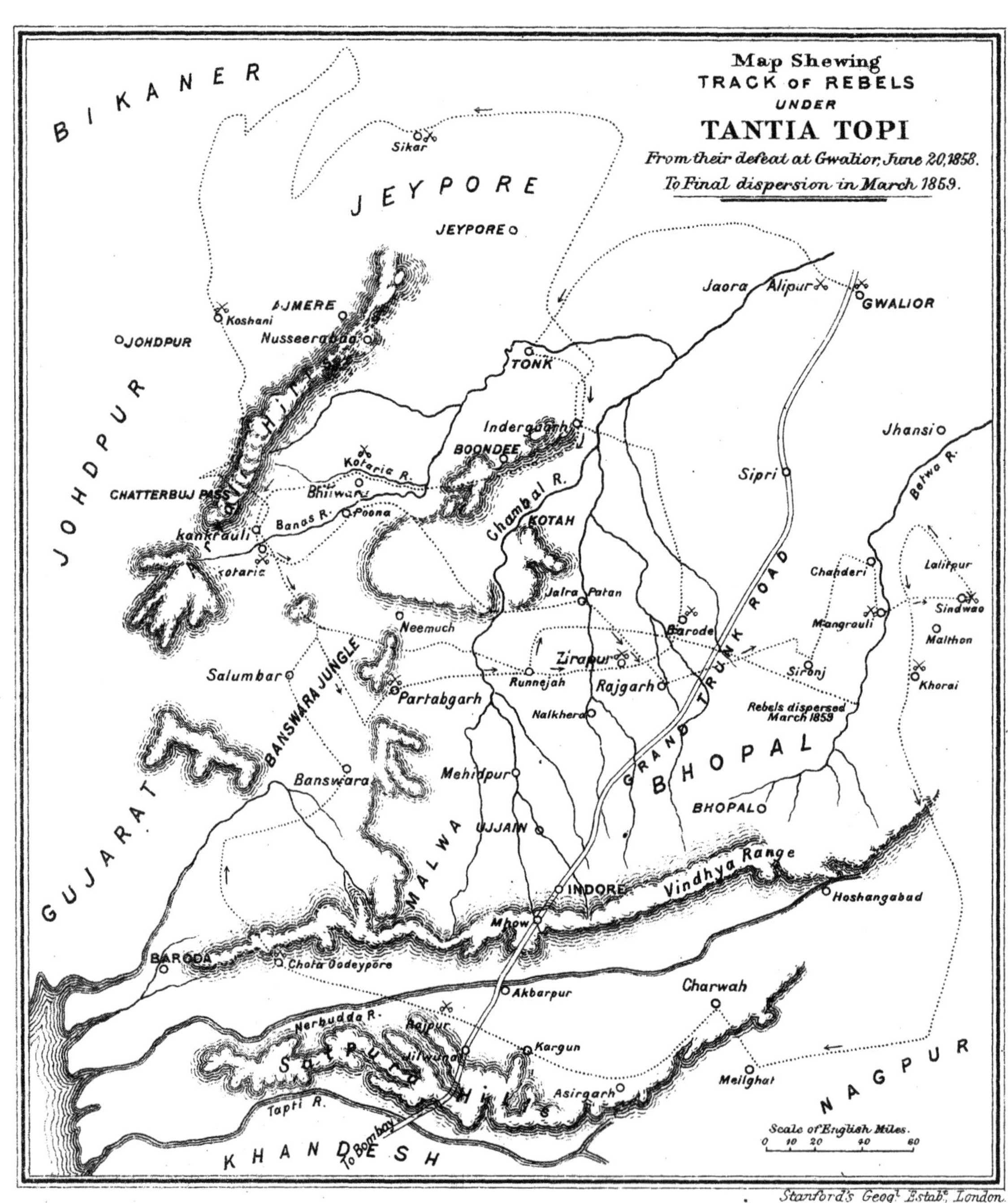

Reproduced by kind permission of Messrs. MacMillan & Co. Ltd., publishers of "History of the Indian Mutiny," by T. Rice Holmes

COURT-MARTIAL PROCEEDINGS OF TANTIA TOPEE.

Sentence: The Court having found the Prisoner Tantia Topee, Guilty as above specified, does now under the provisions of Act XIV. of 1857 sentence him, the said Tantia Topee a resident at Bithoor in the District of Cawnpoor, in the territory of British India, to suffer death by being hanged by the neck until he be dead.

CHARLES R. BAUGH, Captain,
9th Regiment Bombay Native Infantry.
President of the Court Martial.
Sipree, 15th April 1859.

THOS. S. P. FIELD, Captain,
Royal Artillery.
Officiating Judge Advocate General.
Sipree, 15th April 1859.

INSCRIPTION ON NORTH FACE.

SACRED
to the memory of
the Soldiers
of H.M. 86TH ROYAL REGIMENT
who fell gloriously at the battle of the
"BETWA"
on the storming of
"JHANSI,"
on the 1st and 3rd of April, 1858.
Erected to their memory by their
comrades in arms.

CHAPTER XIV.

EIGHTY-THIRD REGIMENT (1860-1912).

Eighty-Third moved to Belgaum—Title of County of Dublin Regiment conferred on the 83rd Regiment—Distribution of Indian Mutiny medals—Establishment augmented and reduced—Improvement in Indian uniform—Ordered to proceed to England—In garrison at Dover and Shorncliffe—Moved to Aldershot—Sent to Sheffield—The Regiment prevented Election Riots at Nottingham and elsewhere—The 83rd ordered to Dublin—Privates Barnes and McCartin on duty at Sligo magazine—Establishment again reduced—Detachments at various Irish towns: Newry, Armagh, Londonderry—Breechloaders first issued—83rd ordered to Gibraltar—New colours presented to the Regiment—Ordered to India—Crossed the Isthmus of Suez by train—Quartered at Deesa—Regiment furnished the guard of honour to H.R.H. the Prince of Wales—83rd sent to Karachi—Ordered to Sukhur on account of Afghan War—Sent to Belgaum—Regiment officially notified that it would be sent to England—Ordered to South Africa—Slight sketch of position of affairs in South Africa—Regiment arrived in Natal—Peace negotiations—Designation of 83rd changed to 1st Battalion Royal Irish Rifles—Returned to England on H.M.S. "Tamar"—Quartered at Dover—Rifle uniform taken into wear—Army Reserve called up on account of Egyptian War—Inspection by H.R.H. the Duke of Cambridge—Army Reserve sent down—Moved to Lydd—Sent to Guernsey and Alderney—Quartered at Portsmouth—Attended Jubilee Review at Aldershot—Crossed to Ireland—Quartered at Mullingar—Improvement in Army cooking—Sent to Fermoy—Changed to Newry—Ordered to Brighton—Mounted Infantry sent to South Africa—Battalion ordered to South Africa, quartered at Ladysmith—Narrative of service of Mounted Infantry in South Africa—Enteric fever at Ladysmith—Battalion sent to India, quartered at Dum Dum and Fyzabad—Various parties sent to the Boer War, to China, to Tibet, and to Somaliland—Delhi Durbar—Tibetan Expedition described—Battalion moved to Meerut—Sent to Burma—Ordered to Kamptee.

1859

THE Mutiny having been suppressed, the 83rd remained at Nusseerabad and Ajmere until the end of February, 1860, when it was ordered to proceed to Belgaum in the southern Mahratta country. It moved in two detachments. The first left on the 17th of February and marched to Cambay, some 390 miles away. It reached that place on the 30th of March, and proceeded by sea to Vingerla, on the Malabar coast. From that port it marched to Belgaum, where it arrived on the 16th of April. The other half of the Regiment did not commence its journey until the 30th of March. It had a trying experience, as cholera broke out, and Lieutenant Colebrook and nine men died from that disease. This party landed at Wagotun, instead of at Vingerla, and marched to Kolapore, where it remained on detachment.

Just previous to this change of station by the 83rd, it received various letters which had been despatched with reference to its title. They are as follow:—

"Howth Castle,

"28th July, 1859.

"SIR,—An application has been made to me by Lieutenant-Colonel Steele, of the 83rd Regiment, in my capacity of Lord-Lieutenant of the county of Dublin, to give the assistance of my sanction and co-operation in forwarding an application from him as Commanding Officer of the 83rd Regiment, that the distinctive appellation of the 'County of Dublin' Regiment may be conferred upon that Corps. Lieutenant-Colonel Steele has suggested that I should communicate with you, as the Colonel of the 83rd, on the subject of his wishes, and I accordingly beg to do so. Lieutenant-Colonel Steele has accompanied that application to me by an extract from the records of the 83rd Regiment, and it most plainly appears that the Regiment was raised in the County of Dublin in the year 1793, under a letter of service granted to its first Lieutenant-Colonel-Commandant, William Fitch, who was killed at the head of the Regiment in the Maroon War, in the Island of Jamaica, three years afterwards. It further appears, from its records, that the 83rd Regiment has seen much active foreign service, and has been distinguished by its discipline and valour in many parts of the world, and recently in the suppression of the Mutiny in India, where it is at present serving.

"Under these circumstances, and having regard to the fact that the Regiment was originally raised exclusively in the County of Dublin, I have much pleasure in expressing my concurrence in the application of Lieutenant-Colonel Steele on the part of his Regiment. I consider that it would be creditable to the County of Dublin that a Regiment raised in it, of whose services it may be justly proud, should be distinguished by its name, and I give this letter to Lieutenant-Colonel Steele, for conveyance to you, in the hope that it may aid in inducing His Royal Highness the Commander-in-Chief to recommend to Her Majesty that the title of the 'County of Dublin Regiment' may be conferred on the 83rd.

"I have, etc.,

"(Sd.) HOWTH.

"To General Sir Frederick Stovin, K.C.B.,
"Colonel of the 83rd Regiment."

The Colonel of the 83rd, General Sir Frederick Stovin, K.C.B., brought this letter to the notice of H.R.H. The Duke of Cambridge, and Her Majesty graciously conferred the title of "County of Dublin" Regiment on the 83rd, just 67 years after the Regiment had been raised in Dublin. The annexed letter authorised this title being adopted:—

"Horse Guards, S.W.,
"29th October, 1859.

"SIR,—By desire of His Royal Highness the General Commanding-in-Chief, I have the honour to acquaint you that Her Majesty has been graciously pleased to authorize the 83rd Regiment, which was raised principally from Recruits obtained in Ireland in 1793, embodied in Dublin in that year, being designated the 83rd 'County of Dublin' Regiment.

"I have, etc.,
"(Sd.) W. F. FOSTER, D.A.G."

1860 On the 27th of November the half of the Regiment at Kolapore was brought in to rejoin headquarters, and on the 11th of February the medals conferred on the 83rd for their part in the suppression 1861 of the Mutiny were issued to them at a garrison parade in Belgaum. Lieutenant-Colonel Steele had been already rewarded by a Companionship of the Bath, which he received on the 12th of May, 1859.

The Regimental Records of the 83rd Regiment were lost in transit to the Depôt companies at Chatham in 1861. Lieutenant Chamley, who was exchanging to the 17th Light Dragoons, was in charge of the records. In passing through London he left them in his cab, and they were never recovered.

On the 24th of July, 1857, the establishment of the 83rd had been greatly augmented, and amounted to twelve companies, of which ten were abroad and two at home. The total was as follows:—

Rank	Number	Rank	Number
Colonel	1	Quartermaster-Serjeant ...	1
Lieutenant-Colonels	2	Paymaster-Serjeant	1
Majors	2	Armourer-Serjeant	1
Captains	12	Hospital-Serjeant	1
Lieutenants	14	Orderly-Room-Clerk	1
Ensigns	10	Colour-Serjeants	12
Paymaster	1	Serjeants	48
Adjutant	1	Drum-Major	1
Quartermaster	1	Corporals	60
Surgeon	1	Drummers or Fifers	24
Assistant-Surgeons	3	Privates	1,140
Serjeant-Major	1		
		Total	1,339

In June, 1861, however, this imposing force was reduced as follows:—

Number of companies the same.

Field Officers	3	Serjeants, etc.	57
Captains	12	Drummers or Fifers	25
Lieutenants	14	Corporals	50
Ensigns	10	Privates	900
Staff Officers	7		
		Total	1,078

The dress was also modified at this time to suit the climate, and is described as a frock of scarlet serge and a wicker helmet, covered with grey linen, with a turban, and this replaced the somewhat unsuitable campaigning garb for the tropics of shell jacket and shako.

Being under orders to proceed to England, the Regiment was permitted to give volunteers to any other regiment serving in India, and 8 serjeants, 10 corporals, 7 drummers, and 463 privates volunteered to remain out there. Previous to the Regiment being thus reduced in numbers, the Commander-in-Chief of the Bombay Presidency, Lieutenant-General Sir William Mansfield, K.C.B., came to Belgaum and inspected the 83rd. He expressed his approbation in no measured terms, saying he had never, in the course of his service, seen a regiment in higher order, and that he should not fail to report accordingly to H.R.H. the Commander-in-Chief.

1862

On the 22nd of January the Regiment turned its face towards home and marched out of Belgaum. It reached Vingerla on the 29th, and remained encamped there until the 5th of February. It then embarked in the hired transport "King Lear." Lieutenant-Colonel Steele, C.B., was in command. No time was wasted, for the ship sailed that evening, touched at the Cape of Good Hope on the 18th of March, obtained fresh supplies, and sailed on the 19th, and reached Gravesend on the 21st of May, 1862. The Regiment disembarked on the same day, and proceeded to Dover by rail, where it was stationed until the 25th of April, 1863, when it marched to take up new quarters at Shorncliffe Camp. Lieutenant-Colonel Steele had left the Regiment on the 29th of July, 1862, and was succeeded by Lieutenant-Colonel Hankey.

Lieutenant E. Brymer was accidentally drowned at Hythe on the 23rd of June, 1863.

1864

On the 28th of April the 83rd was moved by rail from Shorncliffe Camp to North Camp, Aldershot. It is noted in the regimental records that on the 15th of July twenty-eight officers and 500 other ranks of the Regiment proceeded with a flying column to Woolmer and encamped there, returning on the 19th of that month. Evidently manœuvres were not as constant then as now. The Regiment was sent to the same place on the 19th of August to extinguish a fire on the Government property. Travelling by a special train on the South-Eastern Railway, it arrived at Woolmer on the evening of the 19th, and worked hard night and day until the 22nd, by which time the fire was successfully extinguished and the Regiment returned to Aldershot that evening by road.

Up to now the depôt of the 83rd, consisting of two companies, had 1865 remained at Chatham and formed part of the 2nd Depôt Battalion. It was ordered to join headquarters on the 31st of January, thus making it up to twelve companies.

The 83rd left Aldershot in two parties. Headquarters and seven companies went to Sheffield on the 24th of April, and the remaining five companies went to Weedon on the following day, all being conveyed by rail. Whilst at Aldershot the Regiment had won the good opinion of all, and was specially reported on to the Horse Guards by Lieutenant-General Sir John Pennefather, K.C.B., Commanding at Aldershot. Its Brigadier-General spoke most highly of the Regiment, and even the barrack department sent in its special report that never had barracks been left so clean by any troops.

The farewell address of Major-General Russell, C.B., is given in full, as it shows what a General Commanding a Brigade at Aldershot at that time looked for in the regiments under his command:—

"Lieutenant-Colonel Hankey and 83rd Regiment,—When the intimation of the route was first conveyed to me by my Brigade-Major, I felt, to say the least of it, surprised and sorry that I should not have so smart and well-disciplined a Regiment to do my Brigade credit during the ensuing field days in the Long Valley. I can understand that some of you may not regret it, but I assure you I do. Your universal steadiness on parade has given me great satisfaction, and the manner in which you advanced in line and crested the hill not long since, on the Divisional Field Day, was remarked by the Lieutenant-General and admired by all present, and your smartness is a credit to you all—from the senior to the smallest drummer-boy, who draws man's pay.

"Your barrack-rooms are perfect, and their cleanliness is the admiration of the Barrack Department. Your soldier-like appearance in town, and your exemplary conduct while attached to my Brigade at this station has given me the greatest satisfaction and merit my most sincere approbation. Eighty-Third, I wish you well, wherever you may go, and I trust we shall meet again."

From Sheffield two companies were detached to Bradford on the 16th of May, under Major Bray, numbering, however, only 126 of all ranks; whilst from Sheffield also Lieutenant Scott and twenty-two other ranks proceeded to York on the 12th of June.

Serious election riots broke out in Nottingham, and Major Bray was withdrawn from Bradford and given two companies with a strength of five officers and 120 other ranks, and hurried off at 10 p.m. on the 26th of June to Nottingham. They spent the remainder of the night in the Nottingham Exchange, and next day moved into the old cavalry barracks. On the 4th of July one company, under Captain Sprot, was recalled to Sheffield, and another company was brought into Nottingham from Weedon. Captains O'Connor and Karslake now commanded the two companies in Nottingham.

The polling day in Nottingham was the 12th of July, and at 11.30 a.m. the Mayor asked Major Bray to march into the town, as the mob was collecting there in large numbers, and grave trouble was feared. Taking his two companies with him, also a troop of the 15th Hussars, Major Bray, on arrival in the town, found that the mob had assumed such a threatening aspect that the Mayor promptly read the Riot Act and asked Major Bray to clear the streets and to fire upon the crowd, if necessary. With cool judgment, Major Bray at once formed his men up and ordered Captain Ricardo, who commanded the troop of 15th Hussars, to charge. By a series of steady and short advances, following each charge, the streets were gradually cleared of the crowd, and peace was restored. Major Bray, rightly, was much praised for his able dispositions, which obviated bloodshed. A special letter of approval was sent to him from the Horse Guards, by the order of H.R.H. the Duke of Cambridge, Commander-in-Chief.

On the same day a telegram was received from the Mayor of Grantham, saying that the election fever had also broken out in that town, and that he was unable to quell the riots. A very weak company was accordingly sent to him by special train. Captain Johnson was in charge, and had under him Lieutenant Whitlock, one serjeant, and thirty-one rank and file.

However, troops—no matter in what numbers—were to be welcomed with open arms by the Mayor and Corporation of Grantham on that day, and on the arrival of the special train at Grantham the Mayor was waiting on the station platform and marched up at once with the force of thirty-four all told, to the Market Place. Here, however, the mob jeered at the small number of the troops; so the Riot Act was promptly read, and Captain Johnson was asked to order his men to load and clear the streets. A quarter of the company loaded, and then the English common-sense of the mob prevailed, and they wisely went home and left the streets to the military. This company returned to Sheffield on the 14th of July, having bivouacked in the Corn Exchange on the nights of the 12th and 13th, whilst Major Bray brought his two companies back to Sheffield on the 25th of July.

The Mayor of Lincoln also was in trouble over his elections, and on the 15th of July, at 2 a.m., a special train took Captain Gore and forty-four officers and men to that city from Sheffield. The troops arrived about 5 a.m. on the 15th, and remained until the 17th of that month, when, the troubles being over, they returned to headquarters again, with the thanks of the Mayor and Corporation. This party was quartered first in the Militia Barracks, and then in the Corn Exchange.

On the 21st of July the authorities at Rotherham telegraphed about 6.30 p.m. for a military force to be sent at once to quell a riot. Captain Sprot, with forty-six of the Regiment, at once marched off. Captain Sprot was supported by some of the 15th Hussars. Leaving his men at the Railway Station, he went off and personally examined the crowds in the streets, and told the civil authorities that as soon as the public-houses were closed for the night all troubles would cease. He was perfectly right, and by 1.30 a.m. on the 22nd of July he and his men were safely back in Barracks at Sheffield, and Rotherham was quite peaceful.

Lieutenant-General Sprot, in his memoirs, states that the mob was Radical, and the colours of the Radical party at Rotherham were yellow. The mob attacked all other troops until the arrival of the 83rd Regiment's detachment, which, of course, were wearing yellow facings. The mob accepted them as belonging to their party, and made no attempt to attack or resist the men of the 83rd.

The detachment at York had no exciting experiences, and returned to headquarters on the 8th of September.

The detachment at Bradford rejoined headquarters on the 20th of October, but two companies were sent on the 2nd of November to Tynemouth.

1866 On the 30th of January all three detachments of the 83rd left Sheffield, Tynemouth, and Weedon, and embarked at Liverpool for Dublin, on board the "Windsor" and "Trafalgar." The total number embarked consisted of 31 officers, 754 non-commissioned officers and men, and 178 women and children.

On the arrival of the Regiment in Dublin, eight companies went with headquarters to the Curragh, three companies to Armagh, and one to Monaghan. Three days later two companies from the Curragh were ordered to Sligo, whilst on the 25th of February two more companies were taken from headquarters and sent to Boyle.

Political troubles were in full swing in Ireland at this time, and on the 3rd of March, about 1.30 a.m., Privates Barnes and McCarton, who were on sentry over the magazine at Sligo, observed an unusual gathering of people in the field next to the magazine. They promptly warned the commander of the guard, Corporal Davidson, who boldly ran in on the crowd with his guard, scattered the assemblage and captured two of the men, who were handed over to the civil police, and were recognised by the latter as ringleaders in all disturbances. What the exact plot was, is hard to say at this distance of time, but a special letter of thanks was sent to the two sentries and their commander from General Sir Hugh Rose, afterwards Lord Strathnairn, who was Commander-in-Chief in Ireland, thanking them for their conduct in this matter, so the capture was evidently of importance.

On the 26th of March a letter was received from the Horse Guards, directing that the number of companies should be reduced to ten, and that the spare non-commissioned officers of the reduced companies should be retained as supernumeraries until vacancies occurred for them. Lieutenant-Colonel Hankey broke up the two companies on detachment at Sligo, and sent two others from headquarters to relieve them. The rank and file of the companies broken up were distributed amongst the other companies of the Regiment. On the 25th of May a company was sent from the Armagh detachment to Londonderry, and on the 9th of June the Monaghan detachment was withdrawn and a second company was sent to Londonderry; whilst on the 2nd of July the headquarters party, now reduced to a strength of two companies, left the Curragh and went to Enniskillen. On the same day a company was sent from the Armagh detachment to Newry; whilst one company, apparently drawn from Boyle, was also sent to Newry on the 4th of July, and the Londonderry detachment was broken up on the 8th and 9th of August, and a company was sent to Newry and Armagh respectively.

The whole Regiment, less two companies, on the 2nd and 3rd of October, moved to Richmond Barracks, Dublin. The two companies were temporarily left at Enniskillen, but were brought into Dublin on the 5th of December.

On the 27th of December the 83rd received their first breech-loaders. They were converted Enfield rifles of the Snider pattern.

1867

The Regiment was now under orders to proceed to Gibraltar. After being inspected by General Lord Strathnairn, commanding troops in Ireland, and receiving from him the highest praise, the Regiment sent its two depôt companies, via Tilbury, to Colchester on the 27th of March, whilst the eight service companies embarked on the troopship "Himalaya" on the 6th of April, and arrived at Gibraltar on the 12th inst., disembarked the following day, encamped for a week at the North Front, and then relieved the 86th Royal County Down Regiment in the Casemate Barracks.

On the 9th of May the 83rd was presented with new colours on the Almada by the Hon. Lady Airey, wife of Sir John Airey, the Governor of Gibraltar. In her speech on this occasion, Lady Airey mentioned the common tradition in Wellington's Peninsular Army, that the 83rd, celebrated even amongst General Sir Thomas Picton's famous fighting division, was first on the boats crossing the Douro,* first at the night assault on Badajoz, first at the crossing of the River Yardara, at Vittoria; and particularly mentioned Lieutenant Pyne, who, at Talavera, rushed to seize the colours from the hands of the officer who was then carrying them, and who had been sorely wounded;

* There has always been a tradition in the Regiment that a company of the 83rd crossed the Douro in boats early in the day. Professor Oman's researches appear to prove this rumour to be incorrect.

though grievously wounded himself, Lieutenant Pyne manfully refused to surrender his charge until, at the close of the day, he was able to carry the colours to a place of safety.

The Regiment continued its peaceful garrison life at Gibraltar until 1870 the 11th of March, when it embarked on board H.M. troopship "Tamar," en route for India. During the time the 83rd was at Gibraltar it had received 545 recruits, etc., in drafts and in volunteers from other regiments in the garrison. As it was proceeding to India, it was necessary to bring it up to full strength. It embarked with a strength of thirty-four officers and 895 other ranks. The strength of the establishment had been twice changed during its stay at Gibraltar, but at this time recruiting was at a very low ebb, and great difficulties were experienced in filling the ranks of all regiments. One point may strike officers of the present day as rather curious. During the stay of the Regiment in Gibraltar the officers of the 83rd, when in plain clothes, were in the habit of wearing a grey felt hat, turned up at the left side, and that side was held in position by a large regimental brooch-badge, with a long, black cock's feather stuck through it, towering in the air, all looking rather like the hat of a brigand chief on the stage.

An immense concourse of people came to see the 83rd off from Gibraltar, headed by General Sir John Airey, G.C.B., the Governor of the "Rock." His speech of farewell was very complimentary, but may be condensed into one sentence, which he used at the end of it: "I am aware that it is sometimes thought that the words which fall from general officers on such occasions as these are mere stereotyped expressions for general use. I hope you will believe that the sentiments I have expressed emanate from truthful convictions. In my opinion—and I believe, during a period of nearly fifty years' service, I have seen every regiment in the Service—I consider yours unsurpassed, and I doubt much if equalled by any, and it is a gratification and pride again to see you at the head of the Army in the figure of merit for shooting."

So soon as the Governor had ceased speaking, and had been loudly cheered by the men, the Calpé Hounds trotted forward. Colonel Hankey had been the Master, and they had been brought to his farewell parade. Round the Regiment went the smart little pack with their new Master and their huntsman and whips, in their best pink coats, and as they drew away all the bands broke into "Auld Lang Syne," and the Regiment marched steadily on board their transport, not a man absent or drunk.

The "Tamar" took the 83rd to Alexandria and disembarked them there on the 24th of March. They crossed the isthmus of Suez by train and re-embarked at Suez on the 25th of March, on H.M. troopship "Euphrates," and duly arrived at Bombay on the 8th of April. The Regiment remained

COLOURS OF 83RD REGIMENT IN ST. PATRICK'S CATHEDRAL, DUBLIN.

On the left the Colours carried 1848-1867 (including the Central India Campaign); on the right the last stand carried by the regiment 1867-1881

on board until the 10th of that month, when they disembarked and proceeded by rail to Poona, which they reached on the same day.

The establishment of the Regiment had again been altered by the War Office on the 1st of June, 1869, to twelve companies, including two at the depôt. On the 25th of June, 1870, a further letter was received to say that the companies had again been reduced to ten, two of which still acted as depôt companies. The total strength varied but little, being 1,039 of all ranks with the twelve companies, and 1,032 with the ten companies.

The years of this Indian tour passed peaceably on. Inspections came and were successfully met by the Regiment, and complimentary orders were always received and duly noted in the Regimental Orders. Drafts arrived; Lieutenant-Colonel Hankey retired on half-pay on the 19th of November, 1871, after commanding nearly ten years, and was succeeded by Lieutenant-Colonel T. S. Brown.

In the same year, on the 27th of November, 4th of December, and 7th of December, three companies went on detachment to Bombay, Sattara, and Asseenghur respectively. These detachments were withdrawn and again sent out about the same time in the following year, whilst on the 11th of November, 1872, the whole Regiment went to Bombay, to furnish the necessary guards of honour, etc., to the Viceroy, Lord Northbrook, on his arrival in India. After taking part in a Durbar, held to invest the Begum of Bhopal with the Star of India, the 83rd returned to Poona on the 28th of November. Drafts to the total strength of 219 of all ranks joined whilst the 83rd was at Poona. On the 3rd of December, 1873, the Regiment left Poona, to go to a camp of exercise at Chinohwud. It formed part of the 2nd Brigade of the 1st Division in that camp, and at the close of the exercises, on the 29th of December, went to Ahmedabad in two detachments, leaving on the 29th and 30th of December respectively, and arriving on the 1st and 2nd of January. Leaving one company at this place, the remainder of the 83rd then marched for Deesa on the 5th of January, and arrived there on the 15th inst. Seventy-eight men were detached to Mount Abu on the 21st of January, and an incessant flow of officers and men, women and children went to and fro between this station and Deesa.

1872

1873

1874

Two companies were sent to Baroda, on the 20th of October, 1875, and furnished a guard of honour to H.R.H. The Prince of Wales (afterwards King Edward VII.) when he visited that place on the 18th of November.

1875

On the 11th of December one company was sent on in advance of the Regiment, marching from Deesa, and arriving at Porebunder on the 12th of January, 1877. Here it embarked, on the following day, on the I.G.S. "Dalhousie," and on arrival at Karachi was at once sent up to Hyderabad. The remainder of the Regiment left Deesa on the 27th of

1876

December, 1876, being preceded by the Ahmedabad and Baroda detachments on the 20th of December. These two latter detachments arrived at Karachi on the 9th and 13th of January, and were installed in the Barracks with the 56th Foot. This excellent regiment had only one disadvantage, and that was, that it was suffering from an epidemic of small-pox, so the detachments of the 83rd were sent out some fourteen miles from Karachi to Jemedar-ka-Landi, where they were encamped. The 83rd marched down from Deesa in twenty-three days to Mandavic, where it embarked by degrees on the "Dalhousie" steamer, which had to make three trips to carry them all, landing the parties at Karachi on the 22nd, 25th and 28th of January respectively.

The records note that the longest day's march when coming from Deesa was twenty-two miles, which march was performed in seven hours, inclusive of halts, totalling thirty-five minutes in all.

1877 Two more companies were sent on detachment to Hyderabad on the 26th of February and on the 1st of March.

The troubles, which culminated in the series of small wars with Afghanistan, now affected the Regiment, in so far that it was moved to Upper Scinde from Karachi. The headquarters of the Regiment was sent up to Sukkur, leaving Karachi on the 21st of December, 1878, and arriving at Sukkur on the 23rd of that month. The Hyderabad detachment joined the headquarters here on the 5th of January, 1879. An Indian depôt 1879 was formed, consisting of the weakly men and the married families, and an unfortunate officer, Lieutenant Bell, was selected to command it. His depôt consisted of 8 serjeants and 88 privates, with 67 women and 154 children. The whole depôt was sent to Ahmednugger, via Bombay, on the 12th of December, 1878.

The Afghan War did not affect the Regiment in any way beyond this temporary change of stations, and by the 5th of March all the companies had been sent back by degrees to Karachi. Three officers of the 83rd Regiment took part in the Afghan Campaign: Lieutenant Adye, who was recommended for the Victoria Cross on the same occasion as the late Lieutenant-General Sir Edward Leach, V.C., Royal Engineers, received it; Lieutenant Tobin, who led a charge of the Bombay Lancers and had his horse killed and his left arm rendered permanently useless by a sword cut, and only saved his life by shooting his opponent, who attacked him as he lay under his dead horse; Lieutenant Brown, acted as A.D.C. to his father. Whilst at Sukkur the 83rd had been encamped a mile and a half below that town, and when at Karachi the Regiment was partly quartered in the Napier Barracks and partly encamped on the open ground in rear of the Convent. Having been ordered to proceed to Belgaum, the 83rd embarked in two transports, the "Tenasserim" and the "Czarewitch." Sailing from Karachi on the 13th of March, 1879, they landed at Vingerla on the 18th of March.

At Vingerla the Indian depôt rejoined them, and the 83rd marched from

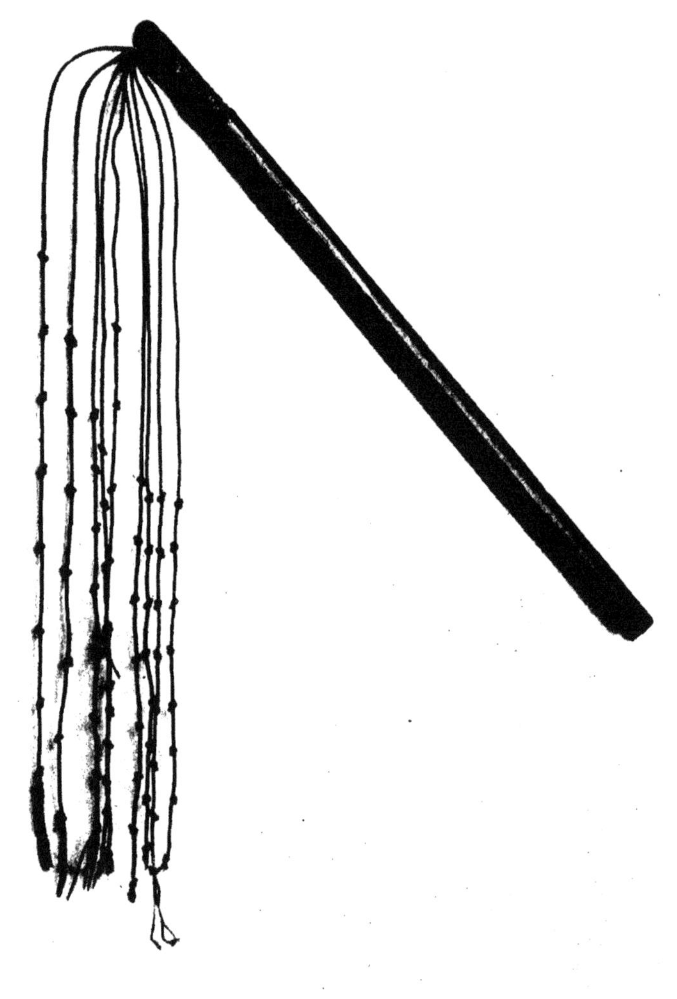

PHOTOGRAPH OF THE LAST "CAT" USED IN THE 83RD COUNTY OF DUBLIN REGIMENT.

Weight of whip - - - five ounces
Length of stick - - one foot five inches
Length of tails - about one foot nine inches
Flogging abolished in the Army, 1879.

this place, en route for Belgaum, on the 21st of March. All the heavy baggage had to be left behind, under the charge of Captain Wyndham's company, as there was no transport available for it, but it finally reached Belgaum on the 1st of April.

On the 1st of December, 1878, a War Office letter was received reducing the establishment from 780 privates to 700, and on the 1st of May, 1879, a further letter was received changing the numbers back to 780 again.

1880 Colonel Brown was appointed to the command of the Scinde District on the 29th of September; so the command of the Regiment devolved upon Lieutenant-Colonel Meurant. Two companies were sent on detachment to Sattara on the 23rd of February. In October volunteers were permitted to join other regiments in India, as the 83rd would shortly be going home. Only four serjeants, four corporals and seventy-four men, however, volunteered to remain behind. On the 26th of November official notification was given to the 83rd that it would proceed home on H.M.S. "Jumna," on the 25th of January, 1881. The Sattara detachment was, therefore, called in on the 3rd of December, and the Regiment prepared to leave India. They were not, however, destined to see Ireland or England so soon as

1881 they expected. On the 3rd of January, 1881, a telegram was received from Army Headquarters ordering the Regiment to proceed on field service to Natal. On the 7th of January the Regiment marched from Belgaum and arrived at Vingerla on the 13th of that month.

On the 14th of January the married families were embarked on the s.s. "Tenasserim," which took them to Bombay, there to be transhipped and sent home to England. They were accompanied by any weakly or time-expired men.

Thus, being "stripped for fighting," as it were, the 83rd went on board the troopship "Crocodile" on the 15th, and landed at Durban on the 30th of January. The 92nd Regiment was also on board the "Crocodile."

A fairly full account of England's dealings in South Africa is given in another chapter of this History. It will, therefore, be sufficient here to give the general outlines of events which called for the despatch of these two regiments to Durban.

The Transvaal had been annexed by the British Government in 1877. It was peopled to a large extent by Dutch farmers, or their descendants, who had moved from Cape Colony in 1837 to avoid British rule and slave emancipation. These settlers had struggled against hardships of every sort, and had met the native owners of the soil again and again in severe combats, and had not always emerged victorious from these fights, in which case there was no emerging at all for the white combatants, only to their friends a knowledge that they had been massacred fighting to the end. Such a people were not likely to accept annexation quietly if it should not happen to suit them. Such an idea appears never to have entered British statesmen's heads.

When the Boers protested, and held meetings, and had interviews with the British Commissioner, and even used threatening language to him, such language was either treated with contempt or replied to in its own vein. It, therefore, came rather as a shock to the British Government when, on the 15th of December, 1880, the Republic was formerly proclaimed at Paardekraal (now Krugersdorp) by Messrs. Kruger, Joubert, and Pretorious. Of military forces in the country to enforce the British rule, there were hardly any. Potchefstroom was attacked on the 16th of December, and on the 20th of that month the Dutch settlers, or Boers, as they called themselves, laid an ambush at Bronkhorst Spruit for two companies of the 94th Regiment, which were marching from Lydenburg to Pretoria. Colonel Anstruther, who was in command of this small party, was told to halt by some Boers, and refused to do so, and a heavy fire was opened on his party, and in a few minutes fifty-six of his men were killed and 101 were wounded, many being hit several times. After this the little force surrendered. The Boers now invaded the British colony of Natal and entrenched a position at Laing's Nek. Major-General Sir George Colley, the Commander-in-Chief in South Africa, marched up with a small force from Maritzburg, and on the 28th of January made an attack with 570 men on 2,000 Boers holding Laing's Nek. Needless to say, he was beaten back, with a loss of 190 men. To anyone knowing the Boers, nothing else could have been expected. Taught from their earliest youth to shoot for food, in the shape of buck and antelope, they were first-rate at judging distance in the open and at concealing themselves, whilst firing most accurately. The British soldier, unless able to close with such a foe, had no chance of success. The Boers took every care that such personal conflicts should be avoided. Their whole system of fighting depended on fire effect. They fought in extended order, kept themselves carefully concealed behind a ridge, with their horses behind them, again under cover, ready to fly if necessary. Their fire was withheld until the enemy approached within the most effective range. If necessary a retreat took place from point to point.

Two days after Laing's Nek, the 83rd and 92nd landed at Durban. The 83rd was extremely weak in numbers, having only a total of 20 officers, 40 serjeants, and 534 rank and file.

On the 1st of February they proceeded inland by rail to a standing camp, named Lilliefontein, where they remained until the 15th of February. On the 8th of that month Colley had fought another unsuccessful fight at Ingogo. Here he was on the march with a small force of 300 men, and was most roughly handled by a much larger force of Boers, who killed and wounded over half his force. Still, he managed to slip away with the remainder of his small party.

The 83rd marched out on the 15th of February, as already mentioned, on its way to the front. Newcastle was its goal. Marching steadily forward, despite heavy rain, the Regiment arrived at Newcastle on the 15th of March.

In the meantime, however, Sir George Colley had lost the engagement at Majuba Hill and, incidentally, his own life. It is so well-known a fight that it is hardly necessary to describe it. It is sufficient to say that Major-General Sir George Colley determined to seize a mountain, called Majuba, which was supposed to outflank the Boer lines at Laing's Nek. Taking 554 men selected from various regiments, he ascended the mountain by night.

Next morning the Boers tried to turn him off the summit which towered some 2,000 feet about their position. Two hundred Boer volunteers crept up the hill under cover of dead ground on the steep hillside. The position had been improperly occupied; the men of all regiments were mixed together. No good positions for bringing effective fire on such an enemy had been occupied. Sir George Colley refused to sanction a bayonet charge. Naturally, therefore, disaster followed. Sir George Colley and ninety-one officers and men were killed, and 134 of all ranks were wounded. A large number of prisoners were taken. Two Boers were killed or died of their wounds. Such was the fight at Majuba Hill on the 27th of February.

Brigadier-General Sir Evelyn Wood, V.C., was the senior officer in the country after Sir George Colley's death. He was on the lines of communications, supposed to be hurrying forward reinforcements, by General Colley's orders. He stated to Her Majesty's Ministers that he held the Boers in the hollow of his hand. Unfortunately, his hands were tied by instructions from England, and though the Government made great profession and sent out reinforcements and General Sir Frederick Roberts, V.C., to command, yet the whole matter ended unsatisfactorily. The heart of the British Government was not in the war.

On the 8th of March an eight days' truce was arranged, Mr. Brand, President of the Orange Free State, being mediator between both parties, with a naturally strong leaning towards the Transvaal. This truce was prolonged from day to day, and on the 21st of March, after a series of meetings and discussions, the three chiefs of the Boers, Messrs. Kruger, Joubert, and Pretorious, agreed to disperse their followers on condition that the independence of the Transvaal was recognised, subject to the suzerainty of the Queen. The 83rd, whilst these negotiations were going on, had arrived at Newcastle, and was encamped on the south side of the Incanda River. On St. Patrick's Day it crossed that river and encamped near Fort Amiel, moving to a fresh camping ground on the Wakerstroom road on the 23rd of March, and again, on the 31st of that month, to a camp at Signal Hill. The ground was generally hilly, but Signal Hill was a large plateau intersected by small ravines.

Two companies were detached from here for wood-cutting fatigue in the Drakensburg Mountains, whilst another company was detailed to work a coal mine, and on the 17th of May the whole Regiment went to the

Drakensburg Mountains until the 28th of that month on a wood-cutting fatigue.

During all the weary time of these protracted and discreditable negotiations, the troops in South Africa hoped against hope that they might be allowed to settle the matter by a fair fight, and various alterations were made to meet the extraordinarily good shooting powers of the enemy. Amongst other things, to make the men less conspicuous, and, therefore, not such good marks, they were ordered to dye their pipeclay belts with coffee.

A draft of fifty-two of all ranks arrived for the Regiment on the 15th of May from the depôt. Whilst the Regiment was encamped here it received the Army Order changing its title to the 1st Battalion Royal Irish Rifles, and the Belfast district was assigned to it for recruiting purposes. It consisted of the three Counties of Down, Antrim, and Louth, and the Regimental Number assigned to that District was the 1st Battalion one of 83rd, whilst the 2nd Battalion, or old 86th Regiment, gave the title of " Royal " to both battalions.

In November the arrangements had been completed with the Boers, and the force about Newcastle was broken up. The 1st Battalion Royal Irish Rifles marched away from the last camp they had occupied there, at Bennett's Drift, on the 7th of November, and arrived at Pietermaritzburg on the 19th of that month. Here it entrained on the 21st and went to Durban, and encamped there until the 25th of January, when it embarked on board the troopship " Tamar," and sailed for England. The 15th Hussars were fellow-passengers on the " Tamar."

1882

Arriving at Portsmouth on the 3rd of January, the Battalion disembarked on the following day, and, accompanied by the married families, who had remained at Portsmouth whilst the troops had been in South Africa, the whole proceeded to Dover. On the 1st of January the strength of the battalion had been again altered from fifty-one serjeants to forty-three, and the establishment of riflemen was reduced from the latest figures (760) to 440. It had only been changed to the former figures on the 1st of May, 1881. The strength being 452, and a draft of 171 of all ranks having arrived from the depôt, volunteers were allowed to go to other corps, and twenty-nine riflemen changed to different regiments. The rifle uniform was taken into use on the 29th of June, 1882.

All soldiers transferred to the 1st Class Army Reserve on or after the 1st of January, 1881, were recalled to the Colours on account of the outbreak of the Egyptian War. They were required to join by the 26th of July, and had to report themselves at the depôt at Belfast. They were then sent across in drafts, and on the 25th of August eighty-four men arrived, and on the 27th thirteen more. The Duke of Cambridge came down shortly afterwards, and inspected the battalion on parade, and, after a searching scrutiny, remarked that he was highly gratified with the efficiency of the battalion, and that it was *perfect* in drill and appearance.

The actual strength of the battalion was extraordinarily weak at the time. For H.R.H. the Commander-in-Chief's parade only 13 officers, 35 serjeants, and 394 riflemen were available.

In this month a slight change was made in the Battle Honours of the Regiment, as it was directed to have the date "1806" added to the honour of "Cape of Good Hope." Of course, many other regiments carried the honour of "Cape of Good Hope" for Kaffir wars, but very few bore it for the actual capture of the Cape from the Dutch.

Following on the inspection by the Duke of Cambridge, an order was received from the Horse Guards on the 10th of September directing that the whole of the Army Reserve men who had rejoined the Regiment should be held in readiness to embark for Egypt. It was further ordered that this party should join the 2nd Battalion Royal Irish Regiment in Egypt, but on the 16th of that month, after the defeat of Arabi Pasha's Army at Tel-el-Kebir, instructions were received that they would not now be required.

A company was sent on detachment to Brighton on the 7th of October, and recalled on the 18th of that month. Immediately after this, orders for demobilization were received, and the procedure was as follows:—

On the 25th of October all men returned to the Reserve who had originally joined the Army Reserve between the 1st of January and the 30th of June, 1881, and on the following 4th of December, all those who had originally joined the Reserve between the 1st of July and the 31st of December, 1881, were sent away, and finally, on the 23rd of December, all men belonging to the Army Reserve who were then serving, were re-transferred to the Army Reserve. The men were not badly treated by the authorities, as they received forty-two days' pay and subsistence money when they left the battalion to join the Reserve.

1883

The first of many drafts to the foreign battalion from the home battalion was sent to join the 2nd Battalion at Bermuda, on the 23rd of February, amounting to 114 of all ranks. The battalion was ordered to Lydd on the 17th of September, to execute its field firing. The distance is twenty-eight miles, and the troops could not complain of being overmarched, as they took three days to march the distance. The field firing was apparently not of a very tedious or extended nature, for it was completed on the 20th, and the battalion commenced its homeward march on the same day, again going over the twenty-eight miles in three days.

A total of 143 riflemen joined the battalion from the depôt during the year 1883, in six drafts; whilst in April, 1884, a draft of seventy-nine of all ranks was sent from the 1st Battalion to join the 2nd Battalion at Halifax, Nova Scotia.

Two companies were sent in April and May by rail to Lydd, to build rifle butts and to lay out a rifle range there. The remainder of the battalion followed on the 19th of May, marching on that date to Shorncliffe; on the

20th to New Romney, and on the 21st to Lydd. The marching is noted in the records as being highly satisfactory. As it did not exceed ten miles on any day, no one could complain of overwork. The battalion at Lydd was rejoined by its two detached companies, and then the whole battalion proceeded to perform a musketry course and a military training course by half-battalions at a time.

1884 The latter was quite a new feature at this time. The total strength at Lydd was twenty officers and 375 other ranks. On the 19th of June the battalion marched back to Dover by the same easy stages.

H.M.S. "Assistance" appeared in Dover Harbour on the 13th of August and took the battalion to Guernsey; four companies were immediately sent on detachment to Alderney.

A further draft was sent to join the foreign battalion on the 22nd of October of a total strength of 134 of all ranks, but the home battalion had received during this year 336 riflemen from the depôt in nine different drafts.

1885 During 1885 nothing of any moment happened to the 1st Battalion. Two drafts were sent to Nova Scotia to the 2nd Battalion, stationed there, one on the 23rd of May, and the other on the 4th of November, being of a strength of 138 and seventy-six respectively.

On the 15th of December the four companies at Guernsey embarked on the "Assistance," and, having taken on board the other four companies at Alderney on the 17th of that month, sailed for Gosport, where they landed on the 18th of December. The total strength was 13 officers, 624 other ranks, 43 women, and 57 children.

1886 At first the battalion occupied the New Barracks, and then changed to Forts Elson and Fareham, moving back again to Gosport on the 8th of October, 1886. On the 15th of January, 1887, the Special Army Orders were brought out, creating the rank of 2nd-lieutenant.

1887 The foreign battalion had nine lieutenants and eight 2nd-lieutenants, whilst the home battalion had only four of the latter rank. At first the unfortunate officers had no badges of rank, to the great confusion of all, but after some years the War Office gave them a badge denoting their rank, as at present.

On the 9th of July, 1887, the battalion attended a Jubilee Review at Aldershot. Six hundred and thirty-six of all ranks were present at this parade. It moved to Aldershot on the 7th of July, camping in Rushmoor Bottom, and returned to Gosport on the 12th of July, having been brigaded with the 2nd battalion of the Rifle Brigade and the 2nd Battalion of the King's Royal Rifles.

A draft of 123 of all ranks went to Gibraltar on the 21st of September to join the 2nd Battalion. The "Assistance" took the battalion to Kingstown, leaving Portsmouth on the 14th December and arriving at Kingstown on the 17th, after a very rough and stormy passage. No one was much the worse for the bad crossing but the regimental coach, when being swung ashore by the bluejackets of H.M.S. "Assistance," slipped and went to the bottom, and had to remain there twenty-four hours, until a diver could be obtained, and then it was brought up, much the worse for its immersion.

The battalion went to Mullingar, and one company was sent to Sligo.

1888 During the year 1888 two drafts went to the 2nd Battalion, then at Alexandria, but only consisted of forty-one of all ranks.

On the 31st of October thirty-two non-commissioned officers and men, under Lieutenant Addison, were sent to Newbridge, to be instructed in the duties of mounted infantry.

Whilst at Mullingar the 1st Battalion was commanded by Lieutenant-Colonel C. J. Burnett, who had succeeded Lieutenant-Colonel Meurant, and had been brought into the Regiment to command a battalion from the 15th Foot. He was most enthusiastic about improving the men's messing, and worked untiringly at this problem. His efforts were irreverently styled "Sheep's heads" throughout the remainder of the army stationed at home, as that delicate morsel formed a greater part of the suggested improvements. There is no denying the fact, however, that Colonel Burnett's well-meant efforts did direct attention to the need for improvement in the scale of living of the men in barracks, and many such improvements as have taken place are largely due to the publicity given by him to this important subject.

One of his efforts was not quite as successful as he might have wished, and this was as follows:—Having been asked to shoot rooks at a neighbouring country estate, the Colonel immediately thought of his men, and returned to barracks with a cartload of young rooks, which he at once distributed to the companies, with a view of having rook pies made up. However, his Irishmen had never partaken of that fare, and, after a hasty consultation as to what was to be done, so as to get rid of the "old crows" and, at the same time, not to hurt the kindly Colonel's feelings, it was decided to throw them out of the back windows of the barracks, which was promptly done forthwith. Lieutenant-Colonel Cutbill succeeded Lieutenant-Colonel C. J. Burnett in command on the 24th of June, 1890. The latter was appointed Assistant-Adjutant-General at Aldershot.

During Colonel Burnett's command Captain Bell organised the first cyclist company in the Army. It was over 100 strong, and was mounted on safety bicycles with solid rubber tyres. It was taken from Ireland to Aldershot during 1889 and was found to be most efficient.

The battalion only had to furnish 136 men to the foreign battalion this year.

1890 Leaving Mullingar on the 1st of July, 1890, the battalion moved to the Curragh Camp for summer drills, leaving the Curragh on the 19th of August, via Dublin and Kingstown. Here it again embarked on the "Assistance," and sailed to Queenstown. One company was sent to Fort Carlisle; the remainder of the battalion went to Fermoy.

1893 The 1st Battalion remained at Fermoy until the 19th of June, 1893, when it moved to Newry. Nothing unusual had occurred whilst at Fermoy. On two occasions the battalion had been taken out to manœuvres at Kilworth—both times in August, viz., in 1891, from the 15th to the 22nd August, and in the following year from the 15th to the 20th of that month. In the same year the drafts to the foreign battalion went up to the large total of 286 of all ranks.

A rifle busby having been sanctioned for all rifle regiments, instead of a green helmet, which resembled a policeman's, the battalion took the busby into wear on Christmas Day, 1891, at Fermoy.

Accommodation was limited at Newry, so a small detachment had to remain in the old barracks at Fermoy, and followed later. One company was sent on detachment to Drogheda on the 9th of October.

Whilst stationed at Newry, the old Glengarry cap, which had been in use since 1874, was abolished, and a field service cap, somewhat similar to a Glengarry, without ribbon tails, was substituted. It had the single advantage that it would double down over the head and button under the chin, and was, therefore, useful for night work in inclement weather.

1894 The battalion left Newry, without regret, on the 16th of October, 1894. The company from Drogheda was brought in on that date, and the whole left Newry for Greenore, and there embarked on the London and North-Western Railway Company's boat for Holyhead. There, two special trains met the battalion, one of which conveyed six companies to Brighton, under command of Lieutenant-Colonel Knox, who had succeeded Lieutenant-Colonel Cutbill on the 26th of June. The other train conveyed two companies to Chichester. Captain Laurie was in charge of this detachment.

This move was the first the battalion had made of late without the unpleasant experience of a crossing in the "Assistance." This somewhat ancient vessel had lately fallen into disuse, and the London and North-Western Railway Company, which was called upon to move troops, instead, provided everything they possibly could for the comfort of all.

The headquarters at Brighton furnished one company for a month to garrison a redoubt at Newhaven.

1895 On the 16th of January the 1st Battalion had the unusual experience of receiving a draft from the foreign battalion, the reason being that the latter, having been ordered to proceed from Malta to India, sent home any men whose Army engagement had nearly expired, or who were not passed as medically fit for India. The draft only numbered fifty-three of all ranks.

On the 25th of May the Chichester detachment marched to Arundel, and next day proceeded to Brighton. On the 28th the whole battalion marched, via Newhaven, Eastbourne, Bexhill, and Appledore, to near Ashford, for a few days' manœuvres, and at the close of these proceeded, also by marching, to Lydd for musketry. All returned to their respective stations by train on the 9th of July.

1896 Lieutenant King-Harman took a party of mounted infantry to South Africa, leaving Brighton on the 27th of April, and its adventures are narrated elsewhere.

Moving to Lydd, by rail, for its annual course of musketry on the 1st of May, the 1st Battalion Royal Irish Rifles then returned to Arundel Park on the 5th of June, where it was joined by three battalions of Sussex Militia and Volunteers, and various exercises were carried out until the 13th of June, when the headquarters of the battalion marched back to Brighton. The distance marched was twenty-two miles, and the heat was very great. Serjeant Bannister, provost-serjeant of the battalion, died from heat apoplexy on this march.

The Chichester detachment remained for four days longer, busily engaged in making the Park resume its wonted appearance. So pleased was its owner, the Duke of Norfolk, with their efforts, that he sent down a present of £10 to the men of the detachment. They politely replied that they considered £10 too much for His Grace to give for such a small service, but that they would gladly accept £5 to drink his health with, and this kindly gentleman's health was duly drunk when the party arrived at Chichester.

Moving to Aldershot by train on the 25th of September, the battalion took over Ramillies Barracks.

Whilst the battalion was stationed here, the Jameson Raid in South Africa took place, and the authorities considered that it was necessary to increase the British Forces in South Africa, as much ill-feeling was naturally engendered between the Transvaal and the British Government by this absurd and most unjustifiable filibustering expedition. The 1st Battalion was one of those selected to proceed to South Africa, and it left Aldershot on

1897 the 24th of April, embarked at Southampton on the hired transport, "Dunera," and disembarked at Durban on the 24th of May, and immediately proceeded to Ladysmith. There were no troops in Ladysmith at this time, excepting Lieutenant King-Harman and his mounted infantry detachment. Their adventures and the reasons for their hasty despatch to South Africa are contained in the following pages:—

Matabeleland and Mashonaland are contiguous countries in Africa, lying to the north of the Transvaal. In 1887, Kruger, the President of the Transvaal, desired to make a treaty of suzerainty over these two countries. He was anticipated by that very able Englishman, Mr. Rhodes, who induced the British High Commissioner to send a Mr. Moffatt, a clergyman, to Lobengula, the chief of the Matabele, who preferred to make a treaty with the English, instead of with the Dutch. Unfortunately, he was unable to control his warlike natives, who murdered some of the Mashonas, a native tribe kept in subjection by the more warlike Matabele. The English traders, who also used these people as their servants, resented this useless butchery, and a war broke out between the Matabeles and the British, as represented by a trading company, known as the Chartered Company. After various engagements a peace was patched up; but the Matabeles were a proud people, and felt that they had not done themselves justice in this war, and only bided their time. Their anticipated opportunity came, when, on the 29th of December, 1895, Mr. Rhodes engineered the Jameson raid, which was an armed invasion of the Transvaal from its western side by a force of some 500 mounted men, under a Dr. Jameson. To get even this small force together, Mr. Rhodes had to take trained men from anywhere he could collect them.

Most of the Matabeleland Police were withdrawn, to take part in this Jameson raid, with the result that the natives rose, hemmed in Bulawayo, and murdered many white settlers. Troops were sent from England to assist in suppressing the rising: among others a battalion of mounted infantry (four companies), under the command of Lieutenant-Colonel E. Alderson, Royal West Kent Regiment. The Irish company in this battalion, commanded by Captain Sir Horace McMahon, Bart., included one section of the 1st Royal Irish Rifles Mounted Infantry. It is difficult to understand why this company should have been commanded by an officer in the Royal Welch Fusiliers, instead of by the officer who had trained it at Aldershot in 1895—Captain F. J. H. Bell, 1st Royal Irish Rifles, who was then Aide-de-Camp to Major-General Burnett. The Royal Irish Rifles section, under Lieutenant King-Harman, and consisting of Colour-Serjeant Sheridan, one serjeant, one corporal, and twenty-five riflemen, left the 1st Battalion at Brighton on the 27th of April, 1896, and joined the mounted infantry battalion at Aldershot. The mounted infantry battalion sailed from Southampton on s.s. "Tantallon Castle" on the 2nd of May, and landed at Capetown on the 19th of that month. By that time, matters had improved in Matabeleland, and it was decided not to send any more troops there, so the mounted infantry were placed in huts at Wynberg, with the idea of soon returning to England.

Many settlers in Mashonaland had joined colonial corps, to help suppress the Matabele rising; so, taking advantage of their absence, and instigated by the Matabele, the Mashonas rose, and the same murderings and pillage took place in Mashonaland as had occurred in Matabeleland. This was about the 20th of June, before the mounted infantry had been sent home, so the "Irish" company and a company formed from the King's Royal Rifles and Rifle Brigade, with a few horses, sailed from Capetown in s.s. "Arab" on the 26th, picked up more horses at Durban, and on the 2nd of July reached Beira, where they were joined by some Royal Artillery with 7-pounders, by some Royal Engineers, Army Service Corps, Royal Army Medical Corps, and by some infantry, with machine guns. The British force was welcomed by the Portuguese Mozambique authorities, who thought its presence would have a good effect on their own natives, who were getting restless.

A camp was formed at Fontesvilla, forty miles up the Pungwe River, and men, horses, and materials, including six months' food supply, were moved up. The means of transport were most primitive. No roads existed; a toy train could only take a few men and horses up each day, and a small tug could convey a few more up the river. The Royal Irish Rifles section remained at Beira unloading horses and supplies until July 11th, when it moved by the toy train up to Fontesvilla, a foul spot, infested with fever and lions.

Here it remained a week, trying to break in the raw horses. From there the troops, in small detachments, were sent by train some 120 miles up to Chirmois, which place was then the rail-head. Another camp was formed at this place. The Irish company moved out of Chirmois on the 20th of July, on its 230-mile march to Salisbury. No proper road existed, merely a villainous veldt track; no bridges spanned the drifts; a large convoy of wagons had to be escorted, so progress was slow. To make matters worse, rinderpest had broken out, and it was impossible to secure many teams of "salted" oxen, viz., oxen which had had the rinderpest and had recovered, so most of the wagons were drawn by mules. Massikessi, a small Portuguese town, on the border, was reached on the 23rd, and the same day the company crossed into Mashonaland. Umtali was reached on the 25th; two days were spent there to re-arrange the convoy and fit the scattered units into their places in the Mashonaland Field Force. Then the march to Salisbury was resumed. Many lions, ostriches, and various kinds of buck were seen on the way.

A halt was made on the 3rd of August, and a portion of the force was sent off to make a night march to attack the kraal of Makoni, the most troublesome chief in Mashonaland, and the leader in the rebellion. The Royal Irish Rifles section took part in this attack, which was most successful. The kraal was captured and burnt, and a large number of natives were killed; but the chief and most of his men took refuge in the caves of the kopje on which the kraal stood, and a column had to be sent three weeks later to capture Makoni. The march was continued, and the telegraph wire which had been cut was repaired. Marandellas was reached on the 9th, and Gatzi's and Magwendi's kraals were destroyed. The Royal Irish Rifles section then pushed on in front of the column, as escort to Colonel Alderson. It passed through swarms of locusts, and entered Salisbury on the 16th. A permanent camp was formed here, and this place became the headquarters of the mounted field force, which numbered about 2,200, namely, 500 British troops, 1,050 Colonials, and 650 natives, together with about 570 horses. With this force the lines of communication from Umtali to Salisbury had to be kept open, some small outlying laagers to be relieved, the villages of Salisbury, Victoria, Enkeldorn and Charter to be garrisoned, and punishment dealt out to the rebellious natives over an area of 114,000 square miles, or over a country about half the size of France.

The problem was a difficult one. There was a great lack of transport animals, owing to the rinderpest, and the country to be operated in was but little known and absolutely unmapped. What was known of this province showed that it was a very difficult country to operate in. Then, to add to these difficulties, the heavy rains were due to begin in December, and all movements of troops would certainly become impossible, whilst the horses were daily dying in large numbers, and were already much too few for the work required from them.

It was, therefore, only possible to clear the country for about sixty miles all round Salisbury, and to establish and garrison posts at points from which expeditions could be pushed farther afield in the following year. The fighting was most uninteresting and disagreeable. The Mashonas never came out into the open, having learnt from their frequent fights with the more robust Matabele to take refuge in the inaccessible caves, over which their kraals were generally built.

The Royal Irish Rifles section did its share of work, and began by a patrol on the 19th of August to Garandoola's kraal, which was burnt. Here Rifleman R. Rainey was hit in the leg by a bullet, fired from a cave at very close range. His leg had to be amputated, but he recovered, and received a pension of 2s. per diem. He was in the battalion football team, and was a fine player. From the 23rd to the 27th of August the Royal Irish Rifles section proceeded on patrol, with some Colonials, to the Hungani River, where several kraals were destroyed and some grain collected. From the 8th to the 14th of September it again moved out, with some more mounted infantry, to Mazöe, twenty-five miles north of Salisbury, where the kraal of a chief, named Amanda, was attacked and burnt. One night some lions got into their laager, but did no harm, beyond eating some sheep, captured from the natives.

From the 18th to the 30th September the whole Irish company, with some Royal Artillery and Colonials, went on patrol again in the district round Mazöe. A skirmish of two days' duration resulted in a chain of rocky hills being captured and the kraals of Gadeera and Chidamba burnt; also some cattle and grain were collected, and some smaller kraals destroyed.

From the 5th to the 24th of October a patrol of all available British troops, with some Colonials, visited the Hartley Hills, fifty miles south-west of Salisbury, and after two days' bickering Mashangombi's and Chena's kraals, on the Umfuli River, were taken and destroyed; a large number of smaller kraals were also burnt, and some grain and cattle captured; but the force lost about twenty killed and wounded. Many of the horses had died, and the men were badly off for boots and clothes, so some of the troops returned to Salisbury in the empty food wagons, and before leaving handed over their horses and boots to those who were to continue the patrol.

The Irish company remained out, and, with some Colonial artillery and dismounted troops, spent from the 24th of October to the 8th of November in Lomogundi's country, sixty miles west of Salisbury. Many kraals were destroyed, but there was little or no fighting, as the natives all moved off into the " fly " country, north, towards the Zambesi, and the horses, etc., could not be taken after them, owing to the presence of the tsetse fly. The famous marble caves of Sinoia—a truly wonderful sight—were visited while on this patrol.

Horses and mules were dying every day, and the survivors were so done up with starvation, that the mounted infantry were dismounted, and the horses led into Salisbury. A patrol of two and a half companies of mounted infantry (one being the Irish company), some Royal Artillery and Colonials, was sent out on the 15th of November to the country occupied by two chiefs (Chiquaqua and Hunzi), thirty miles east of Salisbury. Those rulers at once made their submission, and the patrol returned on the 27th. The Royal Irish Rifles section went out on the 4th and 5th of December fifteen miles down the Charter road, to look for natives who were supposed to have stolen the post-cart oxen, but none could be found.

Major-General Sir F. Carrington, commanding all the troops in Rhodesia, passed through Salisbury, on his way down country, and, as the rains were beginning in earnest, and little more work could be done, he ordered the British troops to leave the country. The newly-raised Rhodesia Mounted Police took over the few remaining horses, and Riflemen R. Baxter, W. Carroll, J. Devlin and J. Moore were allowed to transfer to that corps.

The Irish company marched out of Salisbury on December 8th, down the so-called road it had ridden up some five months before, now turned into a quagmire by the rains, and it reached Umtali on the 18th—a march of 155 miles. A week's halt was made there, and the Irish company marched on Christmas Eve for Bandulas—sixty miles distant—which place they reached on the 27th of December. A good many buck and game birds were shot on the various patrols, especially on the way down country, and provided a welcome change from everlasting bully beef.

Here the railway began, and the section entrained. After three days in the train, in intense heat—115° in the shade—and all having frequently to get out of the train to push it up the gradients—Beira was finally reached on the 29th. Here the s.s. "Pembroke Castle" was waiting, and the troops, going on board, disembarked at Durban on the 3rd of January, 1897. The mounted infantry proceeded at once into huts at Pietermaritzburg, where most of the men suffered from fever, contracted in Mashonaland, and there the Royal Irish Rifles section remained until the 21st of May, when they were sent up to Ladysmith, to prepare the camp for the 1st Battalion, which arrived at that place on the 25th of May, when the mounted infantry section rejoined the battalion.

Ladysmith, then unknown to fame, was only a small dusty village in Natal, when it welcomed the Royal Irish Rifles. The battalion had to camp on a hot and dusty plain, with various hills near at hand, whose names were soon to become household words in England. Here the battalion's tents were pitched, and it settled down for some two years. It arrived under the command of Lieutenant-Colonel Charles Haggard, who had succeeded Lieutenant-Colonel Knox on the 28th of October, 1896, and numbered 20 officers, 593 of other ranks, and 37 women, with 51 children.

The garrison was very shortly increased by the arrival of the 9th Lancers and a brigade division of field artillery.

The mounted infantry section has to be added to the above total, consisting of Lieutenant King-Harman and twenty non-commissioned officers and men.

With reference to Lieutenant King-Harman's section of mounted infantry, the following letter was received on the 27th of May from the Divisional Assistant-Adjutant-General, Natal and Zululand:—

"The Major-General Commanding, Natal and Zululand, requests that you will convey to the Officers Commanding the several Battalions represented in the Mounted Infantry Battalion his high appreciation of the excellent conduct and soldierly qualities displayed by all ranks, both while serving under his command, and also when on Active Service in Rhodesia."

The provisional battalion at Folkestone sent two drafts to the battalion, one, of ninety-five of all ranks, arriving in South Africa on the 28th of December, and the other numbering 114, which joined at headquarters on the 23rd of March, 1898. The establishment of the battalion had been raised from the 24th of April, 1897, from 24 to 28 officers, 38 to 46 serjeants, and from 680 to 880 riflemen. As, however, the Army Order was not published until July, 1897, it was only when these drafts arrived that the battalion actually felt the advantage of the increase in its establishment. For the benefit of those who do not remember the provisional battalion at Shorncliffe, it may be mentioned that it was made up from the trained recruits of such regiments as had both battalions abroad. These recruits were sent here after being trained at the depôt, instead of being sent, as usual, to a home battalion, which in their cases was non-existent.

1898

On the 10th of August the battalion received its first Maxim machine gun.

On the 31st of December medals for Rhodesia, 1896, were presented to Lieutenant King-Harman's mounted infantry section. It shows how quickly the Service now changes when it is known that only eleven all told were still serving with the battalion and received the medal.

During the time that the 1st Battalion served in Ladysmith they furnished one company of mounted infantry, commanded by Captain Noblett. It was formed on the 7th of July, 1897, and disbanded on the 18th of March, 1899, when the company handed over their 119 ponies to the 1st Battalion Leicestershire Regiment at Mooi River. The ponies had suffered much from horse sickness during that time, and the battalion, just before leaving Ladysmith, went through an epidemic of enteric fever, which necessitated the other battalions in camp being sent out from Ladysmith. The Royal Irish Rifles suffered, on the whole, less than the other troops stationed with them, and they remained in Ladysmith despite the outbreak.

1899

Leaving Ladysmith on the 24th of March, the battalion went first by rail to Durban. There it embarked on the R.I.M.S. "Clive," and sailed for Calcutta, en route for Dum Dum. It numbered 13 officers and 631 of other ranks, besides 27 women and 47 children; 7 officers were at home on leave, whilst 91 non-commissioned officers and men remained at Ladysmith, of whom 56 were suffering from enteric fever.

The battalion erected in the cemetery at Ladysmith an obelisk to the memory of those who had died during their tour at that station, numbering one officer (Lieutenant and Quartermaster Cunningham, who had been a great support to the cricket eleven of the battalion) and twenty-three non-commissioned officers and men, also one woman and nine children.

Arriving at Calcutta on the 15th of April, the Royal Irish Rifles disembarked next day, and proceeded on the short railway journey to Dum Dum. Here it found a strong draft from the 2nd Battalion awaiting it, of those who had been left behind when that battalion went home from India in January. The strength of the draft was 388 of all ranks. Six officers also joined here, most of whom had exchanged with other officers of the 1st Battalion to continue serving in India.

This draft brought the strength of the 1st Battalion up to the exceedingly fine total of twenty-four officers and 1,109 of other ranks.

Six companies of the Royal Irish Rifles marched into Calcutta and garrisoned Fort William there on the 18th of September; the other two companies remaining—one at Dum Dum, the other at Barrackpore. Captain Fox Strangways and Lieutenant Eckford were permitted to proceed to South Africa for certain duties on the 18th of September, whilst on the 20th of September Serjeant R. Drysdale and fourteen riflemen, who were trained transport drivers, proceeded on the hired transport "Nurani" to the same destination.

1900

On the 1st of February Captain Noblett and one serjeant (Hewitt) were attached to a squadron of Lumsden's Horse, a body of volunteer mounted infantry, who were sent to South Africa, to take part in the Boer War. Serjeant Hewitt was appointed squadron-serjeant-major. The war in China, known as the Boxer troubles, caused a demand to be made to the battalion for one serjeant and two riflemen for a British field hospital, and for four riflemen as telegraphists, and they duly embarked from Calcutta on the 1st of July for China.

A small draft of one serjeant and four riflemen now arrived from South Africa to rejoin the 1st Battalion. These men had been left behind when the Royal Irish Rifles had left Ladysmith, and, being there when the Boer War broke out, they had the good fortune to serve through the siege of that town, under the command of General Sir George White, V.C.

One man, Rifleman Clydesdale, had been killed by a large shell which destroyed the house opposite to the Ladysmith Post Office, in which house Rifleman Clydesdale was on duty at the time.

Lieutenant-Colonel Haggard retired on the 28th of October, and was succeeded in the command by Lieutenant-Colonel Swaine.

1902 The battalion, under the command of Lieutenant-Colonel Swaine, moved to Fyzabad from Calcutta on the 3rd of February. Despite three drafts from home, it was very much below its original strength. The drafts had arrived on the 10th of January and 17th of April, 1901, and on the 25th of January, 1902, and numbered 99, 43, and 90 respectively of all ranks; yet the wastage had been so great that only 13 officers and 717 of other ranks moved to Fyzabad, with 27 women and 33 children.

The Delhi Durbar, held for the purpose of proclaiming His Majesty the late King Edward VII., took the 1st Battalion Royal Irish Rifles to Delhi on the 18th of November. The military authorities utilized the occasion to hold manœuvres on a large scale, and the battalion did not return to Fyzabad until the 21st of January, 1903. Whilst at Delhi two drafts joined the head-quarters, totalling 129 of all ranks. The following officers, warrant officers, non-commissioned officers and rifleman received Coronation medals on this occasion, according to the regimental records:—

1903 Lieutenant-Colonel Swaine, Major Tobin, D.S.O., Lieutenant Dunn, Serjeant-Major Foster, Bandmaster Williams, Colour-Serjeants Cowden and Elphick, Corporal Verdon, and Rifleman Heron.

The battalion sent one serjeant—Serjeant Bingham—to Somaliland, to take part in the desert warfare in that country.

On the 31st of December a draft of ninety of all ranks arrived from home, followed by a further draft on the 16th of May, 1904, of 120 of all ranks.

1904 These drafts were now arriving from the home battalion in regular order. During the Boer War, whilst both battalions were abroad, the trained men, who had been sent away from the depôt, so as not to clog the training of recruits there, were, for a time, stationed in Sheffield, and afterwards at Londonderry.

A machine gun detachment was now called for to proceed to Tibet on what was politely called a "mission," under the command of General Macdonald. As a matter of fact, there appeared to be little difference between a "mission" and a frontier expedition; but the country was so interesting, being practically unknown to Europeans, that a fuller account than usual is given of the doings of this small party, which was under the command of Lieutenant Bowen-Colthurst. It consisted of Nos. 4277 Serjeant

Lyle, 2732 Rifleman Cochrane, 4342 Rifleman Quinn, 5942 Rifleman Hill, 5808 Rifleman Cunningham, 5985 Rifleman Loftus, and 6011 Rifleman Butler.

The party went off on the 18th of May, and was attached to the 1st Battalion Royal Fusiliers.

The cause of the expedition was briefly as follows:—The Tibetans had followed an immemorial policy of exclusiveness. Warren Hastings had made overtures to them in 1785 without success. Such was the uncompromising spirit of the Lamas, or Monks, who ruled Tibet, that no Viceroy had thought it worth while to renew the attempt to open relations with them until Lord Dufferin sanctioned a commercial mission in 1886. This mission only reached the frontier of Tibet and was withdrawn at the request of the Chinese, who appeared to claim some suzerain rights over Tibet at that time.

This display of weakness incited the Tibetans to such a pitch of vanity that they invaded Sikhim, and established a military post in British territory.

After several vain appeals to China, the Indian Government ordered a military expedition to deal with the matter, and in 1888 the Tibetans were defeated in three engagements and driven back into the Chumbi Valley.

After this the Indian Government attempted to establish some sort of commercial treaty with the Tibetans. These simple creatures argued about the matter for five years, then drew up a treaty with the Indian Government, establishing a trade mart at Yatung, in the Chumbi Valley, and proceeded to build a wall a quarter of a mile from the Custom House, through which no Tibetan or British subject was allowed to pass. Naturally, this checked all intercourse.

Years of negotiations followed, characterized by shuffling and equivocation on the part of the Tibetans and Chinese, and by weak-kneed forbearance and conciliation on the part of the British.

Tibet being nominally under Chinese suzerainty, was able to refer to Pekin on any suggestion being made. As, however, China was quite powerless to deal with the Tibetans, and the Chinese Amban or Regent at Lhasa was practically a prisoner there, these negotiations never came to anything.

In 1901 Lord Curzon was permitted to send a despatch to the Dalai Lama, pointing out that, in view of the continued violation of the frontier by parties of Tibetans, the destruction of frontier pillars, and the restrictions imposed on Indian trade, he would be compelled to resort to more practical measures to enforce the observance of the treaty.

The letter was returned unopened, and the Dalai Lama's answer was to send a mission to St. Petersburg, carrying presents to the Czar.

In 1903 came the announcement that Colonel Younghusband was to proceed as Commissioner to Tibet.

For ten months fruitless negotiations followed, and, except that Bhutanese threw in their lot with the British, no result was achieved. Colonel Younghusband, whose escort consisted of three native regiments—the 8th Gurkhas, 23rd and 32nd Pioneers—decided to advance. A Tibetan army of 2,000 men barred the way at Guru. On the 30th of March, 1904, the first shot was fired by the Tibetan general, and within ten minutes the Tibetans had 700 casualties.

The mission advanced to Gyantse, but no negotiations followed; on the contrary, new Tibetan levies arrived, and severe fighting ensued. Reinforcements were ordered up from India, and with these came the machine gun detachment of the 1st Battalion.

Leaving Fyzabad on the 18th of May, and arriving at Silliguri, by train, on the evening of the 20th, the march of fifty miles to Darjeeling and a 7,000 ft. rise was performed in twenty-eight hours. Chumbi, 110 miles away, was reached on the 2nd of June, the track leading through the Jelap Pass (14,300 feet).

On the 11th of June, the relief force being now all united, and consisting of the following troops:—Half-battalion Royal Fusiliers, Machine Gun Detachment, 1st Battalion Royal Irish Rifles, 7th Mountain Battery, and 40th Pathans, left Chumbi and reached Gyantse, 136 miles away, on the 26th of June, having a skirmish with the enemy at Niani and killing over 300. The Machine Gun Detachment and 1st Royal Fusiliers were in reserve during this action.

On the 28th of June it was decided to disperse another large gathering of Tibetans holding the fort and monastery of Tse-Chen, on the left bank of the Nyang River. The machine gun only fired 420 rounds in this engagement, and the fort was captured by assault with great gallantry by the 8th Gurkhas and 40th Pathans, over 400 Tibetans being killed.

The fighting round Gyantse finished on the 6th of July, when the Jong or fort was assaulted and captured. Over 700 Tibetans were killed in this action, which began at 4 a.m. and lasted until 7 p.m. One thousand four hundred and ninety-two rounds were fired by the Royal Irish Rifles gun section during the day.

On the 14th of July it was decided to advance to Lhasa, the capital of Tibet, 147 miles away. The chief difficulties were the Karo La Pass (16,400 ft. high) and the crossing of the Sanpo, or Brahmaputra River. No opposition was met with, except during the actual crossing of the pass, when fighting took place among the eternal snows, at the record height of 20,000 feet. One hundred and eighty-five rounds were fired from the gun in this encounter, and some seventy Tibetans were killed. This was the last action which the " mission " was engaged in.

The crossing of the Sanpo, 10,000 ft. above the sea, took six days, but the expedition was now only forty-one miles off Lhasa, and on the 3rd of August encamped one mile south-east of the peerless Pota-La, the Dalai Lama's palace.

Next day the detachment formed part of Colonel Younghusband's guard of honour as he rode into the city, through the Pargo Kaling Gate.

For a month negotiations continued. The Dalai Lama had fled, and was deposed, but on the 7th of September the Treaty of Peace was signed in the Pota-La by the National Assembly and by the Chinese Regent.

The clauses of the treaty were as follow:—

1. The boundary pillars to be re-erected.
2. Marts were to be established for trading.
3. Tibet was to pay an indemnity of half a million pounds.
4. Great Britain was to occupy the Chumbi Valley for three years, and finally—
5. Without the consent of Great Britain, no foreign Power whatsoever is to be allowed to concern itself with the administration or government of Tibet, or is to be allowed to construct roads or railways, or telegraphs, or to open mines in Tibet; nor may any portion of Tibet be sold, leased, or mortgaged to any foreign Power.

The last clause was, of course, the most important one. During the signing of the treaty, the machine gun detachment of the Royal Irish Rifles was on guard outside the Pota-La.

All Tibetan prisoners were at once released and given a gratuity of five rupees each; and the return march began on the 23rd of September. A fortnight later, at Gyantse, General Macdonald made his farewell speech to the brigade. On the 21st of October the frontier was crossed at the Jelap Pass, in three feet of snow, and on the 30th of October the detachment marched into Siliguri, having marched the 440 miles from Lhasa in thirty-two days.

So much interest was taken by the battalion in the fortunes of the machine gun detachment, that Lieutenant Bowen-Colthurst's letters to the Colonel were published with Battalion Orders.

Below, a specimen letter, as published in Regimental Orders, is given exactly as published:—

"Gyangtse,

"27th-29th June, 1904.

"Since my last letter on 16th inst. from Phari we have been pushing on steadily towards Gyangtse with only a halt for two days at Kangma. The reason of the halt was to allow of the 40th Pathans (four companies) to join us (two companies of the 40th Pathans have been left at Chumbi and two at Phari), also a force of 600 Thibetans were reported in a monastery 13 miles off on the Lhassa Road—the short cut between Kagma and Baluna. Two companies Royal Fusiliers, four companies 23rd Pioneers, and one section 30th Mountain Battery started at 6.45 a.m. on 23rd inst. to catch them, however, they were not early enough as they found the sangars deserted, although fires were still burning; so they looted the monastery, destroyed about 6,000 rounds of Thibetan ammunition, and secured some considerable amount of forage, grain, timber and a few odds and ends as swords and spears. The same day a force of two companies 32nd Pioneers marched twelve miles through the Red Gorge to Dote and a telegraph line was carried to Dote; if you remember on the first advance of the Thibetan Mission to Gyangtse, the Red Gorge was held by the Thibetans, however, the 8th Ghurkhas went over the hill-tops and drove them out and the Mounted Infantry got at them round the corner and nearly 500 Thibetans were killed. The actual gorge is very narrow and about three quarters of a mile long full of piled up rocks, some with Lamaist paintings of Buddha or other incarnations; painting out here is a common form of worship and owing, I suppose, to the small rainfall these paintings keep very fresh. We left Kangma on the 25th inst. and marched through the Red Gorge sixteen and a-half miles to Saugon; we met no Thibetans, but one of the Mounted Infantry was mortally wounded about nine miles beyond Saugon by a shot fired from a house. The baggage did not arrive till 6 p.m.; on this occasion we had to throw up sangars around the camp.

"I am now going to try and tell you about the first fight we have had. Reveille at 5 a.m. on the 26th, we were shortly on the path, marching towards Gyangtse sixteen miles off; on this occasion I regret to say the Royal Fusiliers formed the rear or third column; we pushed along the first eight miles easily enough, then someone commenced shooting in front; a message was signalled back to 'close up,' which I thought was a good start, we closed up for another mile and found, after a good deal of peering through glasses, that a long hill in front, behind which was a large monastery and two large villages, was held by the Thibetans. The 7th Battery commenced by shelling a small fort on the hill—range about 1,700 yards—hitting it fairly, twice; the 40th Pathans on the left and the Pioneers on the right moved towards the position in a semi-circle. Rather a smart manœuvre at this junction, was, that a portion of the Gyangtse garrison appeared on the skyline of another high range of hills at the back of the villages and monastery. The Thibetans were absolutely surrounded. I got as far to the front as I could but never loosed off, as most of the fighting was house to house and done by the 40th Pathans and Pioneers; although the Gyangtse garrison did a lot of shooting at Thibetans trying to get away up the big valley on the left. The Royal Fusiliers never fired a shot. The 40th Pathans enjoyed themselves immensely and slew about 150, the Pioneers accounted for another 100. Our casualties were two Pathans, two Pioneers and ten wounded, including Major Lye of the 23rd Pioneers.

"The Gyangste Garrison also accounted for a good many Thibetans; the fight was over about 5 p.m., the Thibetans fought with great pluck, a lot of them were left hiding in cellars, it being impossible to get at them. We got into camp that night about a mile from the Mission and about two miles from the Jong, after a fairly stiff day, 17 mile march wading across rivers, etc. However, the next day was a quiet one and I amused myself looking up the Mission Camp. They have built very strong walls, traverses, etc., and are fairly comfortable, this appears a very necessary precaution as the Jong was keeping up a merry fire from Jingalls and I counted five or six balls striking fairly near; these balls weigh three pounds to half a pound, and are made of lead or copper.

" We had heard a rumour that the General intended to capture the fort within five days and on the 28th we commenced operations by moving down the left bank of the Nyang River to clear villages and a large town and monastery. The Royal Fusiliers were in front and advanced through the villages on the left bank very slowly, which were found to be deserted, they got as far as a large monastery where they stopped, but managed to secure some cattle, sheep, chickens, etc. I happened to be on the right flank and was more or less on my own, and shoved along until a couple of Jingalls fired at us from the town. I started by managing to upset the gun mule in a river and nearly drowned him—we had to cross several pretty deep rivers and canals. Sergeant Lyle did some very pretty shooting at 1,100 yds. and I pushed on to about six hundred yds. from the nearest houses; the fight that followed was the prettiest you can possibly imagine. The town was built on the side of a very steep hill and appeared to be full of Thibetans and on top of the hill was a line of sangars and big stone forts; when the Fusiliers were ordered to stop, the Ghurkhas and Pathans pushed up, the Ghurkhas had a good start and climbed up the left edge of the hill (the batteries shelling the sangars), until they got to a large round fort on the very top of the hill, into which they could not climb and the Thibetans inside were shelling them with rocks (this part of the fight was alone worth coming to Thibet to see). However, the Ghurkhas stuck to it and eventually we saw them climb over the edge and bayonet all the Thibetans inside; meanwhile the 40th Pathans had gone straight for the town absolutely regardless of the enemy's fire and had some pretty stiff fighting there too. I turned the Maxim on whenever I saw anybody move and expect we hit one or two (I actually only saw one man fall as Sergeant Lyle got him in the open).

" I hear the Thibetans lost over 400 killed, while our casualties were, Captain Crashter, 46th Pioneers, killed, and two Officers and several natives wounded. We had a long day from 4.30 a.m. to 11 p.m. To-day, the 29th instant, we are sending out parties to collect forage, grain, etc. I have sent out a couple of my mules and am busy refilling belts, etc. All here are very fit.

"(Signed),

" J. C. BOWEN-COLTHURST."

Lieutenant Bowen-Colthurst and all his men rejoined the battalion in capital fettle on the 3rd of November.

Lieutenant-Colonel Swaine went on half-pay on the 28th of October, and was succeeded by Lieutenant-Colonel Brown.

Being ordered to proceed from Fyzabad to Meerut, three companies were sent to Meerut by rail on the 11th of November, as an advanced party, whilst the remainder of the battalion left Fyzabad on the 15th of December, and proceeded on their march of 351 miles to their new station.

1905

At Kumrala two companies ("E" and "F") branched off and marched to Delhi, to be stationed there, arriving at Delhi on the 20th of January; whilst the headquarters, with three companies ("A," "B" and "C") joined the advanced party at Meerut on the 18th of January. Their repose was short, however, for, on the 26th of January, the whole battalion, leaving both stations, took part in the Meerut manœuvres, returning to their respective garrisons on the 13th of February.

One captain (Macnamara) was detached at this time to take charge of a company in the Chinese Regiment in the British Service.

On the 28th of March two companies were sent to Chakrata and one to Landour, but all detachments, including the one from Delhi, were withdrawn by the 7th of November. On the 30th of that month the battalion proceeded to the Rawal Pindi manœuvres. During these manœuvres a review was held in honour of Their Royal Highnesses The Prince and Princess of Wales. The battalion took seventeen officers on this duty, and 637 of all other ranks.

The various detachments were resumed at the close of the manœuvres.

1907

The year 1906 passed peacefully away, and on the 5th of January, 1907, the battalion proceeded by rail to Agra, to take part in a review, in honour of the Amir of Afghanistan. This ceremony over, it returned to its old station on the 19th of January. The draft from the home battalion this year arrived on the 1st of April, and consisted of 116 of all ranks.

1908

On the 15th of April, 1908, the Commander-in-Chief inspected the battalion, and expressed himself as pleased with all he had seen. Brevet-Colonel Brown was placed on retired pay on the 28th of October, and was succeeded by Lieutenant-Colonel O'Leary. This officer took the Royal Irish Rifles from Meerut, on the 12th and 13th of February, 1909, to Maymyo, in Upper Burma, which place was reached on the 20th of February. It remained here until the 23rd of December, 1911, when it moved to Kamptee, arriving there on the 17th of January, 1912. During the summer of 1911, two companies had been stationed in Mandalay.

1909

The battalion lost some of their Officers' Mess plate in December, 1911. It had been sent to England to be repaired and was returning in the P. and O. s.s. "Delhi," which was wrecked in a storm off Cape Spartel, in Morocco. It will be remembered that the Princess Royal of England and her two daughters were nearly drowned on this occasion.

The undermentioned received Coronation Durbar medals, 1911:—

1912 Lieutenant Hutcheson, Lieutenant and Quartermaster Foster, and No. 3531, Serjeant-Major Clark.

On the 28th of October Lieutenant-Colonel O'Leary was placed on half-pay, and was succeeded in the command by Lieutenant-Colonel Laurie.

Officer's Breast Plate, 1832—1855.
Dead gilt plate, bright bevelled edges,
bright gilt mounts.

CHAPTER XV.

EIGHTY-SIXTH REGIMENT (1859-1899).

Eighty-Sixth ordered to England—300 men volunteered to remain in India—Regiment quartered at Portsmouth—Moved to Aldershot—Ordered to Newry—Title recommended to be changed to Royal Irish Light Infantry—Establishment altered—Moved to Dublin—Sent to Gibraltar—Companies "lettered" instead of numbered—Suffered from cholera—Received new colours—Sent to Mauritius—Detained at Port Elizabeth—Twopence per diem increase of pay granted—Journey to Mauritius resumed—Fines for drunkenness introduced—Volunteers from other regiments joined the 86th—Establishment altered—Returned to South Africa—Rank of Ensign abolished, Sub-Lieutenant substituted—86th linked with 83rd Regiment—Forage cap badge altered—Embarked for England—Accident to H.M.S. "Himalaya"—List of officers who returned to England—Another accident to the "Himalaya"—Journey continued in "Tamar"—Sent to Fermoy—First draft to foreign Battalion—Lieut.-Colonel Jerome and Major Adams—Regiment sent to Aldershot—Reserves called out on account of threatened Russian War—Bandmaster Brown—Regiment at Chatham and Dover—Sent to Bermuda—List of officers embarked—Title changed to 2nd Battalion Royal Irish Rifles—Colonel Adams's order on this occasion—Uniform changed to Rifle Uniform—Letter from H.R.H. Princess Louise—Moved to Halifax—General Bradford appointed Honorary Colonel—Battalion embarked for Gibraltar, list of officers embarked—Rank of 2nd-Lieutenant instituted—Band contest at Malaga—Battalion sent to Egypt—Mustapha Pasha Palace—Detachment sent to Cairo—Combat of Gamaizah—Toski Campaign—500 men sent to Assouan—Major Turnbull's death—Detachment returned to Cairo—Battalion quartered at Cairo—Review at Cairo for the Prince of Wales—Battalion sent to Malta and Gozo—Memorial Brass at Downpatrick—Centenary celebrated—General Knowles's farewell speech to the Battalion—Battalion ordered to India—Stationed at Bombay—Death of Lieut. Peck—Battalion assists at checking the Bubonic Plague—Returned to Ireland—Stationed at Belfast—Ordered to mobilize for service in South Africa.

1859 ON the 2nd of February orders were received for the Eighty-Sixth, then stationed at Gwalior, to proceed to Bombay, en route to England. It had remained at Gwalior ever since that stronghold was stormed, on the 19th of June, 1858.

When it reached Mhow, on its way to Bombay, further orders were received that any non-commissioned officers or men who wished, could volunteer for any of Her Majesty's regiments who were remaining in India, and 300 men volunteered to the following regiments:—

7th Regiment		9 privates.
18th ,,		15 ,,
28th ,,	1 serjeant	13 ,,
31st ,,		19 ,,
33rd ,,		7 ,,
51st ,,		12 ,,
54th ,,		1 ,,
56th ,,		23 ,,
57th ,,		25 ,,
72nd ,,		2 ,,
82nd ,,		1 ,,
83rd ,,	1 serjeant	146 ,,
87th ,,		16 ,,
88th ,,		1 ,,
89th ,,		9 ,,
95th ,,		1 ,,
Total		300

The Regiment arrived in Bombay on the 22nd of March, and embarked in three detachments on board the "Genghis Khan," "Kennington," and "Northumberland" on the 18th and 22nd of April and on the 2nd of May respectively.

The first two ships arrived at Gravesend on the 16th of August, and the troops were immediately transhipped on board the "Himalaya," and taken round to Portsmouth, reaching that port on the 20th of August.

The "Northumberland" duly arrived at Gravesend on the 1st of September, and the detachment on board of this ship was disembarked and sent by rail to Portsmouth, arriving there on the 2nd of September. One serjeant and eleven men died during the four months' voyage home.

On disembarkation the regimental state shews: 33 officers and 722 non-commissioned officers and men, being 9 officers and 355 non-commissioned officers and men short of its establishment.

1860 On the 28th of April of this year the Regiment was moved to Aldershot, and became part of the 3rd Brigade, commanded by Brigadier-General Russell, C.B.

It only remained here, however, during the drill season, and on the 19th of September proceeded by train to Portsmouth and embarked for Warren Point on the "Foyle" and "Zephyr," both being steamships, and reached Newry, the Regiment's new station, on the 22nd of September. One company was detached to Enniskillen, and travelled there by rail.

On the 17th of December 300 non-commissioned officers and men, under

Lieutenant-Colonel Stuart, proceeded to Londonderry, for the purpose of keeping the peace in that town on the anniversary of the closing of its gates at the ever-memorable siege. The remainder of the Regiment was in readiness to follow its detachment, but the day passed off without any disturbance.

General Sir Hugh Rose, G.C.B., who had commanded in the Central Indian Campaign, had recommended that the Eighty-Sixth should be known as the "Royal Irish Light Infantry," to commemorate its gallantry in that campaign. After due consideration, this request was refused by H.R.H. The Duke of Cambridge, Commander-in-Chief, for the very sensible reason that endless confusion would be caused between the actual title of the 18th Foot, and the proposed new title of the 86th Regiment. This letter was dated the 14th of August, 1860, and was sent to India, and reached the Royal County Down Regiment at Newry.

1861 On the 18th of May the Regiment moved to the Curragh for the drill season, and on the 1st and 5th of October arrived in two detachments at Dublin, and was accommodated at Ship Street and Royal Barracks. It moved by road on the 30th of September to

1862 Kilkenny and Waterford, having spent most of the summer at the Curragh Camp.

1863 In April the following order changing the establishment was received:—

"General No. 248.

"1st April, 1863.

"His Royal Highness the Field-Marshal Commander-in-Chief has been pleased to direct that all Regiments of Infantry of the Line at Home and in the Colonies (with the exception of those in New Zealand, Australia, China and Ceylon, and the Battalions under orders for India) shall be divided as follows:—

Distribution.	Field Officers.	Captains.	Lieutenants.	Ensigns.	Staff.	Serjeants (exclusive of Schoolmasters).	Drummers.	Corporals.	Privates.
10 Service Companies	3	10	11	9	5	48	21	40	640
2 Depôt Companies	—	2	3	1	—	10	4	10	110
Total	3	12	14	10	5	58	25	50	750

This was followed by a letter dated the 4th of June, shewing how this establishment was to be made up, viz.:—

Colonel	1	Serjeant Instructor of Musketry	1
Lieutenant-Colonel	1	Schoolmaster (to be appointed by Secretary of State) ...	1
Majors	2	Hospital Serjeant	1
Captains	12	Orderly-Room Clerk	1
Lieutenants	14	Colour-Serjeants	12
Ensigns	10	Serjeants	38
Paymaster	1	Drum-Major	1
Adjutant	1	Trained Bandmaster	1
Quartermaster	1	Drummers and Fifers	24
Surgeon	1	Corporals	50
Assistant-Surgeon	1	Privates	750
Serjeant-Major	1		
Quartermaster-Serjeant ...	1		
Paymaster-Serjeant	1		
Armourer-Serjeant	1	Total	929

In May the Regiment moved to the Curragh again for the drill season, and after four months there proceeded, on the 27th of August, to Dublin, being stationed first at Richmond Barracks and then at Beggars' Bush Barracks.

1864 At the end of May the 86th marched to the Curragh, and from there went to Kingstown on the 18th of October, embarked on H.M.S. "Tamar," and sailed for Gibraltar, to relieve the 2nd Battalion of the 9th Foot.

A rough passage was experienced, and one of the crew was drowned, whilst three chargers died on the voyage, but the Regiment duly landed on the 26th of October. Then it was found that the 9th Regiment would not be ready to embark before the 3rd of November, so the 86th Regiment was put up in all the barracks in the fortress by the simple process of "doubling up" with the other regiments. After the 3rd of November it occupied Buena Vista Barracks, Windmill Hill Barracks, and Catalan Bay, having a strength of thirty officers and 738 non-commissioned officers and men.

1865 The present system of "lettering" companies was now introduced, instead of the old system of "numbering," by this circular memo:—

"Horse Guards,

"10th June, 1865.

"His Royal Highness the Field-Marshal Commanding-in-Chief is pleased to direct that, in place of the designation by numbers, the Companies of Infantry will be distinguished in each regiment by letters from 'A' to 'M' (excluding 'J') for all purposes of interior economy.

"On parade they will be numbered from right to left, the company on the right being actually number one."

From July to October cholera existed at Gibraltar, and the military there lost 98, the civilian population lost 424, and the convicts lost 57 by deaths from this disease.

There were at the time four regiments—15th, 22nd, 78th, and 86th—and the Royal Artillery quartered at the "Rock." The 86th only lost seven men, so, on the whole, did not suffer very severely.

Beyond the cholera and a change to the Casemate Barracks, nothing of any importance happened at Gibraltar. It is found noted in the regimental records the hard case of Ensign H. B. Rhodes, who proceeded to England on the 22nd of July, for the purpose of going through a course of instruction at the School of Musketry, Hythe. He arrived at that establishment late, and was not permitted to join. No details are given as to why he was late, whether through the slowness of his ship or from his own sloth. The Horse Guards cheerfully granted him leave until the 22nd of February, 1866, but on the representation of the Colonel of the 86th Regiment that he was suffering from a paucity of officers for duty with the Regiment, it equally cheerfully cancelled the leave, and he was promptly sent back to Gibraltar. Various drafts came and went.

1866

New colours were presented to the Regiment on the 13th of February at Gibraltar, by Lady Airey. Major John Jerome was in command of the Regiment at the time. These were the last colours carried by the Regiment. After it became a rifle regiment the colours were kept for some time in the officers' mess, and were finally deposited in the Cathedral at Downpatrick in 1894.

1867

Orders were received in January that the 86th Regiment was to proceed to the Cape of Good Hope, but their destination was changed to Mauritius on the 24th of February, and on the 18th of April it embarked for that island on H.M.S. "Himalaya," having a strength of twenty-seven officers and 712 non-commissioned officers and men. Leaving the "New Mole" about 10 a.m. on the 19th of April, the transport arrived in Simon's Bay, at Cape Town, on the 14th of May.

Here serious news was received; for the Colonel heard that fever was raging in the Mauritius, and, not wishing to risk his men unnecessarily, he consulted with the military authorities at the Cape as to what course he should pursue. Finally the Regiment sailed from Simon's Bay on the 23rd of May and disembarked at Port Elizabeth on the 25th of May. The 86th remained encamped on the hill above the town until the 21st of September, when it was moved into two large stores, which had been rented by the municipality for the use of the troops. This kindly action was much appreciated by all ranks, as during their four months in tents there had been a succession of rains and storms, which had proved very trying. In particular some companies carrying on their musketry at a place called the "Fisheries," three miles from camp, were greatly inconvenienced by the weather, as the road to the "Fisheries" was completely washed away.

On another occasion even the stores did not save the troops from discomfort, for a torrent rushed through one, which had been a wool store, during a heavy storm, and everyone had to leave the building and withdraw to a safer place.

In September a royal warrant was received granting an increase of pay of twopence per diem to all ranks of non-commissioned officers and men.

On the 19th of November Private Deverell was drowned. Privates Patrick Mahon and John Henehan made most gallant attempts to save him, and attention was called to their honourable conduct in regimental orders of this date.

The fever having somewhat abated in Mauritius, the Regiment now embarked for that island on the 17th of December on H.M.S. "Tamar." Its total strength was 27 officers and 712 non-commissioned officers and men, attended by 65 women and 139 children. The "Tamar" duly arrived at Mauritius on the 30th of December.

1868

Unfortunately, the fever had again commenced to increase in the Island after the Regiment had left Port Elizabeth, and though the troops had been disembarked at Port Louis and placed in barracks there, yet everyone was sent out on detachment as soon as possible—four companies to Cannonier Point, two companies to Point D'Esvay, one company to Grand River, the remainder going to Flacq. Considering the amount of fever, the number of deaths was very small, not exceeding eleven for the year.

On the 11th and 12th of March a very severe hurricane visited the Island and blew down all the tents and a great many of the buildings.

On the 20th of June the depôt companies at home in England were moved from Gosport to Parkhurst, in the Isle of Wight.

A very wise measure was carried out as follows:—All the sickly men of the Regiment were formed into two regular companies, and were sent, to the number of 100, to the Cape, to recuperate their health, on the 11th of August.

1869

Colonel Lowe, C.B., who was then commanding the Regiment, went home on leave this year. His leave was for eighteen months—a rather longer period than is now allowed to commanding officers.

In this year also the first scale of fines for drunkenness was brought out, a most satisfactory reform.

On the 27th of April, the establishment of the Regiment was reduced by two ensigns and raised by the addition of ninety privates. Difficulties were now found throughout the Army in raising sufficient recruits each year, and it was no doubt this paucity of recruits that accounted for the pay being raised by twopence a day in 1867, and also for the short service system, which was so soon to be introduced.

To complete the 86th Regiment, which, of course, suffered in proportion with all others in the lack of recruits, the following volunteers came from different regiments, and the regiments' names are given in full, to show what they were actually called previous to the linked battalion system being introduced:—

32 men from 33rd Duke of Wellington's Regiment.
10 ,, ,, 35th Royal Sussex Regiment.
9 ,, ,, 67th South Hampshire Regiment.
10 ,, ,, 46th South Devonshire Regiment.
7 ,, ,, 87th Royal Irish Fusiliers Regiment.
8 ,, ,, 31st Huntingdonshire Regiment.
5 ,, ,, 1/15th York East Riding Regiment.
3 ,, ,, 2/20th East Devonshire Regiment.
19 ,, ,, 47th Lancashire Regiment.
2 ,, ,, 64th 2nd Staffordshire Regiment.
8 ,, ,, 61st South Gloucestershire Regiment.
2 ,, ,, 1/24th 2nd Warwickshire Regiment.
6 ,, ,, 100th Prince of Wales's Royal Canadian Regiment.

1870

In March a new system of maintaining depôts was started. Instead of being joined with others, and thus formed into depôt battalions, it was directed that they should be attached to infantry regiments serving at home. In pursuance of this order, the depôt of the 86th Regiment was moved from Parkhurst and sent to Belfast, where it was attached to the 1st Battalion of the 18th Royal Irish Regiment, on the 26th of March. They did not, however, remain there long, but were sent to Birr on the 21st of June.

From the 1st of July all enlistments were made without bounty.

On the 11th of July a War Office letter was received, fixing the establishment of the Regiment as follows:—

Service Companies.	Depôt.		Total.
8	2	Number of Companies	10
—	1	Colonel	1
1	—	Lieutenant-Colonel	1
2	—	Majors	2
8	2	Captains	10
10	2	Lieutenants	12
6	—	Ensigns	6
1	—	Paymaster...	1
1	—	Adjutant	1
1	—	Quartermaster	1
1	—	Serjeant-Major	1
1	—	Trained Bandmaster	1
1	—	Quartermaster-Serjeant	1
1	—	Paymaster-Serjeant	1
1	—	Armourer-Serjeant (unless supplied by Corps of Armourers)..	1
1	—	Serjeant-Instructor of Musketry	1
1	—	Hospital Serjeant (unless supplied by the Army Hospital Corps)	1
1	—	Serjeant-Cook	1
1	—	Orderly-room Clerk	1
8	2	Colour-Serjeants	10
32	6	Serjeants	38
1	—	Drum-Major	1
16	4	Drummers and Fifers	20
40	8	Corporals	48
780	92	Privates	872
916	117	Total	1033

Attached—1 Surgeon, 2 Assistant-Surgeons.

1871

On the 13th of July the Regiment left Mauritius, having a strength of twenty officers and 445 non-commissioned officers and men. It again embarked on its old transport, H.M.S. "Himalaya." On the way to the Cape, where they were to disembark the 86th, the vessel had to call at East London, to embark other troops. The weather, however, was so bad that for some days no communication could be effected with the shore and the ship daily ran in towards the shore, and every night, for safety, had to stand out again until the weather moderated somewhat, and surf boats were enabled to come out. On the 31st of July the transport arrived in

Table Bay, and the troops were disembarked on the 1st of August. Here they met the two invalid companies which had been sent away in 1868, and which had since been augmented by drafts amounting to upwards of 300 men from home.

1872 The depôt companies had moved from Birr to the Curragh with the 18th Royal Irish Regiment, but in January they were attached to the 2nd Battalion Royal Fusiliers. On the 13th of January the Royal Warrant was received abolishing the time-honoured rank of ensign.

E. G. Selby Smyth appears to have been the last ensign appointed to the Regiment, having been transferred on the 7th of May, 1870, from the 82nd Foot; whilst on the 19th of March C. Haggard was appointed as sub-lieutenant, from the Royal Military College.

1873 On the 26th of March the Regiment was re-armed with the Martini-Henry rifle.

Following up the various changes in organization which were now taking place in the British Army, is one which has had a great effect on the 86th Regiment. This order is dated 1st of March (G. Order 18), and states that the 86th Regiment is linked with the 83rd or Dublin Regiment and designated the 2nd Battalion 63rd Brigade Depôt. Recruiting for both battalions was to be carried on in the Belfast district. The brigade depôt, when formed, was directed to be stationed at Antrim.

1874 Three companies, consisting of seven officers and 186 non-commissioned officers and men embarked for Natal on the 20th of November, on H.M.S. "Rattlesnake." All three companies, however, returned—two on the 14th of January, and the remaining one on the 29th of January.

1875 On the 1st of April the establishment of the Regiment was reduced to a total of thirty officers and 659 of all other ranks, and on the 1st of April (1875) it was further reduced to thirty officers and 587 of all other ranks.

The forage cap badge of the officers and the tunic collar badge of the non-commissioned officers and men now consisted of a harp and crown, with the motto "*Quis Separabit,*" being granted by two separate orders, dated 30th of June, 1874, and 9th of November, 1874.

On the 13th of January six companies, under the command of Lieutenant-Colonel J. Jerome, embarked on board H.M. troopship "Simoon" for conveyance to England.

The "Himalaya" was to bring the remaining two companies and the 75th Regiment home. Unfortunately, the "Himalaya" broke down, and the six companies were put off the "Simoon," and remained in Cape Town until the "Himalaya" was repaired, when all the Regiment embarked on her and duly set sail for England on the 1st of February.

Previous to leaving Cape Town, volunteers for other regiments stationed there had been called for, and fifty-two non-commissioned officers and men transferred to the 1st Battalion 13th Regiment, five to the 32nd Regiment, and five to the 24th Regiment, which latter relieved the 86th Regiment at Cape Town; and was afterwards cut to pieces at Isandula by the Zulus.

A total of nineteen officers and 619 of all other ranks came home in the troopship, bringing sixty-seven women and 144 children with them. The names of the officers returning to England were as follow:—

Lieutenant-Colonel Jerome.
Captain W. Gray.
Brevet-Major J. Brockman.
Captain Crofton.
,, Chatfield.
,, Brooke.
Lieutenant J. Boulcott.
,, W. Brockman.
Lieutenant H. Stuart.
,, J. Stewart.

Lieutenant and Adjutant J. Spence.
,, F. Graham.
,, R. Knox.
,, C. Loraine.
,, Selby Smyth.
,, C. Haggard.
Paymaster H. F. Luke.
Quartermaster G. Morgan.
Surgeon-Major W. Leach.

Having reached Ascension on the 12th of February, and St. Vincent on the 24th of that month, the voyage was now broken by the discovery of a crack in one of the cranks in the machinery.

The captain of the ship telegraphed to the Admiralty for orders, and was told to make for Madeira, where the Channel Squadron happened to be. On arrival it was found impossible to mend the shaft at the time, so the "Himalaya," escorted by H.M.S. "Sultan," steamed away to Gibraltar. There they were to be transhipped to H.M.S. "Tamar," which was to be sent out from England to bring them home. The "Himalaya" arrived on the 9th of March, and the "Tamar" on the 18th of the same month, and left for Ireland on the 19th of March, and arrived at Queenstown on the 25th of March. Here the Regiment landed next day, and proceeded at once to Fermoy, and was quartered in the New Barracks, changing to the Old Barracks on the 20th of April.

Here they found their depôt companies and those of the 83rd Regiment all being attached to the 33rd Regiment.

On leaving the Cape of Good Hope a special order was published by Lieutenant-General Sir Arthur Cunnyngham, K.C.B., calling attention to the exceptionally good behaviour of the 86th, and on arrival in Fermoy it was inspected by General Lord Sandhurst, G.C.B., Commander-in-Chief in Ireland, who called special attention to the splendid appearance of the men on their return from foreign service. It is noted that on the 29th of June the Regiment was re-armed with the new pattern Martini-Henry rifle, and on the following day received the new valise equipment.

In October eighty-six men volunteered to join the 34th Regiment to make up its strength for foreign service, and in the same month eighteen more men volunteered to join the 35th Regiment for the same reason. This shows the great shortage of recruits at the time.

On the 20th of November the first entry of the since regular drafts from the home battalion to the foreign battalion is noted. It is as follows:— " Twenty-one men (short service transferred to the foreign battalion (83rd)." No one could foresee that this modest draft of twenty-one men would be succeeded by others, sometimes exceeding 200 men a year.

1876

On the 9th of May the Regiment proceeded from Fermoy to the Curragh, and in October a further draft of thirty-five men proceeded to India, to join the 83rd Regiment.

1877

In January Lieut.-Colonel Jerome is noted in the regimental records as having proceeded on leave of absence, and again after five years' service as Lieutenant-Colonel (completed on 6th of March) as having been promoted Colonel, whilst on the 14th of February it is also entered that Major Adams returned from leave, and, in the absence of the Colonel, assumed command. The two brothers Jerome have been already spoken of as gallant soldiers in the Mutiny Campaign, and Colonel J. Jerome shared with Major Adams the reputation of being one of the strictest martinets of latter years. Therefore, it is not surprising to find that when the Regiment moved from the Curragh to Aldershot—on the well-known troopship " Assistance " as far as Portsmouth—it is noted that not a man was drunk or absent.

Whilst the Regiment was at Aldershot, the Russo-Turkish War was proceeding, and the English Government began to see that they would probably be involved in this dispute, and measures were taken to strengthen the Army. In April, 1875, the total establishment of the Regiment had been only 617; on the 1st of October, 1877, it was hurriedly raised to 903 of all ranks, and again on the 1st of April, 1878, to 1,096.

1878

For the first time in the records, regimental transport is assigned to the Regiment. It arrived on the 16th of March, and consisted of eight general service waggons, three ammunition carts, one tip cart, and twenty-seven horses.

The depôt companies of the 83rd and 86th Regiments had all come over to Aldershot with the Regiment; but now, on the 11th of April, they were sent to Belfast, and formed the 63rd Brigade Depôt there. On the 28th of April the Army Reserve having been called out, strong bodies of reservists were sent over to join their old Regiment. Their departure from Belfast was most cheering. Crowds enthusiastically shouted their good wishes, and the men themselves were so anxious to come that those who had been rejected for minor health defects, tried to conceal themselves in the railway carriages, or to hide themselves in the ranks, hoping to be overlooked in the general confusion of departure.

EIGHTY-SIXTH REGIMENT (1859-1899).

The Eighty-Sixth now stood on parade 29 officers, 1,263 non-commissioned officers and men. Thirty recruits arrived from the new depôt on the 5th of June, and the Regiment was now in the most magnificent order.

In the meantime the Treaty of Berlin had been signed and Lord Beaconsfield returned in triumph from the field of his diplomatic labours. The order went forth on the 31st of July disembodying the Army Reserve, and by the 1st of September the spirit of economy was again to the fore, and the establishment was reduced to twenty-five officers and 658 of all other ranks.

1879

On the 4th of January the Regiment moved to Chatham, and both this year and next (1880) they occupied the second place in Army Signalling, being beaten by the Royal Engineers only. Band-Serjeant Brown, of the Regiment, was specially made the bandmaster here, as a reward for his excellent character and good service, greatly to the satisfaction of all.

1880

On the 30th of January the Eighty-Sixth moved to Dover, being partly stationed at the South Front and Citadel.

Leaving Dover on the 30th of October, the Regiment, numbering 22 officers and 693 men, embarked on H.M. Indian troopship for conveyance to Bermuda.

A special report was made to His Royal Highness The Duke of Cambridge, Commander-in-Chief, of the extreme orderliness of the departure of this Regiment from Dover.

The following officers sailed in the " Crocodile " for this colonial tour:—

Lieutenant-Colonel Adams.
Major Mackenzie.
,, Gray.
Captain Crofton Chatfield.
,, Sir Guy Clarke Travers, [Bart.
,, Stuart.
,, Graham.
,, Jackson.
Lieutenant Selby Smyth.
,, Seton.
Lieutenant Gaussen.
,, Swaine.
,, Stewart.
,, Scott.
,, Plummer.
,, Allen.
Captain and Paymaster Luke.
Lieutenant and Adjutant Haggard.
Quartermaster Timothy Cleary.

Having left England on the 31st of October, the " Crocodile " duly anchored off Ireland Island, Bermuda, on the 15th of November, relieving the 1st Battalion of the 19th Regiment there.

1881

On the 30th of June the following Regimental Order was published:

" Prospect, Bermuda,

" 30th June, 1881.

" By General Order 41, of 1881, the 86th Royal County Down Regiment will, from to-morrow (the 1st of July, 1881), be designated as The 2nd

Battalion The Royal Irish Rifles. This change will inaugurate a fourth period in the life of the Regiment. During its first period—from 1793, when it was raised, to 1806—it was known as 'The Shropshire Volunteers,' from 1806 to 1812, as 'The Leinster Regiment' during its third period—from 1812 to this date: that is for seventy years—it has borne its present title.

" The historical records prove that if the fortune of war has not called it to share in those victories and campaigns which have been most bruited abroad, numerous important occasions have offered on which to show not only its endurance and fortitude, but also its valour before the enemy, and on these it has never failed.

" Few Regiments can boast of more varied or more arduous active service than this one experienced from 1796 to 1819, when, during twenty-three years and four months of continuous foreign service it was constantly before the enemy in Egypt, India, Bourbon, and Ceylon, suffering much hardship, in every instance gaining distinction and being repeatedly thanked in General Orders for its gallant conduct.

" In more recent times, during the Indian Mutiny, the siege of Jhansi, the relief of Gwalior, and other operations in Central India, proved that the Regiment was still as ready and as able as of old to do and suffer.

" Though we cannot but feel regret at parting with a designation that has been thus signalized, yet it must be a subject of congratulation to all that the change of title in our case is to one that will stamp us more than ever as a National and Special Corps, and give us the power of earning the same reputation for soldierlike excellence that the existing Rifle Brigade has long held, for it must be remembered that a name is only what our conduct makes it. The Royal County Down has been an honoured and honourable title, because its members have shewn in the field, as in quarters, courage, obedience and endurance, the fruits of sound discipline—rare virtues—which must command respect, and unless we intend that the new name shall become a badge of ridicule rather than of honour, and choose to illustrate in ourselves the old fable of the ass in the lion's skin, the Regiment must resolve, by its recognition of the paramount importance of discipline (the foundation of all military excellence), by its good shooting, by the rapidity of its movements, by its intelligence and resource on outpost duties, by its readiness to face hardships, and, when occasion demands, danger in the field, to justify its claim to the proud distinction that is implied in the title 'The Royal Irish Rifles.'

" As a final testimony to the good behaviour of the old County Down Regiment, and to the estimation in which it was held to the last, the following remarks recently received from Horse Guards on the confidential reports, made on occasion of the last inspection of the Regiment, are published:—

EIGHTY-SIXTH REGIMENT (1859-1899).

"' Horse Guards,

"'8th April, 1881.

"' This Regiment appears to the Field-Marshal Commanding-in-Chief to be in admirable order in every respect, reflecting the greatest credit on its Commanding Officer, Colonel Adams.

"' (Signed) R. B. HAWLEY, Major-General,

"'D.A.G.'

" The Commanding Officer is striving to be as concise as possible, and not from forgetfulness has omitted allusions to many instances of good conduct, evinced during his own command, but he cannot conclude without a brief expression to all ranks of his grateful sense of their willing spirit of co-operation, which has so largely contributed to win the favourable report of every General Officer under whom the Regiment has served, and so reflected credit on himself.

" From the past, the future may be safely augured, for the same spirit will continue to animate the Regiment, which he confidently trusts will yet shine with even greater effulgence as ' The Royal Irish Rifles.' "

The above needs no explanation, for it is the worthy order of a very worthy martinet, on a worthy occasion.

Twenty years before, General Sir Hugh Rose (later Lord Strathnairn) had asked that the Regiment might be created light infantry, for its gallant conduct during the Indian Mutiny. The British War Office moves slowly. The request was declined with regret; but twenty years later the good work and high military efficiency of the Regiment, not only in war, but in peace, was recognised, and honour was conferred with no frugal hand. The title of "Rifles" is only borne by four regiments in the British Service. One in the Peninsula war, Napier states, was composed of " men not accustomed to yield"; its brother-in-arms, originally the Royal Americans, can equal its well-won glories; very great was the honour, therefore, conferred on the 86th Regiment when it joined this illustrious brotherhood, and greatly was it appreciated by the Regiment.

1882

It was some time before the change from red uniforms to green ones was fully carried out. The officers finally changed their garb on the 1st of June, whilst the clothing and equipment of the other ranks did not arrive from England until the 21st of July, and Sunday, the 6th of August, saw the whole battalion parading in its green uniform, which it has now worn for over thirty years.

1883

In the winter of 1883 Her Royal Highness The Princess Louise, whose husband, the Marquis of Lorne, was at that time Governor-General of Canada, stayed at Bermuda, and wrote to the Colonel of the Regiment, pointing out how struck she was with the exceptional smartness of the Regiment on all guards, special notice being drawn to the fact that Her Royal Highness was used to these duties being performed by Her Majesty's Foot Guards, who do not labour under the reproach of want of smartness on such occasions.

On the 2nd of November the 2nd Battalion Royal Irish Rifles left Bermuda in the "Himalaya" for Halifax, Nova Scotia, which was then the last station occupied by British infantry in Canada. At that time two infantry regiments composed the garrison of Halifax with artillery and engineers, but during the time the Regiment was stationed there the garrison was reduced by one infantry battalion; twenty-three officers and 731 of other ranks embarked at Bermuda, and arrived in Halifax on the 6th of November, just in time to enjoy a Canadian winter.

1884 By the end of this year two drafts had arrived, with a strength of five officers and 213 non-commissioned officers and men.

1886 On the 26th of March the Colonel of the Royal Irish Rifles (General The Right Honourable Sir J. Michel, G.C.B.), was promoted to the rank of Field-Marshal.

On the 23rd of May the colonelcy of the Regiment became vacant, by the death of Field-Marshal The Right Honourable Sir John Michel, G.C.B. Lieutenant-General W. H. Bradford was then appointed Colonel of the Royal Irish Rifles in the place of the deceased officer. In June, the battalion was ordered to proceed in the autumn to Egypt, but this destination was changed in September to Gibraltar.

On the 29th of October the battalion embarked on the "Orontes," and proceeded to Gibraltar. The total number embarked was twenty-three officers and 669 other ranks. The following is a list of those officers who embarked on the "Orontes":—

Lieut.-Colonel de Montmorency.
,, Chatfield.
Major Sir G. Clarke Travers, Bart.
,, R. J. Knox.
Captain and Adjutant Selby Smyth.
,, Swaine.
,, Stewart.
,, Rudyerd.
,, Allen.
Lieutenant Welman.
,, O'Leary.

Lieutenant Hallum.
,, Spencer.
,, Cliff.
,, Carstairs.
,, Morphy.
,, Harvey.
,, Lillingston Johnson.
,, Fox Strangways.
,, Rowley.
,, Laurie.
Quartermaster S. McClenahan.

Lieutenant Laurie was the last officer who joined the 2nd Battalion with that rank, for in accordance with a special Army Order, dated the 15th of January, 1887, the establishment of officers was changed from sixteen lieutenants to eight lieutenants, and eight 2nd-lieutenants; and subalterns have ever since joined in the latter rank.

Halifax had been a delightful station, and all were sorry to leave it, the men especially so.

Over 300 of them gave in their names as married without leave, and this was the cause of the loss by desertion of 120 men in the last year of the Regiment's stay in Canada.

On the 9th of November the "Orontes" anchored in Gibraltar Bay at 3 a.m. The voyage had been fairly rough, and on the Sunday afternoon before reaching Gibraltar one of the boilers had cracked, through shortness of water, and great clouds of steam had rushed up from below, driving the stokers out of the stokehold. The fire alarm sounded, and the battalion fell in most steadily on parade at fire quarters.

The troops had to remain on board the "Orontes" until the 13th of November, as those who were moving were not timed to leave Gibraltar until that date. During the whole five days spent on board, alongside the New Mole, the rain poured down in torrents. Coaling operations were also in process, and towards the end of the day a sort of coal dust slush had accumulated on the deck to the depth of three or four inches.

The North Front, including Catalan Bay, the Town Range, and Wellington Front, furnished the barracks for the battalion.

1887 In August the bands of the Royal West Kent Regiment, the King's Royal Rifles and the Royal Irish Rifles went to Malaga, to take part in a band contest. Malaga is a seaport on the east coast of Spain, some sixty miles distant from Gibraltar. H.M.S. "Hecla" conveyed the bands, and all were under the command of Captain Rudyerd, of the Royal Irish Rifles.

Various Spanish regimental bands also competed, including the Marines, the Granada Regiment, the Cuba Regiment, the Bourbon Regiment, etc., not forgetting the Fire Brigade Band, which also entered. Altogether, something like sixty bands competed.

The British bands went ashore about 2 o'clock and met their Spanish rivals, and all marched in procession to the Bull Ring. Here 12,000 people had assembled. Lots were drawn, and the Royal Irish Rifles drew the first place and duly mounted the platform. Each band played first the overture of Weber's opera "Oberon," and then one piece of their own selection. One performer of the Royal Irish Rifles was overcome with nervousness at his strange surroundings, and broke down. The competition lasted until eight o'clock in the evening. The judges awarded the Spanish Marines the first prize. Everyone present concurred in the award. After eight o'clock the Royal Irish Rifles played to a crowded and enthusiastic audience in the Alameda, which was brilliantly illuminated in honour of the occasion. The Royal Irish Rifles band was awarded a gold medal and a diploma, and returned well pleased with themselves two days later, having been fêted throughout by their most courteous Spanish hosts.

The whole battalion was very hard worked at Gibraltar, and the place was unfavourably compared with the last station of Halifax. It was, therefore, with little regret that the headquarters and six companies embarked on the 27th of December on H.M.S. "Euphrates" for Egypt. Arriving at Suez on the 8th of January, the troops went on by train to Ramleh, some six miles from Alexandria, on the coast towards Aboukir Bay, the scene of Nelson's victory of the Nile.

1888

They were quartered in the barracks of Mustapha Pasha, whilst the officers had the palace of that name as their mess.

The palace was very imposing from a distance, but closer inspection shewed much lathe and plaster mixed up with marble slabs, etc. This was accounted for in local tradition by the fact that when nearly completed, the palace had been burnt down by accident, and that the then Khedive, who was a man who knew his own mind, explained to the architect that his (the architect's) own long life depended on the palace reappearing complete in six weeks, and the architect saw this was done somehow.

The remaining two companies arrived from Gibraltar on the steamship "Quetta" on the 21st of January, under the command of Major Sir Guy Clarke Travers, Bart.

On the 2nd of February the headquarters and five companies marched to Ras-el-Tin Barracks, vacated by the 1st Battalion Yorkshire Regiment. These barracks were situated beside the Khedive's palace of that name, and beside the forts, which had been pounded to pieces by the British fleet at the bombardment of Alexandria in 1882. Some of the guns still lay where they had been dismounted by the shells of the fleet, and in teaching the men entrenching much care had to be taken not to disturb the bones of the unfortunate defenders, many of whom had perished at the bombardment and had been buried where they fell.

At the end of February one company joined headquarters; the other two went to a musketry camp at El Mex.

Here a serious complaint was made against the battalion, for Major-General the Hon. R. H. de Montmorency, commanding at Alexandria, received a note from an influential Greek gentleman, stating that his children had been seriously frightened by some soldiers of the Royal Irish Rifles, whom he could not particularly describe, but one was riding a "crimson" horse and the other "one spotted with darkness." On investigation it was found, however, that the Royal Irish Rifles mess tent at the musketry camp at El Mex had been burnt down, and the two culprits were members of the Garrison Board there assembled, being an Engineer officer and one of some other corps, who, after a long board, were hurrying home on a chestnut horse and on a grey horse with black spots. This was the only serious complaint made against the character of the Regiment in the last thirty years.

On the 13th of December ten officers and 450 non-commissioned officers and men were hurried to Cairo at an hour's notice, as the battalions of the Welch Regiment and the King's Own Scottish Borderers, then stationed there, had been hastily sent to Suakin, where Colonel Kitchener (now Field-Marshal Lord Kitchener) was in command, with Osman Digna, and many hardy dervishes, opposed to him outside the town. These ignorant people had dug trenches to approach the town with safety from rifle fire, and it had been found necessary to collect a force to drive them off. The Royal Irish Rifles contributed Lieutenant Fox Strangways, 2nd-Lieutenant Ryan, and thirty-seven non-commissioned officers and men to this force, as part of the British mounted infantry, and Captain McWhinnie and 2nd-Lieutenant Brown, as special service officers to one of the Soudanese battalions at Suakin.

The fighting consisted of a sortie from the town to drive the enemy from the trenches, which were afterwards filled in. The Dervishes left 400 dead in the trenches. 2nd-Lieutenant Brown was hit and knocked off his horse in this sortie, but, fortunately, the bullet struck on the butt of his revolver (borrowed from 2nd-Lieutenant Playfair), and, glancing off, went through his arm. The shooting of the Dervishes was found to be as indifferent as their courage was great, and it was ascertained that their native armourers had "improved" their rifles—most serviceable ones, captured at Baker's defeat at El Teb—by knocking off the backsights of the weapons. Our shooting was little better, for immediately after this the Royal Irish Rifles Mounted Infantry was charged by a body of Dervish horse about their own strength, which in fairly open country was fired upon from a range of 1,000 yards down to a distance of 200 yards, when they decided to retire, and their total loss was one horse wounded.

1889 The detachment at Cairo rejoined headquarters, at Alexandria, on the 7th of January.

There had been no other trouble throughout Egypt during this fighting, though, no doubt, the agents of the Khalifa were very busy at this time, and two cruisers were sent from Malta to Alexandria so soon as Major Wyndham's detachment left for Cairo, to strengthen the British forces.

In July, however, trouble was threatened up the Nile beyond Wady Halfa, where the Khalifa's lieutenant had been directed to move forward, for the purpose of invading Egypt. The lieutenant's name was Wad-el-N'jumi. He was a fine, courageous soldier, but, not unnaturally, of a most cruel disposition. No man who had any relations with N'jumi's forces dared to desert, for if he could not be produced himself, a relation did equally well, and the punishment was either having a hand or foot chopped off. Njumi's forces, when he received the order to advance, consisted of 8,000 fighting men and 8,000 camp followers. Marching round the forts at Wady Halfa, he again struck the Nile, and steadily advanced northwards. In vain the native troops moved out against him. An indecisive engagement at Arguin was the only result, and he still marched northward, despite having lost 500 killed and wounded to the seventy lost by the Egyptians, under Colonel Woodhouse.

On the evening of the 8th of July 501 non-commissioned officers and men of the Royal Irish Rifles left Alexandria by train, with the following officers:—

Lieutenant-Colonel Wyndham.
Captain Hallum.
,, McWhinnie.
Lieutenant and Acting-Adjutant Carstairs.
,, Laurie.
,, and Quartermaster McClenahan.
2nd-Lieutenant Festing.
,, Atkins.
,, Playfair.
,, Jameson.
,, Carter.

The detachment travelled by rail to Assiôut, and there embarked on three steamers and seven barges. The steamers were stern-wheelers, and the barges were attached to them, and troops and stores were piled on them to their full capacity.

In their anxiety to bring as many stores as possible, one company loaded up the landing stage, which was also a barge, and attached it to their steamer, but the mistake was discovered in time.

As a rule, these steamers travelled only by day, as the danger of striking sandbanks at night was very great. Having received orders to push on with all speed, Lieutenant-Colonel Wyndham directed the senior Ries, or pilot, to travel by night; on his objecting two men were posted beside him with loaded rifles, with orders to shoot him if he stopped the steamer unnecessarily. The night travelling then turned out to be as safe as the day, and excellent time was made to Assouan, which was reached on the 15th of July.

During one night's travel the third steamer, under command of Captain McWhinnie, signalled up by lamps that she was sinking fast. The only reply that could be sent was to do the best that they could, beaching their craft, if necessary. The bodies, or hulks of the steamers were only some three feet deep, and were flat-bottomed and only drew some one and a half feet of water. The men of the detachment on board of the sinking steamer were told to break up the decks where they were and to bail for their lives. All through the night this went on, and at dawn the steamer was far behind, but still in view, gamely struggling on.

At Assouan the troops were ordered to wait for the 1st Battalion of the

Welch Regiment, ordered up from Cairo to strengthen the force. The thermometer rose to 115° in the shade.

On the 20th of July the detachment was inspected by the Sirdar of the Egyptian Army, Colonel Grenfell, who promised to do everything in his power to delay the expected engagement until the Royal Irish Rifles joined him, as, indeed, he well might, for they would have been a tower of strength to his plucky black troops. The 11th Soudanese Battalion now marched in to Assouan, with Captain Macdonald (afterwards General Sir Hector Macdonald) in command, having marched from Kosseir to Luxor—almost the identical march made by part of the 86th Regiment in 1801, having taken five and a half days to perform this march.

This battalion went on up the Nile on the 21st of July, leaving the Royal Irish Rifles fretting with rage, but pinning their faith to the Sirdar's promise. The Welch Regiment, having had a slower voyage than the Royal Irish Rifles, arrived about the 25th of July, bringing with them Major Turnbull, Captain Raymond, and Lieutenants Fox Strangways, Weir, and Hall, who had come from leave or other duties to join the detachment of the Royal Irish Rifles.

Captain Raymond had left London so hurriedly to join his Regiment (when he saw in the evening papers that it had gone on service up the Nile) that he had two suits of mufti only with him, one being a shooting suit, and the other a frock coat suit, with tall silk hat, which he had been actually wearing at the time in London. Neither suit was very suitable for campaigning in Nubia at the height of the summer, and, as he was a very tall man, it was difficult to fit him out with khaki from amongst the men. At length Lieutenant Hall's servant, good naturedly, gave up one of his khaki suits to Captain Raymond, and he ceased to command "F" Company in his frock coat. It should be noted, too, that whilst the officers showed such keenness to get to the front, the men were no less anxious to come, and it was twenty-four hours after starting before it was found that one of the men had successfully broken out of hospital and concealed himself in the train so that he might come with his company, thus raising the total to 501, instead of 500 non-commissioned officers and men, as ordered.

However, at last everything was in order, and on the 2nd of August the Royal Irish Rifles were taken round the first cataract on the light railway, which then ran from Assouan to Shellal, and were placed in boats—dahabeeyahs of sorts, which proceeded up the river under sail when the wind was favourable, and by towing when it was adverse. All officers and men took their turns at the towing rope, and so the strange flotilla went south. Some companies had a start of others, but all were straining to be up with the native Egyptian and black troops at Toski before the Dervishes attacked them; but it was not to be.

N'jumi's scouts came back to tell him that white troops were rapidly coming up, and, good general that he was, he attacked the Sirdar and his native army whilst they were drilling on parade before breakfast, and the English officers had only time to serve out extra ammunition and move out to resist the attack. N'jumi fell, and 1,600 of his brave Dervishes died with him, the remainder, wounded and whole, fleeing into the desert away from the Nile, where many of them perished from thirst. The Anglo-Egyptian loss was one Englishman and sixteen Egyptians killed, and 131 British and Egyptians wounded.

N'jumi's scouts had fired on "G" Company's boats at sundown on the 2nd of August and then retired. It was only bravado, for there were but two of these gallant fanatics on the top of the hills looking down on the yellow river, and only two shots were fired, which struck the water near the boats, but it shewed the spirit of the invading army of 8,000 men—willing first to destroy the Egyptian Army, and then to fight the best army in the world.

On the 2nd of August a sad accident occurred. Major Turnbull was standing talking to 2nd-Lieutenant Festing at the stern of the boat about ten o'clock at night, and, not being used to these native craft, happened, without looking, to put his foot into the hole cut at the stern of the boat for the rudder-post. Naturally, he fell forward and overboard. Being a good swimmer, he was in no way disturbed, but called out twice that he was all right. The boat shortened sail and dropped back to him; a cloud passed over the face of the moon, obscuring her light for a moment, and when it had passed Major Turnbull had vanished.

On the 5th of August, Toski having been fought and won, the Royal Irish Rifles came sadly down stream again, the nearest boat having been within twenty miles of the engagement, and three non-commissioned officers and men of the Regiment having been actually present, employed by the Royal Army Medical Corps.

On the way back to Shellal a black dog belonging to the Regiment, which had been brought from Halifax, and was supposed to have a leaning towards the Labrador breed in his obscure pedigree, was found to be barking dismally. As nothing would quiet the animal, steps were taken to ascertain what he was barking at, and it was found to be the body of poor Major Turnbull, floating in the water. It was reverently lifted out and taken to Assouan, where it was interred. The Rev. Joseph Collins officiated at the funeral. He was better known to officers and men as "Tally Ho! Joe," the sporting curate, of the Rev. "Jack" Russell, of Devonshire fame, for whom he had always acted as whip to his pack of hounds.

One hundred men, under Major Stuart, with Lieutenant and Adjutant Harvey and Lieutenant Johnson, left Alexandria on the 26th of July, met the returning battalion twenty miles north of Assouan, and came back with it to Cairo, where all arrived on the 12th of August. The remainder of the battalion, with the married families, had moved up from Alexandria on the

26th of July, under Captain Swaine, and had installed themselves at Abbasiyeh.

Cairo was a pleasant change after Assouan, as the thermometer only went up to 105°. In the six weeks after the 8th of July the Regiment lost one officer and twenty-five men from accident or illness. There was much fever amongst both officers and men, as well as these twenty-six deaths.

The Royal Irish Rifles always thought they had been hardly treated in this campaign. Having made an exceedingly rapid journey up the Nile, they were detained at Assouan for over a fortnight, and were thus too late for the fight at Toski. The Sirdar (now Field-Marshal Lord Grenfell) knew of their anxiety to be present, and of the way in which they were prevented being there. Nevertheless, the Egyptian Medal and Khedive's Star were never granted to the unfortunate battalion, though they had been given in the previous autumn at Suakin. The men always hoped that this omission would be rectified through Lord Grenfell's influence, but this never happened.

The Prince of Wales (afterwards King Edward VII.) came to Cairo on the 1st of November, and at a review took command of all the troops—British and Egyptian. The battalion furnished a guard of honour to him.

Major-General the Hon. Sir J. Dormer mentioned the Toski Campaign to the Regiment on the Queen's birthday parade of the next year (24th of May, 1890), and condoled with them on their extremely unfortunate position in the matter, pointing out that he had brought their very good work to the notice of the Commander-in-Chief, and on the same parade he presented Lieutenant Ryan and thirty-one non-commissioned officers and men with Egyptian medals, some for Gamaizah and some for Toski; whilst Captain McWhinnie and 2nd-Lieutenant Brown received the 4th and 5th Class of the Order of the Medjidie respectively.

1890

The 1888 pattern equipment was issued to the battalion on the 11th of November, 1890.

1891

On the 22nd of February Major-General Walker presented the Khedive's Star to twenty-nine non-commissioned officers and men, who had been employed at Gamaizah and Toski.

The 2nd Battalion Royal Irish Rifles left Cairo on the 21st of March, and embarked on board a hired transport (the "City of Richmond") at Alexandria. Leaving Alexandria at 2 p.m. on the 22nd of March, the ship touched at Cyprus on the 23rd, at 9 a.m., and then proceeded to Malta, where it arrived on the 27th, and four companies, with headquarters, marched to Pembroke Camp, whilst four companies were sent to the neighbouring island of Gozo. In all, twenty-two officers and 822 of all other ranks came from Egypt.

The 2nd Battalion erected a brass memorial tablet to those non-commissioned officers and men who died in Egypt during their stay in that country.

This tablet was placed in the Cathedral at Downpatrick, and was unveiled by Lieutenant-Colonel Wyndham, who commanded the battalion.

The names as given on the tablet are:—

Major Turnbull, Lieutenants Hume and Wynch, Colour-Serjeants Cashen and Garagan, Serjeants Gavin, Connors, and Flack, Lance-Corporals Atkinson, McGuire, and Trainor, Bugler Cooley, Privates Kelly, Bruce, Savage, Coffey, Alexander, Bowers, Murphy, Madden, Gartlane, Johnstone, Hinds, Irvine, Cahill, McHugh, Trueman, Blundell, Mascord, Deary, Woods, Boot, Johnston, Williamson, Gales, and Egan.

The inscription is as follows:—

"This Tablet is erected by the Officers, Warrant Officers, Non-Commissioned Officers and Men, 2nd Royal Irish Rifles, to the Memory of their Comrades, who died whilst serving in Egypt during the years 1888, 1889, and 1890. 'Thy Will be Done.'"

1892 On the 7th of March, 1892, the battalion's quarters were changed to Isola Gate Barracks.

1893 On the 1st of November the centenary of the raising of the Regiment was celebrated by a general holiday and athletic sports, followed by an excellent mess dinner in the evening. This celebration took place at Fort Manoel, where the battalion had moved on the 3rd of October, 1892.

1894 On the 9th of October Major-General C. B. Knowles, C.B., commanding the Infantry Brigade at Malta, took leave of the battalion as it was proceeding shortly to India. He expressed his regret that this was the last time he should see it on parade, as the manner in which the guard duties were performed and the general appearance and turn-out of the men in the town by far excelled those of any other battalion in the command. The 2nd Battalion Royal Irish Rifles embarked on the hired transport "Victoria" on the 18th of November, and proceeded to Bombay, landing there on the 2nd of December. The strength embarked was twenty-six officers and 799 other ranks. The establishment was raised on the embarkation for India to twenty-nine officers and 1,003 other ranks.

The battalion was stationed at Bombay, with detachments at Ahmedabad, Deesa, and Deolali.

1895 It was now inspected by His Excellency the Commander-in-Chief of the Bombay Presidency (Lieutenant-General Nairne). His written report on the troops was:—"This is a very smart and well-drilled battalion, ably commanded, and well officered. Turn-out and drill unusually good."

1896 On the 18th of March the battalion moved to Poona, all detachments being called in. Its strength was then twenty-seven officers and 1,016 of other ranks.

Lieutenant Peck was killed whilst home on leave of absence on the 16th of March. He was cycling with his brother, an officer in the Navy, along the breakwater at St. Heliers, Jersey. His bicycle, catching in a rut, nearly went over the side of the breakwater, but the young officer managed to regain his balance, and quite coolly looked over his shoulder at his brother, who was cycling behind him, remarking: "That was a near thing." At the same moment his bicycle again skidded and threw him on to the rocks below, where he was instantly killed.

1897 A very serious outbreak of the bubonic plague took place in India at this time, causing great mortality, especially amongst the native population of the large towns. It was decided to call for volunteers from British regiments to try and stamp out the disease; 200 were called for from the 2nd Battalion, and 500 men immediately volunteered for this dangerous and unpleasant duty. Finally, 135 non-commissioned officers and men were selected, with five officers. They remained on duty from the 9th of March until the 20th of May, and were thanked for their good work in General Orders by the Commander-in-Chief of the Bombay Presidency.

1899 On the 26th of January the battalion embarked on board the hired transport "Dilwara," and proceeded to England, arriving at Southampton on the 15th of February, and proceeded from that port to Belfast, where they arrived on the following day, and were stationed in Victoria Barracks.

A large draft, consisting of 377 non-commissioned officers and men, had been sent to Dum Dum, near Calcutta, on the 16th of January, to await the arrival of the 1st Battalion from Ladysmith, in South Africa, so the numbers brought back from India in the "Dilwara" were not very large, consisting of fourteen officers and 428 of other ranks.

Troubles now began to loom up in South African politics, and on the 9th of October the 2nd Battalion Royal Irish Rifles was directed to mobilize for service in South Africa.

The menu of the Centenary Dinner of the 2nd Battalion Royal Irish Rifles is attached, as it may be of interest to those in the Regiment in 1993 :—

FORT MANOEL,

1793—1st November—1893.

Hors d'Oeuvre.

ANCHOIS AU CAIRE.

Potages.

TORTUE CLAIR. PUREE D'ASPERGES.

Poisson.

SAUMON DE LA MEDITERRANEE.
GRANDE MAITREE FARCES.

Entrees.

SUPREME DE VOLEILLES A LA ROYALE.
COTELETTES DE VEAU A LA RUSSE.

Rotis.

SELLE DE MOUTON (GAULOIS).
FRICANDEAU DE BOEUF.

Relevee.

FOIE, GRAS EN ASPIC.

Entremets.

GELEE DES FRUITS.
POUDING FRANCAIS A LA BOURBON.
CANAPES AU PRINCE DE GALLES.
POUDING GLACE.

GWELL ANGAU NA CHYWILYDD.

QUIS SEPARABIT.

N.B.—The above menu was printed in the Regimental Printing Shop, which no doubt accounts for the omission of all accents.

The band programme of that night has, unfortunately, been lost.

CHAPTER XVI.

2ND BATTALION ROYAL IRISH RIFLES (1899-1912).

Original two conquests of South Africa—Boomplatz—Great Britain declared Transvaal and Free State independent—Annexed the Transvaal—First Boer War—Events which preceded second Boer War—Reserves called out—List of officers—Voyage to South Africa—Events in South Africa during the voyage—Battalion sent to East London—Position of Boers at Stormberg—General Gatacre's plans for the attack on Stormberg—The advance on Stormberg—Grobler's Commando—The combat at Stormberg—Retreat from Stormberg—Battalion returned to Sterkstroom—Mounted Infantry formed from the Battalion—Sketch of the whole campaign to occupation of Bloemfontein—Battalion moved to Springfontein and Smithfield—Detachment sent to Dewetsdorp—Retreat on Reddersburg—Combat at Reddersburg—Retreat of remainder of Battalion to Aliwal North—Further events of the campaign—2nd Lieut. Soutry at Wolverhoek—"London Gazette" promotions, etc.—Local Mounted Infantry formed—Serjeant Atkins killed—Casualties of the war in the Battalion—Further details about Mounted Infantry—2nd Lieut. Cooke Collis at Dewetsdorp—Corporal Fletcher killed, leading sortie—Surrender of Dewetsdorp—Lieut. Spedding captured—Nos. 1 and 2 Companies on column—Description of Boer tactics—Serjeant Beatson at Wilge River—Colonel Williams's farewell order—Boers pursued to Bothaville—Night march to Hartebeestefontein — Corporal Jackson at Hartebeestefontein — Lieut. Low attacked 250 Boers with 40 men—Serjeant Watson on 2nd July—2nd Lieut. Soutry on 4th July—Nos. 1 and 2 Companies sent to Cape Colony—Commandant Smuts allowed to cross Orange River—Rifleman Fortune's gallantry—The Mounted Infantry made three captures—General Nieuholt's Commando surprised bathing—The pursuit—Hailstorm near Edenberg—Surprise of Schweizer Reneke—Lieut. Low murdered—Surrender of Boer Commandoes—5th Battalion Royal Irish Rifles Mounted Infantry—Promotions, etc., in "London Gazette"—Representative Detachment sent to Australia—The Battalion sent to Dublin from South Africa—Funeral of H.R.H. The Duke of Cambridge—Moved to Aldershot—The Army Football Cup—Sent to Dover—Funeral of H.M. King Edward—Coronation of H.M. King George V.—Quartered at Tidworth—Notes to Chapter XVI.

SOUTH AFRICA became a British possession twice: first by conquest on the 17th of September, 1795. At the Peace of Amiens it was returned to the Dutch, from whom it had been taken, only again to be wrested from them by the expedition under Sir David Baird, in January, 1806, in which campaign the 1st Battalion of the Royal Irish Rifles, under its old name, as the 83rd Regiment, took part, with a detachment of the 2nd Battalion (then the 86th Regiment).

To a nation with such great interests in India when the Suez Canal was not made, the Cape was too valuable a possession to be lightly parted with, as all ships would naturally touch there for water and provisions on their voyages to and from India. Thus, at the final settlement of affairs in 1815, the Cape of Good Hope was retained by England. Great Britain paid £6,000,000 to the Dutch as compensation. It was a small amount, compared to what South Africa was yet to cost her.

The people at the Cape—such of them as were white—were principally Dutch; but there was a strong infusion of French Huguenot and of German blood. All made up a hard, virile race, impatient of control, and well suited to hold their own against all comers.

In 1815 a small rebellion broke out in the Somerset and Tarkastad districts. This was sternly crushed. In 1834 the slaves were to be liberated within four years, by the order of the British Government, and some compensation was paid. In many cases the Dutch farmers refused to accept the compensation, as it was ridiculously small, being only about one-third of the value of the slaves. A great number of the Dutch farmers then decided to leave Cape Colony, and to move, or "trek," as they called it, into the interior of South Africa, far enough away, as they hoped to be beyond the jurisdiction of the British Government. These farmers moved into what is now Natal, into the Orange Free State, and into the Transvaal. Though the great military nation of the Zulus, with sure instinct, objected to their coming, and tried to destroy them, the other parts of the country were almost depopulated from Zulu raids, and the trekkers found but little opposition there. These lands north of Cape Colony soon became the haunt of vagabonds and freebooters, as well as of well-to-do farmers, and these latter often favoured the British nation taking over the government and the protection of their new country. Steadily the line of British influence advanced north to overtake her wandering subjects. Some of the Dutch trekkers, or Boers, as the Dutch called themselves, were opposed to British rule, and in 1848 these persons sent a force which drove the British Resident out of Wynberg. General Sir Harry Smith moved up from Cape Colony with 800 troops and a few loyal farmers from the disturbed district, and met the disaffected Boers on the 29th of August, 1848, at Boomplaatz, some 800 strong, and after a fight drove them northwards, and they did not rally again, but retreated beyond the Vaal River. However, in January, 1852, the British Government decided to renounce their claims to the lands north of the Vaal, and in March, 1854, to the Orange Free State.

Nevertheless, Great Britain was almost forced again to interfere with the land north of the Vaal, as, by constant internal conflicts amongst themselves, and by cruel treatment of the native tribes, not always unavenged, these fair lands had become bankrupt, and generally impotent either to govern or to defend themselves; so, on the 12th of April, 1877, the Transvaal was annexed to the British Empire by proclamation. Again turbulent spirits objected to the rule of the British Government, and a Republic

1877

was formally proclaimed by the malcontents on the 15th of December, 1880. A few small engagements were fought, including Bronkhorst Spruit, Laing's Nek, Ingogo, and Majuba.

1881 More troops were hurried from England, but the heart of the British Ministry was not in the war, and on the 3rd of August, 1881, a humiliating convention was signed, by which the British Government undertook to withdraw from the Transvaal, keeping only shadowy suzerain rights. The Orange Free State, also a Republic, gave its good offices to both sides, but naturally inclined to help the Transvaal Republic.

The tone of the Boers from this time grew truculent. Had they not beaten the finest infantry in the world? Every little matter of friction became a burning question. Still, things might have proceeded without an actual outbreak between the British and the Boers if gold had not been found in the Transvaal. Immediately there was a steady stream of non-Dutch, who proceeded to the gold regions and settled down to assist in the digging of this precious metal. The Boers did not care for this invasion of their country by foreigners, but they would perhaps have come to like the new-comers, or "Uitlanders," as they were called, as they paid heavy taxes into the treasury; but, unfortunately, the Uitlanders objected to paying taxes without sufficient representation, and in 1899 appealed to the Suzerain Power for support in the redress of their grievances. This petition was carefully considered by the Home Government, and repeated attempts were made to arrange matters but without effect. 1899 The English Ministry now became alarmed at the defenceless state of Natal and Cape Colony, if they should be attacked by the Boers, and the troops in those places were gradually increased. On the 9th of October, 1899, an ultimatum was handed to the British Agent at Pretoria demanding, amongst other things: (*a*) that the troops on the border of the Republic should be instantly withdrawn; (*b*) that all reinforcements of troops which had arrived in South Africa since the 1st of June, 1899, should be removed from South Africa within a reasonable time; (*c*) that Her Majesty's troops which were then on the high seas should not land in any port in South Africa.

Failing a satisfactory reply from the British Government within a fixed time, the Republic intimated that the non-receipt of such a reply would be looked upon as a declaration of war. The reply was demanded to be sent by five o'clock on the afternoon of the 11th of October, 1899.

The British Resident's reply was to take his departure on the 12th of that month; 45,000 refugees had already escaped into Natal and Cape Colony.

The President of the Orange Free State was asked whether he also gave the Transvaal ultimatum his concurrence and support. He replied that he did. Thus the British Government found itself plunged into a war with these two hostile Republics.

In one sense they were contemptible opponents. Their male population between 16 and 60 years of age was only 54,000, and their means were small, but practically every man was a trained guerilla fighter, and their very great powers of resistance really came from this, backed up by the fact that their countries were of great extent, and were practically unmapped, and were altogether admirably suited to the tactics of the Boers—mounted riflemen from childhood—whilst in their great distances they were pre-eminently unsuited to the slow movements of infantry, in which, like all civilized Powers, England's strength lay.

There was a further danger, and that was that Cape Colony was ripe for revolt in many places, whilst Natal also had disloyalists amongst its generally most loyal sons; and in any operations of war great dependence must be placed by the invading British Army on the railways. These railways, many hundreds of miles in length, could easily be cut in many places by Boer sympathisers and the invading army might be brought continually to a halt by the failure of supplies, through the destruction of these railways. Again, whilst naturally the two capitals, Bloemfontein and Pretoria, would be points to be made for by the British army, yet when these towns were seized there was nothing to show that the Boer resistance would be thereby weakened, whilst the enemy's forces of mounted riflemen could practically not be cornered, and, though always so dispersed that they could not be caught and crushed decisively, yet, from their superior mobility, they could decline a combat, and yet concentrate at will on any exposed detachment or line of communication, and crush it by superior numbers, disappearing before reinforcements arrived, and repeating this manœuvre in the hope of wearing down by degrees their more powerful but slower-moving enemy.

The actual numbers that the Dutch brought into the field may be taken at between 50,000 and 60,000 from the two Republics, as many lads under sixteen years of age and many men over sixty years of age took part, and to this total may be added some 10,000 rebels and 3,000 foreigners.

On the British side, there were in September in the whole of South Africa six and a half battalions of regular infantry, two cavalry regiments, two companies of garrison artillery, three batteries of field artillery, and a mountain battery, with a few other details.

Ten thousand extra men were ordered out in that month to the Cape from India, England, and elsewhere, but most of them did not arrive until after the war had broken out. Finally, on the 7th of October, the Army Reserves were called out, and Parliament was summoned to meet on the 17th of October, and then everything was bustle and preparation.

It was to meet this nation of born guerilla warriors that the 2nd Battalion Royal Irish Rifles commenced to mobilize on the 9th of October.

There was no hurry. The infantry would be ready long before the ships to convey them to South Africa. There had been many critics of our military system who pointed out that the Reservist might be called out; but as he would probably be a married man, with a fair situation, or a wanderer, such as a fireman on a tramp steamer, it was doubtful if he could be found when he was wanted. This mobilization disproved that theory once and for all, for out of all the Royal Irish Rifles Reservists summoned — 704 in number—only nine failed to put in an appearance. None of these nine absentees were ever heard of again, and it is only fair to think that they were dead.

On the 25th of October, amidst scenes of the wildest enthusiasm, the 2nd Battalion marched from Victoria Barracks to the Great Northern Railway station, and was conveyed to Queenstown by rail, where they at once embarked on the hired transport "Britannic," for South Africa. Twenty-one officers embarked with the battalion and 872 of all other ranks.

The following was the list of the officers who went out at this time:—

Lieutenant-Colonel Eager.	Lieutenant Christie.
Major Seton.	,, Sitwell.
,, Allen.	,, Bradford.
,, Welman.	,, Wright.
Captain Morphy.	,, Spedding.
,, Bell.	,, Daunt.
,, Hall.	,, Stevens.
,, Weir.	2nd-Lieutenant Saunders.
,, Kelly.	,, Davenport.
,, Dimsdale.	,, Rodney.

Lieutenant and Quartermaster Dwyer.

The "Britannic" arrived at Cape Town on the 13th of November. In the meantime events had moved fast in South Africa. The Boers had poured into Natal from the Orange Free State and also from the Transvaal. General Penn Symons had been attacked at Talana, in Natal, on the 20th of October, by some 4,000 Boers. He had three battalions of infantry and three batteries of artillery, also the 18th Hussars.

The first serious engagement of the war was a repulse for the Boers. General Penn Symons was mortally wounded, and Lieutenant Ryan, late of the Royal Irish Rifles, who had settled in South Africa, and was riding with the General during the action, was dangerously wounded, and the force retreated later on to Ladysmith unmolested, the guide being Colonel Dartnell, who had served in the 86th Regiment at the storming of Jhansi. On the same afternoon Elandslaagte, close to Ladysmith, was fought and won. Here Major-General French pounced with the 5th Lancers and Imperial Light Horse, and three infantry battalions, with two batteries, on 1,000 Boers. This had been a sweeping victory. On the 30th of October, however, Sir George White had been driven into Ladysmith after a hard fight, including the disaster of

Nicholson's Nek, where some 1,000 British troops were surrounded by a much larger force and compelled to surrender.

So much for the eastern part of the seat of war. On the western side the Boers had invested Colonel Kekewich and half a battalion of infantry in Kimberley, and a column, 8,000 strong, under Lord Methuen, was being formed on the Kimberley Railway at Orange River, for the purpose of relieving that town.

Mafeking, further north, on the same railway line, was also invested by the Boers. In eastern Cape Colony, the enemy had, on the 13th of November, crossed in small numbers at Aliwal North, and other parties were near Colesberg.

Thus, when the Royal Irish Rifles arrived in Table Bay on the 13th of November, it was considered necessary by the authorities to send them to East London. General Gatacre came on board the " Britannic," with his staff, and the steamer at once sailed for East London, where the troops disembarked on the 16th of November, and entrained for Queenstown on the 17th, and moved to Putter's Kraal on the 22nd of that month.

The battalion had formed part of the 5th Brigade, under the command of Major-General Fitzroy Hart, but this organization was now broken up, and the Royal Irish Rifles remained under the command of Major-General Gatacre, who was to have been in charge of the 3rd Division, to which the 5th Brigade belonged.

General Gatacre was a soldier of boundless energy and great personal courage. Impervious to fatigue himself, he calculated the endurance of his men by his own.

His plans were as follows: The Boers occupied Stormberg Junction, and General Gatacre decided to surprise them from Putter's Kraal. The number of Boers at or near Stormberg did not exceed 2,300, but the danger which urged General Gatacre to the attack was that rebellion was spreading in the Cape Colony, and there was no saying how many rebels might not join the Boer forces at Stormberg.

The position held by the enemy was as follows: Stormberg Junction lies in the centre of a basin, surrounded by hills. The railway runs through these hills, north-east towards Burghersdorp, west towards Rosmead, and south towards Molteno and Queenstown. The Boer laagers were scattered about the basin. The Bethulie Burghers, with rebels, numbered 800 men, and were close to the railway station.

The Smithfield laager, about 700 strong, was on the south-western slope of the Rooi Kop, a hill standing south of the railway station. This commando looked after the nek through which the railway from Molteno came, and had dug trenches and posted two guns here just west of the railway.

The Rouxville burghers were on the western heights, some in schanses (stone-wall enclosures, etc.). They numbered 800 men, and had one gun with them. Commandant E. R. Grobler was in charge.

General Gatacre decided to take some 2,600 men with him, consisting of the Royal Irish Rifles, the Northumberland Fusiliers, three mounted infantry companies, some Cape Police, two batteries of Royal Field Artillery, a company of Royal Engineers, a field hospital, etc.; whilst 400 more men, with four guns, were to join him at Molteno, to be sent to operate on the right flank, under Captain de Montmorency.

The troops were to rendezvous in Molteno before sunset, set out at 7 p.m., and get as far as Goosen's Farm, within two miles of the Nek, soon after midnight. Here they were to rest for two or three hours before making the attack. Shortly before dawn the Royal Irish Rifles were to rush the ridge to the left of the nek, and capture the guns, while the Northumberland Fusiliers seized the under features of the Rooi Kop, on the right. The force was dangerously small for its work.

The party which was to join at Molteno, under Captain de Montmorency, was at Pen Hoek. At midnight on the 8th-9th of December a message was handed to the telegraph clerk at Putter's Kraal, summoning this detachment to join. The clerk omitted to send it. No such palpable precaution was taken as obtaining an acknowledgment of the receipt of the telegram from Captain de Montmorency; so his party did not join the force, as intended.

At 4 a.m. on the 9th of December the infantry began to pack up. The troops were road-making from 5 a.m. to 7 a.m.; breakfast was at 7 a.m.; tents were struck at 11.30 a.m.; and dinner was at 12.30 p.m. Soon afterwards the work of entraining began. The staff-work appears to have been bad—possibly from unavoidable causes. A train-load of mules was allowed to block the line for hours. Though the troops commenced to entrain at mid-day, yet the last troop train did not reach Molteno until 8.30 p.m.

That afternoon General Gatacre arrived at Molteno, and held a consultation with the local inspector of Cape Police there. Here he received a report that the Boers had constructed a wire entanglement in the nek or pass which he meant to attack. This was afterwards found to be inaccurate, but it altered the plans. The General decided not to attack the front of the basin, as he had intended, but to strike in by one of the flanks, and he chose the western side for his point, meaning to strike in on the Boer right.

The ground was intricate, but once there and on the heights, he could command the guns and the whole Stormberg Valley.

At 9.15 p.m. the column marched out of Molteno, the Royal Irish Rifles leading. Just before moving off, Colonel Eager said: "The battalion represents the North of Ireland, which is watching you. I know I have not to ask you to do your duty." Two days' rations of tinned meat and biscuit were carried. The distance to be covered was ten miles. There was a bright moon, which set about midnight. The road at first was quite good; everything looked promising, and the men were in capital spirits. General Gatacre gave the order to fix swords, and the men marched on, carrying their arms in this rather constrained position. The artillery followed the infantry, but with a long interval between, and the wheels of the guns, etc., were wrapped up in raw hide, to deaden the sound. Behind the guns came the mounted troops. Various details, including the Maxim gun detachment of the Royal Irish Rifles, under Lieutenant Wright, had not been informed of the change of plan from the frontal to the flank attack, and they took the direct road to the nek and lost themselves. The guides, in the darkness, missed the right turning, and the force halted at 1 a.m. at a farm, owned by a Mr. Roberts. The guides informed General Gatacre that the distance from the coveted heights was now only one and a half miles; in reality it was three miles away. The Boers had sent out some 600 men that night, probably to beat up Gatacre's left flank. This force was under Grobler, and was laagered three miles farther up the road from the British column, so Gatacre was actually between the two bodies of Boers, who had not the vaguest idea that his outposts were nearer than Molteno. Here indeed was a chance, if he had known of it, to finish up the 600 Bethulie warriors with the bayonet. However, not knowing this, at 2 a.m. the march was resumed. The track—for it could not be called a road—became appallingly bad. Colonel Eager reported to the General that he thought the guide had lost his way. The guide as stoutly protested that he had not. At 3.45 a.m. on the 10th of December the head of the column reached the point which General Gatacre had aimed for. He was at the foot of the heights which formed the western boundary to the Stormberg Basin, and he was on the western side of those heights, in a small valley, which led into the Stormberg Basin.

Everything was as he could wish it. Unfortunately in the dark he did not know that he had arrived there, and his guides did not quite understand his plans, so there was a misunderstanding. The guides thought he wanted to push on by road into the valley, and did not realise that the infantry, facing east, could have climbed straight up the hill and dominated all the Boer camps from these heights with their rifles; so the column toiled along the road, past the heights on their right, until broad daylight came on, still marching in fours, with swords fixed.

Colonel Eager realized the danger, and requested the General's permission to send half a company out as an advanced guard. General Gatacre ordered him not to do so. A few hundred yards to the east of the British force lay one of the Boer laagers, its outpost absolutely unconscious of the presence of the enemy. There was a picquet, with a single Boer sentry on the road which ran through the nek, which the force was now approaching in column of route. To his horror, the sentinel saw this long serpent of marching men drawing near him. He roused his comrades—between ten and twenty in number—and fire was opened. The Dutch poured out of their laager, where most of them had been making coffee, and rushed for the heights. General Gatacre ordered the Royal Irish Rifles to rush through the nek and seize a detached hill just inside it, but it was too late to issue orders then. Everyone had felt that they were called upon to act promptly for themselves in the emergency, and, though three companies ("F," "G," and "H") dashed through the nek for the hill beyond, the remainder of the Royal Irish Rifles formed for attack towards their right flank, and, with the Northumberland Fusiliers prolonging their right, rushed for the summit, led by "C" Company, under Captain Bell. The advance was well maintained, and half the distance had been crossed when the whole force was brought to a standstill by a line of precipices, which rose sheer up for some distance, and was only scalable here and there.

The men laid down under cover, whilst Colonel Eager, Major Seton, Major Welman, and Captain Bell drew together, studied the formation of the ground, and arranged for the forward movement. The three companies who had taken the hill beyond the nek outflanked the Boer position, whilst the mounted infantry had also pushed inside the Stormberg Valley. Everything was in capital train. The General rode up to the three companies on the hill, whilst Colonel Eager, without orders, but wisely comprehending the situation, arranged for the rush to clear the heights. The two batteries—the 74th and the 77th—opened fire on the heights, but, unfortunately, thinking the Royal Irish Rifles were the enemy in the uncertain light, commenced to shell them. The results were instantaneous. The first shell mortally wounded Colonel Eager and severely wounded Majors Seton and Welman, Captain Bell and several riflemen. The next few were equally deadly, and in a few seconds, to the surprise of the Boers, some of whom had been pouring in an ineffectual fire, whilst others were hurrying to the rear, the whole of the infantry who had been lying close under the cliffs, ready to escalade them, were driven down the slope, vainly trying to avoid the deadly shrapnel of their own guns. The officer commanding the Northumberland Fusiliers ordered his battalion to retire to reform it, ready to support either attack. Some of the Royal Irish Rifles, hearing the order, moved with this battalion, assuming that it also applied to them. Some of the Northumberland Fusiliers did not hear it, and remained where they were.

The men who retired first took shelter in the donga at the foot of the hill, but this was enfiladed, so the retirement was continued as far as the small hills across the valley. The movement was carried out in good order, and each part covered the retreat of the others. Arriving at these small hills, one company was told off to hold the heights, whilst the remainder formed in quarter column under cover. General Gatacre had been with the party that held the hill inside the nek. From here he had meant to sweep down the enemy's position, pressing home his attack. With the hills abandoned to the Boers, he saw that this could not be done, so he gave the order to the three companies to retire, which they did, under heavy fire, in good order, and the mounted infantry of the force galloped back, and a new line was formed on a ridge across the road up which the force had marched. This was about an hour and a quarter after the first shot had been fired. Naturally, the noise had drawn in all parties of Boers, even Grobler's detachment. This last commando fired into Gatacre's troops from the rear, and the 77th battery had three guns firing forward and three backward.

In the meantime, some 600 men of the two infantry battalions lay on the hill under the cliffs, keeping up the fight with the Boers. General Gatacre ordered the force he was now with on the ridge to retire. Major Allen, of the Royal Irish Rifles, urged the General to allow him to take up the remaining companies of the Rifles to carry the heights, but General Gatacre refused to let him do this. The remainder, nearer to the enemy, were left to their fate. Afterwards it transpired that the officers and men did not know what was going on, and that they held tenaciously to their ground, expecting that the remainder of the force was moving to make a flank attack. No orders were given, and each party was overpowered in detail. The retreating force, under General Gatacre, was not kept well in hand, and the infantry straggled a great deal. The guns and mounted infantry kept the enemy at a distance, and Lieutenant and Adjutant Sitwell collected the men of the Royal Irish Rifles who were least fatigued and formed an efficient rearguard. About 11 a.m. Molteno was reached; 634 unwounded prisoners (officers and men) were taken by the Boers. The total casualties of the whole force were 28 killed and 61 wounded on the British side, and 8 killed and 26 wounded on the Boer side.

The Royal Irish Rifles loss was as follows:—Twelve non-commissioned officers and men killed and forty-six non-commissioned officers and men wounded; also wounded officers as follow: Lieutenant-Colonel Eager (mortally wounded), Majors Seton and Welman, Captains Bell and Kelly, and Lieutenant Stevens. The following officers were captured: Captain Weir, Lieutenant Christie, and 2nd-Lieutenants Maynard and Rodney, and 216 unwounded non-commissioned officers and men. The battalion, under Major Allen, was entrained that afternoon, with the remainder of the infantry of General Gatacre's force, and was sent down to Sterkstroom.

General Gatacre had, on the whole, bad luck at Stormberg. The idea was sound; but his arrangements were not thoroughly supervised. He surprised his enemy, but, from want of precautions, was not able to use his advantage, and appears to have sent no orders to his troops. That he should have left 600 of them to be made prisoners was also a piece of bad staff work; whilst the crowning calamity was the successful shelling by the British artillery of their own side. On the whole, the force was lucky to have been able to effect their retreat. An enterprising enemy would have stopped it and captured the whole force. The prisoners were sent to Pretoria.

Hard things have been said of the regimental officers; such as that they must have known that their men were left behind on the hill. The officers knew it perfectly well, and reported it, but General Gatacre refused to either wait for the 600 men who were holding their own on the hill, or to let any signal be made to them.

How unexpected his retirement was is but gathered from one incident. After most of the prisoners had been collected by the Boers, a young officer but recently joined went up to a senior officer, who happened to be one of the Royal Army Medical Corps, and asked what orders there were, and whether he should not move the men he was in charge of against the heights. Of course, he was at once taken prisoner by the enemy, who were talking to the Royal Army Medical Corps officer at the time, and his men, without a leader, were also captured. All were eager to fight; all that was required was someone to direct.

On the following day, the 11th of December, General Gatacre came down and inspected the Royal Irish Rifles, making a speech to the men, explaining the reason for the failure of his plans. Probably the men did not understand his explanation; but they did understand full well that they had been repulsed though they had hardly been engaged, and cheered the General after they were dismissed parade, begging to be taken up again to Stormberg to make another attempt.

1900

The battalion marched to Bushman's Hoek on the 23rd of December, returning to Sterkstroom on the 19th of January, 1900. A draft of three officers and 210 non-commissioned officers and men joined on the 17th of this month. "G" Company was formed into mounted infantry on the 21st of January, with Lieutenant Spedding and Lieutenant Saunders as the officers. Two small skirmishes took place on the 26th of December and 24th of January at Cyphergat and Bird's River, respectively, but there were no casualties.

Lieutenant-Colonel Eager died of his wounds on the 3rd of February. At the beginning of March a further draft of six officers and 300 non-commissioned officers and men joined the battalion, and next day "C" Company was converted into a mounted infantry company, under Brevet-Major Festing and Lieutenant Low and 2nd-Lieutenant Davenport. This company received horses and some bridles of sorts, and had to march 100 miles before they were issued with saddles.

On the 9th of March the Royal Irish Rifles entrained at Sterkstroom and steamed northwards to the Orange River, arriving there on the 13th of March. Here they took part in the skirmishing at Bethulie Bridge, across the Orange River, suffering no casualties, and the Boers fled. A train took the battalion up to Springfontein, where it arrived on the 17th, and halted there until the 20th of March, when it marched eastwards to Smithfield, arriving there on the 23rd of March.

In the meantime, the campaign, as a whole, had progressed as follows:—The British troops, under Lord Methuen, had attacked Cronje at Magersfontein, near Kimberley, and had been repulsed, with about 1,000 casualties, on the 11th of December.

This check was met by the Government in England directing a 7th Division to be sent to South Africa.

On the 15th of December General Sir Redvers Buller had attempted to cross the Tugela River in Natal, to relieve Ladysmith, and the attempt had failed, with a loss of over 1,100.

General Lord Roberts was sent out from England to assume command. Nine Militia battalions and the first Yeomanry and Volunteer contingents were sent out, and Colonial forces raised in South Africa itself, and also offers of troops from the Colonies were gladly accepted.

By the end of February, however, the aspect of things had changed. Lord Methuen's force, on the western side of the theatre of war, had been largely reinforced. Lord Roberts himself assumed command of this Army. Kimberley was relieved on the 15th of February. Cronje tried to slip away from Magersfontein towards Bloemfontein, but was headed off at Paardeberg, and finally captured there on the 27th of February, with some 4,000 men.

In front of General Gatacre the Boers had been turned out of many of their positions by General Brabant's mounted Colonial troops, and Stormberg was occupied by Gatacre on the 5th of March; whilst the Boers withdrew north across the Orange River. On the 13th of March Lord Roberts's army entered Bloemfontein, and whilst the troops under Gatacre threatened the Boers from the south, Lord Roberts's victorious forces moved vigorously down from the north, to try and clear the railway line which ran south from Bloemfontein to Springfontein, and there branched to Bethulie and Norvals Pont, as it was on these lines that the main army, under Lord Roberts, depended for their supplies. The Boers who had been south of the Orange River, in front of the Royal Irish Rifles, etc., fled north, and, swinging east, avoided Bloemfontein, and slipped through the gap between the troops in that town and the Basuto border.

Ladysmith, in Natal, had been relieved on the 27th of February by General Buller's force, and the whole of the southern part of the Orange Free State was settling down, and the Royal Irish Rifles were sent to Smithfield, to collect arms from the burghers who had left their commandoes and returned to their farms, and to accept surrenders from all and sundry.

On the 28th of March "A," "D," "H," and "E" Companies, with twenty-five mounted infantry of "C" Company, the whole under the command of Captain McWhinnie, left Smithfield, in consequence of the following message sent to Major Allen, in charge of the Royal Irish Rifles there:—

"Send one company Royal Irish Rifles, with twenty-five men of the mounted infantry of that battalion, to Helvetia, for proclamation duty. . ."

"Send three companies of the Royal Irish Rifles to Dewetsdorp, so as to arrive there on Sunday, the 1st of April."

Leaving "E" Company at Helvetia on the 30th of March, Captain McWhinnie pushed on with the remaining three companies to Dewetsdorp. Before leaving Helvetia, he heard that there was a small party of Boers towards Dewetsdorp, and he, therefore, took the mounted infantry, under Lieutenant Spedding, one march with him, and then, leaving seventeen mounted men at that night's halting place, he directed Lieutenant Spedding and eight men of the mounted infantry to come on to Dewetsdorp with his party.

Captain McWhinnie duly arrived on Sunday morning, the 1st of April, at Dewetsdorp, made all arrangements for the defence of the town, and reported his arrival to the Assistant-Adjutant-General, 3rd Division, via Wepener, by telegraph.

In the town that day Captain McWhinnie was privately informed that a Boer commando had been invited to come to turn him out. That afternoon Lieutenant Spedding was fired upon when scouting north of the town. Reinforcements were sent, and the enemy retired.

At about 5.50 p.m. a number of men were seen on the hills to the west of the town.

The companies had been entrenching during the day, and the men were immediately ordered to take up their positions for defence. However, after every precaution had been taken, it was found that the strangers were "G" Company, which had been made into mounted infantry, and that there was another company of mounted infantry with them belonging to the Northumberland Fusiliers, commanded by Captain Casson, whilst Captain Dimsdale commanded "G" Company, and with him was Lieutenant Saunders.

At midnight on the 1st-2nd of April Captain McWhinnie received the following telegram from the Officer Commanding at Wepener:—

"Your No. 2, Assistant-Adjutant-General, 4th Division, wires:—'Remove your troops immediately to Reddersburg; await further orders by despatch rider.'"

On receiving this message, all the troops were warned that they would probably move at a very early hour.

The officer commanding mounted infantry reported that his horses would not be fit to move before 7 a.m. At 3.30 a.m. the despatch rider duly arrived, and at 5 a.m. the detachment moved off in heavy rain. The state of the weather made the marching slow, as the ox-wagons sank in deeply and compelled the troops to wait for them.

The cause of this telegram was that Commandant Christian De Wet had fought an engagement at Sannah's Post on the 31st of March, and had given Broadwood's force a severe check, and Lord Roberts was, naturally, anxious about the two little detachments of Rifles at Dewetsdorp and Helvetia, the former being only some forty miles south-east of Sannah's Post. Therefore, he telegraphed at once to General Gatacre at Springfontein to get his detachments back towards the line. Unfortunately, General Gatacre did not mention anything about this reverse in his telegram, neither did he lay as much emphasis as he might have done on the need for haste. So far as Captain McWhinnie knew, the whole countryside was settling down and taking the oath of neutrality and surrendering their arms. True, a few misguided men were probably conspiring together, trying to cause trouble, but their more peaceably-disposed brethren would soon get them to disperse.

Having come about four miles from Dewetsdorp, Captain McWhinnie detached Lieutenant Spedding and his mounted infantry to ride to Helvetia, to make sure that that detachment was duly warned, in case of any mistake.

About 9.15 a.m. the officer commanding the mounted infantry reported that his horses could proceed no farther without being fed, and he was allowed to halt there to feed, whilst the duties of advanced, rear and flank guards were taken up by the infantry, and the column toiled on. The mounted infantry had been told to halt for one and a half hours only, and then to trot on to overtake the column. At noon, however, they had not come up. The column, therefore, halted and cooked dinners. An alarm took place whilst this cooking was in progress, men being reported on the neighbouring hills, and Captain Kelly reported that he had heard a shot. The tired infantry were sent out to the tops of the hills, which were found to be untenanted. At 2.45 p.m. the mounted infantry rejoined, having broken the pole of their ambulance, which had caused the delay. The column then resumed its march, and about a quarter to six that evening halted for the night at some dams on Kelly's Farm. At about 2.30 a.m. on the 3rd of April another despatch rider came in with a duplicate of the despatch from Assistant-Adjutant-General, 3rd Division, which had been sent via Smithfield, Helvetia and Dewetsdorp.

The Rifles marched off again at dawn on the 3rd of April, and proceeded without interruption until 10 a.m., when a cloud of dust was reported on the west of the line of march. What had happened was that De Wet had dogged the detachment with a small force all the day before, moving in a valley parallel to its line of march, being joined on the morning of the 3rd of April by reinforcements, bringing up his numbers to some 2,200 men and four Krupp guns. He sent in a message to Captain McWhinnie, stating his numbers, and requesting him to surrender, to spare needless bloodshed. The reply sent to this message was lacking in courtesy, but the blunt language used, urging De Wet to go elsewhere, was perfectly explicit.

So soon as the cloud of dust was seen it was realised that the enemy were upon them, and a position was taken up on a horse-shoe-shaped ridge, standing on a plain of undulating ground, about four miles north-east of Reddersburg. This small range was called Moster's Hoek. It was too large for the force available, but the best that could be obtained under the circumstances. The mounted infantry were ordered to hold the western side of the horse-shoe, and the three infantry companies the remainder. There was no water to be had, excepting from one water-cart, and no food for the men, excepting one box of biscuits. So soon as De Wet received the reply to his request, the firing became general, and the guns, in particular, searched the whole of the horse-shoe with their shells.

About 3.30 o'clock that afternoon Captain McWhinnie personally superintended the taking of some ammunition to the mounted infantry, and asked Captain Dimsdale whether his mounted infantry could hold their portion of the position or required reinforcements. That brave young officer replied "that he could hold it for a month of Sundays." An hour later it was reported that Captain Dimsdale was wounded, and Captain McWhinnie went across again to see him, and found him mortally wounded, whilst Captain Casson, of the Northumberland Fusiliers, was sitting, dead, on some rocks above Captain Dimsdale. Captain McWhinnie, therefore, remained with the mounted infantry until dark, as they then had only two subalterns left. Barclay, of the Northumberland Fusiliers, who was also killed in this fight, and Saunders, of the Rifles. So soon as night fell, a despatch rider was sent off to Bethany, eighteen miles west, on the railway line, to ask for assistance, and the whole position was divided up into four sections, each being allotted either to one of the three infantry companies or to the whole of the mounted infantry. The water was doled out, giving each man somewhat less than a pint, and ammunition bandoliers were replenished. The firing more or less ceased.

At dawn firing recommenced, the opinion being that more men had joined the Boers during the night, and that they had two more guns, but there is no certainty about this matter. The mounted infantry were reinforced by a party of twenty men from "H" Company. At 8 a.m. the enemy were firing on the mounted infantry at a range of thirty yards from one of their sentries, and the British could see nothing to fire at in return. Shortly afterwards the western half of the horse-shoe was rushed from the northern side, where the enemy had assembled in large numbers unperceived, and the mounted infantry then capitulated, some time before 9 a.m. The enemy then proceeded to creep up closer and closer to the localities held by "H" and "A" Companies, and at the same time the position was also being assaulted from the west and south. "D" Company's position was now taken. The enemy's tactics were to bring a very heavy and accurate fire to bear on the British from a distance, whilst they kept pushing in their spare troops to fifty yards from the hostile infantry. How heavy and deadly was this fire may be judged from one incident. Bugler Longhurst had to change from one exposed position to another, less exposed, only five yards away; he moved as quickly as possible, and fell dead at the desired place, with nine bullets through him. Strange to say, Captain McWhinnie strolled calmly about for the last thirty minutes under all this fire without taking cover and remained untouched. "A" Company was then captured, and the assault was continued as before, and also from the east. Hundreds of Boers were now on the position, and the defenders were broken up into two little groups. Seeing that he was outnumbered and overpowered in every way, Captain McWhinnie decided to surrender.

2nd-Lieutenant Soutry, who was on a small knoll of ground with a section, kept up the fight for a little longer, but was also obliged to capitulate.

The loss of the fight may be attributed to the want of guns, to reply to the enemy's artillery, also to the paucity of officers (the mounted infantry had only one officer of three years' service left at the end of the fight for two companies), and also particularly to the overwhelming odds of more than four to one against the British.

The exact casualties of the Royal Irish Rifles were as follow:—Captain Dimsdale and eight non-commissioned officers and men killed; Captain Kelly, Lieutenant Bradford, and twenty-four non-commissioned officers and men wounded. Six officers and 382 non-commissioned officers and men were made prisoners.

Besides these losses of the Rifles, some seventy Northumberland Fusiliers were made prisoners, having also lost two officers killed and about twelve non-commissioned officers and men killed and wounded.

The only chance of escape would have been to have marched towards Bethany during the night, and that would have meant sacrificing all the transport and wounded. About the same time as this surrender took place, the scouts of a relief force were on a ridge overlooking Reddersburg, and they actually heard the last few shots. The force consisted of the Cameron Highlanders, de Montmorency's Scouts, two mounted infantry companies, and three batteries, all under General Gatacre. This officer, however, instead of pushing on at once, decided to retire, and had actually marched back four miles, when he received an order from Lord Kitchener to push forward and occupy the village of Reddersburg. He did so, capturing a couple of Boers, and returned to Bethany next morning, and shortly afterwards was directed to relinquish his command and to proceed to England.

In the meantime, "E" Company was at Helvetia, where it was joined by "F" Company, from Smithfield, and the remainder of the battalion, under Major Allen, was at Smithfield. Major Allen sent a despatch to the Helvetia detachment. This despatch is stated to have arrived just after the mid-day dinner hour on the 3rd of April. Lieutenant Spedding had ridden in the night before from the point four miles south of Dewetsdorp, from which Captain McWhinnie had sent him away, and, after warning the detachment, had

PLAN OF FIGHT AT REDDERSBURG, APRIL, 1900

Kindly photographed by
G. Friedlander, Esq
—— Dover ——

ridden on to Smithfield. The two companies marched out from Helvetia about 3.30 p.m. on the 3rd of April, heading for Smithfield. Time pressed, and the men marched well; half-an-hour's halt only was allowed, and a nek near Smithfield was reached about 1 a.m. on the 4th of April. Here the Smithfield party was met, blankets were got out, and the men slept until 5 a.m., when breakfast was served out, and the troops moved away at 6 a.m., directing their march, via Rouxville, towards Aliwal North. A short halt was made at mid-day for some food, and another halt at nightfall until 10 p.m. The march was resumed until dawn, when Rouxville was reached, and the troops halted until about 9 a.m. (5th of April). Major Allen utilised the time by telegraphing from Rouxville to the General at Aliwal North, giving him some information which had been acquired on the way. This was as follows:— Whilst halted on the evening of the 4th of April, about 9 p.m. a letter arrived from a friend, stating that a very strong body of Boers was close to him, and that they had guns. Major Allen at once ordered all the men to fall in without noise, and stole away, leaving two large fires burning. The enemy, satisfied by seeing these fires, settled down for the night, intending to attack at dawn. It was afterwards known that they were much surprised to find the camp empty. The military authorities at Aliwal North hurried off some mounted troops to reinforce Major Allen. He had left Rouxville at 9 a.m., and, marching hard, with only an hour's halt, passed triumphantly over the Aliwal North bridge, across the Orange River, at 7.30 p.m. on the 5th of April. The loyal townspeople of this little place turned out and cheered the detachment as it marched in. They had also sent out some wagons to bring the tired men in on, but one and all refused to drive, preferring to complete this march on their own sturdy legs. The Boers attacked Major Allen some ten miles north of Aliwal North, but the attack was repulsed with the loss of two men of the reinforcements, who incautiously exposed themselves. The whole march was a very fine performance. The distance actually marched was seventy-five miles, and the time taken was fifty-two hours.

Three-quarters of "C" Company (mounted infantry) were with General Gatacre in his advance on Reddersburg.

On the 15th of April the battalion, now under Major Tobin, moved towards Wepener, where a party of the Cape Mounted Rifles, etc., were surrounded by the Boers. The whole of the relieving force was under the command of General Hart. Wepener was reached on the 20th of April, and the Boers retired. On the 28th of April the Regiment marched to Smithfield, arriving there on the 1st of May. They received here drafts of six officers and 458 non-commissioned officers and men. An extra company ("I") was formed here, commanded by Captain Brennan. In June "D" and "E" Companies were sent to Bloemfontein, whilst "F" Company went to Rouxville. During this month Captain Henty, Lieutenants Hanks and Sheen, with 104 non-commissioned officers and men of the London

Irish Rifles, joined the battalion, and became " K " Company, 2nd Royal Irish Rifles.

On the 6th of August the battalion arrived in Bloemfontein, via Bethulie, and here 100 men were formed into mounted infantry, under Captain Clinton Baker, to take the place of " G " Company, mounted infantry, which had been captured at Reddersburg. In the meantime, the main army, under Lord Roberts, had advanced from Bloemfontein about the beginning of May. Pretoria fell on the 5th of June. General Buller, with the Natal army, had entered the Transvaal during the first days of June. Mafeking had been relieved on the 17th of May. In July 4,000 Boers surrendered to General Hunter in the north-eastern Free State, just south of Harrismith. The Transvaal was finally annexed in August, and the late President of that State, Mr. Kruger, fled to Europe, through Portuguese territory, in September. Despite all these successes, however, the guerilla warfare still went on. During August " F " and " K " Companies were on trek with Colonel White's column. In September headquarters and " B," " D," " E," and " I " Companies, moved to Thaba 'Nchu, where they were joined by " F " and " K " Companies. The whole then operated in the Dornberg district, with General Bruce Hamilton's force, crossing to the west of the railway on one occasion.

In October the battalion was divided as follows:—

" D " and " E " Companies, with Colonel White's column.

" B " and " K " Companies occupied Bulfontein.

" F " and " I " Companies and headquarters returned to Thaba 'Nchu.

" C " and " G " Companies were mounted infantry in the 9th Mounted Infantry Battalion.

" A " and " H " Companies, with all the prisoners of the other companies, who had been released by Lord Roberts's advance, occupied different railway stations on the line to Pretoria, as follows:—

Klip River, under Major Spencer.

Wolverhoek, under Captain Weir.

Natal Spruit, under Captain Kelly.

On the 17th of November, De Wet, with a large force and three guns, attempted to capture a post held by Lieutenant Sheen and forty men of " K " Company. He attacked from daybreak until 9 a.m., when he retired, having suffered considerably in men. " K " Company suffered no loss. At Wolverhoek De Wet crossed the railway line with a force of some 2,000 men. Captain Weir ordered his mounted troops out to harass this party. 2nd-Lieutenant Soutry became detached with five mounted irregular troopers. They were driven into a kraal, and, after losing two men killed, were rushed by a large body of the enemy and taken prisoners. At this time the Boers only retained officers as prisoners, and 2nd-Lieutenant Soutry was ordered to be detained as an officer. With ready wit, one of the irregulars, who had just been released by the Boers, remarked to them: "Call that an officer, why, that is only Soutry"; and the enemy told him to go with his men also, after being disarmed.

1901 In January Lieutenant-Colonel B. R. Hawes was appointed to command the Regiment under peculiar circumstances. Two extra battalions had been ordered to be raised in his regiment, the Munster Fusiliers, and he was selected to command one battalion, and was made a lieutenant-colonel. Then it was decided to raise the Irish Guards instead, and the War Office directed that Colonel Hawes should receive the command of the first infantry battalion which became vacant. Finding that the 2nd Battalion Royal Irish Rifles had no Lieutenant-Colonel since Colonel Eager's death, they sent Colonel Hawes to assume command. "L" Company was formed on the 1st of March. It was said that there were 1,800 men of the 2nd Battalion Royal Irish Rifles at that time in South Africa.

"K" Company (the Volunteer company from the London Irish Rifles) returned to England in April, and Lieutenant Wiley and twenty men of a new section from the same Volunteers joined the battalion. In January the headquarters of the 2nd Battalion Royal Irish Rifles marched to Bloemfontein, returning to Thaba 'Nchu on the 30th of April; whilst on the 27th of that month "H," "E," "I" and "L" Companies went out on trek with Major Massey's column, under Major O'Leary. Four officers and 121 non-commissioned officers and men joined from England by October of this year.

The King was graciously pleased to direct, in the "London Gazette," that the following decorations and promotions should take place:—

Major Allen, promoted to the rank of brevet-lieutenant-colonel.
Captain and Adjutant Sitwell, promoted to the rank of brevet-major.
Lieutenant and Quartermaster Dwyer, promoted to the rank of honorary captain.
Major Tobin was given the Distinguished Service Order.
Lieutenant Spedding was given the Distinguished Service Order.
Lieutenant Saunders was given the Distinguished Service Order.

The following were awarded the Distinguished Conduct Medal:—

Serjeant Rainey.	Rifleman Hanlon.
,, Darragh.	,, Hogg.
Corporal Irvine.	,, Keenan.
Rifleman Wright.	,, McIlhare.

Corporal Boyd had previously received this medal for riding through a cutting under a heavy rifle fire to warn a train of the presence of the enemy.

Whilst at Thaba 'Nchu, Major Sitwell formed a party of "local" mounted infantry. This differed from the ordinary mounted infantry, in that they found their own horses, generally by capture from the enemy, and remained with their battalions, instead of being taken away and formed into regular mounted infantry battalions, such as the 9th battalion of mounted infantry, to which " C " and " G " Companies belonged.

With this local mounted infantry Major Sitwell raided the surrounding country for twenty miles from Thaba 'Nchu and captured isolated parties of the enemy, and generally made himself obnoxious to the Boers. He was assisted by 2nd-Lieutenants Goodman, Maynard, Hogan, and McCulloch.

A description of two of his operations will suffice as showing the good work this mounted infantry carried out, with small means.

By means of his spies, Major Sitwell had ascertained that some twenty-five Boers were in the habit of sleeping in a deserted farmhouse twenty-three miles south of Thaba 'Nchu. Riding out there one night, he surrounded the house, boldly entered it and found his enemies lying asleep on the floor. The detachment outside heard him laughing aloud at the ridiculous scene, and were summoned to enter the house. All would have passed off without incident, only one of the enemy, waking up, saw to his horror a fixed sword-bayonet over him; trying to run away, he was pinned to the ground through the leg and his comrades were awakened by his violent screams. However, all surrendered quietly. Another incident occurred on the 30th of December, when Major Sitwell was riding back from some combined operations with Colonel Pilcher in the Korrannaburg district. He had halted for the night with eighty of his local mounted infantry, at Vlakplaats, some fifteen miles north of Thaba 'Nchu. A small picquet had most fortunately been placed under charge of Serjeant Atkins, on the top of the little kopje overlooking the camp. About 11.30 p.m. the picquet was rushed, and Serjeant Atkins and Riflemen Marshall and Nelis were killed, and the remainder wounded. The firing alarmed the troops bivouacking below. 2nd-Lieutenant Hogan was detailed to carry the hill with twenty-five men. He was promptly shot in the face, and the attack failed. Then Major Sitwell directed 2nd-Lieutenants Goodman and McCulloch to take twenty men each, and climbing up on opposite sides, to drive the enemy off. The Boers were believed to be at least 200 strong, but 2nd-Lieutenant McCulloch cleverly imposed upon them by shouting orders to imaginary captains of companies to bring up many reinforcements, and the Dutch evacuated the kopje and fled.

Besides 2nd-Lieutenant Hogan, 2nd-Lieutenant Goodman was also wounded in the affair, also Riflemen Shanks, Carson and Birch.

Serjeant Atkins had joined the Royal Irish Rifles from the Maltese Police, where he had been employed after some years' service in an infantry regiment. He was an excellent non-commissioned officer, and his death was much regretted by all. 2nd-Lieutenant Hogan had a curious experience with regard to his bullet-wound. The bullet remained in his head just above his mouth, for some years, causing him much pain, and he underwent two operations for the purpose of having it removed, but both were unsuccessful.

However, shortly after the last operation, whilst talking to a fellow-patient, the bullet fell into his mouth.

1902 A fine draft of 149 non-commissioned officers and men arrived on the 5th of March from the 1st Battalion, then in India, to reinforce the 2nd Battalion; whilst a similar number of men were picked from the 2nd Battalion and sent to join the 1st Battalion in India on the 8th of March.

The following were the total casualties suffered by the 2nd Battalion Royal Irish Rifles during the South African War:—

	Officers.	Men.
Killed in action	1	30
Died of wounds	2	2
Wounded	10	105
Died from disease	nil	51
Killed by lightning	nil	1
Killed in collision with armoured train	nil	2
Accidentally shot	nil	1
Drowned	nil	1
Total	13	193

The Royal Irish Rifles remained at Thaba 'Nchu during all the guerilla war with the Boers, until after the conclusion of peace, on the 31st of May.

Shortly after the war commenced, as has been noticed, "C" and "G" Companies were converted into mounted infantry.

"G" Company, under Captain Dimsdale, was formed into a mounted infantry company on the 21st of January. His subalterns were Lieutenant Spedding and 2nd-Lieutenant Saunders. On the 3rd of March Brevet-Major Festing formed "C" Company into a mounted infantry company, with Lieutenants Low and Davenport as subaltern officers. "G" Company was called No. 1 Company, and "C" Company became No. 2 Company. No. 1 Company was working busily a month after the action at Stormberg, and made an attempt to seize Bethulie Railway Bridge, over the Orange River. Despite their efforts, however, the enemy managed to blow the bridge up, but they had to do this prematurely, and left several railway trains behind them, which were promptly seized by the Rifles Mounted Infantry. Captain Dimsdale's Company was (as has already been stated in this chapter) captured at Reddersburg, and he himself was mortally wounded. Lieutenant Spedding was, however, detached with twenty-five men of No. 2 Company, and so escaped being involved in this disaster.

1900 Some forty men were called for on the arrival of "D" and "E" Companies at Bloemfontein from Smithfield, on the 15th of June. These men were formed into a mounted infantry detachment. 2nd-

Lieutenant Cooke-Collis, of the 5th Battalion Royal Irish Rifles, who shortly afterwards joined the 2nd Battalion, was with this party. In September of this year 2nd-Lieutenant Cooke-Collis and some of his men had been sent to Dewetsdorp, but in October were withdrawn, leaving the Orange River Colony Police to hold this town. So soon as the troops left, the Boers swooped down on the town and took the police prisoners. The Royal Irish Rifles Mounted Infantry were ordered back to the town, and a column, under Major Massey, Royal Field Artillery, was sent to Dewetsdorp also, to form the garrison. The column consisted of two guns of the 68th Royal Field Artillery, three companies of the 2nd Gloucestershire Regiment, and one company of the Highland Light Infantry.

The police officer, named Boyle, who, with his men, had been taken prisoners by the Boers, had been set at liberty, after being disarmed. A promise had been extorted from him whilst a prisoner that he would not remove the wives of certain Commandants to the Concentration Camp. When he was back again at his duties in Dewetsdorp he was ordered, from Bloemfontein, to remove these ladies, and, despite his protests, was obliged to carry out the order. The unfortunate 2nd-Lieutenant Cooke-Collis was sent to remove Mrs. Piet Fourie. He was ordered to show her every courtesy due to a Commandant's wife. However, the lady refused to be moved, and 2nd-Lieutenant Cooke-Collis had to carry out his distasteful task by force. On the 16th of November the mounted infantry located De Wet, with 3,000 men and four guns near Thaba 'Nchu. Piet Fourie and Haasbroek were with him, and both determined to have revenge on the British for the removal of their wives.

On the 18th of November De Wet commenced his attack on the town from all points, his first operation being to rush at 3 a.m. the Royal Irish Rifles's cossack post on Lonely Kop, a hill one mile west of the town. One man of this party jumped over the precipice to avoid being captured, but, miraculously, was not killed. The Boers then used Lonely Kop as their gun position during the whole of the operations. On the 20th the water supply of the town was cut off by the enemy, and on the 22nd that part of the position held by the Highland Light Infantry was stormed and occupied by the Boers.

The Royal Irish Mounted Infantry were, on the 23rd, allotted to definite parts of the defence, i.e., to two gun epaulments and to two trenches on either side of No. 1 gun epaulment. 2nd-Lieutenant Cooke-Collis was in this epaulment. The enemy, as usual, throughout sent their attacking party forward, covering their movements by a heavy fire. What the enemy's loss was it is hard to say, but Rifleman Potter accounted for two at a distance of 1,300 yards; both were killed in rapid succession as they tried to rush across a tiny stretch of ground devoid of any cover. Each position had been supplied with water-tanks, but the rifle-fire of the Boers riddled these tanks, and on the 21st, 22nd and 23rd there was no water to be had. On the morning of the 23rd of November the Boers, who had crept up during the night, tried to take No. 2 gun epaulment. Despairing of defeating this move with rifle fire, Corporal Fletcher gallantly led a bayonet charge on the enemy. He was killed ten yards in front of the epaulment, and another rifleman was wounded. The enemy then succeeded in getting into No. 2 gun epaulment, and, working round, came on the rear of No. 1 gun epaulment. Here a fight, of which the Regiment may well be proud, was progressing. This epaulment had two officers in it. Lieutenant Elam, Royal Field Artillery, was shot in the head early in the day, and the defence of the epaulment then devolved upon 2nd-Lieutenant Cooke-Collis. His force consisted of five riflemen and eleven Royal Field Artillery.

It is sufficient to say that when the Boers did finally enter the epaulment they found 2nd-Lieutenant Cooke-Collis badly wounded, with two men lying dead across his legs. All the Royal Irish Riflemen were killed or wounded, and all the gunners were in a similar position but one. That one, the farrier-serjeant, was doing just what his own regiment would have expected of him. He was loading, laying and firing his gun by himself. The total loss of the Rifles was not large—only four killed and five wounded. Dewetsdorp having surrendered, the wounded were left at the town and the unwounded prisoners were hurried away. At the first halting place, Boyle, the officer of the Orange River Colony Police, was taken out and shot by his captors. A certain person was sent to look for 2nd-Lieutenant Cooke-Collis amongst the prisoners, with a view to his being also murdered. Fortunately for 2nd-Lieutenant Cooke-Collis, he was lying severely wounded at Dewetsdorp, and so he escaped from the hands of these scoundrels.

On the 8th of August, No. 1 Company was reconstituted under command of Captain Clinton Baker, and Lieutenant Reeve, of the 4th Battalion Royal Irish Rifles, was given him as a subaltern officer. In the meantime, the men of the original No. 1 Company, taken prisoners at Reddersburg, had been released in June, 1900, and were remounted and sent to Irene and Klip River. These two sections were constantly employed in expeditions, and rejoined No. 1 Company, still under the command of Captain Clinton Baker, in December at Ventersburg Road, on the main railway line in the Free State. Serjeant Rainey and Rifleman Boyd, of the Klip River section, had the distinguished conduct medal conferred on them for their conduct whilst at that place.

No. 2 Company's first experience of mounted infantry work was riding bare-backed horses for some 100 miles, as no saddlery had been issued with the horses, and it had to be picked up at Stormberg. This company was with the Wepener relief force, Corporal McIlhare receiving the distinguished conduct medal for his gallantry in these operations. The company was then stationed on the railway line in the Free State; one section was at Zand

River, under Lieutenant Davenport, and took part in the defence of that place when it was attacked by Commandants Philip Botha and Haasbroek. These stout fighters were repulsed, No. 2 Company losing Riflemen Duddy and Montgomery, whilst Rifleman David Irvine received the distinguished conduct medal for his behaviour in this affair. In August Colonel Sitwell, of the Northumberland Fusiliers, under whose orders No. 2 Company then was, made a reconnaissance to the east of the railway line in the Free State towards a farm called Mooiplaatz. Colonel Sitwell decided to burn this farm for various offences committed against the recognised rules of war by its occupants. Unfortunately for him, some 2,000 Boers were bivouacking within two miles of the farm, and, naturally, when it went up in flames they came out of their laager like a buzzing swarm of disturbed bees, and attempted to surround the small party under Colonel Sitwell. The latter promptly rode as fast as their horses could take them for the shelter of the remainder of the force on the railway line. Colonel Sitwell was, of course, overtaken, and then wisely held every point of vantage so long as he could and retired by alternate parties. He directed Lieutenant Spedding, with a section, to hold a kraal at a farm named Klipfontein. This commanded a drift or ford over a small stream, and Colonel Sitwell retired, with the remainder of his party, to a ridge some 1,500 yards farther back. He told Lieutenant Spedding that he would hold this ridge and signal to him to retire. Unfortunately, Colonel Sitwell found himself unable to hold this ridge, and was promptly hustled off it, and the only signal the unfortunate section received was a heavy and accurate fire poured upon them from the ridge, which commanded the kraal. Twenty-three persons were shut up in this kraal, including the doctor, Lieutenant Nicholson. The kraal was promptly stormed by a party of some 300 Boers, who crept close up, almost unfired at, as anyone moving was shot down by the fire of a strong covering party of the enemy. Still, the fight was gallantly maintained, Lance-Corporal McIlhare again and again running out into the open, so as to be able to fire on the approaching enemy. Finally, a strong party of Dutchmen pushed their rifles over the walls, against the men's helmets, and ordered them to "hands up." The whole party was then stood up against a wall, whilst search was made for Colonel Sitwell, who had, fortunately, escaped. After considering whether they would not shoot the doctor, as they found one more rifle than there were men inside, and they wrongly considered he had used it, their wrath eventually subsided, and they behaved very well to their prisoners. Riflemen Brennan and Thompson were killed at this fight, besides several who were wounded.

Lieutenant Nicholson was shortly afterwards awarded the Victoria Cross for his gallantry.

After keeping this little party prisoners for two days, they were released under promise of proceeding to Ladybrand, many miles distant in the Free State.

2nd-Lieutenant Soutry brought some mounted infantry back to Ventersburg Road in October. They had been with him at Wolverhoek. The men went to No. 2 Company, and 2nd-Lieutenant Soutry took over charge of a section in No. 1 Company. These two companies were now forming part of the 9th Battalion of Mounted Infantry, and the whole battalion was commanded by Major Pine-Coffin. Each company was 153 strong.

1901 In January both companies formed part of Lieutenant-Colonel E. C. J. Williams's column, and they operated with him most successfully in the Heilbron, Frankfort, Parys and Kronstadt districts.

At first all treks were conducted in much the same style. On leaving the carefully-guarded and well-fortified supply depôt on the railway line, the column would push out for some destination known to its chief. The mounted infantry formed the advance, right and left flanks, and rear guards. The convoy and guns and spare troops moved in the middle, as in a frame. The mounted infantry, working in fours, scouted far and wide, generally keeping two or three hundred yards between each group. When any particular point like a hill or house had to be occupied it was "galloped" at once by the force nearest it, supported by troops from behind, whilst the neighbouring groups pushed round on both flanks, still keeping their relative places in the line, so as to bring cross-fire on the enemy's line of retreat. If it were found that the enemy held any locality very tenaciously, it was often sufficient to gallop straight at it, the men shooting off their horses at the gallop. Practically this never failed to turn out the Boers. The accuracy of this fire was amazing. Many of the men were able to hit ant-heaps repeatedly at a distance of 500 yards whilst galloping. The range was not taken but judged by the eye. Of course, constant practice, with unlimited ammunition, accounted for this good shooting.

As the column progressed away from the railway line, little dots, really men on horseback, could be seen riding away, these being the Boer vedettes; then more would be seen riding towards one of the flanks of the frame, and as the column got into difficulties, either through toiling over a drift, or trying to force its way through a defile, the rifle fire commenced in real earnest, and attacks and counter-attacks took place. This went on all day and every day, and, though not mentioned in detail, it may be understood that from January, 1901, to May, 1902, the 2nd Battalion Royal Irish Rifles Mounted Infantry were practically under distant fire every day, with some few exceptions. Therefore, only small details are now and then given to show how matters went.

At the very close of January, returning from Frankfort, on nearing Leuw Spruit, the enemy held up Colonel Williams's column for some time, but the Royal Irish Rifles Mounted Infantry made a wide turning movement, and, after having lost five men wounded and seven horses killed, were able to turn the Boers out of their positions. They lost two more men in these operations. Everyone was very businesslike, generally marching at 4.30 a.m., and halting at 5 p.m., or later, according to the water supply. The object of the movements was to clear farms of inhabitants and supplies. The latter were burnt; the former were brought into concentration camps. On the 26th of February 100 Boers tried to ride down No. 2 Company, then on rearguard, charging down on them, firing from the saddle. However, they were beaten at their own game, as the Rifles counter-attacked them in the same manner, sending their surprised enemy flying headlong.

On the 28th of March an amusing assault took place. Some Boers held a stout kraal very tenaciously, and the mounted infantry worked up to within 900 yards of it. The whole force was advancing, and the staff officer forgot to halt the transport, which at that time was mules entirely. Colonel Williams was in the firing line, which had just been joined by all the reserves, and had given an order for all to mount and charge down on the kraal, when he looked round and saw the mule-wagons joining in the charge, the native drivers of the wagons standing up and uttering wild war cries, whilst they cracked their whips. The kraal was taken at once, the Boers losing four killed and seven wounded; the British lost two men wounded.

On the 12th of April, marching at 2 a.m., the mounted infantry came to the Wilge River, and were challenged by Boers from the opposite side. A Colonial guide replied in Dutch that this was Steinkamp's commando, moving north, but the sentries refused to let them pass until daybreak. The river was sixty yards wide and four feet deep, and the Rifles had to wait. At daybreak the enemy fled. The mounted infantry and 300 of Driscoll's Scouts crossed, and some wagons and a prisoner were run down and captured. Coming back from the raid at 6 p.m., the Wilge was again reached.

No 1 Company was doing rearguard work, and the Boers pressed it hard. Thinking all were over, Lieutenant Saunders galloped down to the river with his section, which had been the rear screen, to try and cross likewise. Captain Clinton Baker had to send him back to the nearest ridge, towards the enemy, to cover the withdrawal of those not yet over. As Lieutenant Saunders reached the top, with twenty-five men, he found 150 Boers charging up the opposite side. A young Commandant, reining up, fired at him three times at fifty yards' distance, whilst Lieutenant Saunders returned the compliment round for round. Both fired from the saddle, and missed every round. Each gracefully retired over his own side of the hill. Serjeant Beatson was confronted by a Commandant, waving his rifle in the air as he charged, shouting "Come on, burghers," and he promptly shot him out of the saddle; whilst Rifleman Fortune performed the same office for the next Dutchman. No one on either side had dismounted, and Lieutenant Saunders now gave the order to gallop to the river. The section galloped down, amidst a hail of bullets. Strange to relate, only two horses were killed, and the Boers were beaten off. Camp was reached at 10 p.m.; the horses had covered sixty miles, and two

Boers had been killed, and four prisoners, with sixteen rifles and five wagons, had been captured.

Lieutenant-Colonel Williams was now transferred to a larger column in the Transvaal, and he took Lieutenant Spedding with him as his staff officer. His farewell order, dated Wolverhoek, 21st of April, 1901, was as follows:—

"Lieutenant-Colonel Williams, on being transferred to another command, desires to express to all ranks his gratitude for the soldierly qualities displayed by them, which has made the command of the force a real pleasure.

"The two companies of the Royal Irish Rifles Mounted Infantry, under command of Captain O. Clinton Baker and Lieutenant Spedding, respectively, have been with him during many treks, and he takes this opportunity of again saying he never wishes to command better mounted infantry than these two companies."

Lieutenant-Colonel Williams had commanded the 3rd Mounted Infantry Corps, so the above order was written by a man who understood what mounted infantry should be. Captain Laurie, having come from England as a special service officer for mounted infantry work, now took over the command of No. 2 Company.

Major (Local Lieutenant-Colonel) Western, of the West Kent Regiment, took over the command of Colonel Williams's column. On the 7th of May some of the mounted infantry, with some of Driscoll's Scouts, took prisoners thirty-two Boers, and on the 12th of May the column moved into the Transvaal, via Viljoen's Drift, near Vereeniging. On the 2nd of June the enemy made a valiant attempt to burn out the camp, setting fire to the dry grass up wind. A hurricane was blowing, and the flames came down twelve feet high. Fortunately, Captains Laurie and Baker had obtained permission from Colonel Western, after some trouble, to burn a belt of grass round the camp; this had been done for the first time that day, and the belt stopped the flames. Needless to say, after this experience, such a precaution was always taken. Parys and Bothaville were all visited during June. Vredefort also was cleared. Here a Boer rode up to within thirty yards of four men of No. 2 Company, who were riding with Captain Laurie. Some fighting was going on at the time, and the country was bushy and hilly. Though actually challenged three times by that officer, he turned and tried to escape. He was buried next day, and his horse remained in the company until the end of the war.

On the 23rd of June Colonel Western's column raided to the south suddenly from Klerksdorp, which it had reached on the 22nd. At daybreak some of Driscoll's Scouts opened proceedings by a chase of twelve miles after some Boers, all of whom they captured, excepting one, and also ten wagons. One lady jumped out of a wagon and sprang on a horse with a man's saddle and disappeared in the distance, waving a rifle; another one complained of her departure, as she had taken her rifle, with which she stated that she had killed three British soldiers. Three days later she was caught driving a Cape cart, and she shot dead at a range of three yards a serjeant-major, who rode up to speak to her. She was sent into Klerksdorp to be tried for murder. No. 2 Company, having reported Boers on the right flank, was ordered to pursue, and did so, picking up a prisoner, five wagons, etc., and chased the remainder to Bothaville. As there were apparently some 500 Boers, it was not considered advisable to continue the chase with only one company, and camp was reached late at night, after sixty miles in the saddle. On returning to Klerksdorp on the 26th of June, it was found that Colonel Hamilton and Colonel Allenby had each brought a small column in, and arrangements were made to clear the country in the direction of Hartebeestefontein. Accordingly, at 8 p.m. on the 27th the three columns started, No. 2 Company leading. As the village was only some twenty miles away, the start to arrive there at dawn was some four hours too early. On arrival near the place No. 2 Company proceeded to surround the village, about an hour before dawn. Whilst doing this, the Officer Commanding the column rode up and ordered the company to push straight through, adding that the map was incorrect, as there was no village there at all. The company pushed on in the dark, four men on foot, with fixed swords, and an officer, leading the way. Two flanking men walked into twenty-seven Boers sitting round a fire in a side street. Who were the most surprised it would be hard to say. At any rate, both sides bolted at once; but the two riflemen returned with some others and picked up five rifles and one suit case, packed. Still pushing forward, in accordance with orders, as day broke, No. 2 Company found itself on a ridge north-west of Hartebeestefontein. In the distance a party of the enemy could be seen seated round their fire; knee-haltered horses were feeding in the close vicinity. Those seated round the fire saw the mounted infantry, and rushed for their horses. They were driven off by a heavy fire at 1,700 yards' range. Captain Laurie then gave the order to mount and gallop down on this party, to secure the horses and, if possible, to take the Boers prisoners. Corporal Jackson and Captain Laurie got some 300 yards in advance of their men, and passing a house were received with a heavy fire at twenty-five yards' range.

Both threw themselves from their horses and rolled over in the long grass, whilst from the doors of a double cottage three men poured in a fire with a Mauser pistol, a Mauser rifle, and a Lee-Speed rifle, whilst two men in the garden alongside used Lee-Metford rifles. A native dashed out of the house, really to try and saddle up some horses. Corporal Jackson, seeing him rushing towards Captain Laurie, shot him dead, whilst another man in the house received a wound in the left arm. Two more riflemen of Corporal Jackson's group crawled up to him through the grass, and the house was rushed, and after a scuffle inside, all were taken prisoners. Corporal Jackson was a very brave man, and was recommended on two occasions for the distinguished conduct medal. Some twenty Boers galloped away from the vicinity of the house. Two more prisoners were taken farther on, and five of the enemy managed to get away, but under a heavy fire at close range, so some may have been wounded. Serjeant Channel took a prisoner that day in an amusing manner. Taking his rifle down hurriedly from his shoulder, a man jumped up out of the long grass with his hands in the air. He had been watching the serjeant from his hiding place, and, seeing the quick motion of the rifle, thought he was discovered, and so surrendered. Captain Baker asked leave to search a kloof. Time pressed, and it was refused. It was afterwards found, from one of the prisoners, that Commandant Liebenberg and sixty men were hiding there as their last refuge. Amongst the captures in the house was Commandant von der Winder, Adjutant of Staats Artillery, and Serjeant von der Merve, of the same force.

Returning to Klerksdorp at once, the column left next day, going west. On the 1st of July, having marched at daybreak, the column was moving forward to a dam to camp at about dusk. All the mounted troops were engaged with the enemy on the banks of the Vaal, some distance off, excepting two sections of No. 2 Company. Some Colonial scouts came back and reported that some Boers were driving cattle away from the dam. Lieutenant Low was sent forward, with 2nd-Lieutenant Davenport, and the two sections of No. 2 Company. Instead of cattle, it turned out to be knee-haltered horses, and there were some 350 Boers. Lieutenant Low, who was a most gallant officer, quickly grasped the situation, and sending 2nd-Lieutenant Davenport with a section to one flank, moved under the ridge to the other flank, and both sections then promptly charged home, yelling and shooting off their horses. The Boers fled. Two were picked up dead, one was a poor old man, with a top hat and umbrella, not to mention a rifle and full bandolier. Some days afterwards their flight was explained by a prisoner as arising from the fact that all their officers were away at a conference at the time.

On the 2nd of July two sections of No. 2 Company were detached seven miles from the column. Five Boers appeared in front of No. 1 Section, and Colour-Serjeant Edwards, who happened to be in command of it, gave chase. After a hunt of a couple of miles 100 Boers appeared and the tables were turned, No. 1 Section riding hard to reach No. 2 Section, which was drawn up on a rise. Some 160 Boers now appeared out of a valley, and galloped down the valley, directly across the front of No. 2 Section, intent on heading off Colour-Serjeant Edwards and his twenty-five men. No. 2 Section opened fire to their front and stopped the rush of the 160 Boers. In the meantime, the enemy were gaining on No. 1 Section, three men, in particular, being within 100 yards of them. Serjeant John Watson suddenly swung his horse round, shot the leading Boer at 100 yards' distance, and galloped into his place. The other two looked after their comrade. No. 1 Section joined No. 2 on the ridge, where the Boers proceeded to attack the whole party, but Captain Laurie, having heliographed a message so soon as No. 1 Section had gone that he required support, Captain Clinton Baker brought No. 1 Company out at a gallop, and the Boers were driven off. On the 4th of July Commander Potgieter tried to ride down the Rifles Mounted Infantry by a dashing charge of 200 Boers in the usual style, but was driven off. Serjeant Reid was saved that day by a hoofpick on his belt turning a bullet.

On the 8th of August the column then was taking part in a combined movement of five columns from the Vaal River to the Modder River, under General Elliot. No. 1 Company was sent forward, about 2 o'clock in the afternoon, to assist a section of some other mounted infantry, which had got mixed up with 500 Boers, and had already lost six men prisoners and one wounded, out of their twenty-five men. 2nd-Lieutenant Soutry was in advance, and he boldly dashed in to thirty wagons, being a mile ahead of everyone else.

A regular game of hide and seek went on. As fast as this party seized a wagon and turned it towards the British, some of the 500 Boers would dash forward to re-capture it; then from all sides people raced forward, some to support the capture, others to re-take it. The weaker party (the mounted infantry) used then to fly back into the big convoy of Boer wagons and seize some more, only again to be attacked by overwhelming numbers. Still, all this delayed the convoy, and the remainder of No. 1 Company came on fast. Then the Boers turned, and, leaving 2nd-Lieutenant Soutry amongst the wagons, attacked No. 1 Company furiously. Colonel Western, who was with No. 1 Company, sent to Captain Laurie to bring him assistance, and that officer galloped out with No. 2 Company, two squadrons of Driscoll's Scouts and a gun. Seeing him coming, the Boers fled, whilst 2nd-Lieutenant Soutry and his section, who had been working with great industry, carried off, despite the Boers' interference, the thirty wagons and nine prisoners.

Colonel Western's column had already marched 26 miles, having been on the move since 3 a. m., when Driscoll's Scouts had captured another laager with some prisoners, and, assisted by No. 2 Company, had beaten off a counter attack and had just brought their capture in as this second fight began.

On the 6th of August 2nd-Lieutenants Soutry and Cooke-Collis had been sent with despatches to another column, and travelled ninety miles, with their fifty men, in twenty-eight hours. Arriving at Bloemfontein on the 11th of August, the two companies of mounted infantry were on the 13th hustled into trains and sent south to Cape Colony to head a party of Transvaalers, under Smuts, who were trying to cross the Orange River.

The mounted infantry held the south bank of the river from Aliwal North to the Basutoland border (Herschel Reserve), with the Connaught Rangers and Lovat's Scouts, until the 3rd of September, when all the troops were marched north across the Orange River, into the Free State. Naturally, Smuts and 600 of his men crossed south by another drift into Cape Colony. When the mischief had been done the two Royal Irish Rifles companies of mounted infantry were sent in pursuit. The chase was not very vigorously pressed for some reason, and ended up in a sleepless night with no food to speak of, and no blankets or great-coats, and bitterly cold weather. On the 13th of September Captain Clinton Baker was taken out on a reconnaissance by Colonel Western, with thirty of his men. He was attacked by Commandant Fouche, with 300 Cape rebels. Serjeant Taylor was shot in the head, and the only chance of saving his life was to get him quickly home to the doctors. Captain Baker conducted his retreat most skilfully, keeping the Boers at bay, though they attacked most resolutely, whilst he and Rifleman Fortune carried Serjeant Taylor in their arms on horseback several miles, getting him home alive, but he died shortly after his arrival in the bivouac. The weather was bitterly cold, and the horses commenced to die of exposure in the lines.

On the 20th of September No. 2 Company crossed the Krai River in flood in obedience to orders. When it was found that no one else could cross, the company was ordered to return by flag signal. In coming back Rifleman Bradshaw's horse was swept away, but Bradshaw was pulled out by an officer and his horse swam ashore.

On the 11th of October, being then in the Free State again, one of the horses of Lieutenant Low's section fell down a gully, estimated to be 150 feet deep. He struck a couple of ledges in falling, and broke his saddle up, but beyond being a little stiff next day, was none the worse for his adventure. The two companies arrived at Thaba 'Nchu on the 27th of November. They were given a warm welcome by the battalion, and also a good many recruits for the rather reduced ranks of the mounted infantry. The two companies only stopped one day with their comrades and then moved off again, but on the 1st of December they obtained their first opportunity to operate at their own discretion, for three days, being sent to reinforce General Thorneycroft's column, thirty miles south of Jammersberg Drift. Hiding both companies in a depression in the ground, observation was kept, and on the 3rd of December patience was rewarded, by sixteen Boers passing near the place. Information of this only arrived from the observation post just two hours after they had passed, as the messenger, Rifleman Cross, lost his way. The Boers were tracked and pursued for twelve miles over the mountains, but escaped, leaving a pack mule and two horses and some ammunition in the hands of the Rifles.

At 2.30 a.m. on the 5th of December the two companies stole out of camp and proceeded to "drive" a ten-mile mountain towards each other. At dawn some of the enemy were seen approaching the mountains, having got inside the observation posts, which had wisely let them pass, and then had closed round them. Nine Boers, two natives, and twenty-one good horses were the result of this lucky meeting.

On the 14th of December No. 1 Company picquetted the veldt paths very cleverly and caught some Boers riding through by night.

The 30th of December saw the Mounted Infantry in the western half of the Free State, riding over the historic field of Boomplaatz, where Sir Harry Smith beat the Boers in 1848.

The column crossed into the Western Cape Colony on the 13th of January. Having been three months without eggs or fresh milk, and these articles of consumption being cheap and plentiful there, were greatly appreciated. At this time a General Nieuwhoudt's commando was being hunted back into the Free State, as it had crossed into the Colony. It was headed in rather a curious way. The column was trekking through a sandy country near the Riet River on the 14th of January, Nos. 1 and 2 Companies doing escort to the guns, when a Colonial scout galloped back to say that the advanced guard was close up to 350 Boers, who were bathing; but as the advanced guard was composed of Yeomanry just out from England, they appeared to be in doubt what to do in the matter. Colonel Western proceeded to mount his horse, blowing a whistle; the pom-pom commenced to gallop forward. For over fifty yards the Colonel failed to get into the saddle, but stuck manfully to his cantering horse. The mounted infantry, wishing nothing better, galloped forward, a company on each side of the gun forming line at three paces interval as they rode. Being very strong, it was a most inspiring spectacle of nearly 800 yards' front of galloping horses. No. 1 Company found the Boers bathing and dashed at them in front, whilst No. 2 Company, seeing the fugitives galloping away, cantered to the east to cut off their retreat and to pin them up in the triangle between the Reit and Orange Rivers. However, it was recalled by the staff officer to do escort to the pom-pom, and the Boers at once slipped away back to the Free State. Their escape was facilitated by a very heavy thunderstorm which commenced as the Rifles reached the Reit River. For twenty minutes nothing could be seen on account of the torrents of rain. Three prisoners, some rifles and horses were taken; it was a lucky escape for the Boers. A stern chase then commenced, winding up on the 19th of January. From 1 a.m. to 4 a.m. on the 18th a rapid march was performed, following on a heavy march of the day before. Leaving again about 6 a.m., twenty-five miles more were done. At 6 p.m., under Colonel Rochfort, another fifty miles were performed, the force halting at 5 a.m. Volunteers were now called for to carry on the pursuit, and the whole of No. 2 Company volunteered to go, and all but twelve of No. 1 Company. These latter had their horses done to a turn. Pushing on at noon, after a scanty meal, called breakfast by courtesy, the force, under Captain Laurie, reached the point where the Boers were understood to be entrenched. It was twenty miles further on; but there was no sign of the enemy. The Rifles then heliographed back that the enemy had flown, but that they were prepared to go ahead again. Colonel Rochfort replied to remain where the force was until daybreak on the 20th, and then to retire. This was carried out, food being obtained at about 12 noon on that day. The reason of the recall was that Colonel Rochfort had heard that Nieuwhoudt had been strongly reinforced.

1902

In February the same hard marching continued often night and day. On the 12th of February, as the column was moving out for a raid from a farm some twenty miles from Edenberg, a hailstorm came on, lasting twenty minutes. The hailstones were over an inch in diameter, and the camp was four inches deep in hail. Every horse stampeded, excepting three, which were killed either by the hailstones or by the lightning—it is not known which. The stream alongside the camp rose from a few inches deep to a raging torrent of five feet in depth. It had to be crossed, however, to collect the horses, and a line was taken across by a party of officers and men holding hands. Four horses were lost, two had to be shot, and twenty were sent back to the hospital on the railway line, having been cut badly by barbed wire.

On the 21st of February two brave native scouts were killed, being surprised at ten yards' distance by some twenty Boers. The two natives most gallantly galloped straight at the party and were both killed; but No. 2 Company, hearing the firing, rushed the hill, which sheltered 250 Boers, and after a long chase brought one home as a prisoner. This was also at the end of a fifty-mile night march.

On the 26th of February Captain Laurie was ordered to proceed to Pretoria to take charge of the 28th Battalion of Mounted Infantry, and he handed over the command of No. 2 Company to Lieutenant Low. On the 10th of March Lieutenant Low made a reconnaissance in advance of his column, and captured three Boers, with many cattle and sheep. He then moved towards his column; failing to find it, he told the company to remain on some rising ground, whilst he rode forward. Meeting two mounted men in British uniform, he spoke to them, and then all three dismounted, and Lieutenant Low gave them a cigarette each, under the impression that they were troopers in some Irregular Corps. The two men had their rifles, whilst Lieutenant Low was unarmed. The two men then said to him: "We are Boers, and you must surrender. The brave little officer promptly refused to surrender, and seized the larger ruffian by the throat. The latter held him, whilst his companion walked up and shot Lieutenant Low in the back, killing him instantly. They then rode away, taking his horse and carbine. One of these men was a Colonial rebel. During the war the names of the two men were sent to Pretoria, but they were not located until peace came, when the Government refused to allow any steps to be taken in the matter. Lieutenant Low was supposed to be the smallest officer in the British Army. He was so small that, after passing well out of Sandhurst, he was refused a commission in the British Army, owing to his shortness of stature. Being in Germany for a year, he was offered a commission in a German cavalry regiment, but before accepting it came over to England, and made another attempt to enter the Army, appearing before a medical board, which passed him at last. He was a brave little gentleman, and his untimely death was greatly deplored by all ranks.

On the evening of the 23rd of March the two companies moved out to form part of a driving line. They travelled ninety miles in some twenty hours, capturing three prisoners and some transport, etc.

On the 15th of April the mounted infantry formed part of a small force under Colonel Rochfort, which surrounded Schweizer Reneke, after a ride of forty-three miles by night, sixty-three prisoners, with much transport, horses and cattle being the result. The following day Colonel Rochfort's party was shadowed by a force of the enemy, but ten more Boers were captured. On the 9th of May Captain Clinton Baker handed over the command of his No. 1 Company to Lieutenant Saunders, as he had been sent to command the 21st Battalion of Mounted Infantry.

No. 2 Company took part in the final drive of the war—right across the Western Transvaal from near Klerksdorp, winding up on the boundary fence of the Transvaal on the 10th of May and closing on Vryberg on the following day. No prisoners were taken by the company, though the force picked up 169 Boers, many of whom were Colonial rebels. A patrol was sent out on the 16th of May on the march back to Bloemhof, commanded by Serjeant John Watson; it consisted of men of No. 2 Company and of some men from the Black Watch and West Kent Mounted Infantry. Having taken a prisoner, they found their retreat cut off by forty Boers. Serjeant Watson took his own men, four in number, straight through the ranks of the enemy, his own horse being hit in two places. The eight men of the other mounted infantry galloped round the flank, losing one man killed.

The local Boer commandos surrendered to Colonel Rochfort at Schweizer Reneke after peace was proclaimed, and out of all his six columns that officer chose one troop of the Royals and one section of the Royal Irish Rifles Mounted Infantry to form his escort. Lieutenant Cooke-Collis commanded this section.

Colonel Rochfort afterwards wrote: "During the whole of my time in South Africa I did not command better or more mobile troops than the two mounted infantry companies of the Royal Irish Rifles, under Captain Laurie and Captain Baker."

Lieutenant-Colonel Western sent the following letter to the Officer Commanding 2nd Battalion Royal Irish Rifles, dated 30th June, 1902:—

"On the two mounted infantry companies, Royal Irish Rifles, rejoining your battalion, I would wish to inform you what excellent troops they have proved themselves in every respect. They have been the instructors of the other units of the column. Their horse-management has been exceptionally good. Though I have not the experience Lieutenant-Colonel Williams has, who informed me as far back as January, 1901, that they were the best mounted infantry he had met, I can honestly say that I have never come across any better, and I strongly suspect Colonel Williams' opinion was a correct one."

By some mistake no names were forwarded to headquarters of the mounted infantry of the Royal Irish Rifles until after the lists for despatches had been finally closed.

On the 13th of November, 1901, 100 non-commissioned officers and men of the 5th Battalion Royal Irish Rifles, who were serving as Militia Reservists with the 2nd Battalion Royal Irish Rifles, were sent back to rejoin their own battalion, which had just arrived in South Africa, having volunteered for service abroad. Amongst those who rejoined this 5th Battalion were several men from the mounted infantry. These joined a company of mounted infantry raised from that battalion by Captain Cole Hamilton, and when Captain Laurie joined the 28th Battalion Mounted Infantry he found this company attached to his command.

On the 31st of March about 200 men of the 28th Battalion Mounted Infantry marched off at 2.35 a.m., towards Boshbult, forming part of a column 1,800 strong, under Colonels Cookson and Keir. After a long march and some fighting, the 28th Mounted Infantry were put to guard the eastern side of the locality, being one and a half miles from the camp. Three thousand five hundred Boers, under the command of General Delarey, charged the camp, but were beaten off, and then 1,200 closed round the 28th Mounted Infantry, which happened to be divided into two parts. After a two-hours fight, the mounted infantry were ordered by Colonel Cookson to fall back upon the entrenched camp. One half of the battalion was at once ridden over by the Boers. The remainder, consisting of the Royal Warwickshire Mounted Infantry and the 5th Battalion Royal Irish Rifles Mounted Infantry, fell back in capital order, the Warwickshire Mounted Infantry forming the firing line, and the Rifles being told off to watch the flanks. The Warwickshire Mounted Infantry behaved very gallantly, losing one half of their number killed and wounded, whilst the Rifles, well handled, by Captain Cole Hamilton, lost no one, though the captain had two bullets through his own soft hat. The despatches as sent home by Lord Kitchener commenced: "The 28th Mounted Infantry behaved splendidly." Earlier in the day a little rifleman, who could not really have been more than 17 years of age, was shot through the arm, but only smilingly turned to his serjeant, remarking, "See my Easter egg?" (It was Easter Monday.) Having tied it up, he remained in the ranks fighting during the rest of the day. Eight officers out of seventeen were killed and wounded on this day alone in this battalion. Next day General Delarey rode off to make peace.

Whilst this was being arranged, however, many further encounters with his men took place. On the 26th of April Lance-Corporal Anderson, of the 5th Battalion Royal Irish Rifles Mounted Infantry, was out on patrol with three men, when he was attacked by four Boers. He mortally wounded one man and killed another man's horse, and then sallied out and captured one prisoner. It was afterwards ascertained that one of the four was a woman dressed as a man.

Many other incidents could be given, but these few will suffice to close the career of the mounted infantry.

On the 5th of July the 2nd Battalion Royal Irish Rifles moved to Bloemfontein, and on the 15th of July the first draft of reservists was sent away. It consisted of 200 non-commissioned officers and men. On the 9th of August, the battalion marched away from Bloemfontein to Modder River station, arriving there on the 17th. It left 570 Reservists at Bloemfontein for despatch to England. Seventy men were sent to India to join the 1st Battalion on the 4th of November.

The following further promotions were made, decorations conferred, and mention in despatches given, in recognition of services rendered during the latter half of the war:—

To be Companion of the Bath, Lieutenant-Colonel B. Hawes.

To be brevet-major, Captain E. C. Bradford.

Mentioned in despatches, Captain G. B. Laurie.

To have the distinguished conduct medal, Riflemen Anderson and Beck.

Colour-Serjeant Crowley, of No. 1 Company, Mounted Infantry, was shortly after the war promoted Serjeant-Major of the 2nd Battalion Royal Irish Rifles, whilst Colour-Serjeant Edwards, of No. 2 Company, was afterwards made quartermaster of the same battalion.

Whilst the South African War was progressing, the Royal Irish Rifles were called upon to furnish a representative detachment to proceed to Australia, to form part of the Imperial Representatives Corps, which was composed of troops of all arms from the United Kingdom and India. This corps attended all the celebrations connected with the inauguration of the Australian Commonwealth, including the administration of the oath to the first Governor-General of that great colony. Here the corps formed a guard of honour to His Excellency.

This party consisted of twenty-three non-commissioned officers and men, and was commanded by 2nd-Lieutenant Ley Becher. They received a warm welcome from all ranks and classes in Australia, but particularly from the "Irish Rifles" of Sydney. This detachment was furnished from the men attached to the details of the 2nd Battalion at Londonderry, and left England in November, 1900, returning again in April, 1901.

"G," or No. 1 Mounted Infantry Company, was now ordered to return to the battalion, as the twenty-nine mounted infantry battalions in South Africa were in process of being reduced to the present number of two. "C" Company, or No. 2 Mounted Infantry Company, had been very greatly reduced by all the Army Reserve men being sent home to England, so every available non-commissioned officer and man was transferred from "G" Company to "C" Company. The result was that "G" Company rejoined the battalion with one colour-serjeant, one serjeant, and one rifleman only on its strength.

The Battalion was now ordered home to Ireland. "C" Company rejoined from the 9th Mounted Infantry Battalion, and the Royal Irish Rifles left South Africa, without regret, early in January, 1903. Landing in England on the 5th of February, it crossed over to Kingstown, in Ireland, on the following day, whence it proceeded to Dublin, and was quartered at first in Richmond Barracks.

Whilst stationed at Dublin it was sent to the Curragh camp, in 1904, for its summer training, marching to Glen Imaal, for field firing. In 1905 and 1906 it went to Kilbride camp, for its musketry training, etc., and went to Maryborough for its manœuvres.

In 1903 its manœuvres were cut short, owing to incessant rain. The battalion was also selected by His Royal Highness The Duke of Connaught to represent the Army in Ireland at the funeral of Field-Marshal His Royal Highness The Duke of Cambridge, and fifty men were sent over to London, under the command of Captain Laurie and Lieutenant Soutry. This party was most hospitably entertained at Wellington Barracks by the Grenadier Guards, and after attending the funeral, returned to Ireland.

The Rifles also furnished a detachment to guard the country residence of the Commander-in-Chief in Ireland. This party consisted of forty men, under Captain Clinton Baker and Lieutenant Becher, and it was stationed at Boyle for over a year.

The battalion remained in Dublin until the 22nd of November, 1906, when, under the command of Lieutenant-Colonel J. Tobin, who had succeeded Colonel Hawes, it moved to Aldershot, where it was quartered in Albuhera Barracks. It numbered twenty-four officers and 545 of all other ranks.

Colonel Hawes had the curious experience of giving up the command of his battalion twice. The point was that he had been appointed to command a battalion of Royal Munster Fusiliers, which had never existed. It was an open question if his period of command counted from the date of his appointment to the command of this battalion, or from the time he was appointed to the 2nd Battalion Royal Irish Rifles. At first, it was considered that it should count from the first appointment, so he resigned the command and left the battalion, but later on was summoned back and replaced in command for a short period.

Lieutenant-Colonel Tobin retired shortly after the battalion arrived at Aldershot, and was succeeded by Lieutenant-Colonel Napier, who had been brought in to the Royal Irish Rifles from the Cheshire Regiment.

1908

Whilst at Aldershot the Regiment won the Army Football Cup. His Majesty The King (then Prince of Wales) attended the match, and presented the cup to Lieutenant-Colonel Napier. The captain of this notable team was Rifleman Edmondson.

The Rifles went on manœuvres in 1907 in Berkshire, in 1908 in Hampshire, in 1909 in Oxfordshire.

It left Aldershot on the 23rd of September, 1909, and went to Dover. Here it was quartered, first at South Front Barracks and at Fort Burgoyne for a year. After that it was moved to the Citadel.

In 1910 the battalion went to Lydd for its musketry training, and to Wells for manœuvres. The weather was not propitious, and after a fortnight's rain the camps were a sight not to be forgotten.

The Rifles also attended the funeral of His Majesty King Edward VII. in London, and lined the streets on that occasion.

1911

In 1911 the battalion was ordered to send a representative detachment of fifty men to London to attend the Coronation of His Majesty King George V. Major Laurie was in command of this detachment, and had Lieutenant Thomas and 2nd-Lieutenant How with his party. Coronation medals were afterwards presented to Colonel Napier, Major Laurie, Lieutenant and Quartermaster Edwards, Serjeant-Major Loakman, and Rifleman Bourne. Major Bradford had been taken for some staff duty on this occasion, and received a Coronation medal also. The detachment was stationed in Whitehall, close to Westminster Abbey, and was actually standing on parade for just over twelve hours that day. It was camped in Regent's Park, and was on duty on the day following the Coronation, when the King drove through London, being stationed beside Charing Cross Station, blocking the Strand there.

The musketry course was carried out at Hythe this year, the battalion being encamped for a month at Dymechurch. There were no manœuvres, owing to a railway strike.

Lieutenant-Colonel Napier handed over command of the battalion to Lieutenant-Colonel Hamilton Bell in November, 1911.

In 1912 the Rifles again went to the Lydd rifle range, and proceeded to Aldershot, and to Hitchin, in Hertfordshire, for their training and manœuvres, returning directly from Cambridge at the close of the Army manœuvres and proceeding to Tidworth, which place they reached about 3.30 a.m. on the 20th of September, 1912, and were quartered in Bhurtpore Barracks there.

NOTES.

The following men received Distinguished Conduct medals during the war for various acts of gallantry:—

2473, CORPORAL J. BOYD, on the 18th of February, 1901, under a heavy fire, rode through a defile, held by the Boers, to warn an advancing train (in which was Lord Kitchener) of the presence of the enemy.

2615, SERJEANT R. RAINEY, for handling his section of mounted infantry on rearguard in a gallant manner, whereby a party on patrol from Klip River, which was being pressed by a superior force of the enemy, was enabled to regain its lines.

2ND BATTALION ROYAL IRISH RIFLES (1899-1912).

3033, SERJEANT J. DARRAGH, while in command of a section, held a post near Thaba 'Nchu, under Lieutenant G. H. Sheen, London Irish, and repulsed De Wet in his attempt to cross the line at that point.

2381, CORPORAL R. IRVINE, for gallantry in the defence of Zand River on the 14th of June, 1900, and of Klipfontein on the 19th of August, 1900.

5356, CORPORAL J. WRIGHT, at Wolvehoek, when De Wet crossed the line east to west, about September, 1900, being pursued by De Lisle, was sent out as a scout by the officer commanding detachment at Wolvehoek, and, in spite of being repeatedly under heavy fire, kept touch with the enemy, avoided capture, and brought back accurate information.

2618, RIFLEMAN J. HANLON, for having, in a reconnaissance near Stormberg, in January, 1900, when Rifleman McLoughlin's horse had been shot, gone back, under heavy fire, and assisted McLoughlin back to his section.

4990, RIFLEMAN J. HOGG, and 2895, RIFLEMAN J. KEENAN.—When Colonel Williams's column was proceeding from Kroonstad to Heilbron, about March, 1901, a man of Driscoll's Scouts was wounded and lost his horse. Hogg, Keenan, and 5194, Rifleman Campbell, alternately carried the wounded man and kept off the enemy, though they were under fire from both sides, being behind the rearguard.

895, CORPORAL G. MCILHARE, for gallant conduct and scouting generally during the operations leading to the relief of Wepener, and also in the defence of the kraal at Klipfontein, on the 19th of August, 1900.

4038, RIFLEMAN R. BECK, for repeated gallantry whilst serving with the local mounted infantry at Thaba 'Nchu.

RIFLEMAN ANDERSON, 5th Battalion.—This man belonged to No. 1 Mounted Infantry Company, and afterwards joined the mounted infantry of the 5th Battalion, and showed great gallantry on all occasions, especially on the 26th of April, 1902, when attacked by four Boers, he having with him only three very young Militia men.

CHAPTER XVII.*

DRESS, COLOURS, AND MEDALS.

Contributors—Dress of 83rd and 86th Regiments when raised—Dress in 1796—Hair worn in a queue in 1799—Queues abolished in 1808—86th Regiment changed to Royal Blue Facings in 1812—Dress of the 83rd Foot in 1826 and in 1829—Cross belts abolished in 1850—Tunic introduced in 1855—Grenadier and Light Companies done away with in 1860—Glengarry cap introduced in 1871—White dress of the Band abolished in 1873—Badges of rank displayed on the shoulder strap—83rd and 86th Regiments both changed into Rifle uniform in 1881—Description of Rifle uniform—Changes made in it to present time—Various Colours carried by the 83rd and 86th Regiments—Colours no longer carried after being created Rifles—First medals issued to the officers of the 86th Regiment in 1801—Five Gold Medals granted to the 83rd Regiment—Peninsular Medals—Indian Mutiny Medals—South African Medals—Minor expedition medals—Victoria Crosses—Distinguished Conduct Medals—Meritorious Service Medals—Regimental Medals.

1793

WHEN the 83rd and 86th were raised in 1793, the uniform of the Army was in a transition stage, passing from the rather complicated dress worn during the eighteenth century to the more simple and more serviceable garb that came into use during the long period of war with revolutionary and imperial France. Officers of line regiments wore a scarlet coat just meeting across the chest, then cut away to show the waistcoat; the skirts of the coat were long, but not very full, were lined with white and turned back, and had laced pockets. The coat was lapelled to the waist with the regimental colour (yellow for both 83rd and 86th), lapels three inches

* The writers of this chapter, Colonel Haggard and Captain Stevens, wish to place on record their acknowledgment of the assistance received from Dr. A. Payne, of Sheffield, and D. Hastings Irwin, Esq., of Buxton, who have been most kind in giving their help in that portion dealing with medals. Mr. Irwin's "War Medals and Decorations" has been freely consulted. The Rev. David Wilson, Succentor of St. Patrick's Cathedral, the Rev. S. Smart, Vicar of Newry, Captain Roger Hall, of Narrow Water, and Philip Crossle, Esq., of Dundalk, have given much assistance in the collection of the somewhat scanty information as to the Colours of the 83rd and 86th. Thanks are also due to A. D. Carey, Esq., Librarian of the War Office, to G. W. Snooke, Esq., of Southsea, to Messrs. Jennens, of Conduit Street (and particularly to Mr. Marshall of that firm), to Messrs. Spink and Son, Messrs. Hawkes and Co., and Messrs. Flight and Son, for information as to uniform and illustrations of badges. Finally, acknowledgment must be made to the writings of the late S. M. Milne, Esq., from which much of the information on the subject of dress has been obtained. Had Mr. Milne lived the chapter would probably have been very much more complete, for application would certainly have been made for that assistance which his unrivalled knowledge of the subject rendered so valuable, and which was so freely given in similar cases.

wide, fastened back with buttons, the buttonholes being embroidered with lace. In the 83rd the buttons and lace loops were in pairs and the officers lace gold. In the 86th the lace was silver, and, as far as can be ascertained, the buttons were at first at equal intervals, but in 1796 the placing of buttons in pairs became general. The collar of the coat was high and open, of the same colour as the facings, and ornamented with a button and laced buttonhole, a relic of the time when the collar was worn turned down and buttoned to the coat. The cuffs were three inches deep with buttons and laced buttonholes. Officers of the grenadier company wore an epaulette of lace and fringe on each shoulder, officers of the battalion companies on the right shoulder only. The waistcoat and knee breeches were white, gaiters of black linen with black buttons reaching to a little below the knee. A crimson sash was worn round the waist over the waistcoat, but under the coat. A gorget hung from gorget buttons on the coat collar by silk ribbons and rosettes of the same colour as the facings. This ornament deserves mention, as it is now almost forgotten. It was a crescent-shaped ornament of gilt or silver, according as the lace was gold or silver, and is generally considered to have been the last remnant of coat armour in general use. Before 1796 it frequently bore the regimental number and crest, but none belonging to either the 83rd or 86th of so early a date has been traced. After 1796 the universal pattern bore the royal arms. The officers in 1793 wore a black three-cornered hat with a black Hanoverian cockade.

The sword, which was straight and had a black leather sheath, was carried in a frog on a white shoulder belt, on which was a breast plate. In the 83rd this ornament was a plain gilt oval, bearing on it "83," surmounted by a crown; in the 86th it was of silver three and a half inches long, bearing on it the crown and number, surrounded by a "French scroll."

The men's dress was similar but the lace was white, with, in the case of the 83rd, one red and one grey stripe, and in that of the 86th two yellow and two black stripes. White cross belts were worn, one supporting the bayonet, the other the cartridge pouch, a plain oval breast plate bearing the regimental number was placed where the belts crossed. The cocked hat, with black cockade, was worn across the head. The grenadiers wore a high fur cap with a white feather. Non-commissioned officers were distinguished by shoulder knots. The serjeant-major and staff-serjeants wore silver lace. This was universal throughout the Army, whether the officers wore gold or silver and remained so until 1855. The serjeants were armed with a sword suspended from a shoulder belt and a halberd with a plain steel spear-head.

1796

The warrant of 1796 made certain alterations in the uniform. The lapels of the coat were to be made either to button over or to fasten close with hooks and eyes all the way to the bottom, and the stand-up collar became very high and roomy to permit of the large neckcloth then coming into fashion. Mr. Milne states that it was probably with this coat that the old gold embroidered buttonhole disappeared. This remained the full dress coat of the officers until after 1815, and is certainly that described in the Records of the 86th, which gives the uniform of the officers in March, 1808, as:—

"Full Dress: Cocked Hat; Lace Coat with Yellow Facings, Silver Lace, full when turned back, and when Buttoned up Lace on left side only, Silver Buttons, Buff Belt; White Nankeen trousers; and gaiters with shoes.

"Marching Order: Scarlet Jacket, with Skirts and no Lace; Blue Pantaloons; Hessian Boots; Round Hat; Black Cockade and Four Inch Feather in Front; Black Sling waistbelt."

The jacket referred to as being worn in marching order had also been introduced for all line regiments in 1796. The skirts were much shorter than in the full dress coat and it was altogether lighter. It is noticeable that the costume described is that which was being worn in India; there was no special uniform for tropical countries at this period.

1799

A warrant, dated the 22nd April, 1799, directed that officers and men of infantry, except the flank companies, should wear their hair queued, tied a little below the upper part of the collar of the coat, ten inches long, with one inch of hair below the binding.

1800

In 1800 the cocked hat worn by the men was abolished, a cylindrical shako of lacquered felt, with a peak, being introduced in its place. The shako had a large oblong brass plate in front, about six inches high; this was of universal pattern, consisting of the royal arms, surrounded by a trophy of arms, standards, trumpets, etc., with a lion underneath. The regimental number was engraved on either side of the lion. A worsted red and white tuft was worn in front, rising from a black cockade. Men of the grenadier company were permitted to wear this shako when not using their fur cap, the tuft in their case being white; the tuft in the light company was green.

The officers retained their cocked hats, and, as noted above, they were still being worn—though in full dress only—in 1808. There is some doubt as to the head dress worn by officers on service at this time. Officers of the marines and of some regiments, notably the 28th, wore a head dress very similar to the modern top hat, with a curled brim and a cockade and feather at the side. In other cases officers wore the same shako as the men, though the plate was often of gilt instead of brass, and with a feather instead of a tuft. There is no definite record for either the 83rd or 86th beyond the "round hat, black cockade, and four inch feather in front," known to have been worn in 1808 by the 86th, and this description would be more or less applicable to both the head dresses described above.

1802

A General Order of 1802 introduced chevrons, worn on the left arm, to distinguish the non-commissioned officers, in place of the shoulder-knot. The serjeant-major had four, serjeant three, and corporal two.

In 1808 a General Order of the 20th July abolished queues and

1808 ordered that the hair should be cut short at the neck; at the same time a sponge was added to the soldier's necessaries, with a view to the cleanliness of his head. About this time a new head dress was introduced, though it probably only came into wear gradually. It was a light felt cylindrical shako, with a black leather peak and a kind of false front, higher than the hat itself; on the left side was a black cockade, with a red and white tuft for battalion companies, white for the grenadier company, and green for the light company; officers of the latter also wore a small silver bugle below the cockade. On the front of the shako was a small scroll pattern gilt plate, with the cipher "G.R.," surmounted by a crown. Officers had a gold cord across the front, with a tassel on the right side, the men a similar cord of white worsted; cap covers of black japanned material were worn on service. This was the shako worn during the greater part of the Peninsula and in the Waterloo campaigns. The picture on the wall of the patio of Government House, Gibraltar, representing the storm of Badajoz by the 83rd Regiment (drawn by Captain Marshman, 28th Regiment, in 1870) is not correct as to uniform; in particular the shako and badge therein depicted were not those worn in 1812.

The dress of the officers on service at this time was a double-breasted jacket with short tails and without any lace in front. Officers of the light infantry company carried the curved light infantry sabre suspended from the shoulder belt by slings; indeed, Mr. Milne states that this sabre was generally carried by all officers on service, in place of the regulation straight sword. A crimson sash was worn outside the coat round the waist.

In accordance with a General Order of February, 1810, the rank of 1810 officers was distinguished by epaulettes. Field officers wore two, the colonel having a crown and star on the strap, lieutenant-colonel a crown, and major a star. Captains and subalterns wore one epaulette on the right shoulder. Officers of the flank companies wore two wings instead of the epaulettes, with a grenade or bugle respectively. The adjutant wore an epaulette strap on the left shoulder, in addition to his epaulette. The paymaster and surgeon wore the regimental jacket, single breasted, but without epaulettes or sash, and carried the sword suspended by a plain waist belt underneath the coat.

The men wore a single-breasted short-tailed jacket, looped across the front with regimental lace set on by pairs; white pewter buttons with the regimental number, and a high collar of the regimental facings open in front, with yellow shoulder straps terminating in a white tuft. The trousers were a blue grey in colour. Cross belts were as previously described, while the knapsack was supported by straps through which the arms were passed, and was secured by a small strap right across the front of the chest. This is the uniform illustrated opposite page 122, but it will be noted that no tuft is shown on the shoulder of the non-commissioned officer or of the private. The details of the sketch have been taken from a coatee in the Musée de L'Armée, Paris, which has no tufts, but it is uncertain whether they have been removed from this particular coatee or were not worn by the 83rd.

1816 Probably as a result of contact with foreign troops during the occupation of France after Waterloo, the distinctively English and very serviceable shako was superseded in 1816 by one more after the style then worn on the Continent.

This was seven and a half inches high and eleven and a half inches in diameter at the top. It had brass chin scales that could be fastened up to the cockade in front when not in use, and an upright feather twelve inches high. The badge was a small plate with the regimental number surrounded by a laurel wreath and surmounted by a crown. That of the officers of the 83rd was of gilt; of the 86th silver with a gilt mount. The officer's shako had also gold or silver lace (according to the plate) two inches broad round the top, and another band three-quarters of an inch broad round the bottom. The grenadier cap, which had fallen into disuse in the war, was now rigidly insisted on, and in 1822 a circular was issued calling attention to the fact that the gorget formed part of an officer's equipment; it, too, had apparently fallen into disuse.

In 1824 the facings of the 83rd were changed from yellow to light yellow, but whether the change was real, or merely a matter of terms, is uncertain. In 1830 the facings are given as again being yellow.

1826 In 1826 the private's coat was altered somewhat in shape, the lace loops across the chest being made to taper gradually to the waist. About this time the officer's dress became almost fantastic, as will be seen from the illustration of an officer of the 83rd opposite page 202. Officers of the light company wore, in addition, whistles with long chains. The trousers were of blue grey, generally very full over the hips; in summer, white trousers were worn. A blue undress frock coat was also introduced for officers, the sash being worn with it.

The pattern of the breast plate had been altered about 1822. In the 83rd the new pattern was a gilt plate with silver mounts, having on it 83 surmounted by a crown and encircled by a laurel wreath. The battle honour "Peninsula" was borne on the belt. That of the 86th was of burnished gilt with silver mounts; the number 86 was worn surmounted by the harp and crown, with the Sphinx superscribed "Egypt" underneath, and the title "Royal County Down."

1828 In 1828 a new shako was introduced, similar in shape to that which it superseded, but an inch and a half lower, and the lace was removed from it, as was also the Hanoverian cockade. A shako plate of universal pattern was ordered to be worn. In form this was a large gilt star six inches by five and a half, surmounted by a crown. Regiments were allowed to place their own devices in the centre; the 83rd adopted an inner star with 83 within a laurel wreath and all the Peninsular battle honours on rays radiating outwards from the centre. That of the 86th was more elaborate, consisting of a silver cut star with gilt laurel wreath, the garter with "Royal County Down," Sphinx and figures, blue enamel behind the centre, and the battle honours "India" and "Bourbon" on either side. The gilt scales to fasten under the chin were retained, and heavy cap lines were now introduced, worn braided across the front of the shako, the loose ends hanging down and terminating in two tassels, these being secured to one of the buttons of the coat. These lines were of gold lace for the officers, white worsted for men of the battalion companies, and green for the light company. An illustration of an officer of the 83rd in this dress is given opposite page 204.

1829 The double-breasted coatee, which remained the full dress of officers till the Crimean War, was introduced by a warrant in February, 1829. It had two rows of gilt buttons down the front, by pairs, a high collar of the regimental facings fastened up in front, on either side of it two loops of lace and buttons. The cuffs were somewhat similar in shape to those now worn by officers in service dress, the flap being ornamented with four buttons and four loops of lace. Large gold epaulettes were worn on both shoulders by all officers except those of the flank companies, who retained gold wings. The length of fringe on the epaulette varied with the rank of the wearer. Except in the tropics, this coatee was worn at mess. A plain scarlet shell jacket was ordered for some stations. White trousers were still worn in the summer, and in the winter a much darker shade of blue grey—practically dark blue. This warrant also authorized a dark blue forage cap, the first of its kind, and replaced the white fatigue jacket hitherto worn by the rank and file by a red one.

1830 In August 1830 the cap lines introduced two years before were abolished, as also the gorget. The feather of the shako was somewhat shortened, and in the light companies replaced by a green ball. The bandsmen were now dressed in white coatees.

1832 About the year 1832 new breast plates were brought into wear by the officers of both regiments, and the pattern then adopted was retained until the abolition of this ornament in 1855. That of the 83rd was a frosted gilt plate with bevelled edges, having on it 83 crowned and the battle honours, including "Cape of Good Hope." That of the 86th was a burnished plate bearing a large harp and crown with 86 superimposed on it. In this year field officers were ordered to wear their swords suspended by slings from a white waist belt, and brass scabbards were introduced. The adjutant was still to wear his sword from the shoulder belt, and his scabbard was to be of steel.

1833 In 1833 the familiar red stripe down the outer edge of the trousers was introduced.

1834 About 1834 the substitution of the ball for the shako feather, first introduced for light companies only in 1830, was made universal, the ball being of white worsted except in the case of the light company, which retained the green.

1836 A General Order of October 10th, 1836, abolished the lace hitherto worn by the rank and file, and plain white tape took its place. The serjeants were given a plain red coatee, with white epaulettes or wings. For the next few years the Army was given a rest from these constant changes of uniform; the dress of a serjeant of the battalion companies of the 86th at this period is given in the plate opposite page 226; it is taken from Cannon's "Historical Record of the 86th."

1844 In 1844 a new shako—the Albert hat—was adopted. It was six and a half inches high, and smaller in circumference at the top than at the bottom. It had a small peak in front, and something rather in the nature of a peak behind. The chin scales were done away with, a simple chain taking their place. Officers wore a raised gilt crown and star plate on the front. In the 83rd the number was in the centre, with the battle honours radiating from it. In the 86th the number was surrounded by a wreath and the title "Royal County Down" on a garter; the harp and "*Quis Separabit*" were surmounted by the Sphinx. The honours "India" and "Bourbon" were on either side of the device.

1850 On the 27th May, 1850, a circular was issued abolishing the old cross belts, the bayonet was to be carried in a frog on the waist belt, and the pouch on a plain shoulder belt. The breast plate for the men was abolished. Both the 83rd and the 86th were at this time serving in India, and the new equipment was long in reaching them, being only taken into use by the 83rd (and probably also by the 86th) during the course of 1856.

1855 Great delay also took place in the introduction of the tunic which, as a result of the Crimean War, replaced the old coatee in 1855. The actual date on which the new dress was taken into wear by the regiments in India cannot be definitely ascertained, probably its introduction was gradual, for officers present in the Mutiny speak of both coatee and tunic being worn. It would seem, however, that the red shell jacket, intended as a fatigue dress, was the uniform most generally worn by the men in 1857-58.

The first tunic was double-breasted, but this was almost immediately replaced by a single-breasted tunic not unlike that still worn. On the first there were two rows of buttons at equal intervals, those on the men's tunic being now of brass instead of, as heretofore, of pewter. The cuff, while remaining the same in shape as hitherto, had now three buttons instead of four. Epaulettes and wings were no longer worn by the officers, and badges of rank were placed on the collar. The sashes of officers and serjeants were worn over the shoulders, a practice continued until after the South African war. With the advent of the tunic also the officers' shoulder belt was done away with—and with it the breast plate—and the sword was suspended by slings from a waist belt similar to that already worn by field officers. At this time also dark blue trousers were introduced for the men, and the light blue grey for summer wear were finally done away with.

In the same year (1855) a new shako was introduced into the Army. Smaller and lighter than the one it followed, it bore some resemblance to the French kepi, the front being lower than the back, and there being a long, straight peak instead of the short sloping peak of the Albert hat. The chin strap was of plain leather.

The dress described was not rigidly adhered to on service at this period, and during the Mutiny there appears to have been an infinite variety of uniform. Thus some regiments stained the white duck summer clothing with mud before going on active service, this being the first introduction of the now familiar khaki. Others wore the red shell jacket—first introduced as a fatigue dress in 1829—with the round forage cap covered with a stiff white curtain. In the 86th the men generally wore shirt sleeves only during the day, and the fatigue jacket at night or in the early morning. The officers wore the shell jacket, and with it a white wicker and linen helmet, specially made for the regiment in Bombay. The dress is shown in the illustration opposite page 260.

1860

In 1860 the grenadier and light companies as such were done away with, and with them disappeared—as regards the 83rd and 86th—the different coloured ball on the shako, the sole remaining distinctive feature of their dress. The experience of the Mutiny had done much to show the suitability, one might almost say the necessity, of a light uniform for service in tropical climates, and in the China war of 1860 a wider use was made of khaki, though it was many years before this became what it is to-day, the regular working uniform of the British soldier in tropical lands.

1861

Although in India itself there was no immediate introduction of khaki, the provision of a dress more suitable than that then in use had received consideration, and "a scarlet serge frock and a wicker helmet covered with grey linen, with a turban round it," were ordered to be adopted in place of the fatigue jacket, and shako then in general use. These were taken into wear by the 83rd, then still in India, in the second quarter of 1861.

1862

In 1862 the shako was once more altered, being made considerably lower. The material was dark blue cloth, with diagonal lines. The plate was reduced proportionately, the number being stencilled and the motto and garter struck, not pierced as heretofore.

1866

In 1866 the black sword scabbard worn by dismounted officers since the raising of the regiments in 1793 was replaced by a steel scabbard for all officers.

1867

In the following year the officers' undress blue frock coat, which had continued, with various modifications in pattern, from 1826, was replaced by a braided patrol jacket.

It is interesting to note as a sign of the unsuitability of the then head dress for wear in a hot climate, and the need for something more suitable, that when, during this year, the 86th embarked at Gibraltar under orders for Mauritius, the men's shakos were handed into store, and the only head dress was the forage cap with pugaree.

1868 In 1868 the old slashed cuff was replaced by a pointed one with, in the case of the officers, gold lace distinctions for various ranks. At the same time chin scales were added to the shako, which was also slightly altered in shape and stiffened, and that of the officers ornamented with gold lace; the plate was of crown and wreath pattern, the number stencilled in the centre of a garter. A gold and crimson sash, gold lace trousers and sword belt were also introduced for levee dress.

1871 The glengarry cap, the first of the folding fatigue caps that continued until after the South African War, was introduced into the Army in 1871, replacing the "Kilmarnock bonnet." The glengarry was not actually taken into wear by the 83rd, then in India, until December, 1875.

1873 The distinctive white dress of the band was abolished throughout the Army in 1873, and about the same time a red serge frock, similar to that introduced in India in 1861, replaced the red shell jacket worn for fatigues since 1829.

1874 By an order dated the 30th June, 1874, permission was granted to the 86th for the harp and crown, and motto "*Quis Separabit*," to be borne on the collar of the men's tunics, and in November the officers were permitted to bear the same device on their forage caps.

1875 In the following year the colour of the uniform of both regiments was changed from red to scarlet. An illustration of the uniform at this period is given opposite page 226.

1880 In 1880 badges of rank were removed from the collar of the officers' tunic and displayed on the shoulder strap, a colonel being distinguished by a crown and two stars, lieutenant-colonel by a crown and star, major by a crown, captain by two stars, and a lieutenant by one. A sub-lieutenant had no badge of rank. A blue cloth helmet, with a gilt spike and mountings, was taken into wear, as also, for the officers, a forage cap, with a semi-circular peak, bearing the regimental number in front.

1881 In the following year came the order combining the 83rd and 86th to form the Royal Irish Rifles. The uniform was changed to green with dark green facings, and the badges of the two regiments combined. The general nature of the uniform then introduced was much the same as it is at present, but the green of the officers' dress was very near akin to black, while the men's tunic had a distinct tinge of blue. The head dress was a black helmet with a black metal helmet plate, similar in design to the officers' belt plate, having in the centre the harp and crown surmounting a Sphinx inscribed "Egypt," and a bugle, with the motto "*Quis Separabit*," the whole enclosed in a wreath of shamrock, on which the battle honours were inscribed, and surmounted by a crown. With the exception of the addition of the honour "South Africa, 1899-1902," to the belt plate, and the substitution of the Imperial for the Royal Crown, the officers' appointments have not altered, but on the rifle tunic then worn the difference in the rank was shown by the lace ornamentation of the cuffs and collar.

The officers' mess dress was a shell jacket looped with braid with a stand-up collar. The sealed pattern had black shoulder cords with badges of rank, but these shoulder cords were never actually worn.

A plain serge frock was worn for fatigues, and later for drill and marching order in temperate climates, that of the officer being cut as a " Norfolk jacket." The forage cap was a round cap without peak.

The first alteration in this dress was the substitution of the busby (1891) for the helmet. When first introduced the plume on the officer's busby was osprey, but later an ostrich plume was substituted.

The " Sam Browne " belt, worn with the khaki uniform, was taken into wear by the officers of the 2nd Battalion in India about 1895, and by the officers of the first battalion on their departure for South Africa in 1897. In both cases the belts were at first black, but during the South African War the brown universal pattern was adopted, the change in colour being first occasioned by an order given by General Gatacre on board the " Britannic " on the voyage from Capetown to East London, November, 1899, that all the black equipment of the regiment should be painted to resemble khaki. This was done, yellow paint being obtained from the ship's stores.

Except for the above the uniform continued with no material change until after the South African War. This campaign, however, showed clearly the advantages of a neutral tinted uniform. The khaki drill worn in India was not suitable, either in colour or in texture, for wear in Europe, and the present Service dress was introduced by an Army Order of the 17th January, 1902.

In this the rank of the officer was no longer indicated by his shoulder strap, but by a very complicated arrangement of braid on the cuff. This was soon found to be unsuitable—gossip at the time stated that officers arriving from home could not get served in the refreshment rooms on the railway in South Africa, as the attendants would not believe their statements as to their rank, and that this fact had much to do with the abolition of this form of distinction—and was superseded by the present slashed cuff, bearing on it the stars and crown similar to those hitherto worn on the shoulder strap. Inconvenience had been found in the war to arise from the fact that the second-lieutenant had no distinctive mark as an officer, hence it was now ordered that a second-lieutenant should wear one star, a lieutenant two, and a captain three.

At the same time many changes were introduced in the dress of the Army as a whole, the object being to provide the soldier with a smart dress for walking out and for ceremonial occasions, and with a workmanlike uniform for service in the field. The principal changes which affected the regiment were a change in the colour of the green uniform, both of officers and men, the new colour being a true green. The green serge frock hitherto worn at home in drill and marching order was done away with except in India, where it was retained in lieu of the tunic for the men, and of the patrol jacket for the officers: this jacket has, however, since been re-established at home as an optional fatigue dress for officers. Changes in the ornamentation of the cuffs and collars of the officers' tunics on promotion were discontinued, and the old pattern braided mess jacket was replaced by a plain green jacket of the universal pattern with a roll collar.

Immediately after the war there was some uncertainty as to the type of head dress to be worn, both in service dress and for walking out. For the latter a plain green cap, somewhat similar in shape to that worn by sailors and firemen, was introduced. It was intensely unpopular, and was generally termed the "Brodrick," from the then Minister of War. In 1905 it was replaced by a green forage cap, with a glazed leather peak, similar to that which had been introduced for officers at the same time as the "Brodrick." This is the walking out head dress still worn.

Just prior to the war, the cap worn on most parades had been the field service cap, a development of the old glengarry, which could, however, be opened out so as to cover the ears and back of the head. Attempt was now made to utilise this with the service dress uniform at home, but though admirable for its original purpose—as a fatigue cap to be carried in the haversack and worn in camp or at night—it was entirely unsuited to daily wear, affording no protection from the sun, and easily falling off. Experiments were then tried with a felt hat similar to that worn in South Africa, but this was soon discarded and the present peaked service dress forage cap introduced in 1905.

In this brief review of the uniform of the regiment considerations of space have prevented detailed description of all the badges and waist belt plates worn at different times by the 83rd and 86th, and by the Royal Irish Rifles. Sketches of many of these have, however, been included in the text. The very recent changes in the introduction of a roll collar on the officers' service dress and green fatigue jackets (1912), and of a white metal badge on the men's forage caps (1913), need no special reference.

Enough has been said to indicate generally the evolution of military dress from the very handsome, but apparently unserviceable, uniform of 1793, illustrated on page 226, to the inconspicuous and unattractive, but thoroughly serviceable, field service dress shown on page 400. It may be that to some the review order of to-day may appear less attractive than some of the costumes illustrated, but consideration will show that the most handsome of these dresses must have been exceedingly uncomfortable to wear on active service, and very difficult to keep clean at all times. Perhaps the most remarkable thing that emerges from any study of the dress of the British Army is the fact that uniform apparently so unsuitable was continually worn by our forerunners on active service in tropical and sub-tropical climates. This fact can only increase our admiration for the men of the past, but possibly it may have some bearing on that appallingly high percentage of deaths and invaliding which must attract the attention of every soldier who studies the details of foreign service in days gone by.

BADGE AND BUTTONS
(Various)

Officer 86th, Silver-plated, before 1812

Rank and File 83rd, Brass, up to 1877

Rank and File 83rd, Brass, up to 1877

Rank and File 86th, Brass, up to 1877

Officer 83rd, Gilt, "Mufti," about 1840

Officer 86th, Glengarry Cap Badge, about 1880, Silver

Officer 86th, Gilt, "Mufti," about 1840

Officer's Servant 86th, Silver-plated, Livery, about 1880

Rank and File 86th, Brass, up to 1877

DRESS, COLOURS, AND MEDALS.

Great difficulty has been experienced in obtaining definite information about the various stand of colours carried by the 83rd and 86th Foot. The Regimental Records have little information on the subject, those of the 83rd in particular being very incomplete; as a result this account is far from perfect. At the time the regiments were raised units received their colours from the colonel, and when a new set was taken into use the old became his property, and often no trace of their subsequent fate can be obtained. The custom now general of laying by old colours in a place of worship having some connection with the regiment was not greatly followed at the beginning of the last century.

There is no record of the actual date of the presentation of the first colours carried by the 83rd and 86th, but they would certainly have been provided by the respective colonels out of the grant for clothing, and must have been similar to those then generally in use. The King's colour was the " union," consisting of the crosses of St. George and St. Andrew, with the number of the regiment in the centre, possibly on a shield surrounded by a wreath of roses and thistles. The regimental colour of both 83rd and 86th was yellow, with the " union " in the upper canton, and the centre similar to that of the King's colour. The size was six-foot six inches by six-foot (on the pole), and they were carried on a pike nine feet ten inches long.

In 1801, on the union with Ireland, it was ordered that St. Patrick's cross should be added to the " union " flag, and the shamrock added to the rose and thistle wreath. There seems to have been a good deal of diversity in the pattern of colours about this time, and Mr. Milne mentions in his " Standards and Colours of the British Army," that in 1806 Mr. George Naylor, then holding the office of " York Herald," was appointed " Inspector of Regimental Colours." He sent out a circular to commanding officers of regiments, desiring them to forward sketches of the colours then in use. Answers were received about the summer of 1807, and greater uniformity subsequently resulted.

There is no trace of the colours carried by the second battalion of the 83rd in the Peninsula. Mr. Milne states that in most cases at this period second-battalion colours were only distinguished from those of the first battalion by having the words " 2nd Batt." placed on the shield under the regimental number; it seems probable that the colours of the 2nd/83rd were no exception. We know that colours were carried throughout the campaign, that after Talavera Lieutenant Pyne, of the Grenadier Company, was specially promoted to a company in the 66th for his distinguished valour " in hastening to the assistance of the colours—in carrying which through this battle officer after officer had been shot down,"* and that " many sergeants also were killed and wounded in protecting the colours,"* but all trace of what became of these, probably the most interesting of all the colours carried by either regiment, on the disbandment of the battalion in 1817 has been lost.

In 1816 the regiments that had taken part in the Peninsular War were

* Bray's " Memories and Services of the 83rd Regiment."

authorised to bear the word "Peninsula" on their colours and appointments. In August, 1819, authority was further granted for the 83rd to assume the honours "Talavera," "Fuentes D'Onor," "Ciudad Rodrigo," "Badajoz," "Salamanca," "Vittoria," "Nivelle," and "Orthes," in "commemoration of the distinguished services of the regiment" on the occasions named. In January, 1827, "Toulouse" was added, in June "Busaco," and in 1836 "Cape of Good Hope."

It is impossible to say whether these honours were inscribed on the colours already in use, or, as was more usual, were inscribed on new colours when these were received. The state of the colours actually being carried probably decided the matter; if they were in good condition the new honours could be added to them, if badly worn this would be impossible. Unfortunately, there is no record as to when new colours were received.

In 1843 new regulations were published directing that honours should not be borne on the Queen's Colour; up to this time they were often borne on this as well as on the regimental colour.

On the 9th May, 1867, the 83rd, then at Gibraltar, were presented with new colours by Lady Airey, wife of the Governor-General, Sir Richard Airey, K.C.B., the whole of the troops in garrison being present. In replying to Lady Airey's address, Colonel Hankey, commanding, referred to the old colours as having been received "nineteen long years ago." This would imply that they had been presented to the Regiment in 1848, when the headquarters were at Kilkenny, or, if after September 12th, at Fermoy. These, then, must have been the colours carried through the Mutiny. They now hang in St. Patrick's Cathedral, Dublin, and it is of them that Mr. Milne writes in his "Standards and Colours," "The wreath worked in its proper colours is of extremely artistic design, very different to most other patterns used at this time, and infinitely superior to that latterly, and, in fact, still in use. Possibly a few stands may have been made with this wreath, but, so far, I have only seen three others like it, viz., the 55th, 62nd, and 66th Regiments."*

The colours presented in 1867 were, in accordance with the order published ten years earlier, considerably smaller than those hitherto carried, being three feet nine inches by three feet (on the staff), and the staff, replacing the old pike, was eight feet seven and a half inches long. The colours, too, now had a fringe. The honour "Central India" was borne on these, as also on those they replaced. They were carried by the regiment until on July 1st, 1881 (in accordance with General Order, No. 41, of the 11th April), it was joined with the 86th, and became the first battalion Royal Irish Rifles. They remained with the battalion a little longer, but in 1882 they also were hung in St. Patrick's Cathedral beside those they had succeeded.

A photograph of these two stand of colours is given opposite page 314.

As stated above, when the office of "Inspector of Regimental Colours" was

* "Standards and Colours of the British Army."

created in 1806, letters were sent to commanding officers desiring them to forward sketches of the colours then in use. "The 86th, its title just* changed from that of the 'Shropshire Volunteers' to that of the 'Leinster Regiment,' sent a coloured sketch of its colours from India." Mr. Milne gives an illustration of this taken from a folio volume in the office of the Inspector of Regimental Colours, and describes it as "the regimental colour of yellow silk, with a neat oval shield in the centre." The shield is red, and bears on it the device

G.R.

LXXXVI.

Regt.

There can be little doubt but that this was the original colour carried in the campaigns of Egypt, in India against the Mahrattas, and in the expedition to Bourbon. Its fellow, the King's Colour, would be that which, under a heavy fire, Private Moore fastened to the flagstaff in the redoubt at Bourbon amid the cheers of friend and foe alike. It is probable that the first battle honour of the 86th, the Sphinx superscribed "Egypt," granted for the expedition of 1801-02, was inscribed on this stand of colours, for a note is appended to the entry in the Records of the extract of the London Gazette (18th May, 1812), which changed the title to Royal County Down, and conferred the badge of the "Irish Harp and Crown" to the effect that the Regiment now bore the Irish Harp and Crown "in addition to the Sphinx and the word 'Egypt' on their colours."

The change of title carried with it the change of the facings and the colours from yellow to blue, and on the 16th September, 1814, a draft from home, under the command of Captain Michael Creagh, joined headquarters at Masulipatam, bringing with it the "new Royal Colours. What became of the old ones is not known. General the Earl of Kilmorey was Colonel at that time, but if the colours passed into his keeping all record of them has been lost, as has been ascertained from his descendants, the present Earl (his great grandson) and General Sir George Wentworth Higginson, G.C.B., his grandson.

In 1823 a letter dated Horse Guards, 15th October, informed the Earl of Kilmorey that His Majesty had authorised the Regiment to bear the words "India" and "Bourbon" on its colours and appointments in consideration of the distinguished conduct of the regiment in India from 1799 to 1819, and in the island of Bourbon in July, 1810.

Possibly it was owing to this addition to the honours that the colonel provided a new set of colours. These were presented to the regiment by his daughter, Lady Emily Needham, at Newry, on the 17th September, 1824. The Needham family was well-known at Newry, which Lord Kilmorey had repre-

* See note on page 73.

sented in four Parliaments, and the "Newry Telegraph" of the 21st September gives an interesting account of the ceremony, as also of the subsequent dinner given by the officers to their colonel, followed by a ball to the Ladies Needham. The account specifically states that the new colours bear the words "India" and "Bourbon."

On the third of December of the same year the old colours were deposited with due ceremony in the Parish Church at Newry, "by order of the Colonel, the Earl of Kilmorey." No inscription was, however, placed in the church with the colours, and in 1894 they were for a time removed, being in a very dilapidated condition. Fortunately, an officer of the Regiment was in Newry at the time, and largely through his efforts a brass tablet was placed beneath the colours when rehung. The inscription on this reads:—

"The colours of the 86th or Royal County Down Regiment were placed in the Parish Church by order of the Colonel, the Right Honourable Earl of Kilmorey, Lay Abbot of Newry and Mourne, on the 3rd December, 1824.

"The colours had been carried at the capture of Bourbon in 1810, on which occasion the King's colour was nailed under a heavy fire to the flagstaff of a captured French redoubt by Corporal John Hall, of the Regiment."

This inscription was not put up without careful inquiry, and it embodied a tradition, but details of the colour carried at Bourbon have since come to light, and though the present condition of the colours in Newry Church renders it very difficult to trace with certainty either their design or colour, yet sufficient remains to cause the writer reluctantly to come to the conclusion that they are not those carried at Bourbon, the decisive proof being that in the regimental colour at Newry the number is superimposed on the union in the upper canton; in the Bourbon colour it was borne on a shield in the centre of the flag.

In March, 1832, the Regiment, then in the West Indies, was informed that permission had been granted for the motto "*Quis Separabit*" to be borne, in addition to the harp and crown. In December the Earl of Kilmorey died, but on March 1st, 1833, when headquarters were at Berbice, in British Guiana, "Lieutenant Colonel Sir M. Creagh, K.H., . . . arrived from England and assumed the command of the Regiment, bringing out the new colours presented by the late Lord Kilmorey." It has been suggested that these colours were not really new but were the "new colours" presented in 1824, which had been sent home to have the addition of the motto approved in the previous year, and it is certainly improbable that a stand of colours would have needed replacement after only nine years of not very strenuous service. On the other hand, the term "new" in the Records is not qualified in any way, and the promptitude with which Lord Kilmorey presented new colours when the honours "India" and "Bourbon" were granted seems to show that it is quite probable that the addition of the motto "*Quis Separabit*" led him to provide another stand.

It is interesting to identify the Lieutenant-Colonel Sir M. Creagh who brought the colours to Berbice in 1833 with the Captain Michael Creagh who had brought the first "royal" colours to Masulipatam nineteen years before.

Whether the colours of 1833 were those presented in 1824 or not, they are undoubtedly the set illustrated in Cannon's "Historical Record," reproduced opposite page 282. They were destined to a long and honourable association with the Regiment, for they were carried throughout the Mutiny, and were only replaced in 1867, when, on February 13th, new colours were presented to the 86th at Gibraltar by Lady Airey, with a ceremony almost identical with that which followed a few months later, when the 83rd received their colours on the same parade ground and from the same hands. It is stated that the old colours had "seen such service that little remained save a few shreds of silk fluttering from the otherwise bare staff." Their condition is hardly to be wondered at when it is remembered that they were carried for thirty-three years. They appear to have suffered greatly in the Mutiny, and the difficulties in the way of repair that were sometimes met with in those days are illustrated by an incident on the homeward voyage on board the "Genghis Khan" in 1859. The colours were handed to the master-tailor to repair, "along with an old pair of red silk pyjamas, out of which to make patches." Fortunately, other material was found.*

These colours now hang above the west door of St. Patrick's Cathedral, Dublin.

The colours presented in 1867 had on them the four honours "Egypt," "India," "Bourbon," and "Central India," and were the last carried by the 86th. When in July, 1881, the regiment became the second battalion Royal Irish Rifles they ceased to be carried on parade. They accompanied the battalion on its wanderings, however, until they were brought home from Malta by Major Selby Smyth in 1894, and on the 5th May, after an impressive service, conducted by the Bishop and the Dean, they were laid up in Downpatrick Cathedral above the brass tablet erected by the officers and men of the second battalion to their comrades who died in Egypt between 1888 and 1890.

When the two regiments were combined to form the Royal Irish Rifles their battle honours and badges were also combined, and though colours were no longer carried all were transferred to the officer's belt plate. One change took place immediately, however, the words "Cape of Good Hope, 1806," being substituted by a General Order of the 1st September, 1882, for "Cape of Good Hope," thus marking more clearly the occasion for which the honour was granted.

One addition and an alteration have since been made to the plate, the words "South Africa, 1899-1902," being granted in commemoration of the South African War, and the Imperial Crown substituted for that hitherto borne on the accession of His Majesty King Edward VII.

* Communicated by the late Major-General Sir J. G. Dartnell, K.C.B., C.M.G.

MEDALS.

From the days of Queen Elizabeth medals have from time to time been conferred by the sovereign as a mark of appreciation of naval and military services. The early medals were usually, though not invariably, bestowed only on a few of the more important commanders, and though during the course of the great war with France at the close of the eighteenth and beginning of the nineteenth centuries medals were often issued to officers, some of whom were in comparatively subordinate positions, yet no general distribution of a military medal to all ranks was made until the issue of that for the Waterloo campaign, announced in the Gazette of April 23rd, 1816. As early as 1784, however, the Honourable East India Company had adopted the custom of conferring medals on their native Indian troops. Such medals were distributed to those who took part in the expeditions to Egypt in 1801, and to the Island of Bourbon in 1810, but no issue was made to the troops of the British Government, which in each case included the 86th Regiment. One of the silver medals granted for the latter expedition is now in the collection of the officers of the 2nd Royal Irish Rifles.

The expedition to Egypt above referred to saw the first medals actually awarded to the officers of either the 83rd or 86th. The Turkish Sultan, Selim III., ordered gold medals termed the " Order of the Crescent," to be struck, and these were issued to British naval and military officers who were permitted by King George III. to accept and wear them. Among others, the officers of the 86th received these decorations.

In commemoration of the victories in the Peninsula two gold medals of different size were struck for each victory and awarded, the larger one to general officers, and the other to field officers, certain staff officers, and to those who actually commanded a battalion in battle in consequence of the death or removal of the original commander. The name of the battle or siege for which the medal was granted was engraved on the reverse, but as many officers became entitled to more than one medal, bars, to be worn above the medal, were introduced for subsequent victories, and it was ordered that no officer should have more than one medal, and no medal more than two bars. When an officer was present on a fourth occasion for which a medal was granted, a gold Maltese cross was awarded in place of the medal, one victory being inscribed on each arm of the cross and bars added for each subsequent action. The 83rd taking part in almost all the great victories of the war, five of these decorations were awarded to the officers. Lieutenant-Colonel Carr, commanding (afterwards Sir H. W. Carr, K.C.B.), received the gold cross with three clasps, but perhaps the greatest interest attaches to the medal won by Lieutenant-Colonel Alexander Gordon for the battle of Talavera. Badly wounded when leading the battalion in its counter charge against the enemy, he was being carried to the rear when a shell burst near and killed both him and his bearers. In recognition of the services of the Regiment in the battle "Sir William Gordon, the brother of Colonel Gordon, received the medal which would have decorated that gallant soldier's breast, had he happily survived the action."* An illustration of this medal, now in the collection of Dr. Payne, appears opposite page 134.

On the 1st of June, 1847, a generation after the campaigns had come to an end; a General Order was issued granting a Military General Service Medal for services between 1793 and 1814. Originally twenty-seven bars were granted, all save six being for operations in the Peninsula, but a General Order dated Horse Guards, 11th February, 1850, added a bar for "Egypt, 1801."

The greatest number of bars awarded with any single medal was fifteen, and this in two cases only. Twelve was the maximum number that could be claimed by any survivor of the 83rd, and the only applicant for this number was Private Thomas Hazlehurst, whose medal, a photograph of which is given opposite page 176, is now, through the generosity of a former commanding officer, in the collection of the officers of the 1st Royal Irish Rifles. Of the large number of officers and other ranks who served with the second battalion of the 83rd between 1809 and 1814 few survived to claim the belated reward in 1847; the medal was granted to twenty-four officers and 183 men only. Of the 86th Regiment only one officer and seven men received the medal with the bar for Egypt.

The issue of the General Service medal in 1847 marked the definite adoption of the system followed ever since of granting these rewards to all ranks taking part in a successful campaign. The next medal of which a general issue was made to either regiment was that for the Indian Mutiny, granted in a General Order of the 18th August, 1858. This was awarded to 34 officers and 892 other ranks of the 83rd, and to 30 officers and 873 others of the 86th.

Although the South African War was the next occasion of a general award of a medal to a whole battalion of the Royal Irish Rifles, yet detachments and individual officers, non-commissioned officers and men have taken part in many of the small wars which have occurred during the last thirty years. Thus detachments of the 1st Battalion Royal Irish Rifles earned the medal granted for services in Rhodesia, 1896, in China in 1900, and in Tibet, 1905. Detachments from the 2nd Battalion earned medals during the operations in Sudan, 1888-1889, and on the North-West Frontier of India, 1897-1898.

Mention must now be made of the special rewards granted to the soldier. First of these stands the Victoria Cross. Instituted in January, 1856, this coveted decoration was won by two officers and two men of the 86th in the operations before Jhansi in April, 1858.

It was awarded to Lieutenant and Adjutant Cochrane, who, at the battle of the Betwa (April 1st) had three horses killed under him in a magnificent

* Bray's "Memoirs and Services of the 83rd Regiment."

charge into the enemy's rearguard when he captured a gun under a heavy fire; and at the storming of Jhansi, on April 3rd, it was won by Lieutenant Jerome and Private Byrne, who together assisted Lieutenant Sewell (when badly wounded) under a very hot fire, and by Private Pearson for his gallantry not only in carrying a wounded comrade out of action under fire, but also in an attack on a number of the enemy, when he himself was badly wounded.

The last survivor of these heroes—Major-General Jerome, V.C.—died on the 25th February, 1901. The cross awarded to Private Byrne is to-day in the collection of the officers, 2nd Battalion Royal Irish Rifles.

Next to the Victoria Cross must come the medal for Distinguished Conduct in the Field. This was instituted in 1854, and was freely bestowed for the Crimean War, but was awarded to seven men only from all the infantry regiments that took part in the Mutiny. Of these seven two belonged to the 86th Regiment, viz., Serjeant-Major Alleyne Wolfe and Quartermaster-Serjeant William McNeill. Colonel Stuart, C.B., who commanded the 86th, in his book, "Reminiscences of a Soldier," refers in warm terms to McNeill's services. He had for a time absolute command of two companies in action, and his Commanding Officer was authorised to offer him a commission as an ensign in recognition of his gallant conduct and exemplary behaviour. McNeill was a married man with three children, and wisely, as his colonel thought, declined the promotion. He was granted the medal for distinguished conduct in the field, with an annuity of £15 in 1860, and was subsequently a barrack serjeant at Portsmouth.

Considerations of space only admit of a brief reference to the medal for "Meritorious Service," instituted on the 19th December, 1845. This is granted with an annuity to non-commissioned officers not below the rank of serjeant, as a reward for distinguished or meritorious service. Ten of these medals have been awarded to men of the 83rd and four to those of the 86th.

All the medals mentioned above were awarded by a Sovereign or by a corporation—the Honourable East India Company having practically the powers of a Sovereign. There remains another class of medals of great interest, viz., regimental medals. These were generally presented by officers or other ranks to certain of their comrades in recognition of bravery, long service, exemplary conduct, good shooting, temperance, etc., at a period before general issues were made to all ranks at the public expense. As it generally occurred that a single example only was struck, specimens are of great rarity and highly prized by collectors. They were generally discontinued when the Long Service Medal was instituted in 1830. Of those relating to the 83rd and 86th Regiments the following have so far come to light:—

83rd Regiment.—(i.) A silver medal, "83rd Regiment of Foot, Peninsula," harp and crown and shamrocks. Inscribed on the reverse: "A reward for military merit, presented to Wm. Hall by his comrades, 1815."

(ii.) A gold medal, on one side, "Award of Merit" in a wreath in relief.

On the reverse " Given to Samuel Holt, Quarter Master, 83rd Foot, for Meritorious Service, 1815."

(iii.) Oval gold medal, inscribed " Ciudad-Rodrigo, 83rd Regiment of Foot," within a wreath; reverse, " A Reward for Distinguished Conduct, 19th January, 1812. J. Wiley, Sergt."

86*th Regiment*.—(i.) A silver medal with " G.R., 86," harp and crown and wreath. Reverse on a shield: " From Lieut.-Colonel R. Crawfurd and the Officers H.M. 86th Foot, to Serjt. Byrne, Best Shot, 1803."

(ii.) A silver medal, " G.R.," surmounted by a crown, with " 86 " below, within a wreath. Reverse: " The gift of Lieut.-Col. Jas. Phillips Lloyd, Commanding His Majesty's 86th Regt. of Foot, to Quartermaster James Carr, as a token of his high regard for him as a soldier, 20th May, 1801."

The various medals distributed in recent years for other than war services hardly call for remark, suffice it that both battalions have received their share of the awards that commemorate the Coronations of the late and present Sovereigns, and the great durbars held in India.

The private collection of medals has increased greatly in recent years, and prices now rule so high that any decoration at all out of the common becomes the object of great competition at the public sales. Unfortunately this affords a strong inducement to the recipients and their families to part with these honourable relics. A brighter side to this picture is afforded when these become much valued objects of veneration in regimental collections, where descendants may rest satisfied that the memory of the services and achievements of those who won them is kept alive, and the medals themselves are guarded in security. The officers of both battalions are fortunate in the possession of collections which include examples of almost all the medals granted to men who have served in the two regiments.

83rd REGIMENT.

Officers on whom Gold Medals, Clasps and Crosses have been Conferred.

Name.	Command at the Time.	Sieges, Battles, &c.	Distinctions.	Remarks.
Blaquiere, J.	Major, 83rd Foot	Orthes	Medal	Died in 1842.
Elliott, Gilbert	Capt., 83rd Foot, Light Companies.	Orthes	Medal	—
Carr, Henry Wm.	Lieut.-Col., 2nd Bn. 83rd Foot.	Fuentes d'Onor, Ciudad Rodrigo, Badajoz, Salamanca, Vittoria, Nivelle, Orthes.	Cross and 3 Clasps.	K.C.B. Died in 1821.
Gordon, Alexr.	Lieut.-Col., 83rd Foot.	Talavera	Medal	Killed in 1809.
Hext, Samuel	Major, 83rd Foot	Badajoz, Orthes, Toulouse.	Medal and 2 Clasps.	C.B. Died in 1822.

From Sir N. H. Nicolas's "History of Orders of Knighthood."

2ND BATTALION 83RD FOOT.

Roll of Officers of the above to whom Her Majesty has granted silver medals under General Order, dated 1st June, 1847, for services in the following battles or actions:—

1, Talavera; 2, Busaco; 3, Fuentes d'Onor; 4, Ciudad Rodrigo; 5, Badajoz; 6, Salamanca; 7, Vittoria; 8, Pyrenees; 9, Nivelle; 10, Nive; 11, Orthes; 12, Toulouse.

	Total Clasps.
Baldwin, C. J., Lieut.—1, 2, 3, 4, 5, 6, 7, 8, 9, 10, 11 ... ... Col., Canad. Mil.	10
Bowles, Chas. P., Lieut.—2, 3, 5 ... ... late Capt., 26th.	3
Carey, Mich., Lieut.—1, 2, 3, 4, 5, 6, 11, 12 ... ... Lieut., H.P., 40th.	8
Colthurst, N., Lieut.—1, 2, 6, 7, 9, 11 (Portuguese.), 8, 13 (Albuhera Service.) ... ... now Colthurst Brabazon. Capt., Unatt.	8
Evans, John, Lieut.—6, 7, 9, 11 ... ... Lieut., H.P., 83rd.	4
Gascoyne, Thos. Bamber, Lieut.—2, 6 ... ... Capt., H.P., 3rd Ceylon Regt.	2
Glasco, John, Asst.-Surg.—1, 3, 4, 5, 6, 7, 8, 9, 10 ... ... Staff Surg., H.P.	9
Irwin, F. C. K. H., Lieut.—1, 3, 4, 5, 6, 7, 8, 9, 10 ... ... Bt.-Lt.-Col., Unatt. Comdg. Troops, W. Australia.	9
Johnston, Francis, Lieut.—1 ... ... late Capt., 83rd.	1
Lane, Ambrose, Lieut.—5, 6, 7, 9, 11 ... ... Lieut., H.P.	5
Matthews, Joseph, Lieut.—4 ... ... late Lieut., 83rd.	1
Maxwell, Wm., Ens.—9, 11, 12 ... ... late Lieut., H.P., Unatt.	3
Mee, Geo., Lieut.—2, 3, 5 ... ... late Capt., 83rd.	3
Neligan, Thos., Ens.—4, 5, 6 ... ... Lieut., H.P., 83rd.	3
Nicholson, John, Lieut.—1 ... ... late Capt., 5th Vet. Batt.	1
O'Niell, Chas., Lieut.—2, 3, 4, 5, 6, 7, 8, 9, 10, 11, 12 ... ... Capt., H.P., 2nd.	11
O'Niell, Wm., Vol. and Ens.—5, 6, 7, 8, 9, 10, 11, 12 ... ... Capt., H.P., Unatt.	8
Ormsby, Jas., Lieut.—2, 3, 6 ... ... Lieut., H.P., 25th Dns.	3

	Total Clasps.
Stephenson, G. A., Asst.-Surg.—3, 4, 5, 6, 7, 8, 9, 10 ... ... Surg., 3rd Dn. Gds.	8
Strangways, W., Lieut.—3, 6 ... ... late Lieut., H.P., 3rd Garr. Batt.	2
Swinburne, F., Lieut. and Adjt.—1, 2, 3, 4, 5, 6, 7, 9, 11, 12 ... Major, 83rd.	10
Vereker, H. T., Lieut.—2, 3, 6 ... ... Lieut., H.P., 62nd.	3
Woodhouse, Robt., Ens.—1, 2, 3, 5 ... ... late Capt., Ret., Full Pay, 38th; deceased.	4
Wyatt, Herbert, Lieut.—4, 5, 6, 7, 8, 9 ... ... late Lieut., 25th.	6
Total Officers 24	

Summary of non-commissioned officers and men of the 2nd Battalion 83rd Foot who were awarded the Silver Military General Service Medal, under General Order dated 1st June, 1847, with a re-capitulation of their clasps.

No. of Men.		No. of Clasps.
43	received	1
18	,,	2
17	,,	3
19	,,	4
24	,,	5
12	,,	6
17	,,	7
11	,,	8
6	,,	9
8	,,	10
7	,,	11
1	,,	12
183	Total Recipients of Medals.	

83RD REGIMENT.

Roll of Claimants for the Medal for Service during the recent operations in India.

Camp, Nusseerabad,
3rd November, 1858

Lieut.-Col. E. Steele.
Major C. W. Austin.
,, J. Heatly (Brevet-Lieut.-Colonel).
Capt. J. S. Moloney.
,, R. R. Wyvill.

Capt. H. de R. Pigott.
,, S. Read, killed in action at Jeerum.
,, T. M. Baumgartner.
,, T. P. Wright.
,, H. S. Cooper, deceased, 13th July, 1858.
,, F. H. D. Marsh.
Lieut. J. Wakefield.
,, T. G. Coote.
,, H. Gandy.
,, C. C. Gore.
,, P. C. Browne.
,, W. Minehear.
,, G. G. Beazley.
,, G. W. W. Wardell.
,, G. M. Onslow.
,, B. Chamley.
,, N. Pennefather.
,, F. Karslake.
Ensign W. F. Anderson.
,, J. Healey.
2nd Class Staff Surg. Edwd. Touch, M.B., present at Defence of Mount Aboo, 21st August, 1857.
Assist.-Surg. H. C. Miles.
Adjutant J. N. Colthurst.
Assist.-Surg. William Sharp.
Lieut. G. L. Huyshe.
Paymaster J. D. Swinburn.
Apothecary James Laurence.
Assist.-Apothecary Baptist Fernandez.
A.-Steward Bernard Thompson, deceased.
Assist.-Apothecary Robert Crawford.
1276, Serjt.-Major John Slater.
1551, Hospl.-Serjt. George Palmer.
2491, O.R. Clerk William Parkhill.
2238, A.-Serjt. Edward Chandler, deceased.
1579, Colour-Serjt. Joseph Anderson.
2724, ,, John Fyfe.
2122, ,, Joseph Gray.
2518, ,, Michael McQuade.
2150, ,, Thomas Maynell.
2705, ,, James Murphy.
2156, ,, Thomas Rees.
2088, ,, James Simpson.
1989, ,, William Taberer.
2055, ,, Patrick Tennison.

Abstract :—

Claimants—31 Officers; 4 Warrant Officers; 50 Serjeants; 19 Drummers; 819 Rank and File.

Extracts from Supplementary Roll of Claimants for Indian Mutiny Medal, 21st March, 1859:—

NAME.	Service for which entitled to Medal.
Captain Edwd. Meurant.	Present at the surprise of and attack on the Rebels at Seekur, on the 21st Jan., 1859, under Lt.-Colonel Holmes.
Qr.-Master Patk. Hayes.	Was also awarded the Medal for the same service.

Extract from Hart's "Army List":—

Captain John Sprot, 83rd Regt.	Served with the Force under Genl. Woodburn in the affair at Aurungabad, in 1857 (Medal), and afterwards as Acting Engineer of the Rajputana Field Force.

86TH REGIMENT.

Roll of officer, non-commissioned officer and men of the 86th Foot who served in Egypt and received the Military General Service Medal with clasp, inscribed, "Egypt, 1801." (Under G.O. of 11th February, 1850.)

*Lieut.-General George Middlemore, C.B., Captain in 1801.
Serjeant John Hopewell, Captain Kyler's (sic) Company.
Private John Burden, Captain Cuyler's Company.
Private William Eyre.
Private Edward Moses.
Private William Jones, Captain Middlemore's Company.
Private John Scott, Captain Cuyler's Company.
Private Anthony Walsh, No. 1 Company.

Roll of officers, etc., 86th Regiment of Foot, who have been employed in the suppression of the Mutiny in India.

Dated at Cantonment,
Gwalior,
17th December, 1858.

Colonel R. H. Lowth.
Major W. K. Stuart.
,, G. Keane (Lt.-Col.).
Capt. J. P. Mayers (Brevet-Major).
,, C. Darby (Brevet-Major).
,, M. Lepper.
,, R. E. Henry.

* Received gold medal for Talavera as Major, 48th Foot; died in 1850.

Lieut. J. R. Stuart.
,, G. S. Nunn.
,, R. F. Lewis.
,, H. E. Jerome.
,, W. Knipe.
,, J. K. D. Mackenzie.
,, A. R. Ord.
,, F. D. Edwards.
,, J. Creagh.
,, J. W. Fry.
,, V. G. Coates, died 27th September, 1858.
,, J. G. Dartnell.
,, G. A. Conran.
,, J. D. Brockman.
Ens. J. W. Sewell.
,, C. Keane.
,, G. Fowler.
,, J. Wells.
,, H. S. Cochrane.
Surgeon T. Stack, M.D., killed at Jhansi, 3rd April, 1858.
,, J. Kellie, M.D.
Assist.-Surgeon T. S. Barry.
Paymaster C. F. Heatley.
Serjt.-Major Jerome Murphy, Ensign 47th Foot.
Qr.-Mr.-Serjeant George Wane.
Paymaster William Thompson.
O.R. Serjeant John Coleman.
Armr.-Serjeant Jonas Warner.
Drum-Major Thomas McDermott.
Colour-Serjeant Francis Clough.
,, James Crosbie.
,, Bernard Farrell.
,, Robert Jackson.
,, William Kelly.
,, William McNeill.
,, Thomas Pickerain, died 19th November, 1858.
,, Matthew Prenderville.
,, John F. Stephens.
,, Benjamin Navan.

Total, all ranks—903.

FOUR RECIPIENTS OF V.C. IN 86TH REGIMENT.

Honorary Major-General

H. E. JEROME, V.C.

Died 25th February, 1901.

(This officer was granted a reward for Distinguished and Meritorious Service, on the 19th of February, 1885.)

THE HISTORY OF THE ROYAL IRISH RIFLES.

Colonel

H. S. COCHRANE, V.C.

(This officer's death was announced in "The Times" of the 21st of April, 1884, but date of death not given.)

Private

JAMES BYRNE, V.C., 86th Regiment.

Died 10th July, 1879.

Private

JAMES PEARSON, V.C., 86th Regiment.

Transferred to 56th Regiment, and later (30th June, 1860), to 83rd Regiment.

Died 23rd January, 1900.

Had annuity of £31 14s. 8d.

THE ROYAL IRISH RIFLES.

List of non-commissioned officers of the above Corps who have been awarded the Meritorious Service Medal with Annuity.

War Office, S.W.,

6th October, 1910.

83rd, Q.M.-Serjt. A. Barber	£20	6-12-47
83rd, Q.M.-Serjt. S. McGahey	£15	14-6-54
86th, Serjt. O. Farrell	£15	14-6-54
86th, Serjt.-Major A. Wolfe	£15	? 1859*
86th, Q.M.-Serjt. W. McNeill	£15	? 1860*
86th, Colour-Serjt. E. Baird†	£10	7-12-85
83rd, Serjt.-Major J. Slater	£10	2-5-88
83rd, Serjt.-Major W. Given	£10	2-5-88
83rd, Colour-Serjt. T. Lynch	£10	2-11-93
83rd, Serjt.-Major G. Linder†	£10	25-9-97
86th, Garr.-Serjt.-Major R. Courtney	£10	1-4-04
83rd, Q.M.-Serjt. J. Linn†	£10	1-4-04
83rd, Serjt. J. Campbell	£10	5-9-04
83rd, Q.M.-Serjt. W. Lynas†	£10	7-3-05
86th, Serjt. P. Currivan†	£10	21-11-08
83rd, Q.M.-Serjt. H. Brown†	£5	11-8-10

* These two medals are for "Distinguished Conduct in the Field."

† Those marked thus are still living.

THE ROYAL IRISH RIFLES.

Warrant Officers, Non-Commissioned Officers and Men, Awarded the Medal for Distinguished Conduct in the Field:—

Name and Rank.	Campaign.	Remarks.
Wolfe, Serjt.-Major A. (86th Regt.)	Central India, 1858–9.	For service with Central India Field Force under Maj.-Gen. Sir Hugh Rose. (With an annuity of £15.)
McNeill, Qr.-Mr.-Serjt. Wm. (86th Regt.)	Central India, 1858–9.	For service with Central India Field Force under Maj.-Gen. Sir Hugh Rose. (With an annuity of £15.)
Gardiner, Serjt.-Major R. (5th Bn.)	South Africa, 1899–1902.	Did most excellent service and set a splendid example in carrying out, not only his own duties but others, which strictly did not belong to his department.
Rainey, Serjt. R. (2nd Bn.) ...	Ditto.	Brought out of fire a disabled comrade.
Darragh, Serjt. J. (2nd Bn.) ...	Ditto.	Behaved gallantly during an attack for 4½ hours against superior numbers.
Boyd, Private J. (2nd Bn.) ...	Ditto.	Helped a wounded comrade under fire. (Promoted Corporal for gallantry in the field.)
Irvine, Corpl. R. (Mtd. Inf.) ...	Ditto.	Conspicuous gallantry on several occasions, notably at Klipfontein, on August 19th, 1900, when retiring from a force of 400 Boers, he went back under a heavy fire and brought in Private Flynn, of the same Company, on his horse, until the second horse was also killed.
Wright, Corpl. J. (2nd Bn.) ...	Ditto.	Distinguished himself when in command of a Mounted Infantry escort, which was attacked by a much superior force.
Anderson, Private R.	Ditto.	Mentioned in Lord Kitchener's despatch of 23rd June, 1902.
Beck, Private R. (2nd Bn.) ...	Ditto.	Gallantly leading 7 men, on occasion of a sudden night attack by Boers, at Vlakplaats, December 30th, 1901.
Hanlon, Pte. J. (Mtd. Inf.) ...	Ditto.	For service with 9th Mounted Infantry.
Hogg, Pte. J. (Mtd. Inf.) ...	Ditto.	For service with 9th Mounted Infantry.
Keenan, Pte. J. (Mtd. Inf.) ...	Ditto.	For service with 9th Mounted Infantry.
McIlhare, Pte. D. (Mtd. Inf.) ...	Ditto.	Conspicuous gallantry on several occasions, notably at Klipfontein on August 19th, 1900, when covering retreat of Col. Sitwell's force, when opposing the enemy's advance at a kraal for 1½ hours, during which time he killed several Boers at close quarters.

AWARD OF MERIT

GIVEN TO

SAMUEL HOLT,

QUARTER MASTER, 83RD FOOT,

FOR

MERITORIOUS SERVICE,

1815.

APPENDICES.

APPENDIX I.

Though these lists are correct to a very great extent, it is known that a few officers' names have been omitted, as they were not shewn in the annual Army Lists, having served only a few weeks with the Regiment.

(Corrected to April, 1913.)

NOMINAL ROLL & COMMISSION LIST OF THE OFFICERS OF THE 83RD FOOT,

ORIGINALLY STYLED COLONEL FITCH'S REGIMENT OF FOOT.

Name.	Ensign or 2nd Lieut.	Lieutenant.	Captain.	Major.	Lieut.-Colonel.	Colonel.	Remarks.
Fitch, William	... ...	... ...	... ...	... ...	28–9–1793	... ...	From 55th Regt.—Lt.-Col. Commandant. Killed in the Maroon War.
Handfield, Charles ...	... ...	... ...	... ...	28–9–1793	... ...	... ...	From 22nd Regt.—Lt.-Col., 89th, 3-12-1793.
Sleigh, William	... ...	... ...	... ...	28–9–1793	15–6–1794	... ...	From 23rd Regt.—Not shown after 1795.
Hansard, Henry Hart	... ...	... ...	28–9–1793	... ...	... ...	... ...	From 69th Regt.—Died, 9-11-1795.
Hay, William	... ...	... ...	28–9–1793	... ...	... ...	... ...	From 43rd Regt.—Died, 14-11-1795.
Staples, William Deane	... ...	... ...	28–9–1793	... ...	... ...	... ...	From 66th Regt.—To 9th Dragoons, 5-7-1794.
Tomkins, Charles ...	... ...	... ...	28–9–1793	... ...	... ...	... ...	From 64th Regt.—To 34th Regt., 31-1-1794.
Browne, Gore	... ...	... ...	28–9–1793	15–6–1794	... ...	... ...	From 35th Regt.—Lt.-Col., W. India Regt., 30-11-1796.
Moore, Francis	... ...	... ...	28–9–1793	... ...	... ...	... ...	From 28th Regt. — Major, 128th Regt., 20-7-1794.
Godley, William	... ...	... ...	28–9–1793	22–8–1794	10–5–1796	... ...	From 41st Regt.—Not shown after 1800.
Foote, Simon	... ...	... ...	28–9–1793	... ...	... ...	... ...	From 41st Regt. — Major, 23rd Regt., 22-4-1794.
Dickson, William ...	... ...	... ...	30–9–1793	... ...	... ...	... ...	From 22nd Regt.—Not shown after 1794.
Legh, ——	... ...	... ...	6–9–1794	... ...	... ...	... ...	Died of wounds received in action, 18-9-1795.
Cane, Maurice	... ...	... ...	7–10–1793	... ...	... ...	... ...	From New Independent Cos.—Not shown after 1802.
Beamish, Thomas ...	... ...	28–9–1793	30–10–1794	1–8–1804	... ...	... ...	Not shown after 1806.
Keane, Edward	... ...	28–9–1793	28–2–1794	... ...	... ...	... ...	Died, 20-8-1796.
Hamilton, Skeffington	... ...	28–9–1793	31–5–1794	... ...	... ...	... ...	Not shown after 1801.
Blunden, Trevor Lloyd	... ...	28–9–1793	... ...	... ...	... ...	... ...	From 17th Regt.—Not shown after 1795.
Napper, John	... ...	28–9–1793	5–3–1796	15–5–1806	... ...	... ...	From h.p., 91st Regt.—Lt.-Col. in Army, 4-6-1813. Died, Dec., 1820.
Gore, Richard	... ...	28–9–1793	14–1–1794	... ...	... ...	... ...	To 55th Regt., 28-2-1794.
Lees, William Eden ...	... ...	28–9–1793	... ...	... ...	... ...	... ...	To h.p. in 1793.

APPENDIX I.

Brunt, Jacob	... ...	28–9–1793	20–8–1796	16–5–1805	13–6–1811	... ...	From Qr.-Mr., 55th Regt.—Adjt., 28-9-1793, 30-9-1802. Retired, 1821.
Davis, Paul	28–9–1793	... ...	... ...	... ...	... ...	... ...	Lieut., Fox's Regt., 25-3-1794.
Staples, John	28–9–1793	... ...	... ...	... ...	... ...	... ...	Lieut., Vere Hunt's Regt., 20-7-1794.
Buchanan, William	28–9–1793	... ...	... ...	... ...	... ...	... ...	Not shown after 1805.
Mathews, Rogerson	28–9–1793	... ...	... ...	... ...	... ...	... ...	Lieut., Vere Hunt's Regt., 20-7-1794.
Brown, Melvill	28–9–1793	8–10–1793	... ...	... ...	... ...	... ...	To W. India Regt., 19-7-1797.
Starke, Julius	28–9–1793	... ...	... ...	... ...	... ...	... ...	Lieutenant, Mountnorris's Regt., 22-5-1794.
O'Connor, Patrick	28–9–1793	... ...	... ...	... ...	... ...	... ...	Lieut., Stratford's Regt., 25-7-1794.
Wilton, Thomas	28–9–1793	19–7–1794	... ...	... ...	... ...	... ...	Died 14-11-1795.
White, Humphrey J.	... ...	... ...	14–6–1794	23–10–1800	... ...	... ...	From 5th Regt.—Died, 27-11-1800.
Corry, Marcus	... ...	... ...	6–9–1794	... ...	... ...	... ...	From 31st Regt.—Not shown after 1795.
Tighe, William George	... ...	... ...	24–9–1794	... ...	... ...	... ...	To 7th D.G., 1-1-1797.
Browne, George Van	... ...	... ...	12–11–1794	... ...	... ...	... ...	From 101st Regt.—Not shown after 1797.
Blaquiere, John	... ...	28–2–1794	21–10–1802	13–6–1811	... ...	... ...	Not shown after Aug., 1814.
Burge, Benjamin	... ...	31–5–1794	... ...	... ...	... ...	... ...	To h.p., 1798.
Benson, Richard	... ...	31–5–1794	... ...	... ...	... ...	... ...	Believed to have died, 17-8-1796.
Gethin, Reed	... ...	14–6–1794	... ...	... ...	... ...	... ...	To 23rd L.D., 12-1-1799.
Smith, Thomas	... ...	15–6–1794	... ...	... ...	... ...	... ...	From 5th Regt.—Died, 8-8-1800.
Armstrong, Richard	... ...	15–6–1794	... ...	... ...	... ...	... ...	From New Independent Cos.—Died, 27-10-1795.
Kent, John	... ...	15–6–1794	... ...	... ...	... ...	... ...	Capt., 13th L.D., 16-7-1796.
Tramasse, Michael Simon	... ...	15–6–1794	... ...	... ...	... ...	... ...	Died, 20-4-1797.
Morris, Francis	... ...	15–6–1794	... ...	... ...	... ...	... ...	From 44th Regt.—Not shown after 1795.
Reeves, William H.	... ...	15–6–1794	... ...	... ...	... ...	... ...	From 53rd Regt.—Died, 2-9-1795.
Corr, Luke	... ...	8–7–1794	... ...	... ...	... ...	... ...	Died, 30-10-1795.

Name.	Ensign or 2nd Lieut.	Lieutenant.	Captain.	Major.	Lieut.-Colonel.	Colonel.	Remarks.
Homan, William J. ...	... ...	31–7–1794	25–6–1803	... ...	... ...	... ...	Not shown after 1806.
Richardson, George Arthur	... ...	31–7–1794	... ...	... ...	... ...	... ...	Not shown after 1799.
Ball, William	... ...	24–9–1794	... ...	... ...	... ...	... ...	Died, 20-8-1800.
Scott, Robert	... ...	24–9–1794	26–7–1797	17–8–1809	... ...	... ...	Retired, 1810.
Hart, Joseph	... ...	11–10–1794	... ...	... ...	... ...	... ...	From 99th Regt.—To 16th Regt., 27-2-1796.
Moreton, J.	... ...	11–10–1794	... ...	... ...	... ...	... ...	From 85th Regt.—Died, 28-10-1795.
Gerrard, Thomas ...	... ...	3–11–1794	... ...	... ...	... ...	... ...	From 108th Regt.—To 17th L.D., 20-6-1799.
Church, James	1794	26–11–1794	... ...	... ...	... ...	... ...	Not shown after 1798.
Brennan, Michael ...	1794	26–11–1794	... ...	... ...	... ...	... ...	Not shown after 1798.
Jackson, George	20–7–1794	... ...	... ...	... ...	... ...	... ...	Not shown after 1795.
Gibson, Thomas	20–7–1794	20–8–1796	... ...	... ...	... ...	... ...	Died, 4-10-1800.
Burke, Thomas	20–7–1794	... ...	... ...	... ...	... ...	... ...	Lieut., 4th Regt., 25-10-1799.
Jackson, Robert	24–9–1794	... ...	... ...	... ...	... ...	... ...	Not shown after 1795.
Knipe, John	24–9–1794	... ...	... ...	... ...	... ...	... ...	Lieut., 59th Regt., 24-5-1797.
Balfour, James	... ...	... ...	... ...	... ...	... ...	18–11–1795	From 77th Regt. — Maj.-General, 3-10-1794; Lt.-Genl., 1-1-1801; Genl., 25-10-1809. Died, March, 1823.
Gibson, Thomas	... ...	... ...	... ...	... ...	27–5–1795	... ...	From 49th Regt.—Not shown after 1803.
Baynes, Sir Edward ...	... ...	... ...	... ...	... ...	20–6–1795	... ...	From 2nd Regt.—Not shown after 1796.
Birch, John	... ...	... ...	1–9–1795	... ...	... ...	... ...	From Fox's Regt.—Not shown after 1796.
Wilson, Christopher ...	... ...	... ...	2–9–1795	... ...	... ...	... ...	From 114th Regt.—Died, 7-6-1801.
Clifford, Miller	... ...	21–2–1795	... ...	... ...	... ...	... ...	Capt., W. India Regt., 22-2-1799.
Brown, Henry	2–2–1795	... ...	... ...	... ...	... ...	... ...	Lieut., 17th Regt., 26-1-1797.
Horridge, William ...	7–2–1795	... ...	... ...	... ...	... ...	... ...	Died, 24-10-1795.
Morris, Francis	21–2–1795	... ...	... ...	... ...	... ...	... ...	Died, 20-8-1796.

APPENDIX I.

Temple, Grenville ...	11-3-1795	24-10-1795	... ...	... ...	... ...	... ...	Not shown after 1798.
Brown, Thomas	1-4-1795	25-12-1796	... ...	... ...	... ...	... ...	Not shown after 1800.
Billings, George	7-4-1795	... ...	... ...	... ...	... ...	... ...	Lieut., 69th Regt., 3-5-1799.
Griffiths, Charles	... ...	... ...	... ...	1-6-1796	... ...	... ...	From 14th Regt.—To 82nd Regt., 9-2-1797.
Vesey, John	... ...	... ...	... ...	14-12-1796	... ...	... ...	From W. India Regt.—Lt.-Col., 52nd Regt., 6-9-1798.
Reynolds, Richard ...	... ...	... ...	3-5-1796	... ...	... ...	... ...	From 49th Regt.—Not shown after 1797.
Carr, Henry William ...	... ...	... ...	4-5-1796	17-9-1807	22-9-1814	... ...	From 68th Regt.—K.C.B. To h.p., 24-10-1816.
Ross, Andrew	... ...	6-7-1796	... ...	... ...	... ...	... ...	From 15th Regt.—Not shown after 1798.
Butler, James	... ...	29-7-1796	... ...	... ...	... ...	... ...	From 62nd Regt.—Not shown after 1797.
Brunt, Abraham	... ...	20-10-1796	3-8-1804	19-8-1813	... ...	... ...	To h.p., 1-1-1818.
Lawton, Philip	2-11-1796	... ...	... ...	... ...	... ...	... ...	Not shown after 1798.—Died in West Indies; date unknown.
Wright, Lachlan ...	25-12-1796	13-9-1798	... ...	... ...	... ...	... ...	Qr.-Mr., 28-9-1793 — 24-12-1796. — Died, 12-12-1801.
Somerfield, Thomas ...	4-12-1798	20-2-1800	19-9-1804	21-12-1820	... ...	... ...	Qr.-Mr., 25-12-1796 — 3-12-1798. — Died, 1833.
Edwards, Benjamin ...	... ...	... ...	... ...	9-2-1797	... ...	... ...	From 2nd Regt.—Not shown after 1798.
McLean, William ...	... ...	... ...	9-2-1797	... ...	... ...	... ...	From 6th Regt.—Not shown after 1798.
Lomax, James	... ...	12-1-1797	... ...	... ...	... ...	... ...	From 62nd Regt.—Capt., W. India Regt., 15-6-1798.
Williams, Thomas ...	... ...	9-3-1797	... ...	... ...	... ...	... ...	From 16th Regt.—Died, 1-12-1800.
Richardson, Joseph ...	... ...	19-7-1797	... ...	... ...	... ...	... ...	From W. India Regt.—Not shown after 1800.
McHugh, Hugh	... ...	26-7-1797	... ...	... ...	... ...	... ...	Not shown after 1800.
McIntosh, Robert Keith	1-11-1797	... ...	... ...	... ...	... ...	... ...	To h.p., 1798.
Hamilton, Henry ...	1-11-1797	... ...	... ...	... ...	... ...	... ...	Lieut., 81st Regt., 20-6-1799.
Skerrett, John Byne ...	... ...	... ...	... ...	29-3-1798	23-10-1800	... ...	From 69th Regt.—Lt.-Col., 10th Bn. of Reserve, 17-9-1803.
Smyth, George Stracey	... ...	... ...	... ...	6-9-1798	... ...	... ...	From 7th Regt.—To 3rd Garr. Bn., 9-5-1805.
Cumberland, Charles ...	... ...	... ...	22-3-1798	... ...	... ...	... ...	From Royal Horse Gds.—Not shown after 1799.

Name.	Ensign or 2nd Lieut.	Lieutenant.	Captain.	Major.	Lieut.-Colonel.	Colonel.	Remarks.
Armstrong, William	… …	… …	5–4–1798	… …	… …	… …	Late 95th Regt.—To h.p., 1803.
Dixon, Robert	… …	11–8–1798	… …	… …	… …	… …	From 89th Regt.—Not shown after 1799.
Lewis, James	… …	22–11–1798	… …	… …	… …	… …	Not shown after 1800.
Hunt, Robert	25–1–1798	… …	… …	… …	… …	… …	Lieut., 40th Regt., 6-11-1799.
Burke, Thomas	8–8–1798	… …	… …	… …	… …	… …	Not shown after 1803.
Farrell, Patrick	25–12–1798	21–3–1800	… …	… …	… …	… …	Died, 26-1-1802.
Law, John	… …	… …	10–5–1799	… …	… …	… …	From 46th Regt.—To R.H. Gds., 24-10-1799.
Cocks, James	… …	20–6–1799	… …	… …	… …	… …	Not shown after 1800.
Fitzgerald, Terence	… …	26–9–1799	… …	… …	… …	… …	From 66th Regt.—Not shown after 1800.
Straubenzee, Frederick	30–8–1799	23–4–1800	8–10–1807	… …	… …	… …	Not shown after Nov., 1814.
Williams, E. H.	12–9–1799	… …	… …	… …	… …	… …	To h.p., 1800.
Creagh, Francis Michael	19–12–1799	21–2–1800	24–3–1806	… …	… …	… …	Major in Army, 12-8-1819.—To h.p., 25-4-1822.
Pike, William	… …	… …	… …	… …	… …	… …	Qr.-Mr., 1-6-1799—24-12-1802.
Smith, George	11–1–1800	20–10–1800	6–6–1805	… …	… …	… …	Not shown after 1806.
Smith, Thomas	7–2–1800	5–2–1801	23–3–1809	… …	… …	… …	To 60th Regt., 28-2-1811.
Hill, Edward	21–3–1800	… …	… …	… …	… …	… …	Died, 30-9-1800.
Ball, John	9–5–1800	12–3–1801	… …	… …	… …	… …	Not shown after 1804.
Lewin, G. R.	30–5–1800	… …	… …	… …	… …	… …	Not shown after 1803.
New, Samuel	5–6–1800	… …	… …	… …	… …	… …	Not shown after 1801.
Oliver, William	12–6–1800	… …	… …	… …	… …	… …	Lieut., h.p., 1802.
Parys, Michael	30–9–1800	… …	… …	… …	… …	… …	From W. India Regt.—Lieut., h.p., 1802.
Graham, James	9–10–1800	… …	… …	… …	… …	… …	Not shown after 1803.
Dunbar, J. Killigrew	… …	… …	… …	11–3–1801	… …	… …	From 69th Regt.—Not shown after 1805.
Grant, Lewis	… …	… …	5–2–1801	… …	… …	… …	From Scotch Bde.—To h.p., 1803.

Lock, John	... ...	... ...	3–7–1801	... ...	... ...	... ...	From 70th Regt.—To h.p., 1803.
Hall, William Henry ...	12–2–1801	21–10–1802	... ...	... ...	... ...	... ...	Capt., 95th Regt., 6-6-1805.
Collins, Henry	12–3–1801	31–12–1802	13–3–1806	... ...	... ...	... ...	Not shown after 1808.
Shearer, Thomas ...	8–10–1801	... ...	... ...	... ...	... ...	... ...	Not shown after 1802.
Le Mesurier, Haviland	... ...	1–10–1802	... ...	... ...	... ...	... ...	From 20th Regt.—Capt., 21st Regt., 25-8-1804.
Campbell, Alexander ...	20–8–1802	25–6–1803	29–3–1809	... ...	... ...	... ...	Died, 1830.
Broad, Richard	1–10–1802	3–9–1803	... ...	... ...	... ...	... ...	Adjt., 1-10-1802—27-8-1804; Capt., 43rd Regt., 28-8-1804.
Bucknall, John Lindsay	8–10–1802	... ...	... ...	... ...	... ...	... ...	Not shown after 1803.
Gibb, William	15–10–1802	5–8–1804	... ...	... ...	... ...	... ...	Not shown after 1805.
Lavery, John	25–10–1802	6–8–1804	... ...	... ...	... ...	... ...	Adjt., 12-10-1804—26-3-1806. — To 13th Regt., 13-5-1808.
Wiley, Daniel	14–9–1804	13–3–1806	11–8–1814	... ...	... ...	... ...	Qr.-Mr., 25-12-1802—6-9-1804.—To h.p., 25-6-1817.
Baird, Joseph	... ...	... ...	... ...	... ...	17–3–1803	... ...	From h.p.—Col., 25-10-1809.—Maj.-Gen., 1-1-1812.
Hutchinson, William ...	... ...	... ...	... ...	... ...	9–6–1803	... ...	From 49th Regt.—To 36th Regt., 2-8-1804.
Nesbitt, Alexander ...	... ...	... ...	16–3–1803	... ...	... ...	... ...	From h.p.—Major, Q.M.G. Dept., 25-3-1805.
Hall, James	... ...	... ...	17–3–1803	... ...	... ...	... ...	From 60th Regt.—To 4th R. Vet. Bn., 26-1-1809.
Holbrooke, Henry ...	... ...	17–9–1803	30–3–1809	... ...	... ...	... ...	Retired, 1810.
Mortimer, Thomas ...	17–2–1803	... ...	... ...	... ...	... ...	... ...	From h.p.—Lieutenant, 58th Regt., 24-3-1804.
Donovan, D. Edward...	31–3–1803	28–8–1804	17–8–1809	... ...	... ...	... ...	Died, 1812.
Venables, Joseph ...	14–4–1803	20–3–1805	13–6–1811	... ...	... ...	... ...	Not shown after May, 1814.
Campbell, Donald ...	12–8–1803	21–3–1805	5–3–1812	... ...	... ...	... ...	To h.p., 20-8-1818.
Vernon, Joseph	3–9–1803	... ...	... ...	... ...	... ...	... ...	Not shown after 1804.
Trotter, William R. ...	... ...	... ...	... ...	28–8–1804	... ...	... ...	From 88th Regt.—Not shown after 1807.
Arnot, Hugo	... ...	... ...	18–2–1804	... ...	... ...	... ...	From h.p.—Not shown after 1808.
Halliday, Joseph Loftus	... ...	... ...	1–8–1804	... ...	... ...	... ...	From 7th Bn. of Res.—To 9th R. Vet. Bn., 26-12-1805.

NAME.	Ensign or 2nd Lieut.	Lieutenant.	Captain.	Major.	Lieut.-Colonel.	Colonel.	REMARKS.
Wilson, James M.	... ...	... ...	2–8–1804	... ...	... ...	... ...	From 10th Bn. of Res.—Not shown after 1805.
Sullivan, James	... ...	... ...	4–8–1804	11–8–1814	... ...	... ...	From 30th Regt.—To h.p., 25-6-1817.
O'Neill, L. G.	... ...	... ...	26–8–1804	... ...	... ...	... ...	From 2nd Bn. of Res.—Not shown after 1808.
Moore, William	... ...	... ...	27–8–1804	... ...	... ...	... ...	From 57th Regt.—To W.I. Regt., 1–5–1805.
Naish, Edward	... ...	... ...	28–8–1804	... ...	... ...	... ...	From 14th Regt.—To h.p., 1807.
Cameron, Allan	... ...	... ...	14–9–1804	13–10–1814	... ...	... ...	From W. India Regt.—To h.p., 25-6-1817.
Collins, Edward	... ...	16–1–1804	... ...	... ...	... ...	... ...	From h.p.—Captain, 21st L. Dgns., 5-9-1805.
Gardiner, Charles	... ...	3–8–1804	... ...	... ...	... ...	... ...	From 6th Bn. of Res.—Capt., N.S. Terr. Inf., 10-10-1805.
Loftus, Henry	... ...	4–8–1804	... ...	... ...	... ...	... ...	From 2nd Bn. of Res.—Not shown after 1808.
Stock, William	31–3–1804	3–5–1805	15–5–1806	... ...	... ...	... ...	Retired, 1810.
Brown, F. G.	21–4–1804	4–5–1805	... ...	... ...	... ...	... ...	Not shown after 1806.
Gapper, Edmund	1–8–1804	5–5–1805	3–6–1812	... ...	... ...	... ...	To h.p., 25-6-1817.
Milne, William	14–8–1804	... ...	... ...	... ...	... ...	... ...	Not shown after 1805.
Pyne, Robert	15–8–1804	6–5–1805	... ...	... ...	... ...	... ...	Promoted Captain, 66th Regt., for gallantry in action, 1809.
Emmett, Edwin Cheere	16–8–1804	7–5–1805	4–6–1812	... ...	... ...	... ...	To h.p., 25-6-1817.
Dennie, Michael	1–9–1804	16–1–1806	... ...	... ...	... ...	... ...	To 7th Garr. Bn., 15-4-1807.
Reynolds, James	7–9–1804	6–6–1805	16–7–1807	... ...	... ...	... ...	Died from wounds received in action, 1813.
Dorrell, B.	7–9–1804	12–3–1806	28–4–1814	... ...	... ...	... ...	Adjt., 7-9-1804—12-5-1814.—To h.p., 25-6-1817.
Beetham, ——	5–10–1804	... ...	... ...	... ...	... ...	... ...	Not shown after 1805.
Townsend, William	18–10–1804	29–10–1806	... ...	... ...	... ...	... ...	To h.p., 15-1-1818.
Trydell, Botet	19–10–1804	30–10–1806	... ...	... ...	... ...	... ...	Capt., 2nd Ceylon Regt., 17-11-1818.
Duke, Charles	26–10–1804	... ...	... ...	... ...	... ...	... ...	Lieut., 1st Garr. Bn., 23-5-1805.
Gunter, Thomas Foster	1–11–1804	... ...	... ...	... ...	... ...	... ...	To 12th Regt., 4-4-1805.
Slater, William	... ...	... ...	... ...	... ...	... ...	... ...	Qr.-Mr., 7-9-1804—26-5-1813.

Name	Date 1	Date 2	Date 3	Date 4	Date 5	Date 6	Remarks
Gordon, Alexander					16–5–1805		From 3rd Garr. Bn.—Killed at Talavera.
Collins, Richard				9–5–1805	17–8–1809		From 1st Garr. Bn.—Killed at Salamanca.
Yonge, Henry			1–5–1805				From W.I. Regt.—To 4th Garr. Bn., 25-9-1807.
Thomas, James			16–5–1805				From 22nd L.D.—Not shown after 1807.
Harris, William			19–9–1805				From 28th Regt.—To h.p., 1809.
Keith, James			3–10–1805				From 77th Regt.—To 65th Regt., 24-9-1807.
Harmer, William John		28–3–1805					From W.I. Regt.—To 2nd R. Vet. Bn., 2-8-1810.
Buchan, George		8–5–1805	11–2–1813				From 10th Regt.—To h.p., 25-6-1817.
Macpherson, William		9–5–1805	19–8–1813				To h.p., 25-6-1817.
Shaw, John	20–3–1805	9–2–1807	21–12–1820				To h.p., 1821.
Hincks, George Powell	4–4–1805	15–6–1806					Died, 1813.
Phillips, William	5–5–1805	10–2–1807	1822				Died, 1822.
Terry, John	6–5–1805	6–8–1807	31–1–1811				To h.p., 24-8-1820.
Robinson, James	7–5–1805	8–10–1807					Not shown after 1809.
Abell, Francis	8–5–1805	11–3–1808					Died, 1822.
Hingston, James	9–5–1805	21–4–1808					Lieut., 1st Foot, 22-4-1807 — 20-4-1808; Capt., Royal African Colonial Corps, 4-1-1824.
Cramer, John Lewis	23–5–1805						Lieut., 3rd Regt., 5-3-1807.
Patullo, William	27–6–1805						Lieut., 6th Regt., 4-3-1807.
Jennings, George	18–7–1805						Lieut., 6th Regt., 19-3-1807.
Smith, James	6–8–1807	23–3–1809					Qr.-Mr., 31-1-1805—5-8-1807.—Died, 1819.
Sheldon, George			7–8–1806				From 78th Regt.—To 1st R. Vet. Bn., 4-8-1808.
Mansergh, George			21–8–1806				From 22nd Regt.—Retired, 1810.
Carnell, P. P.		11–3–1806					From h.p.—To 1st R. Vet. Bn., 5-4-1808.
Hardman, James		2–10–1806	17–11–1814				From 27th Rgt.—To h.p., 25-6-1817.

Name.	Ensign or 2nd Lieut.	Lieutenant.	Captain.	Major.	Lieut.-Colonel.	Colonel.	Remarks.
Cotter, William	16–1–1806	12–3–1808	... ...	... ...	... ...	... ...	To Portuguese Army, 25-10-1814.
Parker, John	13–3–1806	... ...	... ...	... ...	... ...	... ...	Lieut., 61st Regt., 26-6-1807.
Brahan, Henry	27–3–1806	13–3–1808	... ...	... ...	... ...	... ...	Adjt., 27-3-1806—24-7-1811.—Died, 1821.
Fry, George	7–8–1806	15–4–1807	26–7–1810	... ...	... ...	... ...	Killed at Badajoz.
Jackson, James	29–10–1806	12–5–1808	... ...	... ...	... ...	... ...	Capt., 37th Regt., 25-6-1813.
Ormsby, James	30–10–1806	14–3–1808	... ...	... ...	... ...	... ...	Not shown after 1813.
Geddes, William ...	... ...	... ...	19–2–1807	... ...	... ...	... ...	From h.p.—Lt.-Col. in Army, 4-6-1814.—Died, 1822.
Dyer, George Lemon ...	... ...	... ...	24–9–1807	... ...	... ...	... ...	From 65th Regt.—Retired, 1810.
Hext, Samuel	... ...	... ...	25–9–1807	... ...	... ...	... ...	From 4th Garr. Bn.—Major in Army, 21-6-1813.—Died, 1822.
Laird, Charles	... ...	... ...	17–12–1807	... ...	... ...	... ...	From 82nd Regt.—Died, 1813.
McBean, Alexander ...	... ...	30–1–1807	... ...	... ...	... ...	... ...	Died, 1819.
Campbell, Alexander ...	... ...	1–2–1807	... ...	... ...	... ...	... ...	From 1st Foot.—To 1st Foot, 26-11-1807.
Gibson, Alexander ...	... ...	2–2–1807	... ...	... ...	... ...	... ...	From 1st Foot.—To 1st Foot, 21-4-1808.
Richards, Roderick ...	... ...	4–2–1807	... ...	... ...	... ...	... ...	From 1st Foot.—Retired, 1810.
Harding, John	... ...	5–2–1807	... ...	... ...	... ...	... ...	To 11th Regt., 12-9-1808.
Nicholson, John	... ...	8–2–1807	... ...	... ...	... ...	... ...	Capt., York Chasseurs, 6-11-1813.
Cruttwell, James ...	... ...	11–2–1807	... ...	... ...	... ...	... ...	From 97th Regt.—Died, May, 1818.
Elliott, Gilbert	... ...	12–2–1807	14–6–1810	... ...	... ...	... ...	To 32nd Regt., 23-7-1818.
Smith, Thomas F. ...	... ...	26–3–1807	... ...	... ...	... ...	... ...	To h.p., 4-4-1816.
Mee, George	... ...	26–11–1807	8–8–1822	... ...	... ...	... ...	From 1st Foot.—Retired, 1830.
Flood, Francis	11–2–1807	15–3–1808	... ...	... ...	... ...	... ...	Killed at Talavera.
Stewart, Samuel ...	12–2–1807	... ...	... ...	... ...	... ...	... ...	Lieut., 8th Garr. Bn., 31-12-1807.
Muter, Robert	5–3–1807	... ...	... ...	... ...	... ...	... ...	Lieut., 7th Regt., 3-3-1808.
Brock, George	19–3–1807	... ...	... ...	... ...	... ...	... ...	Lieut., 1st Foot, 31-12-1807.

Holland, William ...	26–3–1807	28–9–1808	21–7–1814	... ...	... ...	... ...		To h.p., 25-6-1817.
Rosengrave, Matthias ...	2–4–1807	... ...	... ...	... ...	... ...	... ...		Lieut., 10th Regt., 11-2-1808.
Vereker, Henry	23–4–1807	29–9–1808	... ...	... ...	... ...	... ...		To h.p., 7-3-1822.
Montgomery, John ...	25–6–1807	1–12–1808	... ...	... ...	... ...	... ...		Killed at Talavera.
Capel, William	8–10–1807	... ...	... ...	... ...	... ...	... ...		Lieut., R. Irish Regt., 7-7-1808.
Wood, John	9–10–1807	24–3–1809	... ...	... ...	... ...	... ...		Died, 1810.
L'Estrange, Christopher	31–12–1807	26–3–1809	... ...	... ...	... ...	... ...		Died, 1810.
Wight, Thomas	... ...	... ...	... ...	... ...	... ...	... ...		Qr.-Mr., 8-10-1807—31-7-1809.
Fraser, Robert	... ...	... ...	1–1–1808	... ...	... ...	... ...		From 72nd Regt. — To h.p., 20-3-1823.
Noleken, George	... ...	... ...	11–8–1808	... ...	... ...	... ...		From 3rd Ft. Gds.—To 57th Regt., 30-11-1815.
Baldwin, Connell James	... ...	16–3–1808	... ...	... ...	... ...	... ...		From 87th Regt.—Capt., 50th Regt., 10-2-1820.
Colthurst, Nicholas ...	... ...	29–3–1808	... ...	... ...	... ...	... ...		From 8th Garr. Bn.—To Portuguese Army, 25-10-1814.
Holmes, Richard ...	... ...	30–3–1808	29–8–1822	... ...	... ...	... ...		From 7th Garr. Battn.—To h.p., 15-1-1824.
Dahman, William ...	... ...	31–3–1808	... ...	... ...	... ...	... ...		From 48th Regt.—Killed at Talavera.
Elliott, Thomas	... ...	7–4–1808	... ...	... ...	... ...	... ...		From 52nd Regt.—Died, 1811.
Ramsay, John	... ...	14–4–1808	... ...	... ...	... ...	... ...		From 6th Regt.—Not shown after July, 1814.
Johnston, Francis ...	14–1–1808	27–3–1809	7–4–1825	... ...	... ...	... ...		Retired, 1834.
Cummings, William ...	28–1–1808	28–3–1809	... ...	... ...	... ...	... ...		Died, 1813.
Ferris, Jones	11–2–1808	29–3–1809	... ...	... ...	... ...	... ...		Killed at Fuentes d'Onor.
Richardson, Henry ...	3–3–1808	30–3–1809	... ...	... ...	... ...	... ...		To 13th R. Vet. Bn., 25-1-1813.
Watson, Charles ...	16–3–1808	15–8–1809	... ...	... ...	... ...	... ...		To h.p., 25-12-1818.
Boggie, Thomas ...	17–3–1808	16–8–1809	... ...	... ...	... ...	... ...		Died, 1817.
Irwin, Frederick ...	25–3–1808	17–8–1809	... ...	... ...	... ...	... ...		To h.p., 25-12-1818.
Emslie, John	26–3–1808	8–11–1809	... ...	... ...	... ...	... ...		To h.p., 25-12-1818.

Name.	Ensign or 2nd Lieut.	Lieutenant.	Captain.	Major.	Lieut.-Colonel.	Colonel.	Remarks.
Bowles, Charles	28–4–1808	9–11–1809	... ...	... ...	... ...	... ...	To h.p., 25-3-1817.
Taylor, ——	19–5–1808	... ...	... ...	... ...	... ...	... ...	Not shown after 1809.
O'Neill, Charles	16–6–1808	4–1–1810	... ...	... ...	... ...	... ...	To h.p., 25-12-1818.
Tresilian, John Barnard	6–10–1808	1–2–1810	... ...	... ...	... ...	... ...	Retired, 1812.
Evans, John	26–10–1808	21–6–1810	... ...	... ...	... ...	... ...	To h.p., 25-12-1818.
Carey, Michael	27–10–1808	7–3–1811	... ...	... ...	... ...	... ...	To h.p., 25-12-1818.
Heatley, John	3–11–1808	28–3–1811	... ...	... ...	... ...	... ...	Died, 1813.
Barry, Francis M. ...	22–12–1808	29–5–1811	... ...	... ...	... ...	... ...	To h.p., 25-12-1818.
Powys, Hon. Henry ...	... ...	... ...	26–1–1809	... ...	... ...	... ...	From 52nd Regt.—Killed at Badajoz.
Thompson, Robert ...	... ...	... ...	6–7–1809	... ...	... ...	... ...	From R. Staff Corps.—Died, 1827.
Renwick, Edward ...	... ...	... ...	14–12–1809	... ...	... ...	... ...	From 72nd Regt.—Died, 1831.
Gascoyne, Thomas Bamber	... ...	25–3–1809	... ...	... ...	... ...	... ...	From 6th Garr. Bn.—Capt., Ceylon Regt., 7-2-1822.
Le Toller, Henry ...	16–3–1809	30–5–1811	... ...	... ...	... ...	... ...	To h.p., 25-5-1816.
Cox, Wm. Macarmick	30–3–1809	30–5–1811	... ...	... ...	... ...	... ...	Died, 1820.
Mathews, Joseph ...	6–4–1809	13–6–1811	... ...	... ...	... ...	... ...	To h.p., 25-6-1817.
Broomfield, ——	13–4–1809	28–2–1812	... ...	... ...	... ...	... ...	To h.p., 25-6-1817.
Strangeways, William	15–6–1809	3–6–1812	... ...	... ...	... ...	... ...	To 3rd Garr. Bn., 27-12-1812.
St. Clair, Harry	22–6–1809	... ...	... ...	... ...	... ...	... ...	Not shown after 1810.
Bloxham, Robert ...	27–7–1809	16–8–1810	... ...	... ...	... ...	... ...	Killed at Vittoria.
Shepherd, Henry R. ...	15–8–1809	... ...	... ...	... ...	... ...	... ...	Lieut., 12th Regt., 28-2-1812.
Swinburne, Joseph ...	16–8–1809	4–6–1812	6–10–1825	2–8–1842	... ...	... ...	Adjt., 25-7-1811—19-4-1826; Lt.-Col. in Army, 11-11-1851.—To r.f.p., 4-3-1853.
Woodhouse, R.	17–8–1809	5–6–1812	... ...	... ...	... ...	... ...	To h.p., 25-6-1817.
Green, John	8–11–1809	29–12–1812	... ...	... ...	... ...	... ...	To h.p., 25-6-1817.

Robinson, James	9–11–1809						Lieut., 34th Regt., 2-5-1811.
Holt, Samuel							Qr.-Mr., 1-8-1809—18-9-1816.
Burghersh, John Ld.				20–12–1810			From 3rd D.G.—Lieut.-Col., 63rd Regt., 12-12-1811.
Small, William		2–8–1810					From 71st Regt.—Died, 1813.
Wiley, William	1–2–1810	30–12–1812					To h.p., 25-6-1817.
Crowgey, Beauford	23–8–1810	31–12–1812					To h.p., 4-9-1816.
Widdrington, G. J. W. T.				1–8–1811			From 34th Regt.—Killed at Vittoria.
Alexander, William J.			28–2–1811				From 21st L.D.—Not shown after Aug., 1814.
Lindsay, Thomas	7–3–1811						Killed at Vittoria.
Wyatt, Herbert	28–3–1811	17–6–1813					To Life Gds., 15-10-1816.
Lane, Ambrose	28–5–1811	28–7–1813					To h.p., 25-6-1817.
FitzGibbon, W.	29–5–1811	16–9–1813					To h.p., 25-6-1817.
Brahan, John	30–5–1811	21–10–1813					To h.p., 25-6-1817.
Neligan, Thomas	13–6–1811	11–11–1813					To h.p., 25-6-1817.
Hackett, Isaac	18–7–1811						Killed at Badajoz.
Vavasour, John	25–7–1811	16–2–1813					To 64th Regt., 2-2-1815.
Stevens, Alexander Gordon	24–10–1811	3–3–1814					To h.p., 25-3-1817.
Wykherd, Charles Lodewyk	28–2–1812	28–4–1814					To h.p., 25-3-1817.
Illins, William	30–4–1812	29–7–1813					To 80th Regt., 19-1-1815.
Summerfield, John W.	5–6–1812	8–9–1814					To h.p., 25-3-1817.
Parnall, John	11–6–1812	17–11–1814					To h.p., 25-3-1817.
Macken, Simon	2–7–1812						Died, 1813.
O'Neill, William	12–8–1812	22–5–1818					To h.p., 22-5-1818.
Browne, Luke	31–12–1812	4–8–1818					To h.p., 4-8-1818.

Name.	Ensign or 2nd Lieut.	Lieutenant.	Captain.	Major.	Lieut.-Colonel.	Colonel.	Remarks.
Hamilton, John Potter	... ...	... ...	... ...	... ...	3-6-1813	... ...	From 10th Regt.—To 3rd Foot Gds., 25-7-1814.
Fraser, Thomas	... ...	... ...	4-2-1813	... ...	... ...	... ...	From 3rd Garr. Battn.—To h.p., 25-6-1817.
Stevenson, Arthur J. ...	... ...	25-1-1813	... ...	... ...	... ...	... ...	From 85th Regt.—To h.p., 25-6-1817.
Ormsby, George	... ...	25-11-1813	... ...	... ...	... ...	... ...	From 3rd D.G.—Not shown after March, 1814.
Maxwell, William ...	11-2-1813	... ...	... ...	... ...	... ...	... ...	Lieut., h.p., 20-9-1816.
Nugent, P.	25-2-1813	... ...	... ...	... ...	... ...	... ...	Lieut., 60th Regt., 3-3-1814.
Young, Thomas	15-4-1813	... ...	... ...	... ...	... ...	... ...	Not shown after 1818.
Irwin, Charles	22-4-1813	9-10-1818	... ...	... ...	... ...	... ...	To h.p., 9-10-1818.
Jones, Arthur	5-5-1813	... ...	... ...	... ...	... ...	... ...	Not shown after March, 1814.
Burgess, Francis	6-5-1813	4-8-1814	... ...	... ...	... ...	... ...	To h.p., 25-10-1816.
Grieves, Thomas Peter	17-6-1813	... ...	... ...	... ...	... ...	... ...	To 61st Regt., 25-5-1815.
Nihill, Edward	11-11-1813	17-11-1818	... ...	... ...	... ...	... ...	To h.p., 17-11-1818.
MacNab, ——	16-12-1813	... ...	... ...	... ...	... ...	... ...	Died, 1818.
Dewsnap, George ...	12-5-1814	... ...	... ...	... ...	... ...	... ...	Qr.-Mr., 27-5-1813—11-5-1814; Adjt., 12-5-1814—24-6-1817. To h.p., 25-6-1817.
Dwyer, Robert Henry...	2-3-1814	13-4-1820	... ...	... ...	... ...	... ...	To h.p., 23-5-1829.
Hely, Richard	10-3-1814	... ...	... ...	... ...	... ...	... ...	To h.p., 25-5-1817.
O'Brien, Florence ...	31-3-1814	8-6-1820	... ...	... ...	... ...	... ...	Died, 1825.
Rawlins, ——	28-4-1814	... ...	... ...	... ...	... ...	... ...	To h.p., 25-6-1817.
Warre, Francis	11-8-1814	... ...	... ...	... ...	... ...	... ...	To h.p., 25-6-1817.
Sullivan, James	8-9-1814	... ...	... ...	... ...	... ...	... ...	To h.p., 25-6-1817.
Carmack, G. R.	17-11-1814	... ...	... ...	... ...	... ...	... ...	To h.p., 25-6-1817.
Hunt, William	... ...	... ...	30-11-1815	... ...	... ...	... ...	From 57th Regt.—To h.p., 25-3-1817.
Daly, R. R.	... ...	2-2-1815	... ...	... ...	... ...	... ...	From 64th Regt.—To h.p., 25-1-1817.
Eccles, Cuthbert	... ...	10-8-1815	... ...	... ...	... ...	... ...	From 61st Regt.—To h.p., 25-1-1817.

Gordon, John	27–4–1815	... ...	... ...	... ...	... ...	... ...	To h.p., 6-6-1816.
Laing, James	15–6–1815	... ...	... ...	... ...	... ...	... ...	To h.p., 25-6-1817.
Cochrane, John	7–9–1815	... ...	... ...	... ...	... ...	... ...	Retired, 1817.
Taite, Thomas	7–12–1815	... ...	... ...	... ...	... ...	... ...	To h.p., 25-6-1817.
Hall, William	... ...	... ...	... ...	... ...	... ...	... ...	Qr.-Mr., 12-5-1814. — Died, Sept., 1823.
Cother, Charles	... ...	... ...	... ...	... ...	24–10–1816	... ...	From 71st Regt.–To h.p., 25-12-1818.
Clues, Josiah	... ...	15–10–1816	... ...	... ...	... ...	... ...	From Life Gds.—To h.p., 25-1-1817.
Moloney, John	... ...	7–11–1816	... ...	... ...	... ...	... ...	From 40th Regt.–To h.p., 25-3-1817.
Edwards, Thomas ...	6–6–1816	... ...	... ...	... ...	... ...	... ...	To 50th Regt., 10-7-1817.
Yates, John	... ...	... ...	... ...	... ...	... ...	... ...	Qr.-Mr., 19-9-1816—24-6-1817.—To h.p., 25-6-1817.
Kelly, Richard	... ...	... ...	... ...	1–1–1818	... ...	... ...	From h.p.—Lieut.-Col. in Army, 3-8-1815.—To h.p., 25-11-1828.
Smith, Malcolm Laing	... ...	... ...	18–6–1818	... ...	... ...	... ...	From 3rd W.I. Regt.—Retired, 1825.
Burleigh, John G. C....	22–5–1818	23–3–1821	... ...	... ...	... ...	... ...	To h.p., 17-7-1823.
Driberg, William ...	14–1–1819	8–8–1822	... ...	... ...	... ...	... ...	To Ceylon Regt., 5-8-1824.
Geddes, Robert Graham	13–4–1819	... ...	... ...	... ...	... ...	... ...	Died, 1823.
Young, Brooke	14–4–1819	... ...	... ...	... ...	... ...	... ...	Died, 1823.
Tyndall, Abram	15–4–1819	... ...	... ...	... ...	... ...	... ...	Not shown after 1821.
Smith, Frederick ...	22–7–1819	... ...	... ...	... ...	... ...	... ...	Died, 1823.
Sanderson, Edward ...	... ...	... ...	24–8–1820	... ...	... ...	... ...	From h.p.—Died, 1825.
Wynn, Joseph	... ...	21–12–1820	30–4–1827	... ...	... ...	... ...	From 88th Regt.—To 58th Regt., 22-5-1828.
Young, Aretas Sutherland	13–4–1820	29–1–1824	2–3–1833	... ...	... ...	... ...	To 63rd Regt., 25-12-1835.
Lisle, Robert F. R. ...	8–6–1820	... ...	... ...	... ...	... ...	... ...	Died. 1824.
Cother, Charles	... ...	... ...	... ...	... ...	24–10–1816	... ...	From h.p. *see ante.* — To h.p., 2-12-1829.
Rayson, John	1–2–1821	7–4–1825	2–4–1841	... ...	... ...	... ...	Retired, 19-11-1841.

Name.	Ensign or 2nd Lieut.	Lieutenant.	Captain.	Major.	Lieut.-Colonel.	Colonel.	Remarks.
Law, William Henry ...	... ...	7–3–1822	14–7–1825	2–4–1841	22–12–1848	... ...	From h.p.—Col., 28-11-1854; Maj.-General, 16-5-1856. — To r.f.p., 16-5-1856.
Richardson, William ...	... ...	12–9–1822	... ...	... ...	... ...	... ...	From 45th Regt.—Died, 1823.
Irwin, Charles	... ...	29–8–1822	... ...	... ...	... ...	... ...	From h.p. *see ante.* — To h.p., 5-8-1829.
Hodgson, John	... ...	... ...	... ...	... ...	... ...	20–3–1823	Lt.-Gen., 4-6-1814.—To 4th Regt., 30-9-1835.
Law, Robert	... ...	... ...	20–3–1823	... ...	... ...	... ...	From 1st W.I. Regt.—To Ceylon Regt., 25-9-1824.
Summerfield, John Wm.	... ...	17–7–1823	... ...	... ...	... ...	... ...	From h.p., *see ante.*—Died, 1824.
Auber, Charles	... ...	11–12–1823	... ...	... ...	... ...	... ...	From h.p.—Not shown after 1825.
Johnson, Wm. Stratford	20–3–1823	8–4–1825	... ...	... ...	... ...	... ...	Killed in action in Canada, 13-11-1838.
Watson, Albert	17–7–1823	... ...	... ...	... ...	... ...	... ...	Lieut., 58th Regt., 22-7-1830.
Caulfield, Henry ...	11–12–1823	14–7–1825	... ...	... ...	... ...	... ...	To Ceylon Rifle Regt., 25-12-1828.
Trydell, Botet	... ...	... ...	15–1–1824	3–12–1829	2–8–1842	... ...	From h.p., *see ante.* — Colonel, 20-6-1854; Maj.-Gen., 26-10-1858.
Crofton, Peter	... ...	... ...	25–9–1824	2–3–1833	... ...	... ...	From Ceylon Regt.—To retired f.p., 2-4-1841.
Haggerstone, John ...	... ...	5–8–1824	17–12–1825	... ...	... ...	... ...	From Ceylon Regt. — To h.p., 17-12-1825.
Harrison, John	... ...	28–10–1824	3–12–1829	... ...	... ...	... ...	From 45th Regt.—To h.p., 15-3-1839.
Ainslie, Henry Francis	29–1–1824	7–11–1826	5–4–1831	3–6–1851	... ...	... ...	To r.f.p., 19-1-1855.
Kelly, Robert	25–6–1824	19–12–1826	8–4–1834	... ...	... ...	... ...	Retired, 1839.
Stubbs, John	20–4–1826	25–6–1830	... ...	... ...	... ...	... ...	Qr.-Mr., 8-9-1823; Adjt., 20-4-1826. Died, 19-5-1840.
Burgess, James	... ...	... ...	8–4–1825	... ...	... ...	... ...	From 2nd R. Vet. Battn.—Retired, 1830.
Hodgson, Studholme John	... ...	3–2–1825	30–12–1826	... ...	... ...	... ...	From 45th Regt.—To 39th Regt., 14-11-1827.
Colquhoun, Robert ...	... ...	9–4–1825	16–3–1830	... ...	... ...	... ...	From h.p.—Died, 3-10-1841.
Hotham, Augustus ...	... ...	19–5–1825	19–12–1826	... ...	... ...	... ...	From 40th Regt. — To h.p., 19-12-1826.
Anstruther, Robert ...	... ...	29–12–1825	7–11–1826	... ...	... ...	... ...	From 21st Regt.—To h.p., 7-11-1826.
Bell, William	7–4–1825	... ...	... ...	... ...	... ...	... ...	Lieut., 16th Regt., 26-4-1828.

Atherton, William	8–4–1825						To h.p., 5-2-1829.
Keating, James	9–4–1825						Lieut., 13th Regt., 12-6-1830.
Ball, James	30–6–1825	30–12–1826					Died, 1829.
Kelsall, John	14–7–1825	30–4–1827	4–10–1841	7–5–1854	16–5–1856		Col. in Army, 13-4-1858.—To r.f.p., 13-4-1858.
Pole, John George	14–11–1826						Died, 1829.
Bowles, Henry Stephen George	19–12–1826	3–12–1829					Died, 1832.
Blakeney, George	30–12–1826	16–3–1830					Retired, 1835.
Rusher, John							Qr.-Mr., 20-4-1826—8-6-1838.—To h.p., 8-6-1838.
De Visme, Edward	30–4–1827	8–6–1830					Retired, 1831.
Dundas, Hon. Henry				25–11–1828	3–12–1829		From h.p.—To h.p., 2-8-1842.
Townsend, Edward			22–5–1828	22–12–1848			From h.p.—Died, 2-6-1851.
Garstin, William		25–12–1828	2–8–1842				From Ceylon Rifle Regt.—To h.p., 9-3-1849.
Egerton, Charles Troubridge	26–4–1828	5–4–1831					Retired, 1839.
Caulfield, Henry		14–7–1825	8–6–1830				From Ceylon Rifle Regt., *see ante*.—To 58th Regt., 5-4-1833.
Hamilton, John James Edward		5–8–1829					From 8th Regt.—Retired, 1833.
O'Brien, William	5–2–1829						From Royal Staff Corps.—Lieut., 53rd Regt., 8-6-1830.
Goodrich, James	16–7–1829	23–8–1831					Died, 1839.
Clifford, Hon. Robert	3–12–1829	21–9–1832					Died, 1833.
Emslie, John			15–6–1830				From Ceylon Regt., *see ante*.—Retired, 7-6-1844.
Howard, Hanway		28–9–1830					From 58th Regt.—Retired, 1837.
Grey, George	14–1–1830	29–3–1833	22–3–1839				Retired, 1840.
Pringle, James	16–3–1830						Lieut., 8th Regt., 31-8-1832.
Scott, Henry Murray	8–6–1830						Died, 1832.
Clerk, James	9–6–1830						Lieut., 9th L.D., 24-12-1833.
D'Alton, Edward	13–6–1830	2–8–1833	20–9–1839				Unattached, 12-1-1849.

NAME.	Ensign or 2nd Lieut.	Lieutenant.	Captain.	Major.	Lieut.-Colonel.	Colonel.	REMARKS.
Lloyd, Henry	5-4-1831	8-4-1834	10-4-1840	4-3-1853	... ...	... ...	Died, 6-5-1854.
James, John Taubman	23-8-1831	18-4-1834	... ...	... ...	... ...	... ...	Retired, 1836.
St. Aubyn, Thos. John	24-8-1832	25-9-1835	15-12-1840	... ...	... ...	... ...	Died, Oct., 1846.
Bowles, Francis William	21-9-1832	10-6-1836	... ...	... ...	... ...	... ...	To 94th Regt., 11-8-1837.
Whittingham, Ferdinand	2-11-1832	... ...	... ...	... ...	... ...	... ...	Lieut., 7th Regt., 19-2-1836.
Richardson, John	... ...	... ...	5-4-1833	... ...	... ...	... ...	From 40th Regt.—Retired, 1840.
Coghlan, Roger	... ...	24-12-1833	... ...	... ...	... ...	... ...	From 3rd L.D.—Died, 1834.
Brown, Benjamin Handley	12-4-1833	8-12-1837	19-11-1841	... ...	... ...	... ...	Adjt., 20-5-1840—18-11-1841.—Retired, 19-3-1847.
Labalmondiere, Douglas W. P.	21-6-1833	25-5-1838	7-6-1844	... ...	... ...	... ...	To h.p., 26-4-1850.
Rinzy, Thomas Richards	2-8-1833	14-11-1838	22-7-1845	... ...	... ...	... ...	Unattached, 2-2-1849.
Campbell, Duncan ...	8-4-1834	1-2-1839	3-10-1846	... ...	... ...	... ...	To 90th Regt., 7-5-1848.
Steele, Edward	18-4-1834	15-3-1839	19-3-1847	19-1-1855	13-4-1858	... ...	Retired, 29-7-1862.—Died, 6-8-1862.
Fraser, Hastings	... ...	... ...	... ...	... ...	... ...	30-9-1835	Maj.-Gen., 12-8-1819; Lieut.-Gen., 10-11-1837.—To 61st Regiment, 1-9-1848.
Stubbeman, Denis McCarthy	... ...	... ...	25-12-1835	... ...	... ...	... ...	From 63rd Rgt.—Retired, 22-7-1845.
Dundas, Wedderburn ...	25-9-1835	... ...	... ...	... ...	... ...	... ...	Retired, 1838.
Cary, Septimus Alphonso Frederico.	19-2-1836	20-5-1840	22-12-1848	... ...	... ...	... ...	To 31st Regt., 20-12-1850.
Hamilton, Walter ...	10-6-1836	22-3-1839	... ...	... ...	... ...	... ...	Retired, 30-12-1845.
Durie, William S. ...	... ...	11-8-1837	... ...	... ...	... ...	... ...	From 94th Regt.—Retired, 1838.
Wynniatt, Wenman ...	8-12-1837	20-9-1839	... ...	... ...	... ...	... ...	Died, 14-5-1841.
Hext, Francis John ...	25-5-1838	10-4-1840	... ...	... ...	... ...	... ...	Retired, 9-8-1845.
Austen, Charles Wilson	14-12-1838	15-12-1840	1-12-1848	16-5-1856	26-10-1858	... ...	To 14th Regt., 10-6-1862.
Anderson, David	28-12-1838	15-5-1841	2-2-1849	... ...	... ...	... ...	To 22nd Regt., 17-9-1849.
Imray, Robert	... ...	... ...	... ...	... ...	... ...	... ...	Qr.-Mr., 8-6-1838—6-6-1844. — To 15th Regt., 7-6-1844.

Downman, John Thos.	1-2-1839	19-11-1841	... ...	... ...	... ...	... ...	Retired, 25-9-1849.
Puleston, William Roger	15-3-1839	... ...	... ...	... ...	... ...	... ...	Retired, 1840.
Gage, Hon. William ...	22-3-1839	2-8-1842	9-3-1849	... ...	... ...	... ...	Died, 14-7-1849.
Wallington, John Williams	20-9-1839	7-6-1844	... ...	... ...	... ...	... ...	To 4th L.D., 2-4-1847.
Portal, Robert	10-4-1840	... ...	... ...	... ...	... ...	... ...	Lieut., 3rd Regt., 9-8-1845.
Naylor, James Sadler ...	17-7-1840	22-7-1845	... ...	... ...	... ...	... ...	To 8th L.D., 27-3-1846.
Wills, Wm. Sandford ...	14-8-1840	8-8-1845	... ...	... ...	... ...	... ...	To 5th D.G., 24-12-1847.
Foster, James	15-12-1840	30-12-1845	... ...	... ...	... ...	... ...	To 1st D.G., 16-1-1846.
Spring, Thomas	... ...	2-4-1841	15-7-1849	... ...	... ...	... ...	From 24th Regt.—To 35th Regt., 28-11-1851.
Crowe, John William ...	2-7-1841	3-4-1846	... ...	... ...	... ...	... ...	Retired, 11-5-1849.
Nott, William	19-11-1841	21-5-1846	3-6-1851	... ...	... ...	... ...	Adjt., 19-11-1841—2-6-1851.—Died, 1-9-1858.
Cazalet, George Henry	26-11-1841	... ...	... ...	... ...	... ...	... ...	Lieut., 18th Regt., 20-8-1844.
Lane, Thomas Stewart	19-8-1842	22-5-1846	... ...	... ...	... ...	... ...	Died, 28-6-1848.
Swinburne, John Dennis	7-6-1844	3-10-1846	... ...	... ...	... ...	... ...	Appointed Paymaster, 2-3-1849.
Adams, Thomas	20-8-1844	19-3-1847	4-5-1853	... ...	... ...	... ...	To 78th Regt., 3-10-1854.
Molony, John Sharman	23-5-1845	29-6-1848	15-3-1853	... ...	... ...	... ...	To Staff, 20-4-1860.
Pigott, Henry de Renzy	22-7-1845	19-9-1848	7-5-1854	19-12-1862	... ...	... ...	To 19th Regt., 30-6-1863.
Molony, William Mills	8-8-1845	1-12-1848	19-1-1855	... ...	... ...	... ...	To 22nd Regt., 23-2-1855.
Campbell, John Donald	30-12-1845	... ...	... ...	... ...	... ...	... ...	Appointment cancelled.
Cartmail, J.	... ...	... ...	... ...	... ...	... ...	... ...	Qr.-Mr. (from 15th Regt.), 7-6-1844—27-9-1847.—To 3rd Regiment, 28-9-1849.
Gethin, Sir R., Bart. ...	... ...	16-1-1846	... ...	... ...	... ...	... ...	From 1st D.G.—Retired, 22-5-1846.
Gwyn, Thomas Gabriel Leonard Carew	16-1-1846	5-11-1847	... ...	... ...	... ...	... ...	To 2nd L.G., 28-1-1848.
Bookey, Wm. Kavanagh	3-4-1846	22-12-1848	... ...	... ...	... ...	... ...	Retired, 29-4-1854.
Nunn, William John ...	22-5-1846	22-12-1848	... ...	... ...	... ...	... ...	To 26th Regt., 20-1-1854.

Name.	Ensign or 2nd Lieut.	Lieutenant.	Captain.	Major.	Lieut.-Colonel.	Colonel.	Remarks.
Blackburn, William ...	20–10–1846	... ...	... ...	... ...	... ...	... ...	Qr.-Mr., 91st Regt., 1-8-1848.
Churchill, Ld. Alfred ...	... ...	2–4–1847	... ...	... ...	... ...	... ...	From 4th L.D.—Retired, 19-9-1848.
Crawfurd, R. H. Payne	... ...	24–12–1847	15–3–1853	... ...	... ...	... ...	From 5th D.G.—To 90th Regt., 6-5-1853.
Metge, Stephen Wm. ...	12–2–1847	22–12–1848	... ...	... ...	... ...	... ...	Died, 1856.
Cooper, Herbert Stanley	19–3–1847	22–12–1848	5–3–1858	... ...	... ...	... ...	Died, 13-7-1858.
Colborn, William ...	... ...	... ...	... ...	... ...	... ...	... ...	Qr.-Mr. (from 3rd Regt.), 28-9-1847. Date of death, 2-8-1852.
Stovin, Sir Frederick ...	... ...	... ...	... ...	... ...	... ...	1–9–1848	Maj.-Gen., 23-11-1841; Lieut.-Gen., 11-11-1851; Gen., 14-8-1859. —Died, 16-8-1865.
Read, Samuel	... ...	22–12–1848	16–5–1856	... ...	... ...	... ...	From 28th Regt.—Killed at Jeerum, 23-10-1857.
Mainwaring, Ed. M. H.	... ...	22–12–1848	... ...	... ...	... ...	... ...	From 56th Regt.—Adjt., 11-9-1851—1-8-1856.—Died, 1-8-1856.
Wright, Thomas Parker	... ...	22–12–1848	5–10–1857	... ...	... ...	... ...	From 2nd W.I. Regt.—To h.p., 12-3-1861.
Ellis, James Verling ...	... ...	22–12–1848	24–10–1857	... ...	... ...	... ...	From 66th Regt.—To Ceylon Regt., 8-7-1862.
McKelvey, John Norris	... ...	22–12–1848	... ...	... ...	... ...	... ...	From h.p.—Died, 1856.
Meade, John	... ...	22–12–1848	... ...	... ...	... ...	... ...	From Ceylon Regt.—Capt., 30th Regt., 15-2-1856.
Baumgartner, Mowbray	... ...	22–12–1848	19–8–1856	... ...	... ...	... ...	From 10th Regt. — To B.S.C., 17-11-1863.
Dickinson, Frederick ...	18–8–1848	2–2–1849	13–4–1858	... ...	... ...	... ...	Retired, 16-10-1863.
Hall, William	1–8–1848	15–7–1849	... ...	... ...	... ...	... ...	Adjt., 3-6-1851, till died, 10-9-1851.
Sprot, John	19–9–1848	29–5–1849	14–7–1858	22–1–1867	... ...	... ...	To h.p., 14-9-1867.
Sweeny, Richard Thomas	1–12–1848	3–6–1851	2–9–1858	... ...	... ...	... ...	To h.p., 6-1-1860.
Marsh, Francis Henry Digby	22–12–1848	11–9–1851	... ...	... ...	... ...	... ...	Capt., 21st Regt., 27-3-1858.
Alcock, Ussher William	22–12–1848	12–10–1852	... ...	... ...	... ...	... ...	Retired, 20-7-1855.
Richardson, Marmaduke Nelson	22–12–1848	4–3–1853	... ...	... ...	... ...	... ...	Retired, 10-2-1854.
Rowland, Thomas ...	22–12–1848	29–4–1854	... ...	... ...	... ...	... ...	To 1st Regt., 17-6-1853.
Graham, Thomas	22–12–1848	... ...	... ...	... ...	... ...	... ...	Died, April, 1852.

APPENDIX I.

Lamb, Samuel Burges	... ...	... ...	12–1–1849	... ...	... ...	... ...	From u.l.—To 10th Rgt., 29-9-1851.
Heatley, John	... ...	... ...	17–9–1849	13–4–1858	... ...	... ...	From 22nd Regt.—Brev.-Lieut.-Col., 19-1-1858.—To 69th Regiment, 16-2-1860.
Teesdale, Charles Peregrine	... ...	9–3–1849	... ...	... ...	... ...	... ...	From Ceylon Regt.—Capt., 55th Regt., 10-8-1855.
Sheils, Wm. Cumming	... ...	13–4–1849	... ...	... ...	... ...	... ...	From 67th Regiment. — Retired, 12-10-1852.
Wilson, Sylvester W. F. M.	... ...	11–5–1849	... ...	... ...	... ...	... ...	From 26th Regt.—Capt., 55th Rgt., 31-8-1855.
FitzRoy, William ...	29–5–1849	10–2–1854	... ...	... ...	... ...	... ...	Capt., 63rd Regt., 2-11-1855.
Meurant, Edward ...	12–10–1849	7–5–1854	26–10–1858	14–9–1867	5–10–1879	... ...	See Royal Irish Rifles List.
Murray, James Florence	... ...	... ...	26–4–1850	26–10–1858	... ...	... ...	From 73rd Regt.—To 97th Regt., 7-11-1862.
Bray, Edward William	... ...	... ...	20–12–1850	2–5–1865	... ...	... ...	From 31st Regt.—To 4th Regt., 4-8-1865.
Moore, George Fredk.	... ...	... ...	29–9–1851	... ...	... ...	... ...	From 10th Regt.—To h.p., 9-10-1855.
Cooke, Edward Bowen	... ...	... ...	28–11–1851	29–7–1862	... ...	... ...	From 35th Rgt.—Retired, 19-12-1862.
Huskisson, John Wm.	8–8–1851	19–1–1855	... ...	... ...	... ...	... ...	To 56th Regt., 27-7-1855.
Colthurst, James Nicholas	5–12–1851	11–5–1855	17–11–1863	... ...	... ...	... ...	Adjt., 2-8-1856—23-11-1863.—To 76th Regt., 29-3-1864.
Dunlevie, George ...	16–4–1852	27–7–1855	... ...	... ...	... ...	... ...	To h.p., 28-8-1857.
Holt, Alfred	12–10–1852	... ...	... ...	... ...	... ...	... ...	Lieut., 21st Regt., 9-2-1855.
Hayes, Patrick	... ...	... ...	... ...	... ...	... ...	... ...	Qr.-Mr., 5-11-1852—12-6-1863. — To h.p., 12-6-1863.
Wyvill, Richard Rodes ...	... ...	... ...	6–5–1853	... ...	... ...	... ...	From 90th Regt.—Retired, 21-2-1860.
Villiers, H. P. Villiers ...	... ...	17–6–1853	... ...	... ...	... ...	... ...	From 1st Regt.—Retired, 11-5-1855.
Wakefield, Julian ...	22–4–1853	20–7–1855	6–1–1860	16–9–1868	... ...	... ...	Brev.-Lt.-Col.—Retired, 7-8-1878.
Coote, Thomas Gethin	13–5–1853	10–1–1856	2–5–1865	... ...	... ...	... ...	To 26th Regt., 12-9-1865.
Keogh, Thomas M. ...	... ...	... ...	3–10–1854	... ...	... ...	... ...	From 78th Regt.—Retired, 19-8-1856.
Mylne, Graham	10–2–1854	... ...	... ...	... ...	... ...	... ...	Lieut., 82nd Regt., 22-9-1855.
Jones, Robert Colvill ...	... ...	... ...	23–2–1855	... ...	... ...	... ...	From 22nd Regt.—Died, 4-10-1857.
Forester, Hon. Emilius J. Weld	... ...	... ...	9–10–1855	... ...	... ...	... ...	From h.p.—Maj. in Army, 26-9-1857. —To h.p., 5-3-1858.

NAME.	Ensign or 2nd Lieut.	Lieutenant.	Captain.	Major.	Lieut.-Colonel.	Colonel.	REMARKS.
Keddle, John Shering ...	26–7–1855	18–4–1856	... ...	... ...	... ...	... ...	To 41st Regt., 29-8-1856.
Browne, Peter Clifford	27–7–1855	16–5–1856	17–4–1867	... ...	... ...	... ...	To 23rd Regt., 24-4-1869.
Wardell, George William Henry	16–10–1855	2–8–1856	... ...	... ...	... ...	... ...	Retired, 6-8-1861.
Gandy, Henry	... ...	18–4–1856	21–2–1860	... ...	... ...	... ...	From 26th Regt.—Retired, 15-6-1860.
Gore, Charles Clitherow	... ...	2–5–1856	20–4–1860	7–8–1878	... ...	... ...	From 21st Regt. (See Royal Irish Rifles List.)
Minhear, William ...	... ...	27–6–1856	15–6–1860	... ...	... ...	... ...	From 18th Regt.—Retired, 23-6-1863.
Beazley, George Gant...	... ...	1–8–1856	29–7–1862	... ...	... ...	... ...	From 3rd W.I. Regt.—Hon. Lt.-Col. Retired, 12-3-1881.
Onslow, Guildford Macleay	... ...	29–8–1856	... ...	... ...	... ...	... ...	From 41st Regiment. — Retired, 26-11-1861.
Ivimy, William Henry...	14–3–1856	5–10–1857	23–6–1863	... ...	... ...	... ...	To 60th Regt., 17-10-1863.
Huyshe, Geo. Lightfoot	18–4–1856	19–8–1856	19–12–1862	... ...	... ...	... ...	To R. Bde., 3-3-1865.
Chamley, Braithwaite...	9–5–1856	24–10–1857	... ...	... ...	... ...	... ...	To 17th L.D., 16-2-1861.
Pennefather, Nicholas...	19–8–1856	26–1–1858	17–4–1867	... ...	... ...	... ...	To 34th Regt., 6-11-1867.
Karslake, Frederick ...	19–6–1857	2–9–1858	16–10–1863	5–10–1879	... ...	... ...	See Royal Irish Rifles List.
Anderson, Wm. Forbes	16–10–1857	26–10–1858	... ...	... ...	... ...	... ...	Died, 17-9-1863.
Healey, John	4–12–1857	13–4–1860	... ...	... ...	... ...	... ...	To 66th Regt, 25-2-1862.
Colebrooke, James Robert Alfred	... ...	5–3–1858	... ...	... ...	... ...	... ...	From 26th Regt.—Died, 29-4-1860.
Whitlock, Hy. Cornish	5–2–1858	6–1–1860	22–1–1867	... ...	... ...	... ...	Retired, 6-11-1867.
Murphy, Michael ...	6–2–1858	24–8–1860	13–9–1864	... ...	... ...	... ...	To 58th Regt., 22-8-1865.
Brymer, James Edmund	30–3–1858	24–8–1860	... ...	... ...	... ...	... ...	Died, 23-6-1863.
Blathwayt, William ...	16–4–1858	... ...	... ...	... ...	... ...	... ...	Died, 16-6-1859.
Powys, Littleton Albert	29–10–1858	21–12–1860	20–2–1866	... ...	... ...	... ...	To 59th Regt., 18-12-1866.
Gibb, Reginald Kennett	12–11–1858	6–1–1861	... ...	... ...	... ...	... ...	To 1st W.I. Regt., 9-1-1863.
Ford, Frederick	28–1–1859	... ...	... ...	... ...	... ...	... ...	Retired, 8-3-1862.
Fuller, Henry Albert ...	1–7–1859	26–11–1861	... ...	... ...	... ...	... ...	Retired, 8-9-1865.

Hankey, Aug. Barnard	... ...	... ...	... ...	16–2–1860	29–7–1862	... ...	From 69th Regt.—Col. in Army, 29-7-1867. To h.p., 2-8-1871.
Wright, Fredk. Augustus	3–8–1860	29–7–1862	11–1–1869	... ...	... ...	... ...	See Royal Irish Rifles List.
Davies, Herbert Grant	4–9–1860	19–12–1862	... ...	... ...	... ...	... ...	To 96th Regt., 7-4-1863.
Smith, Charles Lucius	23–10–1860	22–5–1863	15–2–1868	... ...	... ...	... ...	Retired, 30-4-1873.
Sweeney, James Fielding	... ...	... ...	12–3–1861	... ...	... ...	... ...	From 12th Regt.—To h.p., 4-12-1866.
Scott, James George ...	... ...	15–2–1861	14–9–1867	... ...	... ...	... ...	From 17th L.D. — To 5th Regt., 20-11-1867.
Tollemache, Chas. Hay	8–2–1861	23–6–1863	... ...	... ...	... ...	... ...	Died, 22-4-1867.
Gage, John Olphert ...	29–3–1861	18–9–1863	... ...	... ...	... ...	... ...	To 45th Regt., 21-6-1864.
Blunt, George E. Emes	6–8–1861	16–10–1863	8–7–1868	... ...	... ...	... ...	Adjutant, 24-11-1863—7-7-1868.—Retired, 5-1-1870.
Wilson, Henry George	26–11–1861	2–5–1865	... ...	... ...	... ...	... ...	To 1st W.I. Regt., 8-9-1865.
Venables, Thomas ...	... ...	... ...	... ...	7–11–1862	... ...	... ...	From 97th Regt.—Retired, 22-1-1867.
Campbell, Francis Pemberton	... ...	... ...	8–7–1862	... ...	... ...	... ...	From 79th Regt.—To 14th Hsrs., 17-7-1863.
Strickland, Walter Cecil	... ...	25–2–1862	6–11–1867	... ...	... ...	... ...	From 66th Regiment. — Retired, 28-10-1871.
Bridger, Aug. Goring ...	18–3–1862	13–9–1864	... ...	... ...	... ...	... ...	Retired, 2-6-1865.
Church, Henry	29–7–1862	... ...	... ...	... ...	... ...	... ...	To 1st Regt., 26-2-1864.
Bates, Robert	... ...	... ...	... ...	30–6–1863	... ...	... ...	From 19th Regt.—Lt.-Col., 2-5-1865. To r.f.p., 2-5-1865.
Mackenzie, Lawrence ...	... ...	... ...	17–7–1863	... ...	... ...	... ...	From 14th H.—Retired, 13-9-1864.
Stehelin, George Ffrench	... ...	... ...	17–10–1863	... ...	... ...	... ...	From 60th Regt.—To h.p., 1-9-1869.
Thomas, Edwyn	... ...	9–1–1863	... ...	... ...	... ...	... ...	From 1st W.I. Regt. — Retired, 22-5-1863.
Townsend, Thos. Edw. Brackenbury	... ...	7–4–1863	... ...	... ...	... ...	... ...	From 96th Regiment. — Retired, 10-11-1865.
Lyall, James Mitchell ...	30–1–1863	2–6–1865	... ...	... ...	... ...	... ...	Retired, 25-9-1867.
Jackson, Sir Keith George, Bart.	22–5–1863	8–9–1865	... ...	... ...	... ...	... ...	Retired, 1-5-1866.
De Montmorency, Raymond Oliver	17–7–1863	10–11–1865	16–9–1868	... ...	... ...	... ...	See Royal Irish Rifles List.
Blurton, John	18–7–1863	... ...	... ...	... ...	... ...	... ...	To Commt. Dept., 6-3-1867.
Horrocks, Charles ...	9–10–1863	20–2–1866	... ...	... ...	... ...	... ...	Retired, 7-11-1868.

NAME.	Ensign or 2nd Lieut.	Lieutenant.	Captain.	Major.	Lieut.-Colonel.	Colonel.	REMARKS.
Thompson, Wm. Thos.	16–10–1863	22–1–1867	5–1–1870	... ...	... ...	... ...	To 42nd Regt., 29-1-1870.
Copeland, Thomas ...	... ...	... ...	... ...	... ...	... ...	... ...	Qr.-Mr. (from h.p.), 12-6-1863—21-1-1868.—To h.p., 22-1-1868.
O'Connor, Luke E. ...	... ...	... ...	29–3–1864	... ...	... ...	... ...	From 76th Regt.—Died, 10-1-1869.
Browne, Edward ...	... ...	21–6–1864	1–9–1869	... ...	... ...	... ...	From 45th Regt.—To 20th Regt., 9-10-1869.
Disney-Roebuck, F. H. Algernon	26–2–1864	17–4–1867	9–4–1870	... ...	... ...	... ...	To 46th Regt., 20-7-1870.
Butler, Edward Arthur	29–3–1864	17–4–1867	28–5–1870	... ...	... ...	... ...	See Royal Irish Rifles List.
Barney, John Douglas	20–9–1864	17–4–1867	... ...	... ...	... ...	... ...	Retired, 17-4-1869.
Buckley, Edward Percy	... ...	... ...	... ...	... ...	... ...	17–8–1865	Gen., 17-8-1865.
Forster, John Philip Bohun	... ...	... ...	... ...	4–8–1865	... ...	... ...	From 4th Regt.—Retired, 16-9-1868.
Johnson, Edmund Geo.	... ...	... ...	3–3–1865	... ...	... ...	... ...	From R. Bde.—Retired, 28-5-1870.
Fawkes, George Philip	... ...	... ...	12–9–1865	... ...	... ...	... ...	From 12th Regt.—Retired, 29-6-1870.
Fawcett, Alfred	... ...	8–9–1865	... ...	... ...	... ...	... ...	From 1st W.I. Regt. — Retired, 8-7-1868.
Wyndham, Charles John	2–5–1865	23–4–1867	29–6–1870	... ...	... ...	... ...	See Royal Irish Rifles List.
Metcalfe, Booth Hay ...	2–6–1865	14–9–1867	30–11–1870	... ...	... ...	... ...	See Royal Irish Rifles List.
Campbell, Hy. William	8–9–1865	... ...	... ...	... ...	... ...	... ...	Retired, 9-3-1867.
Stevenson, George Newcombe	10–11–1865	25–9–1867	28–10–1871	... ...	... ...	... ...	To 91st Regt., 31-10-1871.
Cartwright, George Skeward	20–2–1866	6–11–1867	11–7–1874	... ...	... ...	... ...	Retired, 10-2-1877.
Wyse, John Francis ...	... ...	... ...	6–11–1867	... ...	... ...	... ...	From 34th Regt.—Retired, 8-7-1868.
Walker, Henry Warren	22–1–1867	... ...	... ...	... ...	... ...	... ...	To 19th Regt., 13-6-1868.
Powell, Thomas Pery...	9–3–1867	15–2–1868	... ...	... ...	... ...	... ...	Retired, Nov., 1875.
Stewart, William	17–4–1867	8–7–1868	10–2–1877	... ...	... ...	... ...	Retired, 9-6-1877.
Cooke-Collis, William...	18–4–1867	8–7–1868	21–4–1877	... ...	... ...	... ...	Adjt., 8-7-1868—27-6-1871. (See Royal Irish Rifles List.)
Gibbs, Thomas Fredk.	19–4–1867	16–9–1868	9–6–1877	... ...	... ...	... ...	Retired, 19-10-1878.
Brooke, Art. Vivian Hen.	20–4–1867	7–11–1868	... ...	... ...	... ...	... ...	To 33rd Regt., 1-12-1869.

Kirwan, Martin Oranmore	8-6-1867	11-1-1869	... ...	... ...	... ...	... ...	Retired, 16-3-1871.
Marriott, Wm. Fredk.	25-9-1867	17-4-1869	... ...	... ...	... ...	... ...	To 41st Regt., 18-12-1875.
Bond, Henry Coote ...	16-10-1867	1-9-1869	9-4-1878	... ...	... ...	... ...	See Royal Irish Rifles List.
Davenport, Charles Talbot	18-12-1867	29-6-1870	7-8-1878	... ...	... ...	... ...	See Royal Irish Rifles List.
Macgregor, Charles Reginald	14-3-1868	... ...	... ...	... ...	... ...	... ...	To 96th Regt., 5-4-1868.
Anderson, James Wm.	25-4-1868	5-1-1870	... ...	... ...	... ...	... ...	To Bo. S.C., 3-7-1874.
Bruce, Edward	10-7-1868	... ...	... ...	... ...	... ...	... ...	To 39th Regt., 17-3-1869.
de Hoghton, William ...	19-6-1868	... ...	... ...	... ...	... ...	... ...	Died, 29-4-1870.
Henderson, J. K. S. ...	... ...	... ...	24-4-1869	... ...	... ...	... ...	From 23rd Regt.—Retired, 9-4-1870.
James, R. H.	... ...	... ...	9-10-1869	... ...	... ...	... ...	From 20th Regt.—Died, 4-4-1871.
Heath, Lewis Forbes ...	... ...	1-12-1869	... ...	... ...	... ...	... ...	From 33rd Regt.—To Ben. S.C., 15-3-1871.
Nuthall, Alfred James Phayre	10-2-1869	28-10-1871	... ...	... ...	... ...	... ...	To Bo. S.C., 9-1-1872.
Bell, James Archd. Robert	17-3-1869	28-10-1871	23-4-1879	... ...	... ...	... ...	To Pay Dept., 12-4-1881.
Hinde, Charles Wm. ...	17-3-1869	... ...	... ...	... ...	... ...	... ...	Lieut., Ben. S.C., 25-10-1871.
Gifford, Hon. Ederic Frederick	17-4-1869	30-11-1870	... ...	... ...	... ...	... ...	To 24th Regt., 26-2-1873.
Hardtman-Berqueley, John Hart	15-6-1869	28-6-1871	... ...	... ...	... ...	... ...	Adjt., 28-6-1871—22-11-1875. — To 107th Regt., 5-1-1876.
Cockburn, George Wm.	... ...	... ...	29-1-1870	... ...	... ...	... ...	From 42nd Regt.—Retired, 30-9-1870.
Stokes, Folliott Stuart Furneaux	9-4-1870	27-10-1871	19-10-1878	... ...	... ...	... ...	Adjt., 22-11-1875—28-12-1878. (See Royal Irish Rifles List.)
Hervey, Charles Gerald Bernard	30-11-1870	28-10-1871	... ...	... ...	... ...	... ...	To Ben. S.C., 20-9-1877.
Brown, Thomas S. ...	... ...	... ...	... ...	... ...	2-8-1871	... ...	From 55th Regt.—Brigr.-General, 5-10-1879.
Thorburn, Chas. James	... ...	... ...	31-10-1871	... ...	... ...	... ...	From 91st Regt. (See Royal Irish Rifles List.)
Cleaveland, George ...	23-9-1871	28-10-1871	... ...	... ...	... ...	... ...	Died, 19-3-1875.
Beresford, George Alexander	30-12-1871	30-12-1871	... ...	... ...	... ...	... ...	Died, 8-6-1875.
Buckland, Philip Arnold	24-4-1872	24-4-1872	... ...	... ...	... ...	... ...	To Ben. S.C., 9-4-1875.

Name.	Ensign or 2nd Lieut.	Lieutenant.	Captain.	Major.	Lieut.-Colonel.	Colonel.	Remarks.
Mayne, Richard Chas. Graham	21–12–1872	21–12–1872	... ...	... ...	... ...	... ...	To Bo. S.C., 1-9-1877.
Brown, Wm. Gustavus	... ...	... ...	... ...	... ...	... ...	29–5–1873	Lt.-Gen., 23-5-1872. (See Royal Irish Rifles List.)
Taylor, Raynsford ...	... ...	... ...	28–5–1873	... ...	... ...	... ...	From 41st Regt.—Retired, 11-7-1874.
Wilkinson, Chas. St. Le	... ...	26–2–1873	... ...	... ...	... ...	... ...	From 24th Regiment. — Retired, 25-11-1874.
Cadell, Hugh Francis	12–2–1873	12–2–1874	... ...	... ...	... ...	... ...	To Mad. S.C., 5-7-1875.
Anketill, Amyatt Wm.	... ...	2–12–1874	... ...	... ...	... ...	... ...	From Militia.—Resigned, 7-6-1875.
Eagar, Henry Averell ...	... ...	2–12–1874	... ...	... ...	... ...	... ...	From Militia. (See Royal Irish Rifles List.)
Walter, Edward Charles Lethbridge	23–5–1874	15–4–1875	... ...	... ...	... ...	... ...	See Royal Irish Rifles List.
Anson, John Hy. Wm.	21–9–1874	21–9–1874	12–3–1881	... ...	... ...	... ...	See Royal Irish Rifles List.
Oakeley, Arthur Henry	... ...	20–11–1875	... ...	... ...	... ...	... ...	From Militia.—Died, 10-10-1876.
Mansfield, Herbert ...	... ...	20–11–1875	... ...	... ...	... ...	... ...	From Militia. — To Ben. S.C., 5-7-1878.
Hogge, John William ...	... ...	18–12–1875	... ...	... ...	... ...	... ...	From 41st Regt.—To Ben. S.C., 23-5-1876.
Marling, Walter Henry	11–2–1875	11–2–1875	... ...	... ...	... ...	... ...	Adjt., 29-12-1878. (See Royal Irish Rifles List.)
Beville, Henry Edward Walter	28–8–1875	28–8–1875	... ...	... ...	... ...	... ...	To Bo. S.C., 14-12-1876.
Warden, Hugh Barwell	28–8–1875	28–8–1875	... ...	... ...	... ...	... ...	To Bo. S.C., 14-12-1876.
Tinley, Gervase Fras. Newport	10–9–1875	... ...	... ...	... ...	... ...	... ...	From u.l.—Lieutenant, Bo. S.C., 5-2-1877.
Read, Hastings	... ...	5–1–1876	... ...	... ...	... ...	... ...	From 107th Regt.—To Ben. S.C., 28-8-1876.
Cole, Duncan	11–10–1876	... ...	... ...	... ...	... ...	... ...	Lieut., Bo. S.C., 11-10-1876.
Baker, Lewis Samuel Hyde	15–8–1877	... ...	... ...	... ...	... ...	... ...	Lieut., Bo. S.C., 16-8-1878.
Alexander, Granville Henry Jackson	... ...	10–2–1877	... ...	... ...	... ...	... ...	From 58th Regt. — To Militia, 20-8-1879.
Graves, Robert Stannus	... ...	31–1–1877	... ...	... ...	... ...	... ...	From 13th Regt. (See Royal Irish Rifles List.)
Enriquez, Albert Dallas	5–9–1877	31–8–1878	... ...	... ...	... ...	... ...	To Ben. S.C., 9-7-1880.
Crawley, Anthony ...	10–11–1877	... ...	... ...	... ...	... ...	... ...	To 48th Regt., 2-3-1878.

Adye, Walter	30–1–1878	19–10–1878	... ...	... ...	... ...	... ...	See Royal Irish Rifles List.
Colborne, Hon. Fras. Lionel Lydston	30–1–1878	30–10–1878	... ...	... ...	... ...	... ...	See Royal Irish Rifles List.
Thompson, Wm. David	1–5–1878	5–3–1879	... ...	... ...	... ...	... ...	See Royal Irish Rifles List.
Brown, John Southwell	1–5–1878	12–4–1879	... ...	... ...	... ...	... ...	See Royal Irish Rifles List.
Westropp, Geo. Ralph Collier	14–9–1878	26–11–1879	... ...	... ...	... ...	... ...	To Bo. S.C., 26-4-1880.
Trant, John Francis ...	21–6–1879	13–3–1880	... ...	... ...	... ...	... ...	See Royal Irish Rifles List.
Burrows, George Vernon	21–6–1879	9–7–1880	... ...	... ...	... ...	... ...	See Royal Irish Rifles List.
Harris, Charles Gerald	9–7–1879	13–9–1880	... ...	... ...	... ...	... ...	See Royal Irish Rifles List.
Monteith, Robt. Wm. Freebairn	6–8–1879	29–9–1880	... ...	... ...	... ...	... ...	See Royal Irish Rifles List.
Barnett, Robert Percy Sanders	13–8–1879	... ...	... ...	... ...	... ...	... ...	See Royal Irish Rifles List.
Kettlewell, Edward Alexander	13–8–1879	... ...	... ...	... ...	... ...	... ...	See Royal Irish Rifles List.
Alban, William Gore ...	13–8–1879	... ...	... ...	... ...	... ...	... ...	See Royal Irish Rifles List.
Meynell, John Jebb ...	... ...	... ...	26–5–1880	... ...	... ...	... ...	From 98th Regt. (See Royal Irish Rifles List.)
Browne, William	14–1–1880	... ...	... ...	... ...	... ...	... ...	See Royal Irish Rifles List.
Wilkinson, Lionel Thomson Verner	19–2–1881	... ...	... ...	... ...	... ...	... ...	See Royal Irish Rifles List.
Tighe, Michael Augustus	23–4–1881	... ...	... ...	... ...	... ...	... ...	See Royal Irish Rifles List.

NOMINAL ROLL & COMMISSION LIST OF THE OFFICERS OF THE 86TH FOOT,

ORIGINALLY STYLED GENERAL CUYLER'S SHROPSHIRE VOLUNTEERS.

NAME.	Ensign or 2nd Lieut.	Lieutenant.	Captain.	Major.	Lieut.-Colonel.	Colonel.	REMARKS.
Cuyler, Cornelius ...	... ...	... ...	... ...	... ...	... ...	30–10–1793	From 55th Regt.—To 69th Regt., 20-6-1794.
Sladden, George ...	... ...	... ...	... ...	... ...	30–10–1793	... ...	From 67th Regt.—Not shown after 1795.
Dickens, Richard Mark	... ...	... ...	... ...	30–10–1793	... ...	... ...	From 44th Regt.—Lt.-Col., 34th Regt., 22-10-1794.
Hardy, Thomas Carteret	... ...	... ...	30–10–1793	... ...	... ...	... ...	From 2nd Dragoons.—Lt.-Col., York Fus., 26-9-1794.
Robinson, Edward ...	... ...	... ...	30–10–1793	... ...	22–10–1794	... ...	From 38th Regt. — Succeeded by Maj. Robert Bell, 15-1-1801.
Hill, Rowland	... ...	... ...	30–10–1793	... ...	... ...	... ...	From 53rd Regt.—Lieut.-Col., 90th Regt., 13-5-1794.
Digby, William Henry	... ...	... ...	30–10–1793	... ...	... ...	... ...	From 7th Regt.—To New Independent Companies, 1-10-1794.
Campbell, Alexander ...	... ...	... ...	30–10–1793	22–10–1794	29–12–1794	... ...	Not shown after 1796.
Bell, Robert	... ...	... ...	30–10–1793	10–12–1794	15–1–1801	... ...	From 23rd Regt.—Local Lieut.-Col., India, 29-4-1802.
Byne, Charles	... ...	... ...	30–10–1793	... ...	... ...	... ...	From 54th Regt.—Appointed Paymaster, 12-4-1799.
Cuyler, George	... ...	... ...	30–10–1793	30–5–1805	... ...	... ...	From 55th Regt. — Lieut. - Col., 28-1-1808.—Inspr. of Mil., N.S.
Nelson, Thomas	... ...	30–10–1793	... ...	... ...	... ...	... ...	Capt., 111th Regt., 7-6-1794.
Houstoun, Hugh	... ...	30–10–1793	... ...	... ...	... ...	... ...	From 53rd Regt.—Maj., 90th Regt., 14-5-1794.
Currey, William Samuel	... ...	30–10–1793	4–9–1795	... ...	... ...	... ...	From 53rd Regt.—Adjt., 5-11-1794—23-6-1795 — Major, 54th Regt., 26-12-1805.
Barnes, Edward	... ...	30–10–1793	... ...	... ...	... ...	... ...	From 47th Regt.—Capt., 99th Regt., 11-2-1794.
Middlemore, George ...	... ...	30–10–1793	15–10–1794	... ...	... ...	... ...	Major, 48th Regt., 14-9-1804.
Semple, William	... ...	30–10–1793	6–1–1796	... ...	... ...	... ...	Killed at Baroach, 25-8-1803.
Pickering, Thomas ...	... ...	30–10–1793	... ...	... ...	... ...	... ...	From 12th Regt.—Not shown after 1795.
Jolley, Charles Edward	... ...	27–11–1793	... ...	... ...	... ...	... ...	Capt., Corsican Regt., 4-4-1795.
Tuffnell, John Charles...	... ...	28–11–1793	... ...	... ...	... ...	... ...	Capt., Dunlop's Regt., 31-5-1794.
Dod, Charles	... ...	29–11–1793	... ...	... ...	... ...	... ...	Not shown after 1794.

Name							Remarks
Gavey, Daniel		11–12–1793	25–6–1803				To 1st R. Vet. Bn., 14-7-1804.
Murray, William	30–10–1793	7–7–1794					Not shown after 1804.
Williams, Walter Carrie	30–10–1793						Lieut., 84th Regt., 2-11-1793.
Burke, James	30–10–1793	8–7–1794	28–9–1804				Died on service, 1809.
St. Clair, William Stirling	30–10–1793						Capt.-Lieut., 110th Regt., 6-6-1794.
Sims, Thomas	30–10–1793						Not shown after 1794.
Coleman, Daniel							Adjt., 30-10-1793—4-11-1794.
Manners, Russell						20–6–1794	To 26th L.D., 25-3-1795.
Baillie, Hugh				29–12–1794			From 97th Regt.—Lt.-Col. in Army, 1-1-1800.—To h.p., 4-2-1802.
Bainbridge, John			1–10–1794				From New Ind. Cos.—Major, Durham Fenc. Inf., 26-2-1795.
Henry, Alexander		6–9–1794					Not shown after 1795.
Macneil, Daniel	1794	11–9–1794					Not shown after 1801.
Morton, William		16–10–1794	30–5–1805				Adjutant, 31-7-1802—8-5-1805.—Not shown after 1806.
Campbell, John	10–9–1794	4–9–1795					Not shown after 1798.
McPherson, Adam	11–9–1794						Lieut., W.I. Regt., 1-7-1795.
Chappelle, John	12–9–1794						Lieut., 7th Regt., 16-4-1795.
Wilson, John Henry	13–9–1794	11–11–1795					Died, 1805.
Bird, John	14–9–1794						Lieut., York Fus., 26-1-1796.
Godfrey, Francis	15–9–1794						Lieut., 17th Regt., 1-10-1795.
Long, John Amherst	16–10–1794	16–6–1795					Not shown after 1803.
Massey, Eyre	6–11–1794	12–8–1795					To 49th Regt., 12-4-1797.
Grinfield, William						25–3–1795	From 3rd Foot Guards. — Died, 19-10-1803.
Orange, William			7–2–1795				From h.p.—To 38th Rgt., 29-8-1798.
Dillon, Charles			2–9–1795				From New Independent Companies. —Not shown after 1802.

Name.	Ensign or 2nd Lieut.	Lieutenant.	Captain.	Major.	Lieut.-Colonel.	Colonel.	Remarks.
Mills, Robert William...	... ...	12–8–1795	19–7–1798	... ...	... ...	... ...	To 36th Regt., 27-6-1805.
Jackson, Richard	... ...	11–11–1795	... ...	... ...	... ...	... ...	Qr.-Mr., 30-10-1793—23-6-1795. — Adjt., 24-6-1795—24-4-1796. To 2nd R. Vet. Bn., 27-2-1806.
Peyton, Reynolds ...	... ...	30–12–1795	... ...	... ...	... ...	... ...	Not shown after 1797.
Page, J. F.	... ...	30–12–1795	... ...	... ...	... ...	... ...	Not shown after 1804.
McLaurin, Laughlan ...	... ...	31–12–1795	7–9–1800	... ...	... ...	... ...	Not shown after 1806.
Macnamara, —— ...	... ...	... ...	... ...	... ...	... ...	... ...	Qr.-Mr., 24-6-1795—24-4-1796.
Lloyd, James Phillips ...	... ...	... ...	... ...	... ...	8–6–1796	... ...	From 46th Regt.—Succeeded by Lt.-Col. Hastings Fraser, 18-4-1805.
McDonald, James ...	... ...	... ...	6–1–1796	... ...	... ...	... ...	From 7th Regt.—Not shown after 1809.
Anderson, John	... ...	1–1–1796	... ...	... ...	... ...	... ...	To 57th Regt., 12-4-1796.
Mann, ——	... ...	2–1–1796	... ...	... ...	... ...	... ...	Not shown after 1804.
Drummond, Peter ...	... ...	6–1–1796	... ...	... ...	... ...	... ...	Not shown after 1807.
Harvey, John	... ...	12–1–1796	... ...	... ...	... ...	... ...	From 91st Regt.—Died, 1805.
Westenra, Warner ...	... ...	20–1–1796	... ...	... ...	... ...	... ...	Retired, 24-3-1802.
Daniel, John	... ...	20–1–1796	... ...	... ...	... ...	... ...	Not shown after 1803.
Brownrigg, J. Castleton	... ...	12–3–1796	... ...	... ...	... ...	... ...	From 95th Regt.—Not shown after 1805.
Hepworth, Peter	... ...	12–4–1796	... ...	... ...	... ...	... ...	From 57th Regt.—Not shown after 1797.
Hudson, Richard ...	... ...	10–5–1796	... ...	... ...	... ...	... ...	From 60th Regt.—To 49th Regt., 5-8-1799.
Masterton, Edward ...	... ...	25–10–1796	... ...	... ...	... ...	... ...	Late 95th Regt.—To 81st Regt., 6-4-1799.
Price, David	... ...	25–10–1796	... ...	... ...	... ...	... ...	From 95th Regt.—Adjt., 6-4-1798—30-4-1802. — Killed in action, 30-4-1802.
Proctor, Ephraim ...	... ...	25–10–1796	... ...	... ...	... ...	... ...	Late 95th Regt.—Died, 1805.
Grant, John	... ...	14–12–1796	7–2–1799	... ...	... ...	... ...	From Scotch Bde.—Not shown after 1810.
Grant, Alexander ...	12–1–1796	3–5–1796	... ...	... ...	... ...	... ...	To 71st Regt., 24-3-1806.
Travers, Robert Otho ...	12–1–1796	25–5–1796	17–11–1804	5–3–1812	... ...	... ...	Not shown after Nov., 1814.

APPENDIX I.

Campbell, Donald ...	20–1–1796	... ...	... ...	... ...	... ...	... ...	Lieut., 51st Regt., 23-9-1799.
Fraser, John	20–1–1796	... ...	... ...	... ...	... ...	... ...	Lieut., 88th Regt., 24-9-1799.
Langlands, George ...	2–2–1796	... ...	... ...	... ...	... ...	... ...	Not shown after 1804.
Campbell, Frederick ...	27–2–1796	... ...	... ...	... ...	... ...	... ...	Lieut., Scotch Bde., 23-3-1800.
Blakeney, William ...	19–4–1796	... ...	... ...	... ...	... ...	... ...	To h.p., 1798.
Young, Wm. Cornwallis	3–5–1796	... ...	... ...	... ...	... ...	... ...	Lieut., 75th Regt., 15-5-1800.
Stuart, James	1–6–1796	... ...	... ...	... ...	... ...	... ...	Not shown after 1801.
Mackay, Geo. Stephen	3–8–1796	19–9–1800	... ...	... ...	... ...	... ...	Not shown after 1805.
de Porbeck, William ...	25–10–1796	19–7–1798	... ...	... ...	... ...	... ...	Not shown after 1801.
McKinnon, John	... ...	... ...	... ...	... ...	... ...	... ...	Adjt., 25-4-1796—5-4-1798.
Hudson, James	... ...	... ...	... ...	... ...	... ...	... ...	Qr.-Mr., 25-4-1796—11-9-1799.
Richardson, James ...	... ...	... ...	25–4–1797	... ...	... ...	... ...	From 23rd Regt.—Not shown after 1807.
Elliott, Freeman Willis	... ...	1–2–1797	... ...	... ...	... ...	... ...	From 10th Hussars. — Not shown after 1801.
Marston, Daniel ...	... ...	12–4–1797	26–2–1805	18–11–1813	... ...	... ...	Retired, 21-10-1821.
Gamble, Samuel	17–8–1797	... ...	... ...	... ...	... ...	... ...	Died, 1802.
Hall, John Joseph Stuart	... ...	... ...	29–8–1798	... ...	... ...	... ...	From 38th Regt. — Succeeded by McLaurin, 7-9-1800.
McMahon, John Markham	... ...	13–9–1798	... ...	... ...	... ...	... ...	From W.I. Regt.—Capt., 6th W.I. Regt., 20-8-1803.
Massey, Luigi	22–3–1798	27–10–1800	... ...	... ...	... ...	... ...	Not shown after 1803.
De Latre, Philip	... ...	6–4–1799	... ...	... ...	... ...	... ...	From 81st Regt. — Capt., Champagne's Regt., 16-6-1803.
McQuarie, Lachlan ...	23–9–1799	24–3–1802	1–12–1806	... ...	... ...	... ...	Not shown after 1808.
D'Aguilar, Chas. George	24–9–1799	1–12–1802	... ...	... ...	... ...	... ...	Adjt., 9-5-1805—29-3-1808.—Capt., 81st Regt., 30-3-1808.
Morrice, Daniel	11–12–1799	9–5–1802	13–2–1807	19–7–1821	... ...	... ...	To h.p., 25-10-1821.
Carr, James	... ...	... ...	... ...	... ...	... ...	... ...	Qr.-Mr., 12-9-1799—14-2-1805.
Carter, Edward	16–1–1800	9–5–1802	9–6–1808	... ...	... ...	... ...	To 4th Garr. Bn., 26-7-1810.

Name.	Ensign or 2nd Lieut.	Lieutenant.	Captain.	Major.	Lieut.-Colonel.	Colonel.	Remarks.
McQuarrie, Laughlan...	... ...	... ...	... ...	12–3–1801	... ...	... ...	From 77th Regt.—Lt.-Col., 73rd Regt., 30-5-1805.
Baird, William	... ...	20–5–1800	20–7–1805	25–12–1813	... ...	... ...	From Scotch Brigade. — Retired, 8-2-1827.
Lanphier, Thomas ...	... ...	20–9–1800	2–4–1806	... ...	... ...	... ...	From 10th Regt.—Lt.-Col. in Army, 12-8-1819.—Retired, 1823.
Wilson, John	... ...	21–9–1800	... ...	... ...	... ...	... ...	Not shown after 1807.
Jenny, Philip	23–2–1800	... ...	... ...	... ...	... ...	... ...	Not shown after 1803.
McLaurin, Neil ...	15–5–1800	1–9–1803	... ...	... ...	... ...	... ...	Died, 1810.
Smith, Emilius Felix ...	1–11–1800	30–5–1805	... ...	... ...	... ...	... ...	Not shown after 1806.
Paton, Robert	2–11–1800	... ...	... ...	... ...	... ...	... ...	To h.p., 1803.
Steel, Henry	3–11–1800	10–1–1804	... ...	... ...	... ...	... ...	Not shown after 1807.
Craufurd, Robert ...	... ...	... ...	... ...	... ...	11–2–1802	... ...	From 60th Rgt.—To h.p., 24-2-1803.
Torrens, Henry	... ...	... ...	... ...	4–2–1802	... ...	... ...	From Surrey Rangers. — To 89th Regt., 19-2-1807.
Weir, John Laing	... ...	... ...	6–5–1802	... ...	... ...	... ...	From 8th W.I. Regt.—To 59th Regt., 25-4-1805.
Richardson, William ...	1–12–1802	29–2–1804	1–8–1810	8–2–1827	... ...	... ...	Died, 2-12-1830
Dowdeswell, William ...	... ...	... ...	... ...	... ...	24–2–1803	... ...	From 60th Regt.—Major-General, 25-9-1803.—To 60th Regiment, 2-6-1808.
Vandeleur, Henry... ...	... ...	12–8–1803	... ...	... ...	... ...	... ...	From 52nd Regt.—Captain, 38th Regt., 14-11-1804.
Creagh, Michael	9–5–1802	28–2–1804	1–1–1810	24–10–1821	... ...	... ...	Lt.-Col., h.p., 31-12-1830.
Nicolson, Robert	5–7–1802	29–4–1804	23–2–1809	... ...	... ...	... ...	Died, 1815.
McLean, Archibald ...	24–9–1802	24–11–1804	5–3–1812	... ...	... ...	... ...	Died on service, Oct., 1818.
Craig, Sir James Henry	... ...	... ...	... ...	... ...	... ...	5–1–1804	From 46th Rgt.—Lt.-Gen., 1-1-1801. To 22nd Regt., 30-10-1806.
Blackall, Andrew	10–1–1804	15–9–1805	... ...	... ...	... ...	... ...	Capt., 4th Ceylon Regt., 14-3-1810.
Jacob, Henry	30–6–1804	9–5–1805	10–3–1814	... ...	... ...	... ...	To 56th Regt., 27-10-1815.
Glover, ——	2–5–1804	... ...	... ...	... ...	... ...	... ...	Not shown after 1808.
McQuarie, Lachlan ...	1–5–1804	16–9–1805	7–9–1816	... ...	... ...	... ...	Adjt., 8-12-1808—24-12-1813. — To h.p., 25-5-1817.

APPENDIX I.

McLaurin, Colin	3–5–1804	17–9–1805	21–10–1818				To h.p., 25-10-1821.
Home, William	24–11–1804	7–1–1806					To h.p., 15-1-1824.
Monro, John Graham	25–12–1804						Killed in action at Bourbon, 8-7-1810.
Fraser, Hastings					18–4–1805		From 10th Regt. — Relinquished command on promotion to Major-General, 12-8-1819.
Sorrell, Thomas Stephen			25–4–1805				From 84th Regt.—Major, h.p., 23-4-1808.
Williams, William			29–8–1805				From 71st Regt.—Major in Army, 4-6-1814. Died at Calcutta, 1817.
Richardson, Charles		24–1–1805	10–2–1814				Capt., 59th Regt., 26-10-1815.
Deane, John		1–8–1805					From 12th Regt.—Capt., 66th Regt., 17-3-1814.
McLean, Alexander		16–9–1805	23–10–1817				From 56th Regt.—Major, 3rd W.I. Regt., 1-9-1840.
Hillhouse, Martin	13–6–1805	1–1–1807					Capt., Bourbon Regt., 2-4-1812.
Strettell, F. E.	1–8–1805	23–9–1809					Retired, 1813.
Leavock, Duncan	27–8–1805						Not shown after 1807.
Macquarie, John	15–9–1805						Not shown after 1809.
Campbell, John	16–9–1805	20–4–1808					Died on passage home, 1819.
Smith, George							Qr.-Mr., 15-2-1805—23-11-1808.
Ross, Sir Charles						30–10–1806	From 85th Regt.—Lieut.-General, 30-10-1805. To 37th Regiment, 25-6-1810.
Shearman, Robt. Marcus			2–1–1806				From 50th Regt.—Died, 1817.
Webb, John		6–3–1806					From 54th Regiment. — To h.p., 25-10-1821.
Wilkins, Martin	7–1–1806						To h.p., 1813.
Carey, Peter				26–3–1807			From 28th Regt. — Lt.-Col., 84th Regt., 18-7-1811.
Stuart, James		2–2–1807	24–11–1813	15–2–1831			From 6th R. Vet. Bn.—Retired, 18-1-1839.
Mercer, Thomas		28–3–1807					Died, 1815.
Cox, George	29–10–1807						Retired, 1810.
Drummond, Edward					2–6–1808		From 60th Regt.—Colonel, 4-6-1813.

Name.	Ensign or 2nd Lieut.	Lieutenant.	Captain.	Major.	Lieut.-Colonel.	Colonel.	Remarks.
Edwards, Wilbraham T.	... ...	... ...	... ...	12-5-1808	... ...	... ...	From 9th Garr. Bn.—Lieut.-Col., Bourbon Regt., 25-1-1812.
Kirkland, David Kennedy	... ...	17-11-1808	... ...	... ...	... ...	... ...	From 27th Regiment.—To h.p., 25-11-1819.
McLean, Francis ...	... ...	22-12-1808	... ...	... ...	... ...	... ...	Capt., Bourbon Regt., 25-1-1812.
White, William Richard	20-4-1808	24-9-1809	... ...	... ...	... ...	... ...	Died, May, 1814.
Mackintosh, Kenneth ...	23-6-1808	1-1-1811	... ...	... ...	... ...	... ...	To h.p., 14-11-1814.
McLean, Archibald ...	... ...	... ...	... ...	... ...	... ...	... ...	Qr.-Mr., 24-11-1808—13-9-1813.
Vanspall, Henry	... ...	1-8-1809	30-1-1823	... ...	... ...	... ...	To 41st Regt., 4-3-1824.
Clarke, George Warde	... ...	7-12-1809	... ...	... ...	... ...	... ...	Capt., 60th Regt., 6-1-1814.
Sandon, John Kedgell	... ...	12-12-1809	... ...	... ...	... ...	... ...	From 1st Ceylon Regt.—To h.p., 13-1-1820.
Cannell, Thomas E. ...	25-5-1809	1-2-1811	... ...	... ...	... ...	... ...	Retired, 1817.
Hodson, John	31-8-1809	25-2-1811	... ...	... ...	... ...	... ...	Died, 1816.
Porter, Robert	7-12-1809	3-3-1812	... ...	... ...	... ...	... ...	To h.p., 11-4-1816.
Grogan, Francis Stubb	21-12-1809	... ...	... ...	... ...	... ...	... ...	Lieut., Bourbon Regt., 27-1-1812.
Needham, Hon. Francis (afterwards Viscount Kilmorey)	... ...	... ...	... ...	... ...	... ...	25-6-1810	Lt.-Gen., 29-4-1802, Gen., 1-1-1812.—Died, 21-11-1832.
Impey, Benjamin ...	... ...	... ...	26-7-1810	... ...	... ...	... ...	From 4th Garr. Battn.—Major, 1-12-1812. Died, May, 1814.
Birkett, James	... ...	1-1-1810	... ...	... ...	... ...	... ...	Adjt., 25-12-1813—1-3-1815.—Dismissed.
Creagh, James	1-1-1810	4-3-1812	7-4-1825	18-1-1839	30-4-1852	... ...	Col. in Army, 30-4-1855.—To r.f.p., 16-9-1859.
Grant, John	15-2-1810	5-3-1812	13-8-1830	... ...	... ...	... ...	Not shown after 1832.
Munro, Donald	14-6-1810	24-11-1813	... ...	... ...	... ...	... ...	To 7th Vet. Bn., 4-12-1820.
Johnson, John	... ...	... ...	... ...	18-7-1811	18-11-1813	... ...	From 77th Regt.—To h.p., 6-7-1826.
Maclachlan, Duncan ...	24-9-1811	10-2-1814	... ...	... ...	... ...	... ...	Died, 1815.
Maclean, Duncan ...	25-9-1811	11-2-1814	... ...	... ...	... ...	... ...	To h.p., 4-4-1822.
Macquarrie, Hector ...	26-9-1811	12-2-1814	... ...	... ...	... ...	... ...	To 48th Regt., 5-3-1818.

Henry, John S.	28-2-1812	13-2-1814					Died, 1816.
Leach, James	4-3-1812	10-3-1814					Adjt., 2-3-1815 — 28-6-1820. — To h.p., 23-11-1820.
Kennedy, Duncan Henry	5-3-1812	30-3-1814					To h.p., 1818.
Hale, Jasper	14-5-1812						Retired, 1813.
Perry, Francis		30-12-1813					From 3rd Garr. Bn.—Capt., R. African Col. Corps, 8-6-1826.
Needham, Francis Henry	8-4-1813						To Gren. Gds., 26-1-1814.
Gould, Pierce Purcell	15-7-1813	31-3-1814					To h.p., 17-1-1822.
Bradford, David	5-8-1813	2-6-1815					To h.p., 4-5-1820.
Norton, Thomas	12-8-1813						Retired, February, 1814.
Caddell, Edward	19-8-1813	11-5-1815					Died, 1819.
Stuart, Henry	2-12-1813	1-9-1816					To h.p., 1818.
Read, Alexander	16-12-1813	7-9-1816					To h.p., 1818.
Law, James	23-12-1813	8-9-1816					To h.p., 1818.
Moreton, Joseph	30-12-1813	20-9-1816					Died, 2-6-1817.
Gill, Richard							Qr.-Mr., 13-9-1813—22-3-1826.—To r.f.p.
Edwards, John Hamilton			11-2-1814				From 60th Regt.—To 46th Regt., 25-7-1818.
Lunn, James			12-2-1814				From 40th Regiment. — To h.p., 11-12-1817.
Carroll, John Doyle		30-3-1814					To h.p., 18-9-1823.
Russell, Andrew	10-3-1814	30-6-1817					To 48th Regt., 24-8-1820.
Holland, John	11-3-1814	21-10-1818	31-10-1834				To h.p., 11-1-1820. — Rejoins, 18-9-1823.—Brev.-Maj., 9-11-1846. Died, 12-2-1847.
Home, Patrick	30-3-1814						Died, Nov., 1818.
Chadwick, James			26-10-1815				From 59th Regt. — Major, h.p., 22-4-1826.
Harrison, George	10-8-1815						To h.p., 25-5-1820.
Dobree, Joseph Philip	24-8-1815						To h.p., 24-8-1815.

Name.	Ensign or 2nd Lieut.	Lieutenant.	Captain.	Major.	Lieut.-Colonel.	Colonel.	Remarks
Dundee, Edward	... ...	11–4–1816	... ...	... ...	... ...	... ...	From 7th Regt.—To 47th Regt., 8-1-1818.
Carroll, Edward	20–9–1816	... ...	... ...	... ...	... ...	... ...	Retired, 1820.
Gammell, William ...	... ...	... ...	11–12–1817	... ...	... ...	... ...	From 104th Regt.—Major, h.p., 29-8-1826.
Williams, Robert ...	1–7–1817	24–10–1821	... ...	... ...	... ...	... ...	To 44th Regt., 18-4-1822.
Carroll, Thomas	20–11–1817	... ...	... ...	... ...	... ...	... ...	To h.p., 26-10-1820.
Grant, Lewis	21–11–1817	30–1–1823	... ...	... ...	... ...	... ...	Retired, 1831.
Humfrey, Benjamin Geale	... ...	... ...	25–7–1818	... ...	... ...	... ...	From 46th Rgt.—To h.p., 19-5-1825.
Lloyd, Henry Vereker...	1–1–1819	... ...	... ...	... ...	... ...	... ...	To h.p., 25-10-1821.
McIntyre, James	22–7–1819	7–4–1825	1–9–1840	... ...	... ...	... ...	Adjt., 8-1-1828—2-8-1830.—To h.p., 6-5-1842.
Nunn, John Oliver Howe	... ...	13–1–1820	22–4–1826	... ...	... ...	... ...	From 46th Rgt.—To h.p., 15-5-1828.
Dolman, John	... ...	4–5–1820	... ...	... ...	... ...	... ...	From h.p.—Adjt., 29-6-1820. Died, January, 1828.
Bunney, Robert	... ...	12–8–1820	24–10–1821	... ...	... ...	... ...	From 48th Rgt.—To h.p., 20-2-1823.
Hart, John	... ...	23–11–1820	... ...	... ...	... ...	... ...	From 35th Regt.—To 4th L.D., 5-7-1821.
Bouverie, James Wm.	29–6–1820	3–7–1823	... ...	... ...	... ...	... ...	To 39th Regt., 14-8-1823.
Close, Frederick	26–7–1820	12–8–1824	... ...	... ...	... ...	... ...	Died, 1826.
Ussher, Richd. Beverley	26–10–1820	8–4–1825	29–8–1826	... ...	... ...	... ...	To h.p., 5-6-1828.
Dalgety, Frederick ...	11–1–1821	23–3–1826	... ...	... ...	... ...	... ...	Paymaster, 69th Regt., 1-3-1839.
Grey, Matthew Robert	... ...	4–4–1822	12–8–1824	... ...	... ...	... ...	From 6th Regt.—To 36th Regt., 8-6-1826.
North, Philip	... ...	18–4–1822	15–2–1831	... ...	... ...	... ...	From 44th Regiment. — To h.p., 13-11-1832.
Alexander, Henry ...	... ...	21–11–1822	13–8–1825	... ...	... ...	... ...	From h.p.—To h.p., 13-8-1825.
Hogg, James	... ...	... ...	20–2–1823	... ...	... ...	... ...	From h.p.—Retired, 1824.
Osborne, William ...	30–1–1823	13–8–1825	... ...	... ...	... ...	... ...	Capt., h.p., 31-10-1826.
Jekyll, Edward	10–7–1823	... ...	... ...	... ...	... ...	... ...	To Gren. Gds., 8-2-1826.
Crawford, Robert	... ...	... ...	12–8–1824	... ...	... ...	... ...	From 41st Regt.—Retired, 1830.

Macdonald, Charles ...	... ...	15–1–1824	... ...	... ...	... ...	... ...	From h.p.—To h.p., 8-4-1826.
Trench, Power Le Poer	12–8–1824	17–12–1825	... ...	... ...	... ...	... ...	To 76th Regt., 19-12-1825.
Wynne, Abraham Wm.	... ...	... ...	8–4–1825	... ...	... ...	... ...	From h.p.—To h.p., 29-3-1827.
Wolseley, Robert Benjamin	... ...	... ...	19–5–1825	... ...	... ...	... ...	From 25th Rgt.—To h.p., 13-4-1832.
Ormond, George	... ...	9–4–1825	... ...	... ...	... ...	... ...	From h.p.—Appointed Paymaster, 26-2-1829.
Halliday, Lewis	7–4–1825	4–5–1826	31–12–1830	... ...	... ...	... ...	Died, 1-7-1844.
Caldwell, William Clark	8–4–1825	29–8–1826	... ...	... ...	... ...	... ...	To h.p., 1-11-1827.
Copinger, Henry	28–4–1825	6–3–1828	... ...	... ...	... ...	... ...	To h.p., 27-2-1829.
Brooke, John Smeaton	13–8–1825	... ...	... ...	... ...	... ...	... ...	Lieut., h.p., 12-12-1826.
Mayne, Robert	22–12–1825	31–10–1826	11–10–1831	... ...	... ...	... ...	Retired, 3-11-1837.
Mallett, John William...	... ...	... ...	... ...	... ...	31–8–1826	... ...	From 89th Regt.—Died, Dec., 1831.
FitzGerald, Thomas ...	... ...	... ...	29–8–1826	... ...	... ...	... ...	From h.p.—To 26th Rgt., 21-2-1828.
Sidley, Hy. Edmond De Burgh	... ...	8–4–1826	11–6–1830	8–4–1842	... ...	... ...	From h.p. — To 50th Regiment, 29-11-1850.
Kearney, Francis ...	... ...	8–6–1826	... ...	... ...	... ...	... ...	From 45th Regt. — Capt., h.p., 27-4-1827.
Gallwey, James	18–2–1826	8–2–1831	... ...	... ...	... ...	... ...	Retired, 1834.
Selway, Benjamin John	23–3–1826	... ...	... ...	... ...	... ...	... ...	Died, 1827.
Davis, Edward	4–5–1826	19–6–1827	... ...	... ...	... ...	... ...	To 35th Regt., 27-6-1827.
Johnson, William ...	22–6–1826	11–6–1830	... ...	... ...	... ...	... ...	To 27th Regt., 22-2-1831.
O'Callaghan, Cornelius	28–9–1826	... ...	... ...	... ...	... ...	... ...	To h.p., 22-5-1828.
Chichester, Arthur Chas.	12–12–1826	... ...	... ...	... ...	... ...	... ...	Lieut., h.p., 10-11-1829.
Martyn, Peter	26–12–1826	... ...	... ...	... ...	... ...	... ...	To 88th Regt., 27-3-1827.
Jerome, Joseph	... ...	... ...	... ...	... ...	... ...	... ...	Qr.-Mr., 23-3-1826—17-3-54. — To h.p.
Barrett, James	... ...	... ...	8–2–1827	31–12–1830	... ...	... ...	From 89th Regt.—To h.p., 8-5-1835.
Versturme, Louis Robt. James	... ...	... ...	29–3–1827	... ...	... ...	... ...	From h.p.—To h.p., 3-7-1828.

Name.	Ensign or 2nd Lieut.	Lieutenant.	Captain.	Major.	Lieut.-Colonel.	Colonel.	Remarks.
Creagh, Jasper	... ...	14-6-1827	16-11-1832	... ...	... ...	... ...	From 27th Regt.—To 95th Regt., 1-9-1837.
Tinné, William Thomas	... ...	20-12-1827	21-12-1832	... ...	... ...	... ...	From h.p. — Adjutant, 3-8-1830—25-10-1832. To 8th Light Dgns., 26-4-1833.
Theobald, Wm. Francis	27-3-1827	15-2-1831	... ...	... ...	... ...	... ...	Retired, 1834.
Whitfield, Abraham ...	25-10-1827	... ...	... ...	... ...	... ...	... ...	Lieut., 33rd Regt., 24-12-1829.
Semple, William	8-11-1827	6-8-1831	... ...	... ...	... ...	... ...	To 18th Regt., 4-11-1837.
Kean, Henry	... ...	... ...	21-2-1828	... ...	... ...	... ...	From h.p.—To h.p., 20-7-1830.
Kirby, Thomas Cox ...	... ...	... ...	15-5-1828	... ...	... ...	... ...	From h.p.—Major, 12-8-1819. To h.p., 13-8-1830.
Baines, Thos. Josephus	... ...	... ...	5-6-1828	... ...	... ...	... ...	From h.p.—Major, 19-7-1821. To h.p., 13-8-1830.
O'Dell, Henry Edward	... ...	... ...	3-7-1828	... ...	... ...	... ...	From h.p.—To h.p., 4-10-1831.
Stuart, William Kier ...	6-3-1828	28-9-1830	8-4-1842	29-9-1854	24-1-1860	... ...	To r.f.p., 20-2-1863.
Phibbs, Owen	25-11-1828	31-12-1830	18-1-1839	... ...	... ...	... ...	Died, 22-10-1842.
Holt, William Foden ...	... ...	26-2-1829	... ...	... ...	... ...	... ...	From h.p.—Capt., h.p., 28-12-1830.
Keogh, Francis Gethings	... ...	27-2-1829	... ...	... ...	... ...	... ...	From h.p.—Capt., h.p., 2-11-1830.
Pearson, John Bonner...	10-11-1829	11-10-1831	4-12-1835	... ...	... ...	... ...	To 89th Regt., 7-10-1836.
Scott, John Walter ...	24-12-1829	... ...	... ...	... ...	... ...	... ...	Died, 1833.
Gibson, George	... ...	... ...	20-7-1830	... ...	... ...	... ...	From h.p.—To h.p., 24-11-1835.
Lowth, Robert Henry...	... ...	... ...	14-8-1830	30-4-1852	10-8-1855	... ...	From h.p.—Col., r.f.p., 24-1-1860.
Carlisle, Robt. Needham	11-6-1830	16-11-1832	... ...	... ...	... ...	... ...	To h.p., 20-10-1837.
Dowman, John	28-9-1830	21-12-1832	... ...	... ...	... ...	... ...	To 40th Regt., 16-8-1839.
Murray, Chas. Thomas	31-12-1830	7-3-1834	... ...	... ...	... ...	... ...	Retired, 1-12-1837.
Servantes, William Thornton	... ...	11-1-1831	... ...	... ...	... ...	... ...	From h.p. —To 77th Regiment, 26-10-1832.
Hutchinson, B. A. S. ...	... ...	22-2-1831	... ...	... ...	... ...	... ...	From 27th Regt.—To 26th Regt., 25-10-1833.
Gilchrist, James	25-1-1831	31-10-1834	... ...	... ...	... ...	... ...	To h.p., 23-4-1841.

APPENDIX I.

Cobbe, Henry C.	15–2–1831	23–1–1835	3–11–1837	... ...	... ...	... ...	To 2nd W.I. Regt., 13-7-1838.
Wilson, Christopher M.	23–8–1831	20–10–1837	... ...	... ...	... ...	... ...	To 40th Regt., 9-11-1838.
Edwards, Joseph	11–10–1831	4–12–1835	11–5–1841	... ...	... ...	... ...	Died, 14-1-1849.
Harris, Lord Wm. Geo.	... ...	... ...	... ...	... ...	... ...	3–12–1832	Major-General, 19-7-1821.—To 73rd Regt., 4-12-1835.
Creagh, Sir Michael ...	... ...	... ...	... ...	... ...	24–1–1832	... ...	From h.p., *see ante.*—To 11th Regt., 7-1-1842.
Bunworth, Richard ...	... ...	... ...	13–4–1832	... ...	... ...	... ...	From h.p.—Died, Oct., 1834.
Fenwick, Horatio	... ...	26–10–1832	17–8–1842	... ...	... ...	... ...	From 77th Regt.—Adjt., 26-10-1832—7-4-1842. To h.p., 5-4-1844.
Gore, Charles William	16–11–1832	3–11–1837	28–12–1841	... ...	... ...	... ...	To 72nd Regt., 18-2-1842.
Moriarty, Henry	21–12–1832	... ...	... ...	... ...	... ...	... ...	Retired, 29-5-1835.
Wigmore, Henry Wm.	... ...	... ...	2–8–1833	... ...	... ...	... ...	From 2nd W. India Regt.—To 95th Regt., 10-12-1837.
Keane, Giles	... ...	25–10–1833	23–10–1842	10–8–1855	... ...	... ...	From 40th Regt.—Colonel, r.f.p., 9-3-1860.
Steele, Robert	13–12–1833	... ...	... ...	... ...	... ...	... ...	From h.p.—Retired, 1834.
Blewitt, Thomas	14–2–1834	... ...	... ...	... ...	... ...	... ...	Retired, 17-6-1836.
Loftus, John	7–3–1834	... ...	... ...	... ...	... ...	... ...	Retired, 22-7-1836.
Middlemore, Robert Frederick	31–10–1834	... ...	... ...	... ...	... ...	... ...	To 91st Regt., 13-11-1835.
Ponsonby, Hon. Sir Fred. Cavendish	... ...	... ...	... ...	... ...	... ...	4–12–1835	Maj.-Gen., 27-5-1825.—To Royal Dragoons, 31-3-1836.
Bouverie, James Wm. ...	... ...	... ...	... ...	8–5–1835	8–4–1842	... ...	From 17th Regt., *see ante.*—To 89th Regt., 14-7-1843.
Rideout, Goring	23–1–1835	1–12–1837	... ...	... ...	... ...	... ...	Retired, 18-6-1841.
Smyth, J. R. Carmichael	29–5–1835	... ...	... ...	... ...	... ...	... ...	Lieut., 1st W.I. Regt., 26-8-1836.
Faulkner, Henry Cole	13–11–1835	18–5–1839	... ...	... ...	... ...	... ...	To 2nd Regt., 26-7-1839.
Cash, Henry Christmas	4–12–1835	18–1–1839	... ...	... ...	... ...	... ...	Retired, 3-11-1840.
Watson, James	... ...	... ...	... ...	... ...	... ...	31–3–1836	Lt.-Gen., 10-1-1837.—To 14th Regt., 24-5-1837.
Hay, Archibald	... ...	... ...	7–10–1836	... ...	... ...	... ...	From 89th Rgt.—Retired, 11-5-1841.
Plunkett, Hon. Edward Sidney	17–6–1836	11–5–1841	... ...	... ...	... ...	... ...	To 95th Regt., 19-6-1841.

Name.	Ensign or 2nd Lieut.	Lieutenant.	Captain.	Major.	Lieut.-Colonel.	Colonel.	Remarks.
Mahon, Maurice Hartland	22-7-1836	3-11-1840	... ...	... ...	... ...	... ...	Retired, 1-4-1842.
Thursby, John Harvey	26-8-1836	18-6-1841	5-4-1844	... ...	... ...	... ...	To 41st Regt., 6-8-1847.
Brooke, Sir Arthur ...	... ...	... ...	... ...	... ...	... ...	24-5-1837	Lieut.-General, 10-1-1837. — Died, 26-7-1843.
Macleod, Arthur Lyttelton	... ...	... ...	1-9-1837	... ...	... ...	... ...	From 95th Regt.—To h.p., 1-2-1839.
Everard, Richd. Nugent	... ...	... ...	8-12-1837	... ...	... ...	... ...	From 95th Regiment. — To h.p., 28-12-1841.
Francklin, Michael Geo.	... ...	4-11-1837	... ...	... ...	... ...	... ...	From 18th Regiment.—To h.p., 20-12-1839.
Macbean, Wm. Forbes	7-7-1837	... ...	... ...	... ...	... ...	... ...	Lieut., St. Helena Regt., 7-1-1842.
Prendergast, Wellesley Bowles	3-11-1837	... ...	... ...	... ...	... ...	... ...	Retired, 4-1-1839.
Edwards, William ...	1-12-1837	28-12-1841	13-12-1847	... ...	... ...	... ...	To 17th Regt., 15-8-1848.
Dickinson, James Esten	... ...	... ...	13-7-1838	... ...	... ...	... ...	From 2nd W. India Regt—Died, 11-2-1843.
Bowen, Hugh Thomas	... ...	9-11-1838	12-2-1843	... ...	... ...	... ...	From 40th Regt.—To 72nd Regt., 30-6-1845.
Stephens, Henry Sykes	... ...	... ...	1-2-1839	... ...	... ...	... ...	From h.p.—Major, 28-6-1838.—To h.p., 26-10-1841.
Bennett, Chas. Luxmore	... ...	26-7-1839	1-4-1842	... ...	... ...	... ...	From 2nd Regt.—Died, 16-8-1842.
Bennett, Richard Baylis	... ...	16-8-1839	... ...	... ...	... ...	... ...	From 40th Regt—Retired, 18-8-1843.
Lane, John	... ...	20-12-1839	... ...	... ...	... ...	... ...	From h.p.—Retired, 16-5-1840
Fraser, Adolphus ...	4-1-1839	... ...	... ...	... ...	... ...	... ...	Retired, 3-11-1840.
Peacocke, Warren Wm. Richard	18-1-1839	1-4-1842	... ...	... ...	... ...	... ...	To h.p., 13-8-1847.
Lecky, Alexander ...	18-5-1839	2-4-1842	15-1-1849	... ...	... ...	... ...	To 2nd Regt., 6-7-1849.
Woodgate, Wm. Harding	... ...	16-5-1840	14-11-1843	... ...	... ...	... ...	From h.p.—Died, 28-2-1854.
Hickey, Edward	... ...	1-9-1840	6-5-1842	... ...	... ...	... ...	From 78th Regt.—To 75th Regt., 3-6-1842.
Monypenny, Robert T. G. G.	25-9-1840	... ...	... ...	... ...	... ...	... ...	Retired, 17-9-1841.
Butler, Charles George	3-11-1840	8-4-1842	15-3-1853	... ...	... ...	... ...	Died, 18-12-1854.
Derinzy, Bartholomew Vigors	... ...	... ...	... ...	... ...	7-1-1842	... ...	From 11th Regt.—Col., 11-11-1851; Inspg. Field Officer, Recruiting District, 30-4-1852.

Brown, Edward	... ...	... ...	26–10–1841	... ...	... ...	... ...	From h.p.—Retired, 1-4-1842.
Lovett, G. W. Molineux	... ...	23–4–1841	... ...	... ...	... ...	... ...	From h.p. — To 26th Regiment, 18-2-1842.
Stoney, George Butler...	... ...	19–6–1841	19–12–1844	... ...	... ...	... ...	From 16th Regt.—To 28th Regt., 5-11-1847.
Holland, Averell Lecky	11–5–1841	6–5–1842	... ...	... ...	... ...	... ...	Retired, 13-8-1847.
Creagh, Chas. Osborne	18–6–1841	5–8–1842	1–3–1854	20–2–1863	3–8–1866	... ...	To 6th Regt., 1-2-1867.
Woulfe, Stephen Roland	17–9–1841	16–8–1842	... ...	... ...	... ...	... ...	To 3rd Regt., 18-8-1843.
Weaver, William Henry	28–12–1841	23–2–1843	29–9–1854	... ...	... ...	... ...	Died, 13-9-1857.
Rattray, William	... ...	... ...	18–2–1842	... ...	... ...	... ...	From 72nd Regt.—Died, 14-11-1843.
Lucas, William	... ...	... ...	3–6–1842	... ...	... ...	... ...	From 75th Regt.—Died, 18-12-1844.
Heatley, Charles Fade	... ...	8–4–1842	... ...	... ...	... ...	... ...	From 13th Regt.—Appointed Paymaster, 17-12-1847.
Woodd, George Leslie...	... ...	8–4–1842	... ...	... ...	... ...	... ...	From 2nd W. India Regt.—Retired, 10-12-1847.
Kirby, Joshua Henry...	... ...	8–4–1842	30–4–1852	1–6–1860	... ...	... ...	From 34th Regt.—To 68th Regt., 23-4-1861.
Kelly, Edward Henry ...	... ...	8–4–1842	... ...	... ...	... ...	... ...	From 51st Regt.—Retired, 9-4-1847.
Barry, John Richard ...	... ...	8–4–1842	... ...	... ...	... ...	... ...	From 15th Regt.—Died, 1-9-1847.
Cox, Henry	... ...	8–4–1842	... ...	... ...	... ...	... ...	From 2nd Regt.—Died, 1-12-1842.
Croker, John Rees ...	... ...	8–4–1842	15–3–1853	... ...	... ...	... ...	From 6th Regt.—Retired, 28-3-1854.
Morrow, D.	7–1–1842	24–2–1843	... ...	... ...	... ...	... ...	Qr.-Mr., 57th Regt., 30-6-1843.
Oldham, John Augustus	1–4–1842	23–6–1843	... ...	... ...	... ...	... ...	To 55th Regt., 11-6-1847.
Archer, Hy. Benjamin	2–4–1842	18–8–1843	... ...	... ...	... ...	... ...	Died, 23-7-1846.
Boyd, James	8–4–1842	13–11–1843	23–1–1855	... ...	... ...	... ...	Adjt., 8-4-1842—25-8-1854; Major, r.f.p., 16-8-1859.
Jerome, John	9–4–1842	14–11–1843	28–3–1854	1–4–1866	6–3–1872	... ...	Retired, 23-5-1877, with hon. rank of Major-General.
Creed, Francis Russell	6–5–1842	21–11–1843	... ...	... ...	... ...	... ...	Died, 29-7-1849.
Matthews, John James	5–8–1842	3–5–1844	... ...	... ...	... ...	... ...	Retired, 23-2-1849.
Bowen, Ralph Cole ...	16–8–1842	... ...	... ...	... ...	... ...	... ...	Died, 22-12-1843.

Name.	Ensign or 2nd Lieut.	Lieutenant.	Captain.	Major.	Lieut.-Colonel.	Colonel.	Remarks.
Maister, John	... ...	... ...	... ...	... ...	... ...	25–8–1843	Lieut.-Gen., 10-1-1837.—Died, May, 1852.
Aplin, Andrew Snape Hamond	... ...	... ...	... ...	... ...	14–7–1843	... ...	From 89th Regt.—Died, 28-9-1854.
Cowper, Francis Boynton	... ...	6–1–1843	... ...	... ...	... ...	... ...	From 3rd W. India Regt.—Retired, 15-10-1852.
Stuart, Edmund Robert	... ...	22–2–1843	... ...	... ...	... ...	... ...	From 39th Regt.—Died, 20-11-1843.
Sparks, Robt. Manners	... ...	18–8–1843	... ...	... ...	... ...	... ...	From 3rd Regt.—Retired, 13-3-1846.
Darby, Charles	... ...	13–10–1843	19–12–1854	3–8–1866	... ...	... ...	From 2nd Regt.—Brev.-Lt.-Col., 25-3-1869. To r.f.p., 8-6-1870.
Beresford, Marcus Wylly de la P.	24–2–1843	19–12–1844	... ...	... ...	... ...	... ...	To 49th Regt., 3-3-1848.
Weaver, Edward Baker	25–2–1843	24–7–1846	... ...	... ...	... ...	... ...	Retired, 10-2-1854.
Gerahty, Digby	4–8–1843	13–3–1846	... ...	... ...	... ...	... ...	To h.p., 7-8-1847.
Porter, William	18–8–1843	... ...	... ...	... ...	... ...	... ...	Died, 29-2-1844.
Baird, Wm. Carpendale	22–12–1843	9–4–1847	... ...	... ...	... ...	... ...	To h.p., 15-10-1847.
Munro, William	... ...	... ...	2–7–1844	... ...	... ...	... ...	From 39th Regt.—To 39th Regt., 12-2-1845.
Robinson, George Wm.	22–3–1844	2–9–1847	14–9–1857	... ...	... ...	... ...	Died, 7-7-1858.
Stuart, John Richardson	23–3–1844	13–12–1847	... ...	... ...	... ...	... ...	Capt., 19th Regt., 9-3-1858.
Mildmay, Arthur Geo. St. John	3–5–1844	13–8–1847	... ...	... ...	... ...	... ...	Retired, 23-6-1848.
Barclay, Walter Channing	6–9–1844	... ...	... ...	... ...	... ...	... ...	Retired, 23-3-1847.
Anderson, Abraham Collis	... ...	... ...	12–2–1845	... ...	... ...	... ...	From 39th Regt.—Major, 20-6-1854; Fort Maj., Edinburgh, 15-10-1858.
O'Brien, Donough ...	... ...	... ...	13–6–1845	... ...	... ...	... ...	From 72nd Regt.—Died, 22-1-1855.
Nunn, Geo. Shearman	9–5–1845	15–1–1849	... ...	... ...	... ...	... ...	Capt., 22nd Regt., 17-3-1858.
Orlebar, Orlando Robt. Hamond	13–3–1846	10–12–1847	... ...	... ...	... ...	... ...	To 28th Regt., 2-2-1849.
Lewis, Ralph FitzGibbon	27–10–1846	30–7–1849	2–7–1858	... ...	... ...	... ...	Major, 23-3-1860.—Retired, 1-2-1868.
Rawlins, Thos. Andrews	... ...	... ...	6–8–1847	... ...	... ...	... ...	From 41st Regt.—Staff-Capt., Chatham Invalid Depôt, 14-9-1855.
Mayers, John Perkins	... ...	... ...	5–11–1847	24–1–1860	... ...	... ...	From 28th Regt.—Lt.-Col., retired, 23-4-1860.
Ellison, Wm. H. Howard	... ...	12–1–1847	... ...	... ...	... ...	... ...	From 17th Regt.—Retired, 27-7-1849.

APPENDIX I.

Baxter, Andrew	... ...	13–1–1847	... ...	... ...	... ...	... ...	From 17th Regt.—Retired, 28-3-1851.
Gordon, James John	... ...	11–6–1847	10–8–1855	... ...	... ...	... ...	From 55th Regt.—To 16th Regt., 23-9-1859.
Beatty, William Henry	... ...	13–8–1847	... ...	... ...	... ...	... ...	From 97th Regt.—Died, 17-8-1854.
Spier, John	... ...	15–10–1847	... ...	... ...	... ...	... ...	From 63rd Regt. — Died, March, 1854.
FitzGerald, Ormond	... ...	17–12–1847	... ...	... ...	... ...	... ...	From Cape Cos.—To 87th Regt., 3-4-1849.
Bowen, Hampden Cole	23–3–1847	... ...	... ...	... ...	... ...	... ...	Retired, 12-7-1850.
Lysaght, Hon. William Henry	9–4–1847	... ...	... ...	... ...	... ...	... ...	Died, 11-3-1852.
Lepper, Maxwell	13–8–1847	23–2–1849	25–9–1855	... ...	... ...	... ...	To 14th Regt., 24-8-1860.
Leet, Edward	10–12–1847	27–7–1849	... ...	... ...	... ...	... ...	To 20th Regt., 13-2-1852.
Scott, Hopton Bassett	24–12–1847	28–3–1851	... ...	... ...	... ...	... ...	To 9th Regt., 31-10-1851.
Welman, Harvey Wellesley Pole	... ...	... ...	15–8–1848	9–3–1860	... ...	... ...	From 17th Regt.—To 10th Regt., 19-3-1861.
Havelock, Hy. Marshman	... ...	23–6–1848	... ...	... ...	... ...	... ...	From 39th Regt.—To 10th Regt., 13-2-1852.
Jerome, Henry Edward	21–1–1848	30–4–1852	... ...	... ...	... ...	... ...	Capt., 19th Regt., 23-7-1858.
Thornton, Chas. Edmund	... ...	... ...	6–7–1849	... ...	... ...	... ...	From 2nd Regt.—To Staff Appointment, 16-3-1855.
Meacham, John	... ...	2–2–1849	... ...	... ...	... ...	... ...	From 28th Regt.—Died, 15-7-1855.
Gardner, Frederick	23–2–1849	7–5–1852	... ...	... ...	... ...	... ...	Died, 26-7-1855.
Cochrane, Hugh Stewart	13–4–1849	15–10–1852	... ...	... ...	... ...	... ...	Adjt., 18-4-1856—23-8-1858; Capt., 16th Regt., 24-8-1858.
Knipe, William	19–10–1849	5–6–1853	26–8–1859	6–3–1872	... ...	... ...	Died, 30-8-1878.
Tudor, Wm. Langley	... ...	... ...	... ...	29–11–1850	29–9–1854	... ...	From 50th Regt.—Colonel, h.p., 10-8-1855.
Jones, Edward Monckton	... ...	1–2–1850	... ...	... ...	... ...	... ...	From Cape Mtd. Rifles.—Capt., 20th Regt., 26-3-1858.
Winniett, William	18–1–1850	... ...	... ...	... ...	... ...	... ...	To 4th Regt., 30-4-1852.
Henry, Robert Edward	12–7–1850	10–2–1854	25–6–1858	... ...	... ...	... ...	To h.p., 10-11-1865.
Hamilton, Thomas Rice	... ...	31–10–1851	... ...	... ...	... ...	... ...	From 9th Regt.—Capt., 3rd Regt., 27-2-1858.
Whitaker, George Thompson	... ...	13–2–1852	... ...	... ...	... ...	... ...	From 10th Regt.—Retired, 7-5-1852.

Name.	Ensign or 2nd Lieut.	Lieutenant.	Captain.	Major.	Lieut.-Colonel.	Colonel.	Remarks.
Adams, George Hewish	... ...	13–2–1852	25–1–1856	8–6–1870	23–5–1877	... ...	From 20th Regt.—Adjt., 26-8-1854—17-4-1856. (See Royal Irish Rifles List.)
Mackenzie, James Kenneth Douglas	28–3–1851	1–3–1854	24–1–1860	23–5–1877	... ...	... ...	See Royal Irish Rifles List.
Parke, Roger	... ...	... ...	... ...	... ...	... ...	26–5–1852	Major-General, 9-11-1846. — Died, 28-4-1854.
Wallace, Robert John...	12–3–1852	... ...	... ...	... ...	... ...	... ...	Retired, 21-7-1854.
Maine, Henry Cracroft	16–4–1852	21–7–1854	... ...	... ...	... ...	... ...	Retired, 19-9-1856.
Brown, Richard Charles	14–5–1852	18–8–1854	... ...	... ...	... ...	... ...	Died, 27-12-1855.
Wilson, Aug. Nicholas	28–5–1852	29–9–1854	9–3–1860	... ...	... ...	... ...	To 4th Regt., 15-1-1861.
Neville, Thomas Josiah	15–10–1852	... ...	... ...	... ...	... ...	... ...	Retired, 22-4-1853.
Fraser, Charles	22–4–1853	19–12–1854	... ...	... ...	... ...	... ...	To 13th Regt., 11-5-1859.
Hay, Lord James ...	... ...	... ...	... ...	... ...	... ...	8–5–1854	Lieut.-General, 20-6-1854. — Died, 18-8-1862.
Ord, Alfred Robert ...	10–2–1854	23–1–1855	... ...	... ...	... ...	... ...	Capt., 65th Regt., 22-4-1862
Edwards, Francis Drewé	28–3–1854	16–7–1855	... ...	... ...	... ...	... ...	To 13th Regt., 1-10-1858.
Mullen, John Fleming W.	21–7–1854	27–7–1855	... ...	... ...	... ...	... ...	To 46th Regt., 18-5-1860.
Hammond, Henry ...	... ...	... ...	... ...	... ...	... ...	... ...	Qr.-Mr., 17-3-1854—15-12-1859.—To 66th Regt., 16-12-1859.
Creagh, James	20–2–1855	10–8–1855	1–6–1860	... ...	... ...	... ...	To 17th Regt., 4-9-1860.
Fry, John William ...	26–6–1855	28–12–1855	20–2–1863	... ...	... ...	... ...	To 91st Regt., 22-3-1864.
Dartnell, John George	27–7–1855	7–11–1856	... ...	... ...	... ...	... ...	Adjt., 5-10-1858—23-5-1859; Capt., 16th Regt., 24-5-1859.
Brockman, Julius Drake	10–8–1855	14–9–1857	30–3–1866	... ...	... ...	... ...	Brev.-Lt.-Col., 1-10-1877.—To Reserve of Officers, 12-3-1881.
Coates, Valentine G. ...	... ...	19–9–1856	... ...	... ...	... ...	... ...	From 9th Regt.—Died, 27-9-1858.
Coran, George Albert ...	... ...	28–11–1856	... ...	... ...	... ...	... ...	From 50th Regt.—Retired, 15-4-1859.
Sewell, Stephen Wm. ...	1–4–1856	28–9–1858	... ...	... ...	... ...	... ...	To 89th Regt., 30-5-1859.
Keane, Charles	1–4–1856	8–10–1858	... ...	... ...	... ...	... ...	Retired, 1-5-1866.
Fowler, George	8–7–1856	3–6–1859	... ...	... ...	... ...	... ...	To 54th Regt., 24-6-1859.

Jackson, Gilbert Sidney	19–9–1856	25–6–1858	1–4–1866	... ...	... ...	... ...	Retired, 21-6-1871.
Stewart, Duncan	... ...	1–10–1858	16–8–1861	... ...	... ...	... ...	From 13th Regt.—Adjt., 3-6-1859—15-8-1861. To 92nd Regiment, 17-11-1863.
Wells, John	23–3–1858	18–10–1859	... ...	... ...	... ...	... ...	Capt., h.p., 16-5-1865.
Saunders, William ...	26–3–1858	9–3–1860	... ...	... ...	... ...	... ...	To 66th Regt., 11-2-1862.
Leadbitter, Matthew Edw.	30–3–1858	14–6–1859	... ...	... ...	... ...	... ...	Retired, 20-12-1864.
Hilliard, Geo. Edw. Anstruther	13–4–1858	14–2–1860	... ...	... ...	... ...	... ...	Retired, 30-1-1863.
Posnett, Richard Jebb	13–7–1858	1–6–1860	6–11–1867	... ...	... ...	... ...	Retired, 8-8-1868.
Jackson, Charles Henry	7–9–1858	20–2–1863	... ...	... ...	... ...	... ...	Capt., h.p., 3-2-1875.
Marshall, Francis Glasse	7–9–1858	16–8–1861	1–2–1868	... ...	... ...	... ...	To 13th Regt., 25-9-1869.
Cullinan, Hy. Valentine	9–11–1858	22–4–1862	... ...	... ...	... ...	... ...	To 94th Regt., 28-4-1863.
Branfill, Benjamin Aylett	... ...	... ...	23–9–1859	... ...	... ...	... ...	From 16th Regt. — Major, h.p., 1-4-1870.
Yardley, Thomas ...	... ...	11–5–1859	... ...	... ...	... ...	... ...	From 13th Regt.—Adjt., 16-8-1861—21-9-1865. Retired, 22-9-1865.
Bowness, Joshua James	... ...	30–5–1859	20–5–1864	... ...	... ...	... ...	From 89th Regt.—Retired, 1-5-1866.
Gray, William Ker ...	... ...	24–6–1859	10–11–1865	31–8–1878	... ...	... ...	From 54th Regt. (See Royal Irish Rifles List.)
Murphy, Jerome Richd.	18–1–1859	... ...	... ...	... ...	... ...	... ...	To 47th Regt., 28-10-1859.
Boulcott, John William	14–6–1859	30–1–1863	9–12–1876	... ...	... ...	... ...	Hon. Major, retired, 12-3-1881.
Philipps, Edw. Berkeley	17–6–1859	20–5–1864	... ...	... ...	... ...	... ...	To 52nd Regt., 18-10-1864.
Sperrin, John Edmund Monro	28–10–1859	... ...	... ...	... ...	... ...	... ...	To 95th Regt., 6-9-1861.
Harris, Edward James	30–12–1859	20–12–1864	8–8–1868	... ...	... ...	... ...	See Royal Irish Rifles List.
Lane, William	... ...	... ...	... ...	... ...	... ...	... ...	From 66th Regiment, — Qr.-Mr., 16-2-1859—22-3-1867. To h.p., 23-3-1867.
Best, Richard Mordesley	... ...	... ...	... ...	19–3–1861	20–2–1863	... ...	From 10th Regt.—To 79th Regt., 13-9-1864.
Fenton, Edward Dyne...	... ...	... ...	24–8–1860	... ...	... ...	... ...	From 14th Regt.—Retired, 30-3-1870.
Lees, Edward John ...	... ...	... ...	4–9–1860	... ...	... ...	... ...	From 17th Regt.—Died, 12-7-1861.
Churchill, John Spencer	... ...	18–5–1860	1–5–1866	... ...	... ...	... ...	From 46th Regt.—Retired, 6-11-1867.

Name.	Ensign or 2nd Lieut.	Lieutenant.	Captain.	Major.	Lieut.-Colonel.	Colonel.	Remarks.
Boulcott, Herbert Chas.	14–2–1860	... ...	... ...	... ...	... ...	... ...	To 19th Regt., 5-5-1863.
Stephens, John Francis	9–3–1860	... ...	... ...	... ...	... ...	... ...	To 99th Regt., 30-11-1860
Shaw, Donald	1–6–1860	24–1–1865	24–3–1869	... ...	... ...	... ...	Retired, 9-12-1876.
Savage, Fred. Stuckeley	... ...	... ...	... ...	23–4–1861	... ...	... ...	From 68th Regt.—Died, 31-3-1866.
Fagan, William	... ...	... ...	15–1–1861	... ...	... ...	... ...	From 4th Regt.—To 12th Lancers, 8-11-1861.
Roe, Robert Edward ...	... ...	... ...	8–11–1861	... ...	... ...	... ...	From 12th Lancers. — To h.p., 1-7-1869.
Gambell, Thomas ...	... ...	11–2–1862	... ...	... ...	... ...	... ...	From 66th Regt.—Retired, 1-7-1867.
Chatfield, Hy. Rowland Spencer	12–3–1861	22–9–1865	9–10–1869	... ...	... ...	... ...	Adjt., 22-9-1865—2-4-1869. (See Royal Irish Rifles List.)
Blackett, Hy. Wise Ridley	16–8–1861	... ...	... ...	... ...	... ...	... ...	To 15th Hussars, 20-5-1864.
Sneyd, Henry Francis	6–9–1861	... ...	... ...	... ...	... ...	... ...	Retired, 26-4-1864.
Michel, Sir John ...	... ...	... ...	... ...	... ...	... ...	19–8–1862	Maj.-Gen., 26-10-1858; Lt.-Gen., 25-6-1866; Gen., 28-3-1874. (See Royal Irish Rifles List.)
Jardine, Manfred Leslie Palmes	8–7–1862	10–11–1865	... ...	... ...	... ...	... ...	To 67th Regt., 22-6-1867.
McGrigor, Rob. Lewis G.	... ...	... ...	17–11–1863	... ...	... ...	... ...	From 92nd Regiment. — Retired, 20-5-1864.
Blake, John Joseph ...	... ...	28–4–1863	... ...	... ..	... ...	... ...	From 94th Regt.—Retired, 12-5-1875.
Travers, Sir Guy Francis Clarke, Bart.	30–1–1863	30–3–1866	30–3–1870	... ...	... ...	... ...	See Royal Irish Rifles List.
Brooke, Richard O'Shaughnessy	17–3–1863	1–4–1866	21–6–1871	... ...	... ...	... ...	To 12th Regt., 12-8-1876.
Griffith, John Griffith Wynn	5–5–1863	... ...	... ...	... ...	... ...	... ...	From 19th Regiment. — Retired, 11-10-1864.
Butt, Thos. Bromhead	... ...	... ...	... ...	... ...	13–9–1864	... ...	From 79th Regt.—Col., 23-5-1864. To h.p., 3-8-1866.
Pike, Francis	... ...	... ...	22–3–1864	... ...	... ...	... ...	From 91st Regt.—Retired, 24-3-1869.
Sidsley, Sydenham Lott	... ...	18–10–1864	... ...	... ...	... ...	... ...	From 52nd Regiment. — Retired, 24-1-1865.
Stuart, Hartwell Harvey	26–4–1864	1–5–1866	7–8–1878	... ...	... ...	... ...	See Royal Irish Rifles List.
Rhodes, Henry Brooke	20–5–1864	1–5–1866	... ...	... ...	... ...	... ...	To 68th Regt., 25-12-1867.
Cooper, George	21–5–1864	... ...	... ...	... ...	... ...	... ...	Succeeded by Ensign Loraine, 30-9-1868.

Stewart, James Douglas	11–10–1864	1–2–1867	31–8–1878	... ...	... ...	... ...	See Royal Irish Rifles List.
Awdry, James Arnold	20–12–1864	6–11–1867	... ...	... ...	... ...	... ...	Retired, 27-6-1874.
Brockman, William Law	... ...	10–11–1865	23–5–1877	... ...	... ...	... ...	From 4th W.I. Regiment.—Adjt., 8-6-1876—10-7-1877. To A.P.D., 1-4-1878.
Graham, Ferguson ...	24–1–1865	1–2–1868	17–8–1879	... ...	... ...	... ...	See Royal Irish Rifles List.
Moore, Regnier Campbell	22–9–1865	15–7–1868	... ...	... ...	... ...	... ...	Retired, 21-6-1871.
Wood, Edward Collins	10–11–1865	... ...	... ...	... ...	... ...	... ...	From 60th Regt.—Retired, 21-8-1866.
Campbell, Colin	... ...	... ...	2–10–1866	... ...	... ...	... ...	From 48th Regiment.—To 7th D.G., 6-5-1868.
Seed, William Henry...	10–4–1866	8–8–1868	... ...	... ...	... ...	... ...	Retired, 17-8-1870.
Bertram, Francis Godfray	1–5–1866	24–3–1869	... ...	... ...	... ...	... ...	Retired, 3-5-1876.
Collingwood, Arthur Burdett	2–5–1866	9–10–1869	... ...	... ...	... ...	... ...	Retired, 29-8-1874.
Jackson, Wm. Hy. Hardwicke	21–8–1866	30–3–1870	13–3–1880	... ...	... ...	... ...	See Royal Irish Rifles List.
Denning, Hamilton Harvey Pitt	16–11–1866	28–10–1871	... ...	... ...	... ...	... ...	Retired, 25-11-1874.
Lowe, Edward W. D. ...	... ...	... ...	... ...	... ...	1–2–1867	... ...	From 6th Regt.—Col., 2-6-1863.—To h.p., 6-3-1872.
Spence, John	... ...	22–6–1867	... ...	... ...	... ...	... ...	From 67th Regt.—Adjt., 3-4-1869—1-12-1875—Capt., 102nd Regt., 12-4-1879.
Howard, Alfred Gordon	... ...	25–12–1867	... ...	... ...	... ...	... ...	From 68th Regt.—Retired, 15-7-1868.
Goldsworthy, Josiah Webbe	1–2–1867	17–7–1870	... ...	... ...	... ...	... ...	Retired, 9-9-1874.
Williams, Henry Plantaganet	6–11–1867	21–6–1871	... ...	... ...	... ...	... ...	Retired, 19-9-1874.
Thomson, William ...	... ...	... ...	... ...	... ...	... ...	... ...	Qr.-Mr., 23-3-1867—6-11-1868.—Paymaster, 102nd Regt., 7-11-1868.
Crofton, Fred. Rob. Cameron	... ...	... ...	6–5–1868	... ...	... ...	... ...	From 7th D.G. (See Royal Irish Rifles List.)
Pattison, Arthur James	1–2–1868	27–10–1871	... ...	... ...	... ...	... ...	Retired, 15-4-1874.
Brookes, Henry	15–7–1868	28–10–1871	... ...	... ...	... ...	... ...	Retired, 2-3-1878.
Knox, Robert John ...	8–8–1868	28–10–1871	12–3–1881	... ...	... ...	... ...	See Royal Irish Rifles List.
Loraine, Chas. Harcourt	30–9–1868	28–10–1871	... ...	... ...	... ...	... ...	Retired, 22-3-1876.
Morgan, William George	... ...	... ...	... ...	... ...	... ...	... ...	Qr.-Mr., 7-11-1868.—To 3rd Regt., 14-6-1876.

NAME.	Ensign or 2nd Lieut.	Lieutenant.	Captain.	Major.	Lieut.-Colonel.	Colonel.	REMARKS.
Flower, Stephen	... ...	... ...	25–9–1869	... ...	... ...	... ...	From 13th Regt. (See Royal Irish Rifles List.)
Selby-Smyth, Edward G.	7–5–1870	28–10–1871	28–4–1881	... ...	... ...	... ...	See Royal Irish Rifles List.
Haggard, Charles ...	27–3–1872	27–3–1872	... ...	... ...	... ...	... ...	Adjt., 11-7-1877. (See Royal Irish Rifles List.)
Seton, Henry James ...	... ...	2–12–1874	... ...	... ...	... ...	... ...	From Militia. (See Royal Irish Rifles List.)
Gaussen, John Samuel	11–2–1875	11–2–1875	... ...	... ...	... ...	... ...	See Royal Irish Rifles List.
Tobin, Frederick John	12–5–1875	12–5–1877	... ...	... ...	... ...	... ...	See Royal Irish Rifles List.
Ware, Charles	... ...	... ...	8–3–1876	... ...	... ...	... ...	From h.p.—Retired, 13-3-1880.
Cutbill, Hy. Duppa Alfred	... ...	... ...	12–8–1876	... ...	... ...	... ...	From 3rd Regt. (See Royal Irish Rifles List.)
Swaine, Arthur Thomas	17–5–1876	17–5–1876	... ...	... ...	... ...	... ...	See Royal Irish Rifles List.
Cleary, Timothy	... ...	... ...	... ...	... ...	... ...	... ...	From 3rd Regt.—Qr.-Mr., 14-6-1876. (See Royal Irish Rifles List.)
Coke, Charles Henry ...	31–1–1877	... ...	... ...	... ...	... ...	... ...	To 105th Mad. L.I., 2-3-1878.
Stewart, Hugh Houghton	15–8–1877	11–10–1877	... ...	... ...	... ...	... ...	From 15th Regt. (See Royal Irish Rifles List.)
Armstrong, Oliver Carleton	11–5–1878	23–4–1879	... ...	... ...	... ...	... ...	See Royal Irish Rifles List.
Mulchinock, Michael Edward	12–6–1878	17–8–1879	... ...	... ...	... ...	... ...	See Royal Irish Rifles List.
Scott-Plummer, C. H....	... ...	7–6–1879	... ...	... ...	... ...	... ...	See Royal Irish Rifles List.
Allen, Edward	13–8–1879	26–4–1880	... ...	... ...	... ...	... ...	See Royal Irish Rifles List.
Buckle, Robt. Alleyne Stewart	13–8–1879	2–10–1880	... ...	... ...	... ...	... ...	See Royal Irish Rifles List.
Welman, Herbert Loftus	17–12–1879	12–3–1881	... ...	... ...	... ...	... ...	From 99th Regt. (See Royal Irish Rifles List.)
Welman, Harvey ...	14–1–1880	28–4–1881	... ...	... ...	... ...	... ...	See Royal Irish Rifles List.
O'Leary, Wm. Evelyn	22–1–1881	... ...	... ...	... ...	... ...	... ...	See Royal Irish Rifles List.
Hallum, Octavius Claude John	22–1–1881	... ...	... ...	... ...	... ...	... ...	See Royal Irish Rifles List.
Formby, Reginald Fred. Robt.	19–2–1881	... ...	... ...	... ...	... ...	... ...	See Royal Irish Rifles List.
Heinemann, Emil Dabney	19–2–1881	... ...	... ...	... ...	... ...	... ...	See Royal Irish Rifles List.
McWhinnie, Wm. John	19–2–1881	... ...	... ...	... ...	... ...	... ...	See Royal Irish Rifles List.
Burke, Wm. Hy. Marsh	23–4–1881	... ...	... ...	... ...	... ...	... ...	To 51st Regt., 29-6-1881

APPENDIX I.

ROLL OF OFFICERS OF THE ROYAL IRISH RIFLES.

Name.	Date of First Commission.	Lieutenant.	Captain.	Major.	Lieut.-Colonel.	Colonel.	Remarks.
Michel, Rt. Hon. Sir John	... ...	... ...	... ...	... ...	... ...	19–8–1862	Gen., 28-3-1874; F.-M., 27-3-1885. Died, 23-5-1886.
Brown, William Gustavus	... ...	... ...	... ...	... ...	... ...	29–5–1873	Gen., 1-10-1877. Died, 27-11-1883.
Adams, George Hewish	... ...	... ...	... ...	... ...	23–5–1877	... ...	To r.f.p., with hon. rank of Maj.-Gen., 19-11-1881.
Meurant, Edward ...	... ...	... ...	... ...	... ...	5–10–1879	... ...	Brev.-Col., 1-7-1881.—To Regtl. Dist., 21-7-1888.
Mackenzie, James Kenneth Douglas	... ...	... ...	... ...	... ...	1–7–1881	... ...	Retired with hon. rank of Colonel, 26-4-1882.
Gore, Charles Clitherow	... ...	... ...	... ...	... ...	1–7–1881	... ...	Retired with hon. rank of Colonel, 10-1-1883.
Gray, William Ker ...	... ...	... ...	... ...	31–8–1878	... ...	... ...	Retired with hon. rank of Lt.-Col., 1-7-1881.
Karslake, Frederick ...	... ...	... ...	... ...	5–10–1879	26–4–1882	... ...	Retired with hon. rank of Maj.-Gen., 12-2-1887.
Crofton, Fred. Rob. Cameron	... ...	... ...	... ...	1–7–1881	19–11–1881	... ...	To h.p., 26-4-1886.—Retired with hon. rank of Maj.-Gen., 21-7-1886.
Harris, Edward James...	... ...	... ...	... ...	1–7–1881	... ...	... ...	Retired with hon. rank of Lt.-Col., 26-7-1882.
De Montmorency, Raymond Oliver	... ...	... ...	... ...	1–7–1881	10–1–1883	... ...	Brev.-Col., 10-1-1887. — To h.p., 10-1-1889. To r.p., 5-6-1889.
Wright, Frederick Augustus	... ...	... ...	... ...	1–7–1881	... ...	... ...	Retired with hon. rank of Lt.-Col., 18-1-1882.
Flower, Stephen	... ...	... ...	... ...	1–7–1881	5–10–1884	... ...	Brev.-Col., 5-10-1888. — To r.p., 5-10-1888.
Chatfield, Hy. Rowland Spencer	... ...	... ...	... ...	1–7–1881	26–4–1886	... ...	Retired with hon. rank of Colonel, 28-3-1887.
Clarke-Travers, Sir Guy Francis, Bart.	... ...	... ...	... ...	1–7–1881	1–7–1888	... ...	To h.p., 1-7-1888.—To r.p., 1-8-1888.
Butler, Edward Arthur	... ...	... ...	... ...	1–7–1881	... ...	... ...	Retired with hon. rank of Lt.-Col., 18-4-1885.
Wyndham, Charles John	... ...	... ...	... ...	1–7–1881	10–1–1889	... ...	To h.p., 10-1-1893. — To r.p., 22-3-1893.
Metcalfe, Booth Hay ...	... ...	... ...	30–11–1870	18–1–1882	... ...	... ...	Retired with hon. rank of Lt.-Col., 31-12-1887.
Thorburn, Charles James	... ...	... ...	31–10–1871	... ...	... ...	... ...	Retired with hon. rank of Lt.-Col., 5-11-1884.
Cutbill, Hy. Duppa Alfred	... ...	... ...	12–8–1876	19–11–1881	28–12–1889	... ...	Brev.-Col., 26-7-1894. — To h.p., 4-6-1894. To r.p., 10-7-1895.
Cooke-Collis, William	... ...	... ...	21–4–1877	18–4–1885	18–4–1892	... ...	To h.p., 18-4-1892. — To r.p., 4-5-1892.
Bond, Henry Coote ...	... ...	... ...	9–4–1878	... ...	... ...	... ...	Died, 27-3-1882.
Stuart, Hartwell Harvey	... ...	... ...	7–8–1878	26–4–1882	2–7–1890	... ...	To h.p., 2-7-1890. — To r.p., 25-11-1891.

NAME.	Date of First Commission.	Lieutenant.	Captain.	Major.	Lieut.-Colonel.	Colonel.	REMARKS.
Davenport, Charles Talbot	... ...	... ...	7–8–1878	... ...	... ...	... ...	To A.P.D., 26-8-1881.
Stewart, James Douglas	... ...	... ...	31–8–1878	26–7–1882	... ...	... ..	Retired with hon. rank of Lt.-Col., 9-12-1885.
Stokes, Folliott Stuart Furneaux	... ...	... ...	19–10–1878	10–10–1885	10–1–1893	... ..	To h.p., 2-9-1895, on appointment to Staff.
Graham, Fergus	... ...	... ...	17–8–1879	9–12–1885	... ...	... ..	Retired, 22-2-1888.
Jackson, Wm. Hy. Hardwicke	... ...	... ...	13–3–1880	14–11–1884	... ...	... ..	Retired with hon. rank of Lt.-Col., 6-10-1886.
Meynell, John Jebb ...	... ...	... ...	26–5–1880	1–7–1888	... ...	... ..	Retired, 2-3-1893.
Knox, Robert John ...	... ...	... ...	12–3–1881	26–4–1886	4–6–1894	.. ...	Retired, 21-10-1896.
Anson, John Hy. Wm.	... ...	... ...	12–3–1881	... ...	... ...	.. ...	Superseded for absence without leave, 19-7-1882.
Selby-Smyth, Edward ... Guy	... ...	... ...	28–4–1881	10–1–1889	2–9–1895	... ...	Adjt., 4-7-1883—3-7-1888.—Retired, 25-11-1896.
Haggard, Charles ...	... ...	27–3–1872	19–11–1881	3–8–1889	28–10–1896	... ...	Adjt., 11-7-1877—31-12-1882; Brev.-Col., 28-10-1900. — To h.p., 28-10-1900. To r.p., 20-2-1901.
Eagar, Henry Averell	... ...	2–12–1874	18–1–1882	28–12–1889	25–11–1896	... ...	Died, 13-2-1900, of wounds received at Stormberg on Dec. 10, 1899.
Seton, Henry James ...	... ...	2–12–1874	26–4–1882	2–7–1890	... ...	... ...	To h.p., 5-10-1901. — To r.p., 27-8-1902.
Marling, Walter Bentley	... ...	11–2–1875	26–4–1882	... ...	... ...	... ...	Adjt., 29-12-1878—16-12-1882.—To 4th Bn., Glos. Regt., 14-3-1885.
Gaussen, John Samuel	... ...	11–2–1875	15–5–1882	... ...	... ...	... ...	To A.P.D., 10-7-1883.
Walter, Edwd. Chas. Lethbridge	... ...	15–4–1875	14–7–1882	... ...	... ...	... ...	Killed driving tandem, 4-5-1885.
Swaine, Arthur Thomas	... ...	17–3–1876	19–7–1882	18–4–1892	28–10–1900	... ...	Brev.-Col., 10-2-1904. — To h.p., 28-10-1904.
Graves, Robert Stannus	... ...	31–1–1877 29–11–1876	17–3–1883	... ...	... ...	... ...	To A.P.D., 9-1-1885.
Tobin, Frederick John	... ...	12–5–1877	1–4–1883	10–1–1893	21–7–1904	... ...	Brev.-Col., 21-7-1907. — Retired, 10-8-1907.
Stewart, Hugh Houghton	... ...	11–10–1877	10–10–1883	... ...	... ...	... ...	Retired, 25-11-1891.
Adye, Walter	... ...	19–10–1878	14–11–1884	2–3–1893	6–2–1904	... ...	Subsequent service in Staff Appointments.
Colborne, Hon. Francis Lionel Lydston	... ...	30–10–1878	15–4–1885	... ...	... ...	... ...	Brev.-Maj., 15-6-1885. Retired, 19-6-1889.
Thompson, Wm. David	... ...	5–3–1879	... ...	... ...	... ...	... ...	To Ben. S.C., 13-9-1880.
Brown, John Southwell	... ...	12–4–1879	15–4–1885	4–6–1894	28–10–1904	... ...	Adjt., 16-12-1882—3-2-1886; Brev.-Col., 28-10-1907. — Retired, 28-10-1908.

APPENDIX I.

Armstrong, Oliver Carleton	23-4-1879	... ...	... ...	... ...	... ...	To Ben. S.C., 29-1-1883.
Scott-Plummer, Charles Henry	7-6-1879	... ...	... ...	... ...	... ...	Resigned, 14-3-1883.
Mulchinock, Michael Edward	17-8-1879	5-5-1885	... ...	... ...	... ...	Resigned, 17-8-1887.
Trant, John Francis	13-3-1880	... ...	... ...	... ...	... ...	To 2nd W. India Regt., 2-8-1882.
Allen, Edward	26-4-1880	19-8-1885	2-9-1895	... ...	... ...	Brev.-Lt.-Col., 29-11-1900.—Retired, 17-9-1902.
Burrows, George Vernon	9-7-1880	... ...	... ...	... ...	... ...	To Mad. S.C., 29-3-1882.
Harris, Charles Gerald	13-9-1880	30-9-1885	... ...	... ...	... ...	To 6th Bn. R.I. Rifles, 10-10-1888.
Monteith, Robt. Wm. Freebairn	29-9-1880	7-9-1886	... ...	... ...	... ...	To A.S.C., 1-4-1889.
Buckle, Robt. Alleyne Stewart	2-10-1880	7-9-1886	28-10-1896	... ...	... ...	Retired, 2-4-1898.
Welman, Herbert Loftus	12-3-1881	24-1-1887	25-11-1896	... ...	... ...	To h.p., 30-11-1901. — To r.p., 15-2-1902.
Welman, Harvey	28-4-1881	... ...	... ...	... ...	... ...	To Bo. S.C., 16-10-1882.
Barnett, Robert Percy Sanders	1-7-1881	... ...	... ...	... ...	... ...	To Bo. S.C., 16-3-1882.
Browne, William	1-7-1881	... ...	... ...	... ...	... ...	To Mad. S.C., 26-4-1882.
O'Leary, Wm. Evelyn	1-7-1881	28-5-1887	30-8-1899	28-10-1908	... ...	Bt.-Col., 6-6-1912.—H.p., 28-10-1912.
Hallum, Octavius Claude John	1-7-1881	4-8-1887	... ...	... ...	... ...	Retired 25-1-1899.
Formby, Reginald Fred. Robert	1-7-1881	... ...	... ...	... ...	... ...	To Ind. S.C., 11-12-1882.
Wilkinson, Lionel Thomson Verner	1-7-1881	17-8-1887	... ...	... ...	... ...	Adjt., 26-5-1890—31-12-1891. — Retired, 20-2-1895.
Heinemann, Emil Dabney	1-7-1881	... ...	... ...	... ...	... ...	Resigned, 23-8-1884.
McWhinnie, Wm. John	1-7-1881	25-10-1887	28-10-1900	... ...	... ...	Retired, 22-9-1906.
Tighe, Michael Augustus	1-7-1881	... ...	... ...	... ...	... ...	To Ind. S.C., 16-10-1882.
Spencer, Archd. Claude Douglas	1-7-1881	31-12-1887	5-10-1901	... ...	... ...	Retired, 9-5-1903.
McQuade, Michael	... ...	... ...	... ...	... ...	... ...	Qr.-Mr., 22-1-1868. Died, 24-2-1886.
Cleary, Timothy	... ...	... ...	... ...	... ...	... ...	Qr.-Mr. (18-3-1874), 14-6-1876 — 16-3-1886. — Retired with hon. rank of Major, 17-3-1886.
Jones, Henry	... ...	... ...	... ...	... ...	... ...	Qr.-Mr. (29-3-1873), 1-4-1882 — 18-3-1884.—Retired with hon. rank of Major, 19-3-1884.

NAME.	Date of First Commission.	Lieutenant.	Captain.	Major.	Lieut.-Colonel.	Colonel.	REMARKS.
Raymond, Hy. Warner	22–10–1881	22–10–1881	1–2–1888	... ...	... ...	... ...	Resigned his commission, 23-12-1891.
Battersby, Hy. Francis	10–5–1882	10–5–1882	... ...	... ...	... ...	... ...	Resigned, 20-3-1883.
Edwards-Hodges, Thomas	9–9–1882	9–9–1882	6–6–1888	... ...	... ...	... ...	To A.P.D., 1-7-1893.
Dunlop, William Hugh	19–2–1879	2–8–1882 9–6–1880	28–5–1887	2–4–1898	... ...	... ...	From 2nd W. India Regt. — To A.S.C., 19-2-1885 — 31-10-1891.— Rejoined, 1-11-1891. — Retired, 30-8-1899.
O'Callaghan-Westropp, George	9–9–1882	9–9–1882	19–9–1888	... ...	... ...	... ...	Retired, 20-2-1889, and joined R.G.A. Mil.
Anderson, Alexander ...	20–12–1882	20–12–1882	... ...	... ...	... ...	... ...	Resigned, 20-3-1883.
Rudyerd, Reginald Burroughs	... ...	... ...	31–10–1883 1–1–1881	... ...	... ...	... ...	From Dorset Regt. — Retired, 21-11-1888.
Hay, Hon. Archd. Fitzroy George	7–2–1883	7–2–1883	... ...	... ...	... ...	... ...	From h.p.—Resigned, 13-3-1883.
Fisher, John	10–3–1883	10–3–1883	... ...	... ...	... ...	... ...	To Norfolk Regt., 7-11-1883.
Cliff, Harold Martin ...	10–3–1883	10–3–1883	21–11–1888	17–9–1902	... ...	... ...	To h.p., 4-9-1889. — To f.p., 19-2-1890.—Retired, 10-12-1902.
Stewart, James Fearnley	10–3–1883	10–3–1883	... ...	... ...	... ...	... ...	To Scot. R., 14-4-1883.
Bruce, Arthur Francis	10–3–1883	10–3–1883	... ...	... ...	... ...	... ...	To Ind. S.C., 1-7-1885.
Carstairs, Albert Joseph Henry	10–3–1883	10–3–1883	19–6–1889	... ...	... ...	... ...	Resigned, 11-5-1892.
Morphy, Henry John...	12–5–1883	12–5–1883	3–8–1889	... ...	... ...	... ...	Retired, 11-10-1902.
Orpen, Charles Henry	12–5–1883	12–5–1883	... ...	... ...	... ...	... ...	Resigned, 29-6-1887.
Beresford, Kennedy ...	12–3–1883	30–5–1883	4–9–1889	10–12–1902	... ...	... ...	From Norfolk Regt. — Retired, 2-11-1911.
Bell, Frederick John Hamilton	25–8–1883	25–8–1883	4–9–1889	9–5–1903	2–11–1911	... ...	Adjt., 26-5-1886—25-5-1890.
Harvey, Chas. Edward Ramsey	25–8–1883	25–8–1883	4–9–1889	6–2–1904	... ...	... ...	Adjt., 4-7-1888—18-1-1892. — Died, 4-9-1912.
Duffy, Lawrence	... ...	... ...	... ...	... ...	... ...	... ...	Qr.-Mr., 20-10-1883—4-11-1884.
Bradshaw, Fredk. Ewart	6–2–1884	6–2–1884	... ...	... ...	... ...	... ...	To Ind. S.C., 19-10-1886.
Lillingston-Johnson, William George	6–2–1884	6–2–1884	4–6–1894	... ...	... ...	... ...	To h.p., 8-7-1891. — Rejoined, 30-5-1894. To h.p., 12-7-1896.
Massey, Godfrey Warburton	6–2–1884	6–2–1884	2–3–1893	... ...	... ...	... ...	Retired, 6-6-1903.
Hume, Arthur Carmichael	6–2–1884	6–2–1884	... ...	... ...	... ...	... ...	Died, 25-4-1888.

Fox-Strangways, Theodore Stephen	14–5–1884	14–5–1884	1–7–1893	21–7–1904	... ...	... ...	Retired, 26-8-1905.
Swabey, Herbert Stephen	23–8–1884	23–8–1884	... ...	... ...	... ...	... ...	Resigned, 2-5-1888.
Naughton, Michael ...	... ...	... ...	... ...	... ...	... ...	... ...	Qr.-Mr., 1-3-1884 — 2-8-1892.—Retired, 3-8-1892, with hon. rank of Lieutenant.
Lane, Thomas Henry ...	... ...	... ...	... ...	... ...	... ...	... ...	Qr.-Mr., 23-4-1884—5-6-1894.—Retired with hon. rank of Captain, 6-6-1894.
Thorpe, Patrick Joseph	... ...	... ...	... ...	... ...	... ...	... ...	Qr.-Mr., 5-11-1884. — Cashiered, 13-12-1893.
Bradford, Wilmot Hy.	... ...	... ...	... ...	... ...	... ...	24–5–1886	General, 1-7-1881.
Rowley, Robert Abercromby Dick	29–8–1885	29–8–1885	1–7–1893	... ...	... ...	... ...	Died, 17-11-1898.
Cole, Hy. Walter Geo.	29–8–1885	29–8–1885	... ...	... ...	... ...	... ...	To Northd. Fus., 16-12-1885.
Laurie, Geo. Brenton ...	2–9–1885	2–9–1885	20–11–1893	28–10–1904	28–10–1912	... ...	
Gray, John Robin ...	25–11–1885	25–11–1885	... ...	... ...	... ...	... ...	To K.R.R.C., 30-1-1886.
Homfray, Herbert Richard	25–11–1885	25–11–1885	... ...	... ...	... ...	... ...	To 1st L.G., 8-12-1888.
O'Brien, Hon. Lucius William	16–12–1885	16–12–1885	... ...	... ...	... ...	... ...	To Rifle Bde., 31-3-1886.
Turnbull, Hy. Francis	... ...	... ...	... ...	13–11–1886 3–11–1885	... ...	... ...	From h.p.—Drowned in Nile near Toski, 2-8-1889. To Mad. S.C., 21-7-1887.
Lowry, James Herbert	30–1–1886	30–1–1886	... ...	... ...	... ...	... ...	
Hanbury, Bernard Kingscote	30–1–1886	30–1–1886	... ...	... ...	... ...	... ...	To R.W. Fus., 26-5-1886.
Smith, Vernon Graham	30–1–1886	30–1–1886	... ...	... ...	... ...	... ...	To K.R.R.C., 5-5-1886.
Evans, Geo. Wm. Wallace D'Arcy	21–7–1886	21–7–1886	13–4–1894	... ...	... ...	... ...	To 20th Hussars, 10-7-1895.
Addison, Arthur Joseph Berkeley	25–8–1886	25–8–1886	20–2–1895	26–8–1905	... ...	... ...	Retired, 29-8-1906.
Burke, Ashworth Peter Mary	25–8–1886	25–8–1886	... ...	... ...	... ...	... ...	Resigned, 22-3-1893.
Pike, Thomas	... ...	... ...	... ...	... ...	... ...	... ...	Qr.-Mr., 7-4-1886. — Retired with hon. rank of Major, 18-9-1903.
McClenahan, Samuel ...	... ...	... ...	... ...	... ...	... ...	... ...	Qr.-Mr., 7-4-1886—14-4-1896.—Retired with hon. rank of Captain, 15-4-1896.
Burnett, Charles John...	... ...	... ...	... ...	... ...	7–5–1887 6–9–1885	... ...	From E. Yorks. Regt.—Appointed to Staff, 23-5-1890.
Despard, Hy. Francis Rushworth	5–2–1887	18–7–1888	2–9–1895	... ...	... ...	... ...	Retired, 16-8-1902.
Ryan, Robert Fenwick	4–5–1887	18–7–1888	2–3–1896	... ...	... ...	... ...	Resigned, 24-8-1898.

NAME.	Date of First Commission.	Lieutenant.	Captain.	Major.	Lieut.-Colonel.	Colonel.	REMARKS.
Blunt, Walter Ernest Onslow Campbell	14–9–1887	21–11–1888	6–10–1896	... ...	... ...	... ...	To A.P.D., 29-4-1898.
Wynch, Alexander Balmair	14–9–1887	... ...	... ...	... ...	... ...	... ...	Died, 21-6-1888.
Carson, Thomas	16–11–1887	21–11–1888	21–1–1897	... ...	... ...	... ...	Retired, 17-1-1906.
Hall, Arthur Robinson Kay	16–11–1887	16–3–1889	1–2–1897	... ...	... ...	... ...	To h.p., 11-5-1905.
Weir, Arthur Vavasour	16–11–1887	3–7–1889	3–2–1897	21–7–1906	... ...	... ...	To Ind. S.C., 18-9-1889.—Returned to Regt., 18-4-1891.
Ponsonby, John ...	16–11–1887	... ...	... ...	... ...	... ...	... ...	To Coldstream Gds., 8-12-1888.
Brown, Howard Benjamin	16–11–1887	3–7–1889	... ...	... ...	... ...	... ...	Resigned, 23-7-1890.
Curzon, FitzRoy Edmund Penn	... ...	... ...	14–11–1888	30–11–1901	... ...	... ...	From Scot. R.—Adjt., 1-1-1892—31-12-1896.—Lieut.-Colonel, h.p., 23-3-1907.
Festing, Arthur Hoskyns	11–2–1888	3–7–1889	15–1–1898	... ...	... ...	... ...	Brev.-Major, 16-1-1898. — Retired, 6–9–1905.
Atkins, William	4–7–1888	3–8–1889	... ...	... ...	... ...	... ...	Capt., Wilts. Regt., 18-7-1896.
Noblett, Louis Hemington	22–8–1888	1–10–1889	15–1–1898	23–3–1907 29–11–1900	... ...	... ...	
Playfair, Alan	22–8–1888	4–12–1889	... ...	... ...	... ...	... ...	To Ind. S.C., 2-1-1890.
Biddulph, Harold Mavromichali	10–11–1888	... ...	... ...	... ...	... ...	... ...	To Rifle Bde., 6-2-1889.
Cary, George Stanley ...	8–12–1888	18–12–1889	... ...	... ...	... ...	... ...	Died, 4-4-1896.
Fox, Brabazon Hubert Maine	8–12–1888	2–1–1890	25–3–1898	29–8–1906	... ...	... ...	Retired, 30-12-1908.
Jameson, William ...	8–12–1888	23–7–1890	... ...	... ...	... ...	... ...	Died, 26-11-1897.
Kelly, Vincent Joseph...	8–12–1888	18–7–1891	2–4–1898	... ...	... ...	... ...	Retired, 12-12-1906.
Westropp, Mich. Seymour Dudley	8–12–1888	2–3–1893	... ...	... ...	... ...	... ...	Resigned, 8-6-1898.
Carter, Godfrey Lambert	8–12–1888	... ...	... ...	... ...	... ...	... ...	To Ind. S.C., 16-9-1890.
Christie, Edgar Jessopp	23–3–1889	2–3–1893	22–2–1900	... ...	... ...	... ...	Adjt., 3-2-1892—2-2-1896.—Retired, 6-6-1903.
Clery, Carleton Buckley Laming	24–4–1889	... ...	... ...	... ...	... ...	... ...	To Ind. S.C., 18-2-1890.
Harvey, Harry Charles	21–9–1889	22–3–1893	... ...	... ...	... ...	... ...	Retired on appointment to Militia, 10-8-1898.
Dimsdale Wilfred Philip	9–11–1889	1–7–1893	22–4–1898	... ...	... ...	... ...	Died of wounds received near Reddersburg, 9-4-1900.

APPENDIX I.

Eckford, Philip Geo. Wright	9–11–1889	20–11–1893	22–2–1900	... ...	... ...	... ...	To Ind. S.C., 12-3-1891.—Rejoined, 2-3-1893.—Retired, 13-3-1909.
Peck, Eustace Marriott	21–12–1889	4–6–1894	... ...	... ...	... ...	... ...	Accidentally killed by fall from bicycle, 16-3-1896.
Baker, Osbert Clinton	1–3–1890	18–11–1894	24–2–1900	22–9–1906	... ...	... ...	Adjt., 1-1-1896—31-12-1899.
King-Harman, Wentworth Alex.	29–3–1890	20–2–1895	... ...	... ...	... ...	... ...	Retired on appointment to Militia, 2-4-1898.
Wilmot-Sitwell, Degge	28–6–1890	27–7–1895	24–2–1900	... ...	... ...	... ...	Adjt., 3-2-1896—24-1-1902; Brev.-Major, 12-10-1901. — Retired, 2-2-1907.
Bradford, Edward Chaloner	29–10–1890	2–9–1895	24–2–1900	30–12–1908	... ...	... ...	Adjt., 25-1-1902—24-1-1905.
Brenan, Hugh Gerald...	17–1–1891	5–1–1896	26–2–1900	... ...	... ...	... ...	To h.p., 8-5-1902.
Knox, Alfred William Fortescue	2–5–1891	... ...	... ...	... ...	... ...	... ...	To Ind. S.C., 2-10-1893.
Kennedy, Herbert Alexander	25–7–1891	2–3–1896	10–4–1900	... ...	... ...	... ...	Retired, 29-12-1906.
Wright, Henry Coram	12–3–1892	17–3–1896	30–11–1901	2–11–1911	... ...	... ...	
Dwyer, John	... ...	... ...	... ...	... ...	... ...	... ...	Qr.-Mr., 23-1-1892.—Retired with hon. rank of Capt., 17-9-1902.
Foley, John	... ...	... ...	... ...	... ...	... ...	... ...	Qr.-Mr., 17-8-1892.—Retired with hon. rank of Capt., 3-5-1907.
Molloy, Gerald Macleay	19–7–1893	18–7–1896	... ...	... ...	... ...	... ...	To Ind. S.C.. 28-10-1897.
Mitchell, James Rankin	9–9–1893	... ...	... ...	... ...	... ...	... ...	To K.R.R.C., 23-12-1893.
Spedding, Charles Rodney	23–12–1893	6–10–1896	1–3–1902	2–11–1911	... ...	... ...	
Will, Duncan Almed Elmsly	7–3–1894	21–1–1897	... ...	... ...	... ...	... ...	To Ind. S.C., 27-4-1897.
Daunt, Richd. Algernon Craigie	7–3–1894	1–2–1897	11–10–1902	5–9–1912	... ...	... ...	
Alston, James Wm. ...	10–10–1894	3–2–1897	17–10–1902	28–10–1912	... ...	... ...	
Cunningham, Joseph ...	... ...	... ...	... ...	... ...	... ...	... ...	Qr.-Mr., 28-2-1894.—Died, 1-8-1898.
Pecknold, Abraham ...	... ...	... ...	... ...	... ...	... ...	... ...	Qr.-Mr., 13-6-1894.—Retired with hon. rank of Major, 24-3-1910.
Carew, Ponsonby May Lynn	... ...	... ...	10–7–1895	... ...	... ...	... ...	From 20th Hussars.—Retired on appointment to Militia, 3-2-1897.
Low, Henry Lawrence	23–1–1895	27–4–1897	... ...	... ...	... ...	... ...	Killed in action, 10-3-1902.
Wilding, Henry	29–5–1895	28–10–1897	... ...	... ...	... ...	... ...	Resigned, 14-9-1898.
Charley, Harold Richard	28–9–1895	13–11–1897	10–12–1902	... ...	... ...	... ...	Adjt., 1-1-1904—31-12-1906.
Sprague, Louis Charles	7–12–1895	27–11–1897	10–12–1902	... ...	... ...	... ...	

NAME.	Date of First Commission.	Lieutenant.	Captain.	Major.	Lieut.-Colonel.	Colonel.	REMARKS.
Macnamara, Charles Carrol	15–4–1896	15–1–1898	10–12–1902	... ...	... ...	... ...	Adjt., 1-1-1900—31-12-1903.
White, Charles James...	5–9–1896	... ...	... ...	... ...	... ...	... ...	To Ind. S.C., 31-7-1898.
Stevens, Reginald Walter Morton	5–9–1896	3–8–1898	2–1–1904	... ...	... ...	... ...	Adjt., 25-1-1905—24-1-1908.
Hallett, Richard Lionel Hughes	5–9–1896	3–8–1898	... ...	... ...	... ...	... ...	To Ind. S.C., 3-12-1900.
Flint, Samuel Kirk ...	18–11–1896	3–8–1898	... ...	... ...	... ...	... ...	Capt., R. Fus., 25-8-1900.
Henderson, Alexander...	... ...	... ...	... ...	... ...	... ...	... ...	Qr.-Mr., 15-4-1896; Hon. Major, 15-4-1911.
Murray, John	7–4–1897	3–8–1898	... ...	... ...	... ...	... ...	Died, 25-10-1898.
Forbes, Alexander Graham	15–5–1897	3–8–1898	... ...	... ...	... ...	... ...	Died, 18-12-1900.
Allgood, Bertram ...	15–5–1897	10–8–1898	6–2–1904	... ...	... ...	... ...	
Dashwood, Robt. Hy. Seymour	15–5–1897	14–9–1898	1–5–1904	... ...	... ...	... ...	Retired on appointment to the Militia, 21-12-1907.
Dunn, Ernest George...	8–9–1897	18–1–1899	16–6–1904	... ...	... ...	... ...	
Saunders, Ernest Howie	8–9–1897	8–12–1899	8–9–1906	... ...	... ...	... ...	To Indian Army, 24-11-1908.
Hughes, Adrian Francis Theodore	8–9–1897	24–2–1900	16–7–1904	... ...	... ...	... ...	To h.p., 7-2-1908.
Baker, Robert Geoffrey	16–2–1898	... ...	... ...	... ...	... ...	... ...	To Ind. S.C., 24-1-1900.
Dixon, Charles Stewart	4–5–1898	24–2–1900	18–7–1904	... ...	... ...	... ...	
Rodney, Lennox Geo. Brydges	4–5–1898	24–2–1900	16–6–1906	... ...	... ...	... ...	
Drage, George Thomas	... ...	... ...	... ...	... ...	... ...	... ...	Qr.-Mr. (Prov. Battn., 13-4-1898), 5-10-1898; Hon. Capt., 13-4-1908.
Monro, Eric William Celestin	4–1–1899	24–2–1900	21–7–1906	... ...	... ...	... ...	
Davenport, Talbot Neville Fawcett	4–1–1899	4–3–1900	... ...	... ...	... ...	... ...	Died, 3-3-1905.
Wallace, Chas. Lyndhurst Willoughby	24–6–1899	4–4–1900	... ...	... ...	... ...	... ...	Resigned, 14-6-1905.
Kirkwood, Andrew Samuel	5–7–1899	4–4–1900	5–7–1908	... ...	... ...	... ...	To Indian Army, 2-11-1909.
Bowen-Colthurst, John Colthurst	12–8–1899	20–8–1900	21–12–1907	... ...	... ...	... ...	
Lanyon, Wm. Mortimer	12–8–1899	14–1–1901	7–2–1908	... ...	... ...	... ...	
Biscoe, Arthur John ...	12–8–1899	1–2–1902	28–6–1908	... ...	... ...	... ...	

Maynard, Percy Guy Wolfe	18–10–1899	1–3–1902	28–6–1908	... ...	... ...	... ...	
Soutry, Trevor Lloyd Blunden	18–10–1899	1–3–1902	28–6–1908	... ...	... ...	... ...	
Nelson, Craig	15–11–1899	... ...	... ...	... ...	... ...	... ...	To Ind. S.C., 23-10-1901.
Cinnamond, Chas. Hy. Lalanne	15–11–1899	10–3–1902	28–6–1908	... ...	... ...	... ...	
Wilson, Robt. Callwell	24–1–1900	... ...	... ...	... ...	... ...	... ...	Resigned, 2-10-1901.
Cooke-Collis, Wm. Jas. Norman	24–2–1900	10–12–1902	28–6–1908	... ...	... ...	... ...	
Goodman, Hy. Russell	7–3–1900	28–5–1903	28–6–1908	... ...	... ...	... ...	Adjt., 1-1-1907—31-12-1909.
Bristow, Samuel Follett	4–4–1900	22–3–1904	... ...	... ...	... ...	... ...	Resigned, 29-11-1905.
Eastwood, William ...	18–4–1900	2–1–1904	28–6–1908	... ...	... ...	... ...	
Stewart, Alexander ...	21–4–1900	... ...	... ...	... ...	... ...	... ...	To R. Garr. Regt., 5-11-1902.
Tisdall, Charles Arthur	21–4–1900	... ...	... ...	... ...	... ...	... ...	To Irish Gds., 10-7-1901.
Becher, Cecil Morgan Ley	21–4–1900	22–3–1904	22–1–1909	... ...	... ...	... ...	
Pilkington, Lionel ...	21–4–1900	... ...	... ...	... ...	... ...	... ...	Resigned, 2-11-1901.
McCulloch, William ...	19–5–1900	1–5–1904	... ...	... ...	... ...	... ...	Resigned, 16-10-1907.
Gaussen, Horace Ash ...	23–5–1900	16–6–1904	... ...	... ...	... ...	... ...	Resigned, 7-4-1906.
Barclay, Leslie Geo. de Rune	19–9–1900	16–7–1904	... ...	... ...	... ...	... ...	Resigned, 6-9-1905.
Hawes, Benjamin Reddie	... ...	... ...	... ...	... ...	21–7–1900 17–2–1900	... ...	From R.M. Fus.—Retired, 21-7-1904.
Robinson, Francis ...	5–1–1901	18–7–1904	... ...	... ...	... ...	... ...	Died, 11-4-1906.
Master, Charles Lionel	5–1–1901	4–3–1905	28–6–1908	... ...	... ...	... ...	Adjt., 25-1-1908—24-1-1911.
Corbett, Regld. David de la Cour	8–5–1901	... ...	... ...	... ...	... ...	... ...	To Indian Army, 6-3-1903.
Haskett-Smith, Wm. Jos. Jer. Saccone	8–5–1901	14–6–1905	8–3–1910	... ...	... ...	... ...	
Hogan, Edw. Michael Angelo John	27–7–1901	14–6–1905	8–3–1910	... ...	... ...	... ...	To h.p., 7-10-1904. — Rejoined, 3-12-1904. To Indian Army, 27-7-1910.
Stokes, Folliott Arthur	19–10–1901	17–1–1906	... ...	... ...	... ...	... ...	To Indian Army, 20-3-1904.—Rejoined, 2-10-1906. Resigned, 6-3-1907.
Barton, Francis Hewson	19–10–1901	... ...	... ...	... ...	... ...	... ...	To Indian Army, 18-11-1903.

Name.	Date of First Commission.	Lieutenant.	Captain.	Major.	Lieut.-Colonel.	Colonel.	Remarks.
Giles, Stanley Edmund Hercules	29–1–1902	... ...	... ...	... ...	... ...	... ...	To A.S.C., 1-1-1904.
Forbes, Hon. Bertram Aloysius	29–1–1902	17–1–1906	8–3–1910	... ..	... ...	... ...	
Durant, Hugh Norcott	29–1–1902	7–4–1906	11–3–1910	... ...	... ...	... ...	Adjt., 25-1-1911.
Chatterton, Geo. Alleyn	30–4–1902	12–4–1906	... ...	... ...	... ...	... ...	Retired on appointment to Special Reserve, 2-4-1910.
O'Sullivan, Arthur Moore	23–7–1902	9–7–1906	11–3–1910	... ...	... ...	... ...	Adjt., 1-1-1910—31-12-1912.
Templeton, George ...	... ...	... ...	... ...	... ...	... ...	... ...	Qr.-Mr., 17-9-1902; Hon. Captain, 17-9-1912.
Smyth, Charles Devaynes	28–1–1903	9–7–1906	13–5–1910	... ...	... ...	... ...	Retired, 22-11-1912.
Parsons, Alfred Henry...	28–1–1903	... ...	... ...	... ...	... ...	... ...	To Indian Army, 20-5-1905.
Foster, Harry William	... ...	... ...	... ...	... ...	... ...	... ...	Qr.-Mr., 21-10-1903.—Hon. Lieut.
Smyth, Pierson Florence John	3–12–1904	28–4–1907	... ...	... ...	... ...	... ...	Died, 29-1-1913.
Whelan, John Percy ...	... ...	8–7–1905	8–3–1910	... ...	... ...	... ...	From R. Garr. Regt.
Drought, John Bertie Armstrong	28–1–1905	27–4–1907	... ...	... ...	... ...	... ...	
Kelgan, Peter Daniel ...	10–5–1905	... ...	... ...	... ...	... ...	... ...	Died, 28-5-1905.
Martyr, John Francis ...	20–5–1905	27–4–1907	... ...	... ...	... ...	... ...	
Scott, Gerald Chaplin...	20–5–1905	28–4–1907	... ...	... ...	... ...	... ...	To Indian Army, 6-1-1913.
Kenny, Edward Courtenay	16–8–1905	... ...	... ...	... ...	... ...	... ...	To Indian Army, 24-9-1907.
Ludlow-Hewitt, Edgar Rainey	16–8–1905	24–9–1907	... ...	... ...	... ...	... ...	
Graham, Fergus Reg. Winsford	29–11–1905	5–12–1907	... ...	... ...	... ...	... ...	
Hutcheson, Norman Heber	23–5–1906	21–12–1907	... ...	... ...	... ...	... ...	
Mansergh, Robert Otway	23–5–1906	7–2–1908	... ...	... ...	... ...	... ...	
Umbers, Thomas Ralph	23–5–1906	24–10–1908	... ...	... ...	... ...	... ...	To h.p., 5-1-1910. — Rejoined, 27-1-1912.
Butler, Henry Halpin Cavendish	29–8–1906	... ...	... ...	... ...	... ...	... ...	Resigned, 2-10-1907.
Waller, Edmund de Warrenne	29–8–1906	28–6–1908	... ...	... ...	... ...	... ...	To Indian Army, 22-7-1909.
Galwey, Arthur Wm. ...	7–11–1906	28–6–1908	... ...	... ...	... ...	... ...	

Napier, Henry Edward	... ...	... ...	... ...	... ...	2-11-1907	... ...	From Cheshire Regt. — To h.p., 2-11-1911.
Tee, Charles Clifford ...	... ...	27-4-1907	... ...	... ...	... ...	... ...	From R. Warwick Regt.
Wright, Allan O'Halloran	29-5-1907	22-1-1909	... ...	... ...	... ...	... ...	Adjt., 1-1-1913.
Peebles, Arthur Edmondson	8-7-1907	1-4-1909	... ...	... ...	... ...	... ...	
Varwell, Ralph Peter ...	9-10-1907	1-4-1909	... ...	... ...	... ...	... ...	
Dillon, Stephen Searle	9-10-1907	22-4-1909	... ...	... ...	... ...	... ...	
Olive, Chaloner Cary	2-11-1907	22-7-1909	... ...	... ...	... ...	... ...	Died, 31-7-1909.
Knox, Utred Frederick	20-11-1907	1-8-1909	... ...	... ...	... ...	... ...	
Cornwall, Arthur ...	... ...	... ...	... ...	... ...	... ...	... ...	Qr.-Mr. (Northd. Fus., 7-4-1900), 4-5-1907.—To R. Inniskg. Fus., 3-4-1909.
Merriman, Arthur Drummond Nairne	... ...	... ...	20-5-1908 5-2-1903	... ...	... ...	... ...	From Manchester Regt.
Gifford, Herbert Llewellyn	... ...	... ...	8-8-1908 21-1-1903	... ...	... ...	... ...	From R. Garr. Regt.
Gillespie, Andrew Allan	8-2-1908	... ...	... ...	... ...	... ...	... ...	Resigned, 27-11-1909.
Hutcheson, Robert Barratt	22-2-1908	1-1-1910	... ...	... ...	... ...	... ...	
Thomas, Edmd. Meysey	27-5-1908	8-3-1910	... ...	... ...	... ...	... ...	
Newport, Charles Johnstone	27-5-1908	5-1-1910	... ...	... ...	... ...	... ...	
Browne, Dominic Augustus	19-9-1908	8-3-1910	... ...	... ...	... ...	... ...	
Ffrench, Raymond Patrick Thomas	19-9-1908	11-3-1910	... ...	... ...	... ...	... ...	
Reynolds, Thos. James	... ...	... ...	15-5-1909	... ...	... ...	... ...	From R.I. Fus.
How, William Fitzherbert	3-3-1909	1-4-1910	... ...	... ...	... ...	... ...	
Foley, James Walter ...	22-5-1909	7-6-1910	... ...	... ...	... ...	... ...	
Gartlan, Gerald Ian ...	18-9-1909	7-6-1910	... ...	... ...	... ...	... ...	
Gavin, Robert Fitz Austin	18-9-1909	15-4-1913	... ...	... ...	... ...	... ...	
Cowley, Victor Leopold Spencer	6-11-1909	... ...	... ...	... ...	... ...	... ...	
Dawes, Chas. Reginald Bethel	11-12-1909	... ...	... ...	... ...	... ...	... ...	

Name.	Date of First Commission.	Lieutenant.	Captain.	Major.	Lieut.-Colonel.	Colonel.	Remarks.
Clements, George Stuart	25–12–1909	... ...	... ...	... ...	... ...	... ...	
Teele, William Beaconsfield	9–4–1910	... ...	... ...	... ...	... ...	... ...	
Burges, Wm. Armstrong	20–4–1910	... ...	... ...	... ...	... ...	... ...	
Sullivan, Hy. Rupert Fiennes	20–4–1910	... ...	... ...	... ...	... ...	... ...	Resigned, 23-3-1912.
Rodwell, Reginald Mandeville	5–10–1910	... ...	... ...	... ...	... ...	... ...	
Norman, Geoffrey Schuyler	5–10–1910	... ...	... ...	... ...	... ...	... ...	
Whitfield, Arthur Noel	5–10–1910	... ...	... ...	... ...	... ...	... ...	
Swaine, Henry Poyntz ...	5–10–1910	... ...	... ...	... ...	... ...	... ...	
Edwards, George Wm.	... ...	... ...	... ...	... ...	... ...	... ...	Qr.-Mr., 30-3-1910.—Hon. Lieut.
Matthews-Donaldson, Chas. Lionel Grey	25–3–1911	... ...	... ...	... ...	... ...	... ...	
Adeley, Gerald Graham	11–1–1913	... ...	... ...	... ...	... ...	... ...	
Innes-Cross, S. M. ...	5–2–1913	... ...	... ...	... ...	... ...	... ...	

APPENDIX II.

SUCCESSION LIST OF COLONELS

OF THE

83RD AND 86TH REGIMENTS (NOW ROYAL IRISH RIFLES).

Name of Colonel.	Regiment.	Date of Appointment.
William Fitch	83rd	28-9-1793
Cornelius Cuyler	86th	30-10-1793
Russell Manners	86th	20-6-1794
William Grinfield	86th	25-3-1795
James Balfour	83rd	18-11-1795
Sir J. H. Craig	86th	5-1-1804
Sir Charles Ross	86th	30-10-1806
Francis, Earl of Kilmorey	86th	25-6-1810
John Hodgson	83rd	20-3-1823
Lord Harris	86th	3-12-1832
Hastings Fraser	83rd	30-9-1835
Hon. Sir F. C. Ponsonby	86th	4-12-1835
James Watson	86th	31-3-1836
Sir A. Brooke	86th	24-5-1837
John Maister	86th	25-8-1843
Sir F. Stovin	83rd	1-9-1848
Roger Parke	86th	26-5-1852
Lord James Hay	86th	8-5-1854
Sir John Michel	86th	19-8-1862
Edw. Pery Buckley	83rd	17-8-1865
William Gustavus Brown	83rd	29-5-1873
W. H. Bradford	R.I.R.	24-5-1886

SUCCESSION OF COLONELS

OF THE

83RD AND 86TH REGIMENTS OF FOOT,

SUBSEQUENTLY

THE ROYAL IRISH RIFLES.

WILLIAM FITCH, 83RD REGIMENT.

Appointed 28th September, 1793.

WILLIAM FITCH commenced his military career on the 16th of August, 1775, as an Ensign in the 66th Foot, then stationed in Ireland, but transferred almost immediately to the 65th Foot, in which he obtained his next step in rank on the 16th of December, 1778.

When, in 1779, Spain joined France in the war which was being carried on by the latter country against England, a 90th Regiment was raised for the second time, the first regiment bearing that number having been disbanded in March, 1763. To this regiment Lieutenant Fitch was appointed as a Captain, his commission bearing the date of November 30th, 1779. The Peace of Versailles, in September, 1783, was, according to the custom of the time, followed by the disbandment of the regiment, its officers being placed on half-pay. During the brief period of his late regiment's existence, Captain Fitch visited for the first time the part of the world where he was eventually to meet his death, fighting the rebellious Maroons.

On the 31st of May, 1788, Captain Fitch returned to the active list on the roll of officers of the 61st Regiment, with which he performed a second tour of duty in Ireland. He transferred to the 58th Foot on the 30th of November, 1790, and gained the rank of Major on the corresponding date in the following year. From the middle of 1791 his next junior brother company commander was the Honourable Arthur Wellesley, subsequently the Duke of Wellington.

Having exchanged into the 55th Regiment in April, 1792, his commanding officer was Colonel Cornelius Cuyler, the gentleman who raised the 86th Regiment, eventually to be so closely linked to Fitch's future 83rd.

Further events in the life of Colonel Fitch will be found related in the present volume.

CORNELIUS CUYLER, 86TH REGIMENT.

Appointed 30th October, 1793.

GENERAL SIR CORNELIUS CUYLER, Bart., was born at Albany, in North America, on the 31st of October, 1740, and was appointed Ensign in the Fifty-Fifth Foot on the 31st of May, 1759, and, immediately proceeding to North America, joined his regiment before the fort of Ticonderoga, on the west shore of Lake Champlain, in July of the same year, in time to take part in the reduction of that post. In 1760 he served at the reduction of Isle-aux-Noix and at the capture of Montreal, which completed the conquest of Canada. In 1764 he was appointed Captain in the Forty-Sixth Foot, with which corps he served two years on the frontiers of North America, one year at New York, and eight in Ireland. On the breaking out of the American war in 1775, he was appointed first Aide-de-Camp to Lieutenant-General Sir William Howe, who proceeded to Boston with reinforcements. In 1776 he was promoted to Major in the Fifty-Fifth, but continued to perform the duty of first Aide-de-Camp to Sir William Howe, then Commanding-in-Chief in North America, and served at the reduction of Long Island, the capture of New York, and the battle of White Plains. He also accompanied the expedition to Pennsylvania in 1777, and served at the battles of Brandywine and Germantown, and in November of that year he succeeded Colonel Meadows, who was removed to the Fifth Foot, in the Lieutenant-Colonelcy of the Fifty-Fifth, which corps he commanded in the retreat from Philadelphia to New York, 1778, and was at the battle of Freehold, under Lieutenant-General Sir Henry Clinton. In November of that year he proceeded, with his regiment, to the West Indies, and was engaged in the capture of St. Lucie. He performed the duties of Adjutant-General to the troops in the West Indies, under Major-General Christie in 1781. He afterwards performed the duties of Quartermaster-General in the West Indies until 1784, when he returned to England, and took the command of his regiment, then in Ireland. In 1787 he was appointed to the situation of Quartermaster-General in the West Indies, which he held till 1792, when he succeeded to the command of the forces in the Windward and Leeward Islands. He commanded an expedition against Tobago, and, having captured the principal fort by storm, on the morning of the 15th of April, 1793, the island submitted. Returning to England soon afterwards, he was promoted to the rank of Major-General, and appointed Colonel of the Eighty-Sixth Regiment, then first raised, and styled "Cuyler's Shropshire Volunteers." He was also placed on the staff in Great Britain, and in April, 1794, he obtained the appointment of Lieutenant-Governor of Portsmouth; in June of that year he was removed from the Eighty-Sixth to the Sixty-Ninth Regiment. In June, 1796, he was appointed Commander-in-Chief in the West Indies, with the local rank of Lieutenant-General; in January, 1798, he was promoted to the rank of Lieutenant-General in the Army, and returned to England in May following. In June he was appointed to the command of the Sussex district; and in January, 1799, he was nominated Commander-in-Chief in Portugal—the Government of that country having refused to ratify a treaty of peace with France, and agreed to receive British troops into the ports; he returned to England in November, the greater part of the troops being sent to the Mediterranean. The rank of General was conferred on this distinguished officer in 1803; he was also appointed Governor of Kinsale, and in July, 1814, he was further rewarded with the dignity of Baronet.

General Sir Cornelius Cuyler, Baronet, died at St. John's Lodge, Herts., on the 8th of March, 1819, after an honourable service of sixty years.

Sir Cornelius Cuyler married Anne, daughter of Major Richard Grant, at the Church of St. Julian, Shrewsbury, on the 18th of November, 1786, and had issue—three sons and four daughters. One son died at the age of six years. The other two both served at the battle of Waterloo—Charles, Captain in 69th Foot; Augustus, Ensign in 2nd Coldstream Guards, Aide-de-Camp to General Sir George Cooke.

RUSSELL MANNERS, 86TH REGIMENT.

Appointed 20th June, 1794.

THIS officer was appointed Cornet in the Royal Regiment of Horse Guards in May, 1755; Captain in the Seventh Dragoons in February, 1758; and in April, 1760, he was promoted to the Lieutenant-Colonelcy of the Twenty-First Dragoons, or Royal Foresters. He served in Germany, under Prince Ferdinand of Brunswick, and, at the peace in 1763, when the Royal Foresters were disbanded, he was appointed Lieutenant-Colonel of the Second Dragoon Guards. On the breaking out of the American war, in 1775, he was appointed Colonel of the Nineteenth Light Dragoons, then newly raised; in 1777 he was promoted to the rank of Major-General, and, in 1782, to that of Lieutenant-General; in 1783 his regiment was disbanded. The Colonelcy of the Eighty-Sixth Foot was conferred on Lieutenant-General Manners in 1794; in 1795 he was removed to the Twenty-Sixth Light Dragoons, and in 1799 he was promoted to the rank of General.

On the 23rd of May, 1800, as General Manners was riding, accompanied by two other gentlemen, in a post-chaise, to Cambridge, he was stopped by two highwaymen, who demanded his money, when he shot one dead on the spot, and the other rode off. In September of the same year, he was residing at Southend, for the benefit of his health, and, having a presentiment of his approaching death, he set off for London, alone, to obtain medical advice, but he was taken ill on the road, and died at an inn at Billericay, in Essex, on the 11th of September, 1800.

WILLIAM GRINFIELD, 86TH REGIMENT.

Appointed 25th March, 1795.

WILLIAM GRINFIELD was appointed Ensign in the Third Foot Guards, in 1760; he was promoted to the rank of Lieutenant and Captain in 1767, to that of Captain and Lieutenant-Colonel in 1776, and in 1782 he was promoted to the rank of Colonel in the Army. In 1786 he obtained the commission of second Major in his regiment. He commanded the first battalion of the 3rd Foot Guards, under His Royal Highness the Duke of York, in Flanders, and evinced great personal bravery and ability on several occasions, particularly at the siege of Valenciennes, and at the recapture of the post of Lincelles, on the 18th of August, 1793, for which he received the thanks of the Duke of York in General Orders. He had been appointed Lieutenant-Colonel of the 3rd Foot Guards a few days before this action occurred, and in October of the same year he was promoted to the rank of Major-General.

In 1795 he was rewarded with the Colonelcy of the 86th Foot; in 1798 he was promoted to the rank of Lieutenant-General, and at that eventful period he was called upon to transfer his services to the West Indies, with the important appointment of Commander of the Forces in the Windward and Leeward Islands. On the renewal of hostilities with France in 1803, he commanded an expedition against St. Lucie, and, having captured the fort of Morne Fortuné by storm on the 22nd of June, the island submitted. He landed on the island of Tobago on the 30th of June, and, by a spirited advance upon Scarborough, forced the French General,

Berthier, to surrender. He captured the islands of Demerara and Essequibo from the Dutch on the 19th of September, and Berbice in a few days afterwards. On the 25th of September he was promoted to the rank of General. He died at Barbadoes on the 19th of October, 1803, of the yellow fever, surviving his wife only three days. It is recorded that a short time before he left England for the West Indies, Mrs. Grinfield's brother died, leaving them £20,000; the General, finding two cousins of the deceased were left unprovided for, observed to his wife, that, as themselves possessed an ample fortune, he purposed making provision for the unfortunate relatives; she readily assenting, he sent for them, and divided the whole legacy between them.

JAMES BALFOUR, 83RD REGIMENT.

Appointed 18th November, 1795.

JAMES BALFOUR, a son of Robert Balfour, of Balbirnie, in the county of Fife, was born in 1743.

The early years of his military life were spent in the 6th Foot, in which regiment he attained the rank of Lieutenant in June, 1762, becoming a Captain on the 22nd of February, 1762, and Major on April 16th, 1777. After serving with his regiment at Gibraltar, in England, and in Scotland, he embarked with it in October, 1772, for the West Indies. The main object with which the regiment was sent westward was to assist in reducing the turbulent Caribs of St. Vincent to submission. The area of operations was of a difficult nature, and the arduous task imposed upon the troops was not accomplished till February, 1774, involving untold hardship, in consequence of which many lives were lost among both officers and men.

The Sixth remained on the scene of its late exploits till after the outbreak of the American War of Independence. Ordered to North America in 1776, the regiment landed at New York, and joined the army commanded by Sir William Howe. It was found, however, that the health of the men had been too greatly impaired by their prolonged stay in an insalubrious climate, for which reason the few men remaining fit for service were assigned to other corps, and the regiment sent home, arriving in England in March, 1777.

In 1779 the 99th Regiment—the second to bear that number—was raised by public subscription for service in the Island of Jamaica, then menaced by possible invasion. To the command of this corps Major Balfour, who already held the army rank of Lieutenant-Colonel since November 29th, 1779, was gazetted on the 2nd of June, 1780, and embarked with it for Jamaica. When, however, the emergency which had called the regiment into existence had passed away, the corps was promptly disbanded, its commanding officer sharing the fate of most of the other officers in being placed on the half-pay list, on which he remained from 1783 till October, 1787.

The occasion of his being restored to the active list was the raising of a third 77th Regiment—its two predecessors having been disbanded in 1763 and 1783 respectively. This revived unit was destined for immediate service in East India, where it was to share in the toil and glory of much eventful service during Lord Cornwallis's campaign against Tippoo Sultan, 1790-91, at the reduction of Ceylon in 1795, the Mysore war of 1799, including the storming and capture of

Seringapatam, and many additional operations in Gujarat and the Deccan. Lieutenant-Colonel Balfour was promoted to Colonel in the Army on the 18th of November, 1790, and commanded with distinction the 1st Brigade of General Sir Robert Abercromby's Army in the operations against Tippoo Sultan. As a mark of distinction for the eminent services of the troops under his command, Sir Robert was eventually appointed to succeed Lord Cornwallis as Commander-in-Chief of the Forces in India.

Promoted Major-General on the 3rd of October, 1794, Balfour vacated the command of the 77th Regiment on September 1st, and was appointed Colonel of the 83rd Regiment on the 18th of November, 1795. He gained the ranks of Lieutenant-General and General on the 1st of January, 1801, and the 25th of October, 1809, respectively, dying in 1823, at the ripe age of 80 years.

SIR JAMES HENRY CRAIG, K.B., 86TH REGIMENT.

Appointed 5th January, 1804.

JAMES HENRY CRAIG obtained a commission of Ensign in the 30th Foot on the 1st of June, 1763, and he served with his regiment several years at the fortress of Gibraltar. In March, 1771, he was promoted Captain in the 47th Foot, with which corps he served in the American war. The 47th were at Boston when hostilities commenced; they took part in the actions at Concord and Bunker's Hill in 1775, and in 1776 they served in Canada. In December, 1777, Captain Craig was promoted Major in the 82nd Regiment, then serving in America, and in 1781 he obtained the Lieutenant-Colonelcy of that corps, from which he was removed, in 1783, to the 16th Foot. In 1790 he was promoted to the rank of Colonel, in 1794 to that of Major-General, and in 1795 his services were rewarded with the Colonelcy of the 46th Foot. In 1801 he was advanced to the rank of Lieutenant-General, and was removed to the 86th Regiment in 1804. On the 25th of March, 1805, he was appointed Commander-in-Chief in the Mediterranean, with the local rank of General; he was also honoured with the dignity of a Knight of the Bath, and nominated Governor of Blackness Castle; in 1806 he was removed to the 22nd Regiment. The services of General Sir James Craig were afterwards transferred to British North America, of which country he was appointed Governor, with the local rank of General in Upper and Lower Canada, dated the 21st of August, 1807. In 1809 he was removed to the Colonelcy of the 78th Highland Regiment, or Ross-shire Buffs. On the 1st of January, 1812, he was promoted to the rank of General in the Army, which he only held a few days, his decease occurring on the 12th of the same month.

SIR CHARLES ROSS, BART., 86TH REGIMENT.

Appointed 30th October, 1806.

CHARLES ROSS, son of Admiral Sir Lockhart Ross, Bart., of Balnagown, who signalised himself during the Seven Years' War, obtained a commission of Cornet in the 7th Dragoons, in January, 1780, and in May, 1784, he was promoted Captain in the 3rd Irish Horse, now 6th Dragoon Guards, in which corps he remained three years, when he was advanced to the commission of Major in the 37th Foot. On the 16th of March, 1791, he was promoted to the Lieutenant-Colonelcy of his regiment, and he performed the duty of Commanding Officer several years with reputation to himself and advantage to the Service. He afterwards took an active part in raising the 116th Regiment, but this corps was disbanded in 1796. On the 18th of June, 1798, he was promoted to the rank of Major-General, and to that of Lieutenant-General in October, 1805. In December of the same year he was appointed Colonel of the 85th Foot, from which he was removed, in October, 1806, to the 86th, and in June, 1810, he was appointed to the 37th Regiment. He was endowed with many amiable qualities, which rendered him an ornament to his country; he was eminently useful in every relation which connected him with society, particularly courteous in public life, and affectionate and valuable as a friend. He died at Balnagown Castle, in the county of Ross, on the 8th of February, 1814.

The Honourable FRANCIS NEEDHAM, 86th Regiment.

Appointed 25th June, 1810.

The Honourable Francis-Jack Needham, third son of John, tenth Viscount of Kilmorey, choosing the profession of arms, procured a commission of Cornet in the 18th Dragoons on the 17th of December, 1762; in February, 1765, he was removed to the 1st Dragoons, in which corps he obtained a Lieutenancy in 1771, and in May, 1774, he was promoted Captain in the 17th Dragoons. He accompanied his Regiment to North America in the spring of 1775, and served at Boston, under Lieutenant-General Gage; he also served at the capture of Long Island, under General Sir William Howe, and received, with his regiment, the thanks of the Commander-in-Chief, for his conduct at the battle of Brooklyn. He also served in the actions at White Plains, and in the Jerseys; afterwards proceeded to Philadelphia, took part in several skirmishes in Pennsylvania, and in covering the retreat to New York, in the performance of which service he was engaged at Freehold. He was subsequently stationed in the lines in front of New York, where he was taken prisoner by the Americans. In August, 1780, he was promoted Major in the 76th Highland Regiment, then serving in America, with which corps he shared in the contest until the peace. In February, 1783, he was promoted to the Lieutenant-Colonelcy of the 104th Regiment, and six weeks afterwards he was appointed Captain and Lieutenant-Colonel in the 1st Foot Guards; he was nominated Aide-de-Camp to the King in 1793, with the rank of Colonel. In 1794 he was appointed Adjutant-General of the expedition to the coast of France, under Lieutenant-General the Earl of Moira; and in 1795 he was appointed third Major in the 1st Foot Guards, promoted to the rank of Major-General, and placed on the home staff. He was subsequently detached, Second-in-Command to Major-General Doyle, with Monsieur Compte D'Artois and his suite, to take possession of Isle Dieu, which place the troops maintained so long as the Navy could afford them protection. An appointment on the staff of Ireland was next conferred on him, and he commanded a body of troops during the rebellion in 1798; he was at the battle of Arklow, on the 9th of June, and commanded a division at Vinegar Hill, on the 21st of June. He continued on the staff of Ireland until April, 1802, when he was promoted to the rank of Lieutenant-General. He had previously been appointed Lieutenant-Colonel in the 1st Foot Guards (21st of August, 1801), and in April, 1804, he obtained the Colonelcy of the Fifth Veteran Battalion. In 1806 he was elected Member of Parliament for Newry, and he sat for that borough in four Parliaments. He was appointed Colonel of the 86th Regiment in 1810, and took great interest in the reputation and welfare of his corps; in 1812 he was promoted to the rank of General. On the decease of his brother Robert, in 1818, he succeeded to the dignity of Viscount Kilmorey. Large and influential estates in Ireland were bequeathed to him by a distant relation; and in January, 1822, he was advanced to the dignity of Earl of Kilmorey and Viscount Newry and Mourne, in the County Down, Ireland. This excellent and patriotic nobleman died at his seat of Shavington, in Shropshire, on the 21st of November, 1832, much regretted, particularly by his numerous tenants in Ireland, to whom he had evinced great kindness.

JOHN HODGSON, 83RD REGIMENT.

Appointed 20th March, 1823.

JOHN HODGSON was a son of Field-Marshal Studholme Hodgson, by his wife Catherine, second daughter of Lieutenant-General Sir Thomas Howard, and was born in 1757.

He was educated at Harrow and eventually entered the 4th King's Own Regiment of Foot, of which his father was Colonel, his Ensign's commission being dated May 20th, 1779. He offers one of the rare examples of an officer of his time spending all the years of his service as a regimental officer with the one regiment throughout. His subsequent regimental commissions as Lieutenant, Captain, Major, and Lieutenant-Colonel were dated November 27th, 1780, September 24th, 1787, May 14th, 1794, and September 1st, 1795, respectively.

About the time of Hodgson's first appointment the King's Own were stationed in the island of Antigua, whence, in 1780, they returned to England, and shortly afterwards proceeded to Ireland for a seven years' tour of duty.

In May, 1787, the regiment embarked for Nova Scotia and Newfoundland, remaining there till ordered to Canada in 1794. While in Nova Scotia, Captain Hodgson discharged, for about five years, the duties of Fort-Major at Halifax.

On the passage home of the regiment in the autumn of 1797, under Lieutenant-Colonel Hodgson's command, the transport carrying the regimental headquarters was chased by the French privateer "La Vengeance." There was no hope of escape, and after a show of resistance, in which several officers and men were wounded, the colours were sunk into the sea before the surrender of the vessel. The officers and men concerned were carried into captivity and detained in France for above a year before being exchanged.

In 1799 the King's Own, now consisting of three battalions, were included in the ill-fated Walcheren expedition, during which Colonel Hodgson was one of the many officers wounded in the sanguinary action ensuing upon the attack of the enemy's position between Beverwyck and Wyck-op-Zee, on October 6th, 1799.

APPENDIX II.

When the Treaty of Amiens (March, 1802), came in view, leading to the disbandment of the second and third battalions of the King's Own, Colonel Hodgson went upon the half-pay list.

Subsequently he filled the posts of Governor of Bermuda and of Curaçoa, which latter appointment he held up to the date of the restoration of the island to the Dutch in 1814.

Having been promoted Colonel (25th September, 1803), Major-General (25th July, 1810), and Lieutenant-General (4th June, 1814), General Hodgson was successively appointed to the colonelcies of the 3rd Garrison Battalion, of the 83rd Regiment (20th March, 1823—30th September, 1835), and of his old corps, the 4th King's Own, becoming a full General on the 10th of January, 1837.

Two sons of his marriage with Catherine Krempion, a sister of the Countess of Terrol—Studholme John Hodgson and John Studholme Hodgson—eventually became general officers in the Army.

General Hodgson died at his London residence on the 14th of January, 1846.

WILLIAM GEORGE, LORD HARRIS, C.B., K.C.H., 86th Regiment.

Appointed 3rd December, 1832.

This distinguished officer was the son of General the first Lord Harris, and entered the Army as Ensign in the 76th Regiment of Infantry on the 24th of May, 1795; was promoted Lieutenant in the 36th Regiment on the 3rd of January, 1796, from which he was removed to the 74th Highlanders on the 4th of September following, and joined in India in 1797.

Lieutenant Harris served at the battle of Mallavelly on the 27th of March, 1799, and during the campaign under his father, Lord Harris, which led to the capture of Seringapatam, and was in nearly all the affairs—outposts, and in the storming party on the 4th of May, 1799, which carried that fortress, where Lieutenant Harris was one of the first to enter the breach, for which he was commended on the spot by Major-General (afterwards Sir David) Baird.

Being sent home with the captured standards, Lieutenant Harris had the honour of presenting them to His Majesty King George III., and was promoted to a company in the 49th Regiment on the 16th of October, 1800, which he joined at Jersey, and, embarking with it towards the end of the year for England, was wrecked on the passage off Guernsey.

Captain Harris afterwards accompanied his regiment in the expedition to the Baltic, under the command of Admiral Parker and Vice-Admiral Nelson, and was present in the frigate "Glatton" in the desperate action of Copenhagen on the 2nd of April, 1801.

In 1802 Captain Harris embarked with the 49th Regiment for Canada, and served in the upper province for two years; being then appointed to a Majority in the 73rd Regiment, he proceeded to join that corps in India, and on his way out he was employed in the capture of the Cape of Good Hope in January, 1806, and was present at the action of Blueberg.

The Seventy-Third having quitted India previously to his arrival, he returned to England the same year, and found he had succeeded to the lieutenant-colonelcy of that regiment.

Upon the formation of the second battalion of the Seventy-Third, which was placed on the establishment of the Army from the 24th of December, 1808, Lieutenant-Colonel Harris was appointed to the command of it, and zealously applied himself to perfecting its discipline, and rendering it efficient in every respect. In 1813 Lieutenant-Colonel Harris embarked on a particular service with the second battalion of the 73rd Regiment, but afterwards joined the expedition to Stralsund, in Swedish Pomerania, under Major-General Samuel Gibbs. On arrival Lieutenant-Colonel Harris was selected to take the field with his battalion and place himself under the orders of Lieutenant-General Count Wallmoden, and was present in the action of the Gorde (in which he highly distinguished himself), under that commander on the 16th of September, 1813.

In November, 1813, the second battalion of the Seventy-Third re-embarked in the Gulf of Lubec for England; but, on arriving at Yarmouth, it was ordered, without landing, to join the army of General Sir Thomas Graham (afterwards Lord Lynedock) in Holland. During the winter campaign before Antwerp, rendered more difficult in consequence of the severity of the weather, Lieutenant-Colonel Harris had the honour of carrying the village of Merxem by storm, under the eyes of His Majesty King William IV. (then Duke of Clarence), and, during the remainder of the operations, was employed as Brigadier-General.

After the peace of 1814, when Antwerp was delivered up, Colonel Harris, to which rank he had been promoted on the 4th of June, 1814, was quartered in that town, and remained in the Low Countries with his battalion during the remainder of the year 1814 and the early part of 1815.

On the return of Napoleon from Elba, Colonel Harris joined the army of the Duke of Wellington, and his battalion was appointed to the brigade commanded by Major-General Sir Colin Halkett, and took part in the stubborn contest of the 16th of June, 1815, at Quatre-Bras, assisted in covering the retreat on the 17th, and on the 18th of June, at Waterloo, bore a gallant part in the complete defeat of Napoleon in that memorable battle. Colonel Harris, late in the afternoon, received a shot through the right shoulder, from which severe wound he continued to suffer at times for the remainder of his life. On retiring on half-pay a testimony of admiration and regard was presented to him by the officers of his battalion in the shape of a splendid sword.

On the 19th of July, 1821, Colonel Harris was advanced to the rank of Major-General.

Major-General the Honourable William George Harris was employed on the staff of the Army in Ireland from the 17th of May, 1823, until the 24th of June, 1825, when he was appointed to the command of the Northern District of Great Britain, which he retained until the 24th of July, 1828, and contributed materially in quelling the disturbances in the manufacturing districts.

On the decease of his father, Lord Harris, in 1829, he succeeded to the title, and from that period lived in retirement at Belmont, the family seat, near Faversham, in Kent.

On the 3rd of December, 1832, Major-General Lord Harris was appointed Colonel of the 86th Regiment, and was removed to the Seventy-Third on the 4th of December, 1835.

In January, 1837, Lord Harris was promoted to the rank of Lieutenant-General. His decease occurred at Belmont, after a short illness, on the 30th of May, 1845.

Lord Harris was a Knight Commander of the Royal Hanoverian Guelphic

Order, a Companion of the Bath, and a Knight of the Order of William of Holland.

[Reprinted, by permission of the Controller of H.M. Stationery Office, from "Historical Record of the 73rd Regiment," by Richard Cannon, Esq., Adjutant-General's Office, Horse Guards.]

HASTINGS FRASER, C.B., 83RD REGIMENT.

Appointed 30th September, 1835.

ENSIGN HASTINGS FRASER was appointed to the 74th Regiment on the 9th of April, 1788, about six months after its formation as a regiment of Highlanders, and landed at Madras in June, 1789. He was promoted Lieutenant on the 3rd of November, 1790, and obtained his company command on December 7th, 1797.

His first chance of seeing active service occurred in 1790, when his regiment entered upon the three years' operations against Tippoo Sultan, being present at the siege and storming of Bangalore, the defeat of Tippoo in his strong position before Seringapatam, the capture of the fort of Penagra, and finally the siege of Seringapatam.

In June, 1793, the 74th Regiment formed part of the force sent against Pondicherry, which achieved its object with the capitulation of the fortress on the 22nd of August, after a three weeks' siege.

Captain Fraser next, in 1797, sailed with the projected expedition against Manilla, and had reached Penang, when the troops were recalled in consequence of threatened fresh hostilities in Mysore.

War with Tippoo Sultan re-opened in 1799, during which Captain Fraser took part in the action of Mallavelly, March 27th, and the storming and capture of Seringapatam on the 4th of May.

In 1800 he was engaged in the operations against the Southern Polygars, entailing great exertions and loss of life, though no distinct record exists of the results achieved.

Captain Fraser was promoted Major of the 46th Regiment on the 28th of July, 1802, and in September, 1804, he joined the 10th Foot on promotion to the rank of Lieutenant-Colonel.

He exchanged into the 86th Regiment in April, 1805, and actually commanded it for the long term of thirteen years, namely, from 1806 till promoted Major-General, on the 12th of August, 1819.

The 86th Regiment was selected to form part of the force sent against the island of Bourbon in 1810, and at the eventual surrender of the French garrison Colonel Fraser had the gratification of being able to congratulate his officers and men on the particularly distinguished part taken by them in the operations, which earned for the regiment the royal approbation and authority to bear the word "Bourbon" on the colours, and brought to its Commanding Officer the reward of a Companionship of the Order of the Bath.

During the years intervening between 1811 and 1819 the 86th were engaged in many arduous services, in which, however, they being mainly the work of detachments, Colonel Fraser took no personal part, and his active career came to a close with the surrender of his command on arrival home in the autumn of 1819.

During his tenure of the colonelcy of the 83rd Regiment—from September 30th, 1835, to August 31st, 1848—he was promoted Lieutenant-General, ranking as such from the 10th of January, 1837, and on September 1st, 1848, he was transferred to the 61st Regiment.

He died on the 6th of October, 1851.

The Honourable Sir FREDERICK CAVENDISH PONSONBY, K.C.B., G.C.M.G., K.C.H., 86th Regiment.

Appointed 4th December, 1835.

Honourable Frederick Cavendish Ponsonby, third son of Frederick, third Earl of Bessborough, was appointed Cornet in the 10th Dragoons, in 1800, and rose, in 1803, to the commission of Captain in the same corps, from which he exchanged to the 60th Regiment, in 1806. In 1807 he was appointed Major in the 23rd Light Dragoons, at the head of which corps he distinguished himself at the battle of Talavera, in 1809, and was promoted, in 1810, to the Lieutenant-Colonelcy of the Regiment. In 1811 he served under Lieutenant-General Graham, at Cadiz; and at the battle of Barossa, in March of that year, he attacked, with a squadron of German Dragoons, the French cavalry covering the retreat, overthrew them, took two guns, and even attempted, though vainly, to sabre Rousseau's battalions. On the 11th of June, 1811, he was appointed Lieutenant-Colonel of the 12th Light Dragoons, at the head of which corps he served under Lord Wellington, and distinguished himself, in April, 1812, at Llerena, in one of the most brilliant cavalry actions during the war. At the battle of Salamanca he charged the French infantry, broke his sword in the fight, and his horse received several bayonet wounds. He repeatedly evinced great judgment, penetration, and resolution in outpost duty, and was wounded in the retreat from Burgos, on the 13th of October, 1812. At the battle of Vittoria he again distinguished himself; his services at Tolosa, St. Sebastian, and Nive were also conspicuous; and on the King's birthday, in 1814, he was promoted to the rank of Colonel in the Army. He commanded the 12th Light Dragoons at the battle of Waterloo, where he led his regiment to the charge with signal intrepidity, received sabre cuts on both arms, was brought to the ground by a blow on the head, was pierced through the back by a lancer, plundered by a tirailleur, ridden over by two squadrons of cavalry, and plundered a second time by a Prussian soldier; but afterwards recovered of his wounds. His services were rewarded with the following marks of royal favour:—Knight Companion of the Order of the Bath, Knight Grand Cross of the Order of St. Michael and St. George, Knight Commander of the Hanoverian Guelphic Order, a cross, a Waterloo medal, Knight of the Tower and Sword of Portugal, and Knight of Maria Theresa of Austria. In January, 1824, he was nominated inspecting field officer in the Ionian Islands; he was promoted Brigadier-General on the staff of those islands, in March, 1824; and in June, 1825, he was advanced to the rank of Major-General. He was removed to the staff of Malta, and retained the command of the troops in that island until May, 1835. In December, 1835, he obtained the Colonelcy of the 86th Regiment, from which he was removed to the Royal Dragoons in the following year. He was an ornament to his profession. In him, military talent was united with the most chivalrous bravery, calm judgment, cool decision, resolute action, and modest deportment. He died on the 10th of January, 1837.

[The services of Sir Cornelius Cuyler, Russell Manners, William Grinfield, Sir James Craig, Sir Charles Ross, the Hon. Francis Needham, Earl of Kilmorey, and the Hon. Sir Frederick Cavendish Ponsonby are reprinted, by permission of the Controller of H.M. Stationery Office, from "Historical Record of the 86th Regiment," by Richard Cannon, Esq., Adjutant-General's Office, Horse Guards.]

SIR JAMES WATSON, K.C.B., 86TH REGIMENT.

Appointed 31st March, 1836.

JAMES WATSON, son of Major James Watson, of the Royal Invalids, was born in 1772.

The story of his regimental career is in reality a fragment of the history of the 1st Battalion 14th Foot, for, true to his first choice, he remained with his regiment from the day of first joining it till promotion to higher command necessitated the severance of former ties.

His first commission (in the 64th Foot) was dated the 24th of June, 1783, but his tender years dictated the necessity of his being kept on half-pay till, at the riper age of fifteen, he joined the 14th Foot as Ensign, becoming Lieutenant on the 18th April, 1792; Captain, March 11th, 1795; Major, December 9th, 1802; Lieutenant-Colonel, May 15th, 1808.

His first active service was during the campaign in Flanders and the retreat to Bremen, 1794-95, and during the two years following he took part in the expedition to the West Indies, which resulted in the capture of St. Lucia, St. Vincent, and Trinidad.

In June, 1807, he embarked with his battalion for India, in which country he served for upwards of twenty-seven years, not counting occasional breaks, as in 1810, when he assisted at the capture of the Isle of France, and in 1811, when he took part in the reduction of Java, including the storming and capture of the formidably fortified position of Cornelius, in spite of the great numerical superiority of the defending Gallo-Batavian troops. The distinguished services of Colonel Watson on the latter occasion were commended in orders, and he was rewarded with a gold medal.

When, in 1813, it was decided to send an expedition against the pirates of Sambas, on the western coast of Borneo, Colonel Watson was entrusted with the command, and after a series of sharp conflicts, he succeeded in bringing the Sultan and his adherents to reason.

The next enterprise in which he was engaged, the last while in command of his regiment, was the capture in 1817, of the town and fortress of Hatrass, which was held by a refractory Hindoo Zemindar, named Dyaram.

Colonel Watson soon after received the appointment of Brigadier-General, in which capacity he took part in the Pindaree and Mahratta wars, and was engaged at the capture of Dhomone, of Mundhla (where he personally led the storming party), and of the fortress of Asseerghur, receiving the medal subsequently awarded to the troops engaged.

Promoted Major-General in 1821, he held a divisional command in India from 1830 to 1837, acting meanwhile as Commander-in-Chief during the temporary absence of Lord William Bentinck.

He was appointed Colonel of the 86th Regiment on the 31st of March, 1836, became a Lieutenant-General in 1837, and was transferred to his old corps, the 14th Regiment, in May of the latter year.

In 1839 he was nominated a K.C.B., and attained the rank of full General on the 11th of November, 1851. He died on the 12th of August, 1862.

Sir ARTHUR BROOKE, K.C.B., 86th Regiment.

Appointed 24th May, 1837.

Arthur Brooke, born in 1772, was the third son of Francis Brooke, of Colebrooke, co. Fermanagh. Entering the Army in 1792 as Ensign in the 44th Foot, he remained with that regiment to the date of his promotion to Major-General's rank in August, 1819. In 1793 he was promoted Lieutenant, and as such served with Lord Moira's Division in Flanders in the years 1794 and 1795. In the latter year he was promoted Captain, and took part in Sir Ralph Abercromby's operations in the West Indies, where his regiment remained till 1798. Having subsequently been employed during the Egyptian campaign of 1801, he purchased his Majority in 1802, and a Lieutenant-Colonelcy in June, 1804. The regiment under his command proceeded to Malta in 1805, whence, in 1808, it embarked for Sicily and took part in Sir John Stuart's operations against Naples, returning to Malta in August, 1811. Promoted Colonel in 1813, he accompanied Lord William Bentinck to the east coast of Spain, and, being senior Colonel, took command of the brigade of which his regiment formed part, greatly distinguishing himself in every encounter with Suchet's troops. At the conclusion of the treaty of peace with France in April, 1814, Brooke received orders to march his own regiment and certain other units from Lord Bentinck's force to Bordeaux, there to embark for an expedition against the United States, under the command of Major-General Robert Ross. He himself was assigned to the command of one of the three brigades, into which the force was divided, the brigade in question consisting of his own regiment and the 4th Foot, the latter commanded by his brother, Francis. The victory of Bladensburg was mainly due to the skilful flank movement executed by Brooke's command. After the subsequent burning of the Capitol and other public buildings of Washington, the expeditionary force re-embarked at St. Benedict and sailed for an advance on Baltimore. Upon landing, on September 12th, 1814, at North Point, about thirteen miles from Baltimore, a skirmish ensued, in which General Ross received a mortal wound, the supreme command thus devolving on Colonel Arthur Brooke. A strong force of Americans was routed, and on September 13th Baltimore appeared to be at the mercy of the British Commander. When, however, it was found that naval co-operation was impracticable, owing to the harbour entrance having been blocked by vessels sunk for that purpose, the troops re-embarked, and the armament eventually proceeded to the West Indies, where Brooke handed over the command to Major-General Sir John Keane. He subsequently rendered distinguished services in the operations against New Orleans and Mobile, and after his return home was nominated a C.B. in a gazette dated June 4th, 1815. Promoted Major-General in 1819, he received the appointment of Governor of Yarmouth in 1822, was created a K.C.B. in 1833, promoted to Lieutenant-General's rank on the 10th of January, 1837, and appointed to the Colonelcy of the 86th on the 24th of May of the same year. He died in London on the 26th of July, 1843.

JOHN MAISTER, 86th Regiment.

Appointed 25th August, 1843.

John Maister joined the Service in 1793, being appointed Ensign in the 54th Regiment on the 13th of November of that year.

Promoted Lieutenant on the 14th of January, 1794, he exchanged into the 61st Regiment in the following April, and after his promotion, on March 30th, 1795, to the rank of Captain, he transferred to the 20th Foot, then in Jamaica, in September, 1795, just a few months too late to have taken part in the fighting against the Maroons.

He returned to England with his regiment in 1796, and was engaged in ordinary garrison routine till, in the summer of 1799, the 20th Regiment was ordered to join the expedition to Holland under His Royal Highness the Duke of York. Thus he took part in the fruitless Helder operations of five weeks' duration. Having passed unscathed through the action of Krabbendam, on September 10th, and the first battle of Egmont-op-Zee, on the 2nd of October, he was wounded in four places during the second battle of Egmont-op-Zee, on the 6th of October, 1799.

On the 20th of June, 1801, he was promoted Major, but in consequence of a reduction of establishment in 1802 he was, in July, 1803, absorbed into the 39th Regiment. With the latter he served till, on the 20th of August, 1807, he obtained a Lieutenant-Colonel's commission in the second battalion of the 34th Regiment, then stationed in Jersey. The battalion was ordered for service in the Peninsula in 1809, landing at Lisbon on the 4th of July, and eventually joined Sir Rowland Hill's division of Wellington's army. Thus Lieutenant-Colonel Maister had the gratification of leading his battalion into battle at Busaco, on the 27th of September, 1810, and to command it during the withdrawal into the famous lines of Torres Vedras.

Before hostilities re-opened in 1811, he returned to England, probably owing to reasons of impairment of health, and took up the not unimportant duties of supervising the preparation of reinforcements for the two battalions on foreign service, the regimental depôt then being at Beverley.

He was promoted Colonel on the 4th June, 1814, and subsequent to the disbandment of the second battalion in April, was placed on half-pay on the 25th of June, 1817.

The ranks of Major-General and Lieutenant-General were conferred upon him in July, 1821, and January, 1837, respectively, and he was appointed Colonel of the 86th Regiment on the 25th of August, 1843. In that capacity he continued till his death, in May, 1852.

SIR FREDERICK STOVIN, G.C.B., K.C.M.G., 83RD REGIMENT.

Appointed 1st September, 1848.

FREDERICK STOVIN, son of James Stovin, of Whitgift, near Howden, Yorkshire, was born in 1783. Having obtained an Ensign's commission on March 22nd, 1800, in the 52nd Foot, he served with it in Pulteney's expedition to Ferrol, being one of the few officers engaged in actual fighting. He was promoted Lieutenant on January 7th, 1801, and on June 24th of the following year obtained a company in the 62nd Foot, from which, on July 9th, 1803, he was transferred to the 28th Foot. With the latter regiment he served in Ireland till 1805, when he was employed as Brigade-Major in Lord Cathcart's expedition to Bremen. In 1807 he took part in the siege and capture of Copenhagen, and in the year following served under Sir John Moore in Sweden, and subsequently in Spain during the Corunna campaign. Having joined the staff of General Alexander Mackenzie Fraser in the capacity of Aide-de-Camp, he was engaged in the Walcheren expedition of 1809, being present at the capture of Flushing. He then rejoined the 28th Foot for regimental duty, and in January, 1810, proceeded to Gibraltar, thence, in April, to Tarifa, where he distinguished himself in a sortie against an old convent, from which the French were dislodged. Subsequently he acted for a few months as Brigade Major at Gibraltar, but ill-health necessitated his return to England in September. In July of the following year (1811) he returned to the Peninsula as Aide-de-Camp to Sir Thomas Picton, being present at the capture of Ciudad Rodrigo and Badajoz. Eventually he was appointed Assistant-Adjutant-General to Picton's (the Third) Division, and continued serving with it to the end of the war, taking part in the battles of Salamanca, Vittoria, Pyrenees, Nivelle, Orthes, and Toulouse. For his meritorious services he was promoted Brevet-Major on April 27th, 1812, Brevet-Lieutenant-Colonel on August 24th, 1813, and was awarded the gold cross with two clasps. Appointed Deputy-Adjutant-General to the expeditionary force operating against the coasts of the United States in 1814, he was present in the unsuccessful attack on New Orleans, during which he was wounded. He was nominated K.C.B. on January 2nd, 1815. His promotion to a Majority in his regiment, the 28th Foot, was dated May 9th, 1816, and on September 2nd, 1819, he obtained the Lieutenant-Colonelcy of the 92nd Highlanders, which regiment he commanded in Jamaica from October, 1820, to the middle of 1821, when, in August, he exchanged into the 90th Light Infantry. He held the command of the latter whilst it was stationed in the Ionian Islands till he was placed on half-pay, on April 23rd, 1829, receiving for his services the distinction of K.C.M.G. Promoted Colonel in the Army on July 22nd, 1830, and Major-General on November 23rd, 1841, he was given the Colonelcy of the 83rd Foot on September 1st, 1848, becoming Lieutenant-General on November 11th, 1851, General on August 14th, 1859, and being nominated G.C.B. on May 18th, 1860. He died at St. James's Palace on August 16th, 1865, the fact of his residence in the Palace precincts being explained by his having been a Groom-in-Waiting to the Queen from 1837 to 1860, in which latter year he was made an extra Groom to Her Majesty.

APPENDIX II.

ROGER PARKE, 86TH REGIMENT.

Appointed 26th May, 1852.

ROGER PARKE joined the 39th Foot as an Ensign on the 30th of June, 1795, and remained with the regiment throughout the long period of thirty-one years' service as a regimental officer. The dates of his successive regimental commissions were: Lieutenant, October 1st, 1795; Captain, October 9th, 1800; Major, February 25th, 1808. Though he also held the army rank of Lieutenant-Colonel from the 4th of June, 1814, yet he was regimentally still a Major at the time of his being placed on half-pay on the 18th of May, 1826.

Experience of active service came to him early in his career, since his regiment proceeded to the West Indies towards the end of the year, in which he joined it, and assisted at the seizure of Demerara, Essequibo, and Berbice.

Returning to England in 1803, the first battalion, in which Captain Parke served—a second battalion had been raised that year—embarked for Malta in March, 1805. In November of the same year the flank companies, Captain Parke's included, were detached for service at Naples, under Lieutenant-General Sir James Craig, and subsequently proceeded to Sicily, whence they returned to Malta in February, 1806.

On promotion to his Majority, Captain Parke was posted to the second battalion, which was selected to join Wellington's army in the Peninsula, and landed at Lisbon on the 2nd of July, 1809, in time to take part in the battle of Busaco. Subsequently, Major Parke was present at the expulsion of the French from the fortress of Campo Mayor, at the investment of Badajoz, and at Albuhera. At Arroyo dos Molinos, where upwards of 1,000 prisoners, including General Brun, were taken, together with the guns and baggage of an entire division; he commanded with distinction the light companies of his brigade, which were formed into a special battalion, and received special mention in Sir Rowland Hill's despatch to the Duke of Wellington.

The first battalion having meanwhile arrived in the Peninsula, the second transferred to it all its men still effective for field duty, after which the skeleton battalion embarked for home, where it arrived in March, 1812. For his services in Spain Major Parke received the Peninsular medal, with a clasp for Albuhera.

While on the half-pay list, Lieutenant-Colonel Parke was promoted Colonel on the 10th of January, 1837, and Major-General on the 9th of November, 1846. In May, 1852, he was appointed Colonel of the 86th Regiment, and he died on the 28th April, 1854.

LORD JAMES HAY, 86TH REGIMENT.

Appointed 8th May, 1854.

LORD JAMES HAY, second son of the seventh Marquess of Tweeddale, by his wife, Hannah Charlotte, daughter of the seventh Earl of Lauderdale, was born in 1788. He was appointed Ensign in the 52nd Light Infantry on the 23rd of January, 1806, being promoted Lieutenant on the 6th of August, 1807. As such he served in the Peninsular war, taking part in the battle of Vimiera, in 1808.

Though his battalion did not arrive on the battlefield of Talavera in time, he himself was serving with a detail attached to General Stewart's Brigade, and thus went through the two days' struggle.

On the 8th of February, 1810, he was promoted Captain in the 4th West India Regiment, but could not have actually joined that corps for duty, since he was present at the battle of Busaco and Fuentes d'Onor, September 27th, 1810, and May 3rd-5th, 1811, respectively.

He exchanged into the Grenadier Guards, his commission as Lieutenant and Captain being dated June 27th, 1811, and remained with that corps till finally placed on the half-pay list on the 26th of November, 1830. His next step of regimental promotion (to Captain and Lieutenant-Colonel) bears the date of March 26th, 1818. He took part in all the main fighting in the Peninsula during the year 1813, being present at Vittoria, the battle of the Pyrenees, at Nivelle and Nive, and eventually received the Peninsula medal with a clasp for each of the eight battles in which he had fought.

On the organization of the Duke of Wellington's army for the Waterloo campaign Lord James was nominated for the post of extra Aide-de-Camp to Lieutenant-General Sir Charles Colville, and on the death of Ensign Lord James Hay, eldest son of the seventeenth Earl of Erroll, who was killed at Quatre-Bras, he succeeded him in the appointment of Aide-de-Camp to Major-General (afterwards Sir Peregrine) Maitland, who commanded the brigade composed of the second and third battalions of the Grenadier Guards. He was present at the battles of June 16th, 17th, and 18th, and during the subsequent advance on Paris witnessed the taking of Peronne. After the withdrawal of the main portions of the allied armies, he remained on duty with the army of occupation in France, returning to England in November, 1818, with the third battalion of the Grenadier Guards.

While on the half-pay list, Lord James received the further steps in rank of Colonel (November, 1846), and Major-General (June, 1854), and was appointed Colonel of the 86th Regiment on the 8th of May, 1854.

Major-General Lord James Hay died on the 18th of August, 1862.

Sir JOHN MICHEL, G.C.B., 86th Regiment.

Appointed 19th August, 1862.

John Michel, eldest son of General John Michel, of Dewlish and Kingston Russell, Dorset, was born on the 1st of September, 1804, and educated at Eton. He was gazetted to an Ensigncy in the 57th Foot on the 3rd of April, 1823, passing thence through the 27th to the 64th Foot, which latter regiment he joined at Gibraltar, and in which he obtained a Lieutenancy on the 28th of April, 1825. Having purchased an unattached company on December 12th, 1826, he exchanged back to the 64th at Gibraltar on February 15th following. In February, 1832, he entered the senior department of the Royal Military College, Sandhurst, passing out in November, 1833, with a first-class certificate, whereupon he rejoined for regimental duty in Ireland, exchanging to the 3rd Buffs, then serving in Bengal, in February, 1835. In May, 1840, he purchased a Majority in the 6th Foot, and in April, 1842, a Lieutenant-Colonelcy in the same regiment. He commanded the latter at home and at the Cape until 1854. During the Kaffir war of 1846-47 he commanded a brigade, and during part of the war of 1852-53 was at the head of the 2nd Division of the army operating in the Waterkloof district, receiving a medal for his services, and being made a C.B. Promoted Brevet-Colonel on January 20th, 1854, he was assigned to the command of the York recruiting district, exchanging, however, to half-pay, 98th Foot, on appointment to the post of Chief of the Staff with the Turkish contingent in the Crimea. At the end of the Crimean war he received the 2nd Class of the Medjidie and the Turkish medal. In 1856-57 he was in command of a brigade at Fort Beaufort, Cape of Good Hope, at a time of a threatened Kaffir rising. Being on his way to another command in China, he was wrecked in the Straits of Sunda, on July 10th, 1857, and carried to Singapore. Thus his services were diverted to India, where he was placed on the Bombay staff on the 18th of February, 1858. On the organization of a division to form the Malwa field force, Michel was placed in command, and was eminently successful in his operations against Tantia Topi, Rao Sahib, and other obdurate rebel leaders, which ended with their defeat and the capture of Tantia Topi. Michel, who had been promoted Major-General on October 26th, 1858, was rewarded with the K.C.B., and received the Indian Mutiny medal. He continued in command of the division so brilliantly led by him in the field, his headquarters being at Mhow, till, towards the end of 1859, he was selected for the command of the 1st Division of the Army under Sir James Hope Grant, proceeding to the north of China. He was present in the action at Sinho and at the occupation of Pekin, on October 12th, 1860. On the 18th of October the summer palace was burned by his division as an act of retribution for the treacherous treatment meted out to Mr. (afterwards Sir Harry) Parkes and other captives in the hands of the Chinese. For his services in the China war he was rewarded with the G.C.B., and also received a medal with a clasp for the Taku Forts. On the return to India of the main portion of Sir Hope Grant's army, Sir John Michel assumed command of the troops left in North China, returning home in January, 1862, on the further reduction in the strength of the troops. He was appointed Colonel of the 86th Regiment on the 19th of August, 1862, became a Lieutenant-General on June 25th, 1866, and General on March 28th, 1874. When, in the autumn of 1873, regular manœuvres were held in England for the first time, Sir John was selected to direct them. From 1875 to 1880 he held the command of the troops in Ireland, where his social qualities and ample means assured him great popularity in all circles. The rank of Field-Marshal was conferred on him on the 27th of March, 1885. His death occurred at his seat, Dewlish, Dorset, on the 23rd of May, 1886.

EDWARD PERY BUCKLEY, 83RD REGIMENT.

Appointed 17th August, 1865.

EDWARD PERY BUCKLEY, of New Hall, County of Wiltshire, was born on the 7th of November, 1796, entered the Grenadier Guards as Ensign on the 24th of June, 1812, and served with his regiment in the field from March, 1813, till the end of the Peninsular war in the following year. He was thus present at the passage of the Bidassoa, the battles of the Nivelle and Nive, and the investment of Bayonne, the medal subsequently awarded him bearing the Nivelle and Nive clasps.

On the 23rd March, 1814, he was promoted Lieutenant and Captain, and as such served with the second battalion of the Grenadier Guards at Quatre Bras and Waterloo, and during the advance on Paris, was also present at the taking of Peronne. He returned to England with his battalion in the middle of January, 1816, receiving the brevet rank of Major on the 19th of July, 1821.

A break in his regimental service occurred in September, 1826, upon going on the half-pay list as Lieutenant-Colonel, unattached. He rejoined his regiment as Captain and Lieutenant-Colonel on the 12th of April, 1827, and finally closed his active career by retiring upon half-pay on the 9th of November, 1830.

Lieutenant-Colonel Buckley was promoted Colonel in November, 1841; Major-General in November, 1851; Lieutenant-General in October, 1858; General on the 17th of August, 1865, the date of his appointment to the Colonelcy of the 83rd Regiment.

General Buckley had the distinction of having been Equerry to the Queen for the long span of twenty-one years. He died on the 28th of May, 1873.

WILLIAM GUSTAVUS BROWN, 83RD REGIMENT.

Appointed 29th May, 1873.

SAVE for a short interval of about eleven months, William Gustavus Brown spent all the years of his career as a regimental officer in the 24th Regiment, in which he was appointed Ensign on the 7th of July, 1825. He became Lieutenant on May 11th, 1830, and Captain on May 10th, 1844.

He took part, with the 24th Regiment, in the second Sikh war, being present in command of his light company at the passage of the Chenab, the action of Sadoolapore, and the battle of Chillianwallah, where he was wounded. For his share in the latter battle he received special promotion to the 39th Regiment, with which he was present at Goojerat as a Major. Soon after, however, he purchased a vacant Lieutenant-Colonel's commission in his former corps, being taken on its strength on the 21st of December, 1849, and was advanced to the brevet rank of Colonel on November 28th, 1854.

During the Indian Mutiny the 24th Regiment was stationed at Rawal Pindi, and though engaged in much harassing work in the Jhelum district, it did not take part in the fighting in the main theatre of action.

In July, 1858, Colonel Brown surrendered his battalion command to take up the duties of Brigadier at Calcutta, followed by similar employment, as a first-class Brigadier, at Delhi.

After his return from India he held the command of the 2nd Infantry Brigade at Aldershot from November, 1861, to January, 1863, when he was appointed to the command of the British troops in China, which he held till July, 1864. He had been promoted Major-General on the 29th of the preceding April, and attained to the rank of Lieutenant-General on May 23rd, 1872.

His appointment to the Colonelcy of the 83rd Regiment was dated the 29th of May, 1873, and he was subsequently promoted full General on October 1st, 1877. He died on the 27th of November, 1883.

APPENDIX II.

WILMOT HENRY BRADFORD, The Royal Irish Rifles.

Appointed 24th May, 1886.

Wilmot Henry Bradford, born on the 6th of February, 1815, was appointed Ensign in the Eightieth Foot on May 24th, 1833, and promoted Lieutenant on the 26th of August, 1836. In the following month he exchanged into the Rifle Brigade, in which corps he was promoted Captain on August 27th, 1841, and Major on August 8th, 1851. At the commencement of the war against Russia in 1854 he was serving with the second battalion, and proceeded with it to the Crimea, taking part in the battle of the Alma and the siege of Sebastopol, and was eventually awarded the Crimean medal, with two clasps, the Turkish medal, and the 5th Class of the Medjidie. On promotion to the rank of Lieutenant-Colonel, on December 29th, 1854, he was posted to the third battalion, but remained on duty with the second battalion till the following February, and in August, 1855, exchanged into the Royal Canadian Rifle Regiment, in the command of which he remained till May 11th, 1863. He was promoted Colonel in December, 1859, Major-General in March, 1868, Lieutenant-General in October, 1877, and retired on the 1st of July, 1881, with the honorary rank of General. Field-Marshal Sir John Michel, G.C.B., having died on the 23rd of May, 1886, General Bradford was selected to succeed him in the colonelcy of the Royal Irish Rifles.

2ND BN. ROYAL IRISH RIFLES CHARITABLE FUND.

THE ORIGIN OF GENERAL FRASER'S DONATION, OR THE BENEVOLENT FUND OF THE 86TH REGIMENT.

IN the year 1808, when the 86th Regiment was stationed at Goa, in the East Indies, the Commanding Officer, Lieutenant-Colonel Hastings Fraser (now General) presented the Regiment with the sum of 11,300 sicca rupees, equal to about £1,412 sterling, to be placed in the Government Funds, in the joint names of the Lieutenant-Colonel of the Regiment and the two next senior officers present, one being a regimental Major.

The interest accruing thereon to be expended as the Commanding Officer may direct, in relieving and assisting wives and families of deserving soldiers of the Regiment, assisting in clothing and instructing their children, and for relieving such cases of distress as may be brought to his notice.

In the year 1827, when the Regiment was stationed at the Island of Trinidad, West Indies, it appears that the late Colonel W. Mallet sanctioned some of the officers, non-commissioned officers and privates to become subscribers of a day's pay yearly, to be added to the interest of the above donation, and laid out for charitable purposes, but which was done away with by Lieutenant-Colonel Sir Michael Creagh, when he assumed command of the Regiment, and the sums so subscribed were returned, as will appear by the following Regimental Orders, issued by him :—

REGIMENTAL ORDERS,

By Lieutenant-Colonel Sir M. Creagh.

Berbice, 28th May, 1833.

The Benevolent Fund of the 86th Regiment being found sufficient to meet the purposes for which it was originally intended without the contribution of the non-commissioned officers and privates, Sir Michael Creagh directs a committee, composed of the following officers, shall assemble to-morrow morning, for the purpose of handing over to the captains of companies the sums hitherto subscribed by these individuals since the arrival of the Regiment in this country.

Major BARRETT, *President.*
Captain SIDLEY }
Lieutenant DALGETTY } *Members.*

APPENDIX III.

Regimental Orders by Sir Michael Creagh.

Demerara, 2nd January, 1834.

With reference to Regimental Orders of the 28th May, 1833, on the subject of the subscriptions (for the years 1827, 1828, 1829, 1830, 1831, and 1832, of the non-commissioned officers and privates to the Regimental Benevolent Fund), Captain Sidley, as Treasurer, will be good enough to pay over to the captains of companies the amount of the sums subscribed, with an explanatory company list, that the amount of each man's subscription for the above years may be credited to his account.

Lieutenant-Colonel Sir Michael Creagh (with the advice and concurrence of the two next senior officers), in consequence of the interest being reduced and other inconveniences, directed that the above sum of 11,300 sicca rupees should be transferred from the Bengal to the English Funds. The necessary documents were prepared accordingly and transmitted to India by Messrs. Cox and Co., the Regimental Agents.

It is to be clearly understood that no individual whatever has any claim on the Fund. It must also be understood that the interest only can be expended, and to encroach on the principal would be acting in direct opposition to the expressed wishes and intentions of the donor.

(Signed) M. Creagh,
Lieutenant-Colonel Commanding,
86th Regiment.

Letter 129. Received 2nd March, 1849, at Deesa,

War Office, 12th January, 1849.

Sir,—It having been represented to the Secretary at War that regimental funds exist in the Regiment under your command, for other purposes than those recognised by Her Majesty's Regulations, I am directed to request you will furnish me, for the information of the Secretary at War with a detail of the circumstances under which such funds originated, the regulations under which they have accumulated and are administered, the purposes to which they are applicable, and the balance which at the present time remains disposable, and in what manner that balance is invested or secured for the intended objects of the funds.

I have, etc.,
(Sd.) L. Sullivan.

Officer Commanding,
86th Regiment of Foot.

[*True Copy*]
(Sd.) C. F. Heatley, Paymaster,
86th Royal Regiment.

Camp, Deesa,
24th March, 1849.

SIR,—I have the honour to acknowledge the receipt of your letter (No. 116828), dated 12th January, 1849, and in reply beg to state that the only fund existing in the 86th Regiment not recognised by Her Majesty's Regulations is a Benevolent Fund, the origin of which is as follows:—

In the year 1807, when the Regiment was stationed in the Portuguese Settlement of Goa, under command of Lieutenant-Colonel (now Lieutenant-General Hastings Fraser, C.B., 61st Regiment), that officer established a canteen, for the purpose of supplying the soldier with a better description of liquor from Bombay than the native contractor usually supplied. In about two or three years the profits of this canteen amounted to about Eleven or Twelve Thousand Rupees, which was by the existing regulations and customs of the Service in India the property of the Lieutenant-Colonel, or the Commanding Officer of the station. Lieutenant-General Fraser, however, in the most disinterested manner, declined appropriating this money to his private use and presented it to the Regiment as a Fund for Benevolent Purposes, the principal to be lodged in Government Securities in the name of the Commanding Officer and the two next seniors, the interest accruing thereon to be expended as the Commanding Officer for the time being should direct, in relieving and assisting the wives and families of deserving soldiers of the Regiment, assisting in clothing, and at that period instructing the children, and for relieving all such cases of distress in the Regiment as might be brought to his notice. These instructions of the liberal donor have been carefully followed up during the past forty years, with the greatest advantage to the recipients, more particularly on home service, when undertaking long marches, embarking for foreign service, during sickness, and on occasion of the death of one or both parents, leaving offspring totally unprovided for, as also in cases where the soldier, from the extent of his family, is found unable to support them on his pay. The existence of this fund has, I believe, been brought to the notice of every Inspecting-General, as well as to that of the several Commanders-in-Chief since the year 1819, and although not recognised by regulation, its object and mode of application has been generally approved.

I beg leave to annex a statement of the Benevolent Fund up to 31st December, 1848, shewing a credit of about £2,245 11s. 3d., in explanation of which I offer the following remarks:—

1st. From the formation of the fund to the year 1820, or later, the money was funded in India at a high rate of interest, during which period the Regiment, being in this country, the demands upon the fund were limited.

2nd. Since the return of the Regiment to India, in 1842, the demands have been very limited, an allowance being granted by Government to the families of soldiers and to the Canteen Fund being made applicable to some cases of Benevolence formerly defrayed by the Benevolent Fund.

ESTIMATE OF BENEVOLENT FUND TO 31ST DECEMBER, 1848.

	£	s.	d.
Invested in 3½% English Funds	1,499	9	0
Balance in Cox & Co.'s hands to 31st December, 1848 ...	181	14	11
Balance in Paymaster's hands to 31st December, 1848 ...	564	7	4
Total Credit of the Fund	£2,245	11	3

APPENDIX III.

The balance in hands of the Regimental Agents and the Paymaster will be invested in a New Loan, expected to be opened in this country immediately.

In conclusion, I beg to add that no officer or soldier has ever contributed a fraction to this fund, the accounts of which are regularly audited at the expiration of each quarter by the Commanding and the two next Senior Officers present with the Regiment, in whose names the funded property is invested, and also that no other than the Commanding Officer has any control over the distribution of this fund.

I have, etc.,

(Sd.) James Creagh, Major,

Commanding 86th Regiment.

[*True Copy.*]

(Sd.) C. F. Heatley, Paymaster,

H.M. 86th Regiment.

No. 119582. War Office, 26th November, 1850.

Sir,—With reference to your letter of the 24th March, 1849, reporting that there existed from the year 1807 in the 86th Foot a Charitable Fund, formed from the extra price of spirits sold in the canteen, which then amounted to about £2,245, I am directed to call your attention to the enclosed Act 12 & 13 Victoria, cap. 71, the fourth section of which prescribes that all funds of this nature shall be paid into the Bank of England, to the account of the Paymaster-General, with the object that such funds shall thenceforward be applied only with the knowledge and consent of the Commander-in-Chief and Secretary at War, and that the present Trustees shall be relieved from responsibility.

Under the provisions of this Act, it becomes necessary that all the securities in which the capital of this Charitable Fund have been invested should be sold, and the proceeds, in sterling money, together with any other sums in the hands of the Agents or Paymaster, paid into the Bank of England, to the account of the Paymaster-General.

My letter of the 6th June, 1849, expressed to you the satisfaction of the Secretary at War with the manner in which this Benevolent Fund had been administered by the Regiment, and I am to acquaint you that, although the Act prescribes that the capital of the fund shall be paid into Public Purse, it is in no respect intended to limit the discretion of the Officer Commanding for the time-being, in making such charitable donations to the wives, widows and children of the soldiers, or to discharged soldiers themselves, as he may deem expedient, the only condition being that he shall transmit quarterly to the Secretary at War a separate detailed statement of the several sums he may have ordered the Paymaster or Agent to pay and charge against the public.

During the period the Regiment is serving in India, it is not required that any statement of the existing Canteen Fund should be made to the Secretary at War. Upon the return, however, of the Regiment to this country, any unexpended balance of such fund should be reported to the Secretary at War, that it may be paid into the Bank of England and added to the Charitable Fund already existing.

I am, etc.,

(Sd.) L. SULLIVAN.

[*True Copy.*]

(Sd.) C. F. HEATLEY,

Paymaster, 86th Regiment.

No. 119581/10 War Office, 30th November, 1852.

SIR,—I am directed to acquaint you that the Lords Commissioners of Her Majesty's Treasury have decided upon allowing interest upon the Regimental Charitable Funds paid into the Bank of England under the Act 12 & 13, Victoria, cap. 71, at the rate of £3 15s. sterling per cent. per annum, that being the rate of interest allowed upon the deposits of soldiers in the Regimental Saving's Bank.

I am accordingly to request that at your earliest convenience, you will prepare and transmit to me a certified account, shewing, on the one hand, the sums passed to the credit of the Regimental Charitable Fund, and, on the other, the amount of the donations therefrom charged against the public, specifying the date on which each sum was so credited or charged to the public, and also the particular Pay List or other channel through which the credit or charge was accounted for.

The amount to be now transmitted will commence with the date when the fund was transferred to the custody of the public, and will terminate on the 31st March, 1852. Interest at the above-mentioned rate of £3 15s. sterling per cent. per annum should be calculated upon the balance unexpended on the last day of each quarter, and charged in the said account.

A similar account in continuation of the above should be prepared to 31st March in each subsequent year, and be despatched so as to arrive at this office on or before the 1st December of that year.

I am, etc.,

(Sd.) A. HAWES.

[*True Copy.*]

(Sd.) C. F. HEATLEY,

Paymaster, 86th Regiment.

APPENDIX IV.

REGIMENTAL SONG OF THE 86TH OR ROYAL COUNTY DOWN REGIMENT.

I.

Come fill your glasses to the brim,
Ye lads that love renown,
And toast unto that gallant Corps
That wears the Harp and Crown,
I mean the " Bourbon Heroes," the Royal County Down.

II.

When George the Third of " *Bourbon* " heard,
He answered with a smile:
" 'Egad! they are a gallant band,
Those lads of Erin's Isle,
Their colour shall be Royal blue,
They'll wear the Harp and Crown,
And be called the Bourbon Heroes, or the Royal County Down."

III.

Come fling our Colours to the breeze
And let them flutter free,
That " Central India " on our banner
Proudly we may see,
For " *Gwalior Fort* " and " *Jhansi* " to
The bravery redounds
Of Ireland's favourite fighting sons—
The Royal County Downs.

IV.

Now fill your glasses once again,
And toast with three times three :—
Here's " Honour "—it's our password,
And we shall honoured be :
" Honour " is the password that
Has led us to " Renown ";
Here's " Honours " to the men who added
*Bourbon** to the Crown.

* The 86th Royal County Down Regiment captured Bourbon from the French in 1810.

It is not known by whom this song was composed, but the 86th Regiment is stated to have marched off parade singing it in Gibraltar on the occasion of their new colours being presented to them by Lady Airey.

REGIMENTAL MARCHES, Etc.

So far as can be traced, the original Regimental March of the 83rd Regiment was "Garry Owen."

About the year 1879 this was changed to the present Regimental March, which is entitled "Off! said the Stranger."

This tune is also the Regimental March of the Durham Light Infantry.

The 86th Regiment marched past about A.D. 1845 to the tune of the "Kinegar Slashers."

This was changed about 1875 to "Patrick's Day."

When the two Regiments were amalgamated in 1881 as the "Royal Irish Rifles," the Regimental March of the 83rd was retained for both, as it was known that the 18th Royal Irish Regiment used "Patrick's Day" as their Regimental March.

The words of "Off! said the Stranger" are given in this Appendix.

It has also been the custom for nearly thirty-three years for the band of the 83rd Regiment to play the Russian National Anthem before winding up their programme. This is believed to have come in at the same time as the Regimental March was changed. Both incidents occurred during the time Lieutenant-Colonel Meurant was in command.

Another custom is for the band to play "God Bless the Prince of Wales" previous to closing their programme in the same manner as the Russian National Anthem. This is believed to have originated from the time in 1875, when the Regiment furnished a guard of honour to the late King Edward VII. (then Prince of Wales).

Yet another regimental custom of the 1st Battalion Royal Irish Rifles is that of drinking the King's health daily after dinner, seated.

When what was called the Regent's Allowance was granted to companies in the reign of George the Third the money was expended on cheapening the cost of wine to the officers. The habit then arose in many regiments of drinking the King's health daily, and to avoid any inconvenience it was drunk seated. The 1st Battalion is one of the very few regiments which has carried out this custom to the present day.

APPENDIX V.

WORDS OF THE REGIMENTAL MARCH

OF THE

ROYAL IRISH RIFLES.

Off! said the stranger, off, off, and away,
And away flew the light bark, o'er the silv'ry Bay,
We must reach ere to-morrow the far distant wave,
The billows we'll laugh at—the tempest we'll brave;
The young roving lovers their vows have been given
Unsmil'd o'er by mortals, but hallow'd in heav'n;
She was Italy's daughter, I knew by her eye;
It wore the bright beam that illumines her sky.

Off! said the stranger, off, off, and away,
And away flew the light bark, o'er the silv'ry Bay,
And she has forsaken her palace and halls
For the chill breeze and the light which falls
O'er the pure wave from the heav'ns above,
And their guiding star was the bright star of love;
Off! said the stranger, off, off, and away,
And away flew the light bark, o'er the silv'ry Bay,

Note.—The above is a portion of "The Light Bark," written by Miss A. Mahony, composed expressly for Madame Vestris by J. Craven. It was written previous to 1830.

OFFICER'S BREAST PLATE, 1832-1855.
Burnished gilt plate, dead gilt mounts.

INDEX.

INDEX.

INDEX.

G.

INDEX.

INDEX.

INDEX.

INDEX.

INDEX.

Q.

R.

INDEX.

INDEX.

Y.

Z.

www.ingramcontent.com/pod-product-compliance
Ingram Content Group UK Ltd.
Pitfield, Milton Keynes, MK11 3LW, UK
UKHW052105270726
14058UKWH00005BA/657